Joan Ribas Gispert
Coordination Chemistry

Joan Ribas Gispert

Coordination Chemistry

WILEY-VCH

WILEY-VCH Verlag GmbH & Co. KGaA

The Author

Prof. Joan Ribas Gispert
Department of Inorganic Chemistry
University of Barcelona
Av. Diagonal 647
08028 Barcelona
Spain

All books published by Wiley-VCH are carefully produced. Nevertheless, authors, editors, and publisher do not warrant the information contained in these books, including this book, to be free of errors. Readers are advised to keep in mind that statements, data, illustrations, procedural details or other items may inadvertently be inaccurate.

Library of Congress Card No.:
applied for

British Library Cataloguing-in-Publication Data
A catalogue record for this book is available from the British Library.

Bibliographic information published by the Deutsche Nationalbibliothek
Die Deutsche Nationalbibliothek lists this publication in the Deutsche Nationalbibliografie; detailed bibliographic data are available on the Internet at <http://dnb.d-nb.de>.

© 2008 WILEY-VCH Verlag GmbH & Co. KGaA, Weinheim

Composition   SNP Best-set Typesetter Ltd., Hong Kong

Printing   Betz-Druck GmbH, Darmstadt

Bookbinding   Litges & Dopf GmbH, Heppenheim

Printed in the Federal Republic of Germany
Printed on acid-free paper

ISBN: 978-3-527-31802-5

Contents

Coordination Chemistry. Joan Ribas Gispert
Copyright © 2008 WILEY-VCH Verlag GmbH & Co. KGaA, Weinheim
ISBN: 978-3-527-31802-5

Preface

During my academic life at the Department of Inorganic Chemistry of the University of Barcelona, I have taught Coordination Chemistry for several years. In my University, this course is optional, and offered to those students that, having passed the general (and essential) topics of Chemistry (Inorganic, Organic, Analytical and Physical Chemistry), decide to deepen in this field of Inorganic Chemistry. Since the beginning of my career as a Professor I missed a textbook for students on this topic that would help them both, on the theoretical aspects as well as on the more descriptive ones. A revision of the relatively modern text books reveals the presence of only one book written in English that may be useful for students in this respect. Its title is *"Physical Inorganic Chemistry. A Coordination Chemistry Approach"*, by S.F.A. Kettle (1996). This is indeed a very good book, however, – as may be inferred from the title – it fails to treat many important aspects of Coordination Chemistry.

The lack of a formal, more comprehensive, textbook, gave me the idea to do the effort to write a systematic book fulfilling the purposes stated above, in Spanish language. This book, with the title *"Química de Coordinación"* was published in 2000 (Ed. Omega-Universitat de Barcelona). The present new book is not a translation from the Spanish language to English of the above mentioned work, but rather a complete and updated revision. I have deliberately omitted some chapters (such as that devoted to Coordination Chemistry of lanthanides) and reduced some of the more complicated aspects related to Quantum Chemistry. For example, in the Spanish book, I wrote a separate Chapter on Spin-Orbit coupling, which I have now included in the Chapter on Crystal (Ligand) Field. Also, a very different perspective is now given in the Chapter devoted to Supramolecular Chemistry. I have now adopted a more restricted vision owing to the tendency of many authors (including myself) to use the term Supramolecular in our research papers in such a manner that the first meaning (from J.M. Lehn) has been completely changed and lost. Some other issues have been added in this English version; chapters devoted to polynuclear complexes, clusters with metal-metal bond, biocoordination chemistry or metal-organic frameworks (MOFs) have been included.

The present book – like the parent Spanish book – is addressed to students who already have a background in Inorganic Chemistry, Organic Chemistry, and Fundamentals of Spectroscopy, Quantum Chemistry and Group Theory. This book

Coordination Chemistry. Edited by Joan Ribas Gispert.
Copyright © 2008 WILEY-VCH Verlag GmbH & Co. KGaA, Weinheim
ISBN: 978-3-527-31802-5

has been, thus, thought for graduate students or for advanced undergraduate students.

An important aspect to be considered when trying to write a modern book on Coordination Chemistry is the treatment of this field made by almost all textbooks of general Inorganic Chemistry. Important concepts such as bonding, thermodynamic stability, kinetics and mechanisms, are generally well treated in graduate Inorganic textbooks. Thus, this new book is thought for readers that already have a minimum background on these subjects. However, there are some issues of modern Coordination Chemistry that are scarcely treated in textbooks of Inorganic Chemistry, such as Photochemistry, Magnetic properties of complexes (polynuclear systems are generally ignored), applications of Electronic Paramagnetic Resonance, Mixed-Valence compounds, Supramolecular Coordination Chemistry, the so-called Metal-Organic Frameworks (MOFs, of increasing importance), etc. All these issues are treated in this book in separated Chapters.

With this book I have, thus, tried to fill in many gaps that are currently empty in related books for graduate and undergraduate students. This needed to be accomplished keeping in mind unavoidable space restrictions. Neither the editor nor the author can propose to the students an excessive number of pages. For this reason, the more "classical" aspects of Coordination Chemistry have been relatively shortened, trying, however, to emphasize the most important ones. For example, I decided to join in a Chapter the part devoted to the thermodynamics and non-redox mechanisms. These subjects are well known by many students. On the contrary, I have emphasized the redox mechanisms in a separated Chapter, because they are a preliminary requirement for understanding the Mixed-Valence compounds.

In the spirit expressed above, I propose at the end of many chapters selected references to help readers interested in a particular topic, to find additional information. My honest advice is that the reader should explore these references to complete the discussion given in the text, mainly for some descriptive and graphical aspects, for which the original paper will always be the best source. I am thinking, for example, about the Chapters devoted to Polynuclear complexes, Clusters, Photochemistry, Supramolecular Coordination Chemistry or Metal-Organic Frameworks. In these Chapters the help of the referenced literature is of paramount importance. With this idea in mind I have tried to choose, when possible, the most recent literature, because the reader usually can find there other references to similar papers published previously. In my opinion this task is very pedagogical for the student. Furthermore, when possible, I have indicated references to Reviews, very useful for the reader to acquire a larger vision of a particular issue.

Finally, this book is not intended to encompass all aspects that are related to Coordination Chemistry, in its more general and wide sense. In many Universities around the world, some parts that can be "formally" considered as derived from Coordination Chemistry are separated matters in the curriculum proposed to students, such as Organometallic Chemistry and Bioinorganic Chemistry. I have embraced in this book the same general tendency.

Acknowledgments

A number of colleagues have read either chapters or parts of different chapters of the manuscript or have given me their opinion about the meaning of some problematic terms, ideas, papers published by them, etc. I can assure that their suggestions and criticisms have been invaluable. Of course all the errors and obscure passages that remain after all revisions have to be attributed to the author.

These colleagues and friends are (in alphabetical order):

Dr. Guillem Aromí, Departament de Química Inorgànica, Universitat de Barcelona (Spain)

Dr. Stuart R. Batten, School of Chemistry, Monash University (Australia)

Dr. Roman Boca, Department of Inorganic Chemistry, Slovak Technical University, Bratislava, (Slovakia)

Dr. Jaume Casabó, Departament de Química, Universitat Autònoma de Barcelona (Spain)

Mr. Jordi Casabó, Departament of Mathematics, Universitat Politècnica de Catalunya, Barcelona (Spain)

Dr. Montserrat Corbella, Departament de Química Inorgànica, Universitat de Barcelona (Spain)

Dr. Dante Gatteschi, Laboratory of Molecular Magnetism, Polo Scientifico Universitario; Dipartimento de Chimica, Università di Firenze, Firenze (Italy)

Dr. Xavier Giménez, Departament de Química-Física, Universitat de Barcelona (Spain)

Dr. Thorfinnur Gunnlaugsson, School of Chemistry, University of Dublin, Trinity College, Dublin (Ireland)

Dr. Andreas Hauser, Department de Chimie Physique, Université de Genève, Genève (Switzerland)

Dr. Richard M. Hartshorn, University of Canterbury, Christchurch (New Zealand)

Dr. Carlos Lodeiro, CQFB-REQUIMTE, Departamento de Quimica, Universidade Nova de Lisboa (Portugal)

Dr. Miguel Julve, Departament de Química Inorgànica/Instituto de Ciencia Molecular (ICMol). Universitat de Valencia, Paterna (Valencia, Spain)

Dr. Robert J. Lancashire, Department of Chemistry, University of the West Indies, Kingston (Jamaica)

Dr. Francisco Lloret, Departament de Química Inorgànica/Instituto de Ciencia Molecular (ICMol). Universitat de Valencia, Paterna (Valencia, Spain)

Dr. Manel Martínez, Departament de Química Inorgànica, Universitat de Barcelona (Spain)

Dr. Jane Nelson, Department of Chemistry, Loughborough University, Loughborough. (United Kingdom)

Dr. Juan Carlos Paniagua, Departament de Química-Física, Universitat de Barcelona (Spain)

Dr. Mark Pedersen, Center for Computational Materials Science, Naval Research Laboratory, Washington, D.C. (USA)

Dr. Spiros Perlepes, Department of Chemistry, University of Patras, Patras (Greece)

Dr. José Antonio Real, Departament de Química Inorgànica/Instituto de Ciencia Molecular (ICMol). Universitat de Valencia, Paterna (Valencia, Spain)

Dr. Jan Reedijk, Coordination and Bioinorganic Chemistry, Leiden Institute of Chemistry, Leiden (The Netherlands)

Dr. Jordi Ribas-Ariño, Departament de Química-Física, Universitat de Barcelona (Spain)

Dr. Néstor E. Katz, Instituto de Química Física, Facultad de Bioquímica, Química y Farmacia, Universidad Nacional de Tucumán (Argentina)

Dr. Pascual Román, Departamento. Química Inorgánica, Facultad de Ciencia y Tecnología, Universidad del País Vasco, Bilbao (Spain)

Dr. Jean-Pierre Sauvage, Laboratoire de Chimie Organo-Minérale, Université Louis Pasteur-CNRS, Institut Le Bel, Strasbourg (France)

Dr. Vassilis Tangoulis, Department of Chemistry, University of Thesalloniki (Greece)

Dr. Boris Tsukerblat, Department of Chemistry, Faculty of Natural Sciences, Ben-Gurion University of the Negev, Beer-Sheva (Israel)

Dr. Margherita Venturi, Dipartimento di Chimica "G. Ciamician", Università di Bologna (Italy)

Dr. Michel Verdaguer, Professeur Emèrite, Université Pierre et Marie Curie, Paris (France)

I also acknowledge Drs. Guillem Aromí, E.Carolina Sañudo and Nuria Aliaga (Department of Inorganic Chemistry, University of Barcelona) for their continuous help in many questions concerning the English language. Their help has been unforgettable.

Most of the 3-D-structural diagrams in the book have been drawn using Crystal Maker 1.2.1 for Windows XP, with coordinates accessed from the Cambridge Crystallographic Base. However, Prof. S.R. Batten has gently supplied some very complicated structures on interpenetrated networks. An special mention to him, as well as another special mention and acknowledgment to Ms. Adelaida Pàmies, a very good friend, who designed and drew the most beautiful Figures (Borromean rings, Catenanes, etc) impossible to do – at least for me – with ChemDraw or Crystal Maker.

Special Acknowledgments

I would like to express, finally, a very particular acknowledgment to some special colleagues and friends. First I would like to invoke my first-cousin, Prof. Jaume Casabó, now emeritus Professor at the Universitat Autònoma of Barcelona. He gave me (at the early 1970s) a modern vision on Coordination Chemistry in a moment when in Spain, for political reasons, Science (in capital letters) was in urgent need of a renaissance. Jaume, with his pedagogical attitude, opened a new and different world of Coordination Chemistry to me.

During my life as a researcher I had the opportunity to work with many expert scientists, in different fields, but always related to Coordination Chemistry: Dr. René Poilblanc and Dr. Patrick Cassoux, in the Laboratoire de Chimie de Coordination, CNRS, Toulouse (France) (1980); Dr. Olivier Kahn (*in memoriam*) in the Université of Paris-Sud, Orsay (France) (1986). I would like to make a special mention of Olivier (with his colleagues in Orsay in the early 1986, Michel Verdaguer, Jean-Jacques Girerd and Yves Journaux) who helped me to enter in the not always easy field of molecular magnetism. I remember Olivier as a great scientist, friend and pedagogue, with enthusiasm and passion for spreading the science.

In Orsay, I met Dante Gatteschi, deeply Italian, but also universal. I have always liked very much to speak to Dante in Italian language. How many times we have spoken not only about magnetism and e.p.r., but also about football!

Later, I met Dr. Marc Drillon (Strasbourg, France), and Dr. George Christou (University of Indiana, Bloomington (USA) (1992). With them I learned many important issues on molecular magnetism.

I am indebted to my closer collaborators in the Department and, mainly, to my colleague and friend, Dr. Montserrat Corbella. She was my graduate and undergraduate student some years ago. I taught her the fundamentals of Inorganic Chemistry. Then, she became specialist in Bioinorganic Chemistry, mainly in Models on Biocoordination and she helped me with extraordinary kindness to write the Chapter devoted to Biocoordination Chemistry. Her clear ideas and enthusiasm were invaluable to me.

I am indebted to my wife, Maria Rosa, and my son, Jordi, who demonstrated great patience when finding me often – may be too often – busy with the writing of this manuscript.

Last but not least, I am also indebted to great musicians who "accompanied" me in the writing of the manuscript. The "Missa Solemnis", the piano and violin concerts of Beethoven, the operas "Norma" and "I Puritani" of V.Bellini, "Tannhauser" and the final part of "Tristan und Isolde" etc, have filled my mind during the hard task of giving form, writing, revising, changing, abbreviating, drawing, etc, the manuscript. It is a good experience to be alone, trying to understand the "mystery" of the spin-orbit coupling, with the music of Beethoven or Mozart filling the ambient. You can try ... good luck!

Barcelona, January 2008
Joan Ribas

Introduction: Definitions, History, Nomenclature

Definitions: Complexes, Coordination Compounds

The compounds studied in this context will be called "coordination compounds", or "complexes". The term "complex" dates from the end of the 19th century. By that time, a distinction was made between first order compounds, formed by atoms, and higher order compounds, formed by a combination of molecules (these are sometimes also called "molecular compounds").

A special class of higher order complexes was the historically important cobalt and platinum ammine complexes. These complexes formed the archetypes of Werner's coordination theory. Werner called them coordination compounds, consisting of a central metal atom (or central ion) and a certain number of coordinated molecules or ions. In the context of the compounds he studied, Werner already used the terms complex and coordination compound (exactly: *coordination unit*), mostly as synonyms, although strictly speaking, the term "complex" was the more general one. In those coordination compounds which were studied at the turn of the century, '*ligands*' were stable, even in their dissociated form [1]. Therefore Werner's complexes were higher order compounds also in the original meaning of the term. Today, other ligand systems are gaining importance for which no dissociation equilibriums in solution should be formulated as these ligands are unstable when dissociated from the metal. There is, however, no reason not to call these compounds "complexes" or "coordination compounds"!

Summary   In modern coordination chemistry, the terms "complex" and "coordination compound (unit)" are largely used as synonyms and are applied to almost any molecular compound of transition metals (*although not necessarily*), formed by a central metal atom and its coordinated set of ligands. It may be a cation, an anion or non-ionic.

IUPAC Recommendations (2005): *Each coordination compound either is, or contains, a coordination entity (or complex) that consists of a central atom to which other groups are bonded. A coordination compound is any compound that contains a coordination entity. A coordination entity is an ion or neutral molecule that is composed of a central atom, <u>usually that of a metal</u>, to which is attached a surrounding array of other atoms or groups of atoms, each of which is called a ligand.* [2].

The Historical Development of Coordination Chemistry

Any attempt to summarize the history of coordination chemistry is complicated by the fact that there is no clearly defined beginning. It was not until the middle of the 19th century that coordination compounds began to be the object of chemical investigation. Coordination chemistry became an independent discipline by the end of the 19th century through Alfred Werner's work. Therefore we will start by trying to trace the history of coordination chemistry up to Alfred Werner by highlighting several individual compounds which were fundamental for the early scientific development.

The Development of Coordination Chemistry Before Alfred Werner's Theory

Perhaps the earliest known of all coordination compounds is the bright red alizarin dye, a calcium aluminum chelate compound of hydroxyanthraquinone. It was first used in India and known to the ancient Persians and Egyptians long before it was used by the Greeks and Romans. Alizarin was mentioned by Herodoto in ca. 450 B.C.

The first scientifically documented proof of the formation of a coordination compound (1597) is the description of the tetramminecopper(II) complex, $[Cu(NH_3)_4]^{2+}$, by the physician and alchemist Andreas Libavius. He observed a blue color caused by the effect of a solution of $Ca(OH)_2$ and NH_4Cl on bronze (copper–tin alloy).

In the case of "Prussian blue", $Fe_4[Fe(CN)_6]_3$, for the first time a complex compound was isolated and used as a color pigment (Diesbach and Dippel, 1704). Its preparation was kept secret for 20 years, until in 1724 it was published by John Woodward (based on Diesbach's instructions). The preparation required a mixture of potassium hydrogen tartrate, potassium nitrate and charcoal, as well as dried and carefully pulverized bovine blood, calcinated iron(II)sulfate, potassium alum and hydrochloric acid, and the preparation was extremely laborious. A closer revision of Diesbach's prescription soon showed that the use of potassium alum was unnecessary, but not, however, the use of an iron salt, which was recognized rather early as a basic ingredient of the color-giving component.

At the time of the French Revolution a certain "Citoyen Tassaert" (1798), reported the formation of a brown solution after reaction of cobalt nitrate or chloride with an excess of aqueous ammonia, although he was not able to explain this observation. As is known today, this was the hexaammine cobalt(III) complex $[Co(NH_3)_6]^{3+}$, which is why Tassaert is considered by some historians to be the discoverer of this substance class.

At the beginning of the 19th century, a whole series of complexes was synthesized, and for the first time a purposeful, preparative methodology was established. Louis-Nicholas Vauquelin (1763–1829) discovered $[Pd(NH_3)_4][PdCl_4]$ (Vauquelin's salt). The corresponding platinum compound, $[Pt(NH_3)_4][PtCl_4]$ was discovered by Heinrich Gustav Magnus (1802–1870) and is known as Magnus' Green salt. William Christoffer Zeise discovered the first organometallic compound, $K[Pt(\eta^2-C_2H_4)Cl_3]\cdot H_2O$, (Zeise's salt) in 1825. According to some authorities, the first metal ammine to be isolated in the solid state was the reddish yellow $[Co(NH_3)_6]_2(C_2O_4)_3$, described in 1822 by Leopold Gmelin. Gmelin also discovered several cyano derivatives, such as $K_3[Fe(CN)_6]$, $M_3[Co(CN)_6]$ and $M_2[Pt(CN)_4]$.

Two extremely important platinum(II) compounds were discovered in 1844: *cis*- and *trans*-$[PtCl_2(NH_3)_2]$. Werner, in his first paper on the coordination theory, discussed these compounds and considered them to be geometric isomers with a square-planar coordination. The explanation was finally given by I.I. Chernyaev in his famous *trans*-effect (1926).

In 1852 Oliver W. Gibss began to collaborate with Frederick A. Genth on an investigation which has since become famous in the annals of coordination chemistry. In 1856 they reported the preparation, properties, analytical data and reactions of 35 salts of some cobalt-ammine cations. Since many of the compounds were colored, the next step was to name these compounds on the basis of color: $CoCl_3\cdot 6NH_3$ (yellow, *Luteo* complex); $CoCl_3\cdot 5NH_3$ (purple, *Purpureo* complex); $CoCl_3\cdot 4NH_3$ (green, *Praseo* complex), $CoCl_3\cdot 4NH_3$ (violet, *Violeo* complex), $CoCl_3\cdot 5NH_3\cdot H_2O$ (red, *Roseo* complex). Addition of $AgNO_3$ to a freshly prepared solution of these salts, resulted in different amounts of precipitated AgCl (3 for *Luteo*, 2 for *Purpureo*, and 1 for *Praseo* and *Violeo*). Another kind of experiment provided useful information about the number of ions present in a solution of these different complexes. One early observation was that two or more complexes having the same chemical composition but different chemical and physical properties sometimes existed. These compounds were called *isomers*.

Early Theories of Coordination Compounds [3]

Several hypotheses and theories were proposed to account for all these experimental facts. The development of a structural theory for organic compounds predated that for coordination compounds. At the time people began to consider the structure of complexes, the concept of the tetravalency of carbon and the formation of carbon–carbon bond chains in organic compounds was already well recognized. This concept had a marked influence on the chemists of that time. No doubt it influenced Blomstrand and his student Jorgensen. Therefore a chain structure was used to account for the additional six ammonia molecules in $CoCl_3\cdot 6NH_3$.

Amongst chemists preparing complexes in the 19th century, Sophus Mads Jorgensen was probably the most productive. He systematically synthesized a large number of complexes, introduced, with others, the chelate ligand ethylenediamine into coordination chemistry and created the basis for Alfred Werner's theory of coordination chemistry. For a long time, both scientists were opponents on conceptual questions and originated a scientific dispute which turned out to be extremely fertile.

Werner

Werner was the first inorganic chemist to be awarded the Nobel Prize in Chemistry (1913), at the age of 26. His greatest contribution to coordination chemistry came as an inspiration in 1893 [4]. Three of his most important postulates are:

- The number of groups attached to an atom (something that he referred to as its *secondary valency*) need not equal its oxidation number (*primary valency*). The chemistry of the cobalt(III)-ammonia compounds could be rationalized if in them cobalt had a primary valency of three, as in $CoCl_3$, but a secondary valency of six, as in $[Co(NH_3)_6]Cl_3$. The term secondary valency has now been replaced by *coordination number* and primary valency by *oxidation state* but Werner's ideas otherwise stand largely unchanged.

- Every element tends to satisfy both its primary and secondary valencies. Werner postulated that in the series of cobalt-ammonia-chloride compounds, the cobalt exhibits a constant coordination number of 6 and, as ammonia molecules are removed, they are replaced by chloride ions which tend to act as covalently bound to the cobalt rather than as free chloride ion. For Werner, to describe the complex chemistry of cobalt, one must therefore consider not only the oxidation state but also its coordination number. Werner thus formulated these salts as $[Co(NH_3)_6]Cl_3$, $[CoCl(NH_3)_5]Cl_2$, and $[CoCl_2(NH_3)_4]Cl$.

  Realizing that these formulations implied a precise statement of the number of ions formed in solution, Werner chose, as one of his experimental studies, measurements of the conductivities of a large number of coordination compounds.

- The secondary valence is directed towards a fixed position in space and therefore can be treated by applications of structural principles. By means of the number and properties of the isomers obtained, Werner was able to assign the correct geometric structures to many coordination compounds long before any direct experimental method was available for structural determination. Werner's method was that used previously by organic chemists to elucidate the structure of organic compounds. Werner postulated that the six ligands in a complex such as $[Co(NH_3)_6]^{3+}$ were situated in some symmetrical fashion with each NH_3 group equidistant from the central cobalt atom. Three such arrangements are possible: a planar hexagon, the trigonal prism and the octahedron. In every case Werner investigated, the number of isomers found was equal to that expected for an octahedral complex. In spite all possible difficulties in synthesizing some of the isomers, Werner was correct in his conclusion concerning the octahedral geometry of coordination number 6 for cobalt(III). His results furnish negative evidence, not positive proof. However, he was able to prove definitely that the planar and the trigonal prism structures cannot be correct. The proof involved demonstrating that complexes of the type $[M(AA)_3]$ are optically active. He was also correct in his assignment of square-planar geometry for the four-coordinate complexes of palladium and platinum, from the fact that two isomers had been isolated for compounds of formula $[MA_2B_2]$.

It is worth noting that the first crystallographic confirmation of Werner's assignment of octahedral geometry to Pt(IV) complexes, $(NH_4)_2[Pt^{IV}Cl_6]$, was published in 1921, some 20 years after his theories were completed. The square-planar complexes $(NH_4)_2[Pd^{II}Cl_4]$, $K_2[Pd^{II}Cl_4]$, $K_2[Pt^{II}Cl_4]$ were confirmed the next year, as well as the tetrahedral complex $K_2[Zn^{II}(CN)_4]$.

The ability of Werner and others to assign the correct structures from indirect data and logic was hailed by Henry Eyring, the former American Chemical Society President: *"The ingenuity and effective logic that enabled chemists to determine complex molecular structures from the number of isomers, the reactivity of the molecule and of its fragments, the freezing point, the empirical formula, the molecular weight, etc., is one of the outstanding triumphs of the human mind"* [5].

Note For the reader interested in exploring the historical aspect in depth, one book is essential: '*Werner Centennial*', published in 1967 by the American Chemical Society as No 62 in their Advances in Chemistry Series [6]. This book contains over 40 chapters on historical and current (in 1967) chemistry, including chapters devoted to the Werner–Jorgensen controversy and so on.

Historical Steps Forward

Following the general theory on the covalent bond of G. N. Lewis (1916), the first attempts to interpret Werner's views on an electronic basis were made in 1923 by N.V. Sidgwick and T.M. Lowry. Sidgwick's initial concern was to explain Werner's coordination number in terms of the sizes of the sub-groups of electrons in the Bohr atom. He soon attempted to systematize coordination numbers using the concept of the 'effective atomic number' (EAN). He considered ligands to be Lewis bases which donated electrons (usually one pair per ligand) to the metal ion, which thus behaves as a Lewis acid. Ions tend to add electrons by this process until the EAN (the sum of the electrons on the metal ion plus the electrons donated by the ligand) of the next noble gas is achieved. Today the EAN rule is of little theoretical importance. Nevertheless, it is extremely useful as a predictive rule in one area of coordination chemistry, that of metal carbonyls and analogs.

In 1927 W. Heitler and F. London proposed the quantum treatment of the H_2 molecule, which was developed by Linus C. Pauling for coordination compounds. Pauling (1931) developed the valence bond theory, with the so-called hybrid orbitals, to explain the bonding in coordination compounds. Some predictions (later verified) were made, such as that the paramagnetic compounds of Ni(II) should show tetrahedral or octahedral coordination and diamagnetic compounds should show square planar coordination.

In 1926 I.I. Chernyaev pointed out the general regularity of what he called the *trans*-effect in order to describe the influence of a coordinated ligand on the practical ease of preparing compounds in which the group *trans* to it had been replaced.

In 1929 Hans Albtrecht Bethe proposed the crystal field model for ionic solids. Between 1930 and 1940 several physicists, in particular J. H. Van Vleck, developed new aspects of this theory, the so-called Theory of the Crystal (Ligand) Field. Since 1950, both theories have been developed by chemists to explain the bonding in coordination compounds. In 1938 Ryutaro Tsuchida published the 'spectrochemical series' based on the results of his measurements of the absorption spectra of cobalt complexes.

In 1937 in his Thesis "On the Stereochemistry of metals with coordination number four" Jensen confirmed the suggestion of Werner regarding the planar configuration of a series of Pt(II), Pd(II) and low-spin Ni(II) complexes. He pioneered the use of dipole moment measurements for stereochemical problems in coordination chemistry.

Important Milestones

In 1957, Sir Ronald S. Nyholm, a reputed researcher in the field of Coordination Chemistry, propagated the concept of *"Renaissance of Inorganic Chemistry"*, which stimulated a greater enthusiasm for research in coordination chemistry [7].

Reaction Mechanisms The classical book of F. Basolo and R.G. Pearson, *Mechanisms in Inorganic Reactions* (1967) summarizes all efforts devoted to this subject until that date. In 1983 and 1992, Henry Taube and Rudolph A. Marcus, respectively, were awarded the Nobel Prize for their investigation on the mechanism of redox processes (inner sphere and outer sphere mechanism, respectively; see Chapter 12). Also a crucial area of study has developed over the last few decades, in parallel to the above studies of redox chemistry, in relation to the understanding and applications of mixed-valence complexes (see Chapter 13).

Metal–Metal bond It was not until 1957, with the determination of the structure of $[Mn_2(CO)_{10}]$, that unequivocal evidence for metal–metal bond formation in metal carbonyls was obtained. However, complexes with a metal–metal bond were not the object of serious studies until 1963, when they became the focus for two different important laboratories (headed by F.A. Cotton and W.I. Robinson, respectively). It is very instructive to read the Introduction of the book *Multiple Bonds Between Metal Atoms*, by Cotton and coworkers, on the history of these very important discoveries [8]. At that time the quadruple M–M bond had reached the category of 'classical'. More than 40 years later, the first metal–metal quintuple bond was discovered and studied theoretically (2005–2006) (see Chapter 6).

Molecular Magnetism Although the concepts describing the magnetic properties of d^n ions had been developing since the first theory on Crystal Field (associated with such as Van Vleck), and these had been exploited by various researchers up to the 1970s, it can be said that the 'renaissance' of molecular magnetism dated from the conference '*Magneto-Structural Correlation in Exchange Coupled Systems*' (1983). At that meeting the different concepts and ideas of physicists and chemists on the topic were brought together for the first time. The interested reader can consult the book (now a classic) published from this conference, for a quick overview of the ideas existing at that time [9]. The most recent advances in this area are developed in Chapter 10 of this book.

Supramolecular Coordination Chemistry The concept of molecular recognition through crown ethers, cryptands, and similar ligands was introduced by Donald J. Cram, Jean-Marie Lehn and Charles J.Pedersen in the 1960s. They were awarded the Nobel Prize in 1987 for their research on this subject. The term 'Supramolecular Chemistry' was coined by Lehn in 1978. The diffuse concept of 'self-assembly' lies within this modern term. The issue is discussed briefly in this book (see Chapter 14).

Organometallic Chemistry   The discovery of ferrocene (1951) and the Ziegler–Natta catalyst (1953) were crucial milestones in the field of organometallic chemistry (as a part of coordination chemistry). With the discovery and interpretation of the structure of ferrocene (by G. Wilkinson and E. O. Fisher), it can be said that modern organometallic chemistry was born. The arrival of the Ziegler catalyst (in Germany) and its application by G. Natta (in Italy), was followed by more than 200 papers in 5 years. It is a paradigmatic example where basic research, driven sometimes by curiosity, became suddenly part of an industrial process with great worldwide importance.

Biocoordination Chemistry This new branch, born from general coordination chemistry, started with the discovery by A. D. Allen and C. W. Senoff in 1965 of the first dinitrogen complex, $[Ru(NH_3)_5(N_2)]^{2+}$, which hinted at the important goal of the fixation of N_2 to produce NH_3. The serendipitous discovery of *cis*-$[PtCl_2(NH_3)_2]$ (cisplatin) as a powerful anticancer drug by B. Rosenberg in 1965 was another milestone of coordination chemistry and its application to the life sciences. Currently, biocoordination chemistry is one of the most prolific and important areas derived from coordination chemistry (see Chapter 17).

Photochemistry of Coordination Compounds   The application of physical chemistry in photochemistry has been known for many decades. It was not, however, until 1970, when V. Balzani and V. Carassiti wrote the first monograph *Photochemistry of Transition Metal Complexes* [10], that the field gained extraordinary relevance and coordination photochemistry and photophysics became a prime topic in Inorganic Chemistry forums. Indeed, this is perhaps the area of coordination chemistry with the most publications in the highest impact journals. The possibility of obtaining, for example, photomolecular machines as well as H_2 production from water and sunlight are top priorities in current research.

Metal–Organic Frameworks (MOFs)   Finally, a new strong area has emerged in the current literature (since the beginning of this decade). Study of the so-called MOFs offers almost infinite possibilities and applications.

Although it is undoubtedly still in its infancy, the number of publications in this field has increased exponentially in the past three to four years. This subject will be treated in Chapter 16.

Nomenclature of Inorganic Chemistry (Coordination Compounds). Extract from IUPAC recommendations 2005

Names of Ligands in Coordination Entities

Anionic Ligands   The rule *now* used, without exception, is that anion names ending in 'ide', 'ite' and 'ate' respectively are changed to end in 'ido', 'ito' and 'ato', respectively, when modifying the ligand name for use in additive nomenclature. *All anionic ligands, upon coordination to a metal, have the simple change in the ending from e to o.*

Certain simple ligands have historically been represented in names by abbreviated forms: fluoro, chloro, bromo, iodo, hydroxo, hydro, cyano, oxo, etc. Following the rule stated above, these are now fluorido, chlorido, bromido, iodido, hydroxido, hydrido, cyanido, oxido.

Neutral Ligands   They are used without modification (even if they carry the ending 'ide', 'ite' or 'ate'). Examples: $MeCONH_2$ = acetamide (*not* acetamido).

Charge Numbers, Oxidation Numbers and Ionic Properties

The key point is: *Arabic numerals are used to designate charge on an atom or groups of atoms, while Roman numerals are used to indicate the (formal) oxidation state of an atom.*

It is also important to distinguish between names and formulae: In formulae the (formal) oxidation state is given as a roman number *in superscript*, e.g. $[Co^{II}Co^{III}W_{12}O_{42}]^{7-}$; $[Mn^{VII}O_4]^-$; $Fe^{II}Fe_2^{III}O^4$. In names the formal charge can be given at the end, e.g. $[CuCl_4]^{2-}$ = tetrachloridocuprate(II).

The charge can also be used e.g. sodium tetranitratoborate(III) can also be written as sodium tetranitratoborate(1–); $K_4[Fe(CN)_6]$ as potassium hexacyanidoferrate(II) or potassium hexacyanidoferrate(4–) or tetrapotassium hexacyanidoferrate; $Na[PtBrCl(NH_3)(NO_2)]$ is sodium amminebromidochloridonitrito-κN-platinate(1–).

"Naked" Ions

In speech and in line formulae iron(2+) and iron(II), and even Fe(II) or Fe^{2+} can be used. It is not totally correct, for these ions with *definite charge*, to write Fe^{II}, Cr^{III}, etc. When it is used (such as occurs in many journals) it indicates not a naked ion, but an ion in a chemical formula: it is more a formal oxidation state than a charge.

It is important to note that for certain atoms, when written as naked but without true ionic charge (such as molybdenum(VI) or similar), it is possible to write: molybdenum(VI), Mo(VI) or even Mo^{VI} (with a preference for the full element name, molybdenum(VI)), and indeed many journals do this. In such cases, Mo^{6+}, V^{5+}, etc. would be required if the ion is naked but, since there is a definite implication that it is part of a complex, that does not necessarily apply.

The name of a monoatomic cation is that of the element with an appropriate charge number appended in parentheses: Na^+ = sodium(1+); Cu^{2+} = copper(2+); Cr^{3+} = chromium(3+).

For anions the rule is the same: homopolyatomic anions are named by adding the charge number to the stoichiometric name of the corresponding neutral species: O_2^- = dioxide(1–) (superoxide accepted); C_2^{2-} = dicarbide(2–) (acetylide accepted); N_3^- = trinitride(1–) (azide accepted), S_2^{2-} = disulfide, Pb_9^{4-} = nonaplumbide(4–).

Summary **Arabic numerals** are crucially important in nomenclature and their placement in a formula or name is especially significant. As right subscripts they indicate the number of individual constituents (atoms or group of atoms), unity is not indicated, e.g. $[Co(NH_3)_6]Cl_3$. As right superscripts they indicate the charge, unity is not indicated, e.g. NO^+, Cu^{2+}, etc.

Roman numerals are used in formulae as right superscripts to designate the formal oxidation state, e.g. $Fe^{II}Fe^{III}_2O_4$, $[Mn^{VII}O_4]^-$. In names they indicate the formal oxidation state of an atom, and are enclosed in parentheses immediately following the name of the atom being qualified, e.g. $[Fe(H_2O)_6]^{2+}$ = hexaaquairon(II); $[FeO_4]^{2-}$ = tetraoxidoferrate(VI).

Plus and minus signs, + and –, are used to indicate the charge of an ion in a formula or name, e.g. Cl^-, Fe^{3+}, Cu^+, Cu^{2+}, As^{3-}, $[Co(CO)_4]^-$.

Ionic charge is indicated by means of a *right upper index*, as in A^{n+} or A^{n-} (*not* A^{+n} or A^{-n}). If the formula is placed in enclosing marks, the right upper index is placed outside the enclosing marks.

Ordering of Metal and Ligands in Coordination Entities

Ligands are listed in *alphabetical order* (multiplicative prefixes indicating the number of ligands are not considered in determining that order): $[CoCl(NH_3)_5]Cl_2$ is pentaamminechloridocobalt(2+) chloride.

In the formulae, ligands are ordered alphabetically according to the abbreviation or formula used for the ligand, irrespective of charge. Thus, CH_3CN, MeCN would be ordered under C and M, respectively. Single letter symbols precede two letter symbols: CO precedes Cl.

In the Recommendation of 1990, charged ligands were cited before neutral ligands. The currently recommended formula for the anion of Zeise's salt is now

$[Pt(\eta^2-C_2H_4)Cl_3]^-$ whereas in the older recommendations it was $[PtCl_3(\eta^2-C_2H_4)]^-$ because chloride is anionic.

Enclosing Marks

Chemical nomenclature employs three types of enclosing marks, namely: braces { }, square brackets [ ], and parentheses ( ). In formulae, these enclosing marks are used in the following nesting order: [ ], [( )], [{( )}], etc. Square brackets are normally used only to enclose entire formulae; parentheses and braces are then used alternately. In formulae, the coordination entity is enclosed in square brackets whether it is charged or uncharged: $[Co(NH_3)_6]^{3+}$, $[PtCl_4]^{2-}$, $[Fe_3(CO)_{12}]$.

Hyphens

Hyphens are used in formulae and in names. Note that there is no space on either side of a hyphen. They are used (i) to separate symbols such as μ (mu), η (eta) and κ (kappa) from the rest of the formula or name: $[\{Cr(NH_3)_5\}_2(\mu\text{-}OH)]^{5+}$, μ-hydroxido-bis(pentaamminechromium)(5+) and (ii) to separate geometrical or structural and stereochemical designators such as *cyclo, catena, closo, cis, trans* from the rest of the formula or name.

Elisions

In general, in compositional and additive nomenclature *no elisions* are made when using multiplicative prefixes: tetraaqua (*not* tetraqua), etc. However, *monoxide*, rather than monooxide is an allowed exception through general use.

Specifying Donor Atoms

The *kappa convention* (κ) is used to specify bonding from isolated donor atoms to one or more central atoms. Single atoms are indicated by the italicized element symbol preceded by a Greek kappa, κ, e.g. $[NiBr_2(Me_2PCH_2PMe_2)]$ = dibromido[ethane-1,2-diylbis(dimethylphosphane-κP)]nickel(II). Simple examples are thiocyanato-κN for nitrogen-bonded NCS and thiocyanato-κS for sulfur-bonded NCS.

In certain cases the kappa convention may be simplified: donor atoms of a ligand may be denoted by adding only the italicized symbol(s) for the donor atom (or atoms) to the end of the name of the ligand, when there is no possibility of confusion. Thiocyanato-*N* and thiocyanato-*S*, and nitrite-*N* and nitrite-*O* are good examples.

The *eta convention* (η) is used for any cases where the central atom is bonded to contiguous donor atoms within one ligand. It specifies the bonding of contiguous atoms of a ligand to a central atom, i.e. the number (hapticity) of *contiguous* ligating atoms that are involved in bonding to one or more metals. Most examples are organometallic compounds, to designate the hapticity of a ligand. The number of contiguous atoms in the ligand coordinated to the metal is indicated by a right superscript numeral, e.g. η^3 ('eta three' or 'trihapto'). Thus, it is used only when there is more than one ligating atom, the term η^1 is not used. The contiguous atoms are often the same element but need not be.

The η-convention may also be used for ligands in which σ-bonds are coordinated in a side-on fashion, such as the H–H in complexes of dihydrogen (η^2-H$_2$). η^2-peroxido is another non-organometallic case, indicating that both oxygen atoms are linked to the same metal ion. Owing to the confusion in the literature relative to η and κ conventions, it is important to note that the η convention is valid *only* when the coordinating atoms are contiguous, that is, when they are directly bonded to each other. Otherwise, the κ convention should be applied. *The η convention should be used wherever there are contiguous ligating atoms.*

Bridging Ligands

Bridging ligands are indicated by the Greek letter μ appearing before the ligand symbol or name and separated from it by a hyphen. If the bridging ligand occurs more than once, multiplicative prefixes are employed, as in tri-μ-chlorido, etc. Bridging ligands are listed in alphabetical order together with the other ligands, but in names a bridging ligand is cited before a corresponding non-bridging ligand. In formulae, bridging ligands are placed after terminal ligands of the same kind. Example: $[Cr_2O_6(\mu\text{-}O)]^{2-}$ = μ-oxido-hexaoxidodichromate(2–)

The bridging index 2 is not normally indicated. The kappa convention is used together with μ when it is necessary to specify which central atoms are bridged, and through which donor atoms.

Multiplicative Prefixes

These are listed in the table below.

1	mono	7	hepta (heptakis)
2	di* (bis)	8	octa (octakis)
3	tri (tris)	9	nona (nonakis)
4	tetra (tetrakis)	10	deca (decakis)
5	penta (pentakis)	11	undeca (undecakis)
6	hexa (hexakis)	12	dodeca (dodecakis)

* In the case of a ligand using two donor atoms, the term 'bidentate' rather than 'didentate' is recommended because of prevailing usage. The prefixes bis, tris, tetrakis, etc. are used with composite ligand names or in order to avoid ambiguity.

References

1 The term ligand was first proposed by Alfred Stock when speaking in Berlin on borane and silane (1916), but it did not come into extensive use among English-speaking chemists until the 1940s and 1950s, largely through the

popularity of Jannik Bjerrum's doctoral dissertation.

2 Connelly, N.G., Damhus, T., Hartshorn, R.M., Hutton. A. T. (Eds.), **2005**, *Nomenclature of Inorganic Chemistry. IUPAC Recommendations 2005*, Royal Society of Chemistry, Cambridge. (I thank Prof. Richard Hartshorn for his kind help in some cases where there could be some ambiguity.)

3 Kauffman, G.B., Sophus Mads Jorgensen (1837–1914). A chapter in coordination chemistry history, *J. Chem. Educ.* **1959**, *36*, 521–527.

4 One night in 1892, Werner awoke at 2 am with the solution to the problem of the constitution of 'molecular compounds', which had come to him like a flash of lightning. He arose from his bed and wrote without interruption. By 5 pm on the following day he had finished his most famous paper, 'Beitrag zur Konstitution Anorganischer Verbindungen (Contribution to the Constitution of Inorganic Compounds).

5 Eyring, H. Trends in Chemistry, *Chem. Eng. News.* Jan. 7, **1963**, *41*(1), 5.

6 'Werner Centennial', Kauffman, G.B. ACS Advances in Chemistry Series, Vol. 62, **1967**, American Chemical Society, Washington, DC.

7 Nyholm, R.S., The Renaissance of Inorganic Chemistry, *J. Chem. Educ.* **1957**, *34*, 166–169.

8 Cotton, F.A., Murillo, C.A., Walton, R.A. (Eds.) (**2005**) *Multiple Bonds Between Metal Atoms*, 3rd edn., Springer Science, New York.

9 Willet, R.D., Gatteschi, D., Kahn, O. (Eds.) (**1985**) *Magneto-Structural Correlation in Exchange Coupled Systems*, NATO ASI Series, Series C: Mathematical and Physical Sciences, Vol. 140, Reidel, Dordrecht.

10 Balzani, V., Carassiti, V. (**1970**) *Photochemistry of Coordination Compounds*, Academic Press, London.

Part One Structure and Bonding

1
Bonding in Coordination Compounds

1.1
d Wavefunctions

The wavefunction of an electron, in polar coordinates (Figure 1.1), is expressed by the formula $\phi_{n,l,m} = R(r)_{n,l} Y(\theta,\phi)_l^m$, where $R_{n,l}$ is the radial part and Y_l^m the angular part. Symmetry operations only alter the angular part, regardless of the value of n (the principal quantum number).

Y_l^m corresponds to what are known as spherical harmonics, which can be broken down into two independent parts, Θ_l^m and Φ_m, which in turn depend on the angles θ and ϕ of the polar coordinates, $Y_l^m = \Theta_l^m \Phi_m$. Φ_l^m are the standard Legendre polynomials, which depend on $\sin\theta$ and $\cos\theta$; and $\Phi_m = (2\pi)^{-1/2}e^{im\phi}$.

The wavefunctions of the orbitals s, p, d and f are expressed as follows: Y_0^0 refers to an s orbital; Y_1^0, Y_1^1, Y_1^{-1} to the three p orbitals; Y_2^0, $Y_2^{\pm1}$, $Y_2^{\pm2}$ to the d orbitals; and Y_3^0, $Y_3^{\pm1}$, $Y_3^{\pm2}$, $Y_3^{\pm3}$ to the f orbitals. The mathematical expressions of these functions have an imaginary part, $\Phi_m = (2\pi)^{-1/2}e^{im\phi}$. Given that $e^{\pm im\phi} = \cos(m_l\phi) \pm i\sin(m_l\phi)$, one usually works with linear combinations of the orbitals, which enable the imaginary part to be suppressed. Those functions in which the imaginary part has been suppressed are known as *real wavefunctions* of the atomic orbitals. Using the mathematical expressions that relate the polar to the Cartesian coordinates ($r^2 = x^2 + y^2 + z^2$; $x = r\sin\theta \cos\phi$; $y = r\sin\theta \sin\phi$; $z = r\cos\theta$) it is possible to determine the equivalence between the real wavefunctions in polar and Cartesian coordinates, which are those conventionally used to 'label' the d orbitals (xz, yz, xy, x^2-y^2, z^2) (Table 1.1).

1.2
Crystal Field Effect on Wavefunctions

1.2.1
Qualitative Aspects

In 1930 Bethe and Van Vleck studied the effect of isolating a Na^+ cation by placing it inside an ionic lattice, such as NaCl. They sought to determine what happens

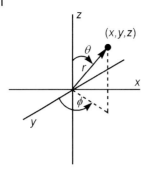

Figure 1.1 Relationship between cartesian and polar coordinates.

Table 1.1 Imaginary and *real* angular wavefunctions of d orbitals [1, 2].

Y_2^m	Imaginary function	Real function (combination)	Real function		Real function[a]	Orbital
			Normalizing factor	Angular function		
Y_2^0	$(5/8)^{1/2}(3\cos^2\theta - 1)(2\pi)^{-1/2}$	$\|0>$	$(\sqrt{5}/\pi)/4$	$(3\cos^2\theta - 1)$	$2z^2 - x^2 - y^2$	d_{z^2}
Y_2^1	$(15/4)^{1/2}\sin\theta\,\cos\theta\,(2\pi)^{-1/2}\,e^{+i\phi}$	$2^{-1/2}[\|1> + \|-1>]$	$(\sqrt{15}/\pi)/2$	$\sin\theta\,\cos\theta\,\cos\phi$	xz	d_{xz}
Y_2^{-1}	$(15/4)^{1/2}\sin\theta\,\cos\theta\,(2\pi)^{-1/2}\,e^{-i\phi}$	$2^{-1/2}[\|1> - \|-1>]$	$(\sqrt{15}/\pi)/2$	$\sin\theta\,\cos\theta\,\sin\phi$	yz	d_{yz}
Y_2^2	$(15/16)^{1/2}\sin^2\theta\,(2\pi)^{-1/2}\,e^{+2i\phi}$	$2^{-1/2}[\|2> + \|-2>]$	$(\sqrt{15}/\pi)/4$	$\sin^2\theta\,\cos2\phi$	$x^2 - y^2$	$d_{x^2-y^2}$
Y_2^{-2}	$(15/16)^{1/2}\sin^2\theta\,(2\pi)^{-1/2}\,e^{-2i\phi}$	$2^{-1/2}[\|2> - \|-2>]$	$(\sqrt{15}/\pi)/4$	$\sin^2\theta\,\sin2\phi$	xy	d_{xy}

a Cartesian coordinates.

to the energy levels of the free ion when it is placed inside the electrostatic field, known as the *crystal field*, which exists in the crystal. It was known that, prior to being subjected to the crystal field, the energy levels of the free ion are degenerate. They demonstrated that, depending on the symmetry of the crystal field, this degeneracy is lost; at the same time they developed a theory which they applied to 3D ionic solids. However, 20 years were to pass before chemists applied the theory to coordination compounds. The essential idea of the model applied to complexes is to assume that the coordination sphere of the anions or ligands surrounding a metal ion behave as a set of negative point charges which interact repulsively with respect to the electrons of the central metal cation. The two single electrons of a ligand act like a negative point charge (or like the partial negative charge of an electric dipole), which undergoes a repulsive effect with respect to the electrons of the d orbitals of the central metal ion. The theory is very simple, easy to visualize and correctly identifies the importance of the symmetry of the orbitals. The bond is thus essentially electrostatic: there is cation(+)/ligand(−) attraction but, at the same time, repulsion between the ligands and the electrons of the central cation.

In this first chapter, the theory will only be used to explain the bonding in coordination compounds. The concept of term, multiplet, state, etc. will be dealt with in Chapter 8. Group theory is of great use in both situations.

Starting from a cyclic point group it is possible to obtain the irreducible representations for any value of the quantum number l, by means of the following formula:

$$\chi(\alpha) = \left[\sin\left(l + \frac{1}{2}\right)\alpha \right] / \sin(\alpha / 2) \tag{1}$$

(α = angle of rotation corresponding to the cyclic group; $\alpha \neq 0$)

This formula can be used to calculate the *qualitative* splitting of the atomic orbitals for any crystal field of a given symmetry (Tables 1.2 and 1.3).

It can be seen that the p orbitals do not split under the effects of octahedral or tetrahedral fields. However, if the symmetry is reduced they do split. Obviously, the s orbital never splits, regardless of the symmetry.

For the two main symmetries (O_h and T_d) the splitting caused by the crystal field can be visualized intuitively by drawing the d orbitals together with the ligands. Figure 1.2 illustrates the splitting of d orbitals for octahedral symmetry, while Figure 1.3 does the same for tetrahedral symmetry.

In an octahedral complex (O_h) the six ligands are situated along cartesian axes (Figure 1.2) whose origin is the metal ion. Therefore, the ligands are close to the

Table 1.2 Splitting of the atomic orbitals in O_h symmetry.

Orbital	l	$\chi(E)$	$\chi(C_2)$	$\chi(C_3)$	$\chi(C_4)$	Irreducible representation
s	0	1	1	1	1	a_{1g}
p	1	3	−1	0	1	t_{1u}
d	2	5	1	−1	−1	$e_g + t_{2g}$
f	3	7	−1	1	−1	$a_{2u} + t_{1u} + t_{2u}$

Table 1.3 Splitting of the orbitals in T_d, D_{4h}, D_3 and D_{2d} symmetry fields.

	T_d	D_{4h}	D_3	D_{2d}
s	a_1	a_{1g}	a_1	a_1
p	t_2	$a_{2u} + e_u$	$a_2 + e$	$b_2 + e$
d	$t_2 + e$	$a_{1g} + b_{1g} + b_{2g} + e_g$	$a_1 + 2e$	$a_1 + b_1 + b_2 + e$
f	$a_2 + t_2 + t_1$	$a_{2u} + b_{1u} + b_{2u} + 2e_u$	$a_1 + 2a_2 + 2e$	$a_1 + a_2 + b_2 + 2e$

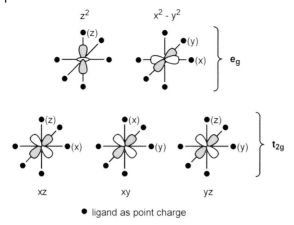

● ligand as point charge

Figure 1.2 Representation of the d orbitals and point charges in an O_h crystal field.

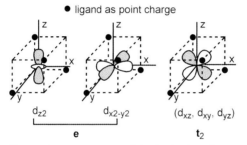

Figure 1.3 Representation of the d orbitals and point charges in a T_d crystal field.

d_{z2} and d_{x2-y2} orbitals, whose symmetry label is e_g. The other three d orbitals (d_{xy}, d_{xz}, d_{yz}) are further away from the ligands. Their symmetry label is t_{2g}. If we consider the repulsions between the d orbitals and the ligand electrons, it is logical to assume that these repulsions are greater between the ligands and the e_g orbitals than between the same ligands and the t_{2g} orbitals (which are directed toward the bisectors of the planes *xy, xz, yz*). Therefore, a simple logical deduction tells us that the five d orbitals have split into two groups: e_g and t_{2g}. This same logic enables us to deduce that the potential energy of the three degenerate t_{2g} orbitals is less than the energy of the two degenerate e_g orbitals.

A tetrahedral field also splits the five d orbitals into two sets, t_2 (d_{xy}, d_{xz}, d_{yz}) and e (d_{z2} and d_{x2-y2}). It can easily be seen that the splitting of the d orbitals must be less than and the reverse of what occurs in the octahedral case. Indeed, with tetrahedral geometry none of the ligands moves toward the central ion according to the direction of the orbitals. In this case the t_2 orbitals are closer to the ligands, as they are directed toward the mid-point of the edges, whereas the e orbitals are directed towards the center of the faces (Figure 1.3).

1.2.2
Quantitative Aspects

Group theory does not tell us anything about the relative energies of the different groups of orbitals. In order to determine the corresponding energies it is necessary to use energy calculations with the corresponding operators.

1.2.2.1 Energy Operator of the Ligand Field

The potential created by six point charges (assuming O_h geometry) at a point x,y,z is:

$$V_{(x,y,z)} = \sum_{i=1}^{6} ez_1 / r_{ij}$$

where r_{ij} is the distance from the charge i to the point x,y,z. $1/r_{ij}$ can be written according to the abovementioned spherical harmonics.

1.2.2.2 Effect of V_{oct} on d Functions

The energy calculation is complicated and beyond the scope of this book. Further information can be found in specialized books, such as that of Figgis and Hitchman [1]. The final result for the energy involved in splitting the d orbitals is $10Dq$, where $D = 35ze/4a^5$ and $q = 2e<r>^4/105$. Therefore, $Dq = (1/6)(ze^2<r>^4/a^5)$. If point charges representing the ligand atoms are replaced by point dipoles, which provides a more realistic representation of ligands such as water or ammonia, a similar expression for Dq is obtained: $Dq = 5\mu<r>^4/6a^6$. In both expressions ze is the charge of the anion, a is the internuclear distance between the metal and the anion or dipole and $<r>$ is the average distance of the d orbital electron from its nucleus.

The splitting of the d orbitals, with their corresponding energies, is shown in Figure 1.4. Although the functions $|1>$ (xz) and $|-1>$ (yz) have been written separately, they can be combined linearly. Any linear combination is permitted, as in a cubic field they are degenerate functions. The linear combinations of $|2>$ and

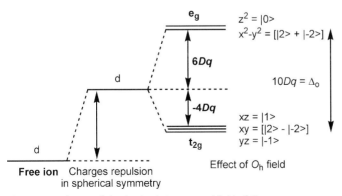

Figure 1.4 Splitting of the d orbitals in a crystal field of O_h symmetry.

|−2> are *necessary* because the corresponding orbitals belong to different symmetries. They cannot be indicated as |2> and |−2>.

The energy difference $10Dq = \Delta_o$ is termed the *crystal field splitting parameter*. The fact that there are only two types of orbitals (e_g and t_{2g}) means that if the separation between them is $10Dq$, the energy of the t_{2g} orbitals will be $-4Dq$ and that of the e_g orbitals $6Dq$, in order to fulfil the principle of energy conservation.

1.2.2.3 Crystal Field Splitting Parameter

The parameter $10Dq$ varies according to the identity of the ligands and the central ion. On the basis of many empirical observations, ligands can be grouped into what is known as the *spectrochemical series* according to the intensity of the crystal field they create, from lesser to greater energy. As regards some of the main ligands, the order of the spectrochemical series is:

$$I^- < Br^- < S^{2-} < SCN^- < Cl^- < NO_3^- < N_3^-, F^- < OH^- < ox < O^{2-} < H_2O < NCS^-$$
$$< CH_3CN < NH_3 \approx py < en < bpy < phen < NO_2^- < PPh_3 < CN^- < CO$$

From the point of view of crystal field theory, the relationship $F^- > Cl^- > Br^- > I^-$ is logical, as the smallest anion has more repulsion energy (q_1q_2/r) with decreasing r. However, crystal field theory cannot explain why an anionic ligand such as OH^- creates a weaker field than H_2O, or why CN^- and CO are among the ligands with the strongest field. This is one of the theory's weak points, and the explanation for it is to be found in the theory of molecular orbitals.

The values of $10Dq$ also depend on the metal ion. In general, the most important variations are as follows:

- $10Dq$ increases with the oxidation number. For example, for the divalent ions of the first transition series the value of Dq varies between 700 and 3000 cm⁻¹, whereas for trivalent ions it varies between 1200 and 3500 cm⁻¹.

- $10Dq$ increases upon moving down a group. If, for an M^{3+} ion of the first transition series, Dq has a value of 1200–3500 cm⁻¹, then the value for the second and third series will be 2000–4000 cm⁻¹. In general, the value of $10Dq$ increases by 50% when moving from the first to the second transition series, and by 25% when passing from the second to the third series.

On the basis of a wide variety of experimental data, it is possible to calculate empirically the value of $10Dq$, as the part corresponding to the metal and the part corresponding to the ligand can be parametrized: $10Dq = M\Sigma n_i L_i \times 10^3$ (cm⁻¹) [3, 4].

1.2.2.4 Weak and Strong Fields. Crystal Field Stabilization Energy

Placing one, two or three electrons in the d orbitals of an octahedral complex does not present a problem, as their placement is necessary $(t_{2g})^1$, $(t_{2g})^2$ and $(t_{2g})^3$. The problem arises when we want to place a fourth electron (or more). The new electron may be placed in two different sites: it may remain in the t_{2g} orbitals and pair with an existing electron, which results in energy destabilization due to the pairing energy of the electrons, P; alternatively, it may jump to the (free) e_g orbitals, and thus it will not have to pair, although it will have to overcome the positive energy $10Dq$. Naturally, which of the two phenomena occurs will depend on the *relative* values of $10Dq$ and P. If $10Dq >> P$ the electrons will tend to be paired in the t_{2g} orbitals; if $10Dq << P$ the electrons will tend to jump to the e_g orbitals. The former case is known as *strong field* or *low spin*, whereas the latter is referred to as *weak field* or *high spin*. This situation occurs for d^4, d^5, d^6 and d^7 configurations. The electron configurations d^8, d^9 and d^{10} only have one possibility, as is the case for d^1, d^2 and d^3.

Figure 1.5 illustrates these possibilities for all configurations. For each one of the configurations the figure in brackets is what is termed the *crystal field stabilization energy* (CFSE). It can be calculated by multiplying the number of electrons in the t_{2g} orbitals by $-4Dq$ and the number of electrons in the e_g orbitals by $6Dq$. The pairing energy, P, is always positive. The final energy calculation will be given by

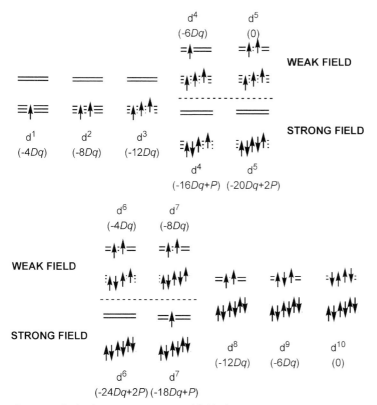

Figure 1.5 d-orbital occupancy in a crystal field of O_h symmetry (for both weak and strong field).

adding the value of P as many times as there are numbers of electron pairs in the configuration. There is one important detail to bear in mind here: if, in the four configurations in which a strong or weak field is possible, the energies of the two fields are mathematically equal, the two configurations will be in equilibrium. Any small external perturbation (temperature, pressure, etc.) may easily result in a switch from one configuration to the other. This phenomenon, known as *spin transition*, is of great importance when studying the magnetism of this series of complexes.

Remark: for the calculation of the pairing repulsion energy it is necessary to take into account the pairing energy in the spherical shell, before the application of the crystal field [5]. For $d^6 = P$; $d^7 = 2P$; $d^8 = 3P$; $d^9 = 4P$ and $d^{10} = 5P$.

All the above is valid for elements of the first transition series. Complexes of the second and third series generally adopt the low spin (strong field) configuration because, even with ligands that create a weak field, the field produced by the ion is sufficiently intense.

1.2.2.5 Splitting of d Orbitals in a Tetrahedral Field

$V_{\text{tet}} = -(4/9)V_{\text{oct}}$, and therefore $Dq_{\text{tet}} = -(4/9)Dq_{\text{oct}}$. This relationship can be verified by means of the angular overlap model of molecular orbital theory (see below). As a consequence, tetrahedral complexes are exclusively weak field complexes and there is no need to distinguish between two types, as in the case of octahedral complexes.

1.2.2.6 Splitting of d Orbitals in a Tetragonally-distorted Octahedral Field. Square-planar Complexes

According to group theory, when a crystal field of D_{4h} symmetry is applied, the d orbitals split into $a_{1g}(z^2) + b_{1g}(x^2 - y^2) + b_{2g}(xy) + e_g(xz, yz)$ (Table 1.3). Let us begin by studying the effect of a tetragonal distortion on the relative energies of the t_{2g} and e_g orbitals of a complex with O_h symmetry. From the point of view of group theory, the new point group will be D_{4h}, regardless of whether the distortion takes the elongated or compressed form. An elongation with respect to the z axis will result in the stabilization of all the orbitals with a z component, as the increasing distance between the two ligands will lead to lower ligand–electron repulsion energy. In contrast, in the case of compression the energy diagram will be the other way round: the orbitals with a z component will become more unstable due to greater repulsion. These effects are shown, for both cases, in Figure 1.6.

If the distortion in the form of elongation is large enough the two ligands will separate in the *trans conformation*, giving rise to a square-planar complex (D_{4h}). Figure 1.6B shows the energy diagrams, starting from an elongated octahedral complex. It is demonstrated that $\Delta_{\text{sp}} \approx 1.3\Delta_o$.

Although the same reasoning can be applied to other geometries it is generally the case, except for octahedral, tetrahedral and square-planar geometry, that molecular orbital theory offers a much more realistic approach. Even in the case of the above-mentioned complexes, crystal field theory ignores what are properly covalent

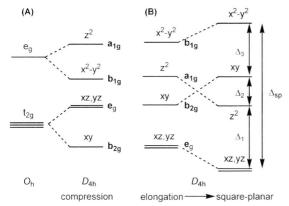

Figure 1.6 Compression or elongation of an O_h crystal field. Extrapolation to a square-planar geometry.

bond interactions between the metal ion and the ligands, and thus it tends to be replaced by *ligand field theory* (see Chapter 8).

1.3
Molecular Orbital Theory

As for non-coordination compounds, two theories can be applied to explain covalency: valence bond theory and molecular orbital theory. In the first the electrons are assumed to be localized: it is thus easy to explain the geometry, and it is necessary to introduce the 'artificial' concept of *hybrid* orbitals. In all books on General and Inorganic Chemistry, the valence bond theory applied to complexes with the typical hybridizations sp^3, sp^3d, sp^3d^2 etc. is explained, with more or less detail, emphasizing its current limitations. Thus, the interested reader can consult any of these books. Anyway, this concept of hybridization in coordination compounds has undergone a 'renaissance' since the Hoffmann definition of the 'isolobal analogy' concept, which will be treated in Chapter 6.

In molecular orbital (MO) theory it is assumed that the bond between the central ion and the ligands is essentially covalent, and produced by the overlap of the s, p, and d orbitals of the central ion and the ligand group orbitals of adequate symmetry.

In this approach it is necessary to distinguish between qualitative aspects (symmetry of the MOs formed) and quantitative ones (energy of the MOs). With regard to the qualitative aspects it is once again necessary to turn to group theory, and there is really no difference between determining the nature and symmetry of the molecular orbitals of a simple molecule such as water and those of a coordination compound. The only difference lies in the fact that in a complex the d orbitals *necessarily* play a role, as they are the most important in this type of molecule.

As clearly indicated by Y. Jean in his recent book [6], there are four stages to the general procedure for constructing the MOs of an ML_n complex: (i) find the appropriate point group symmetry; (ii) determine the symmetry properties of the orbitals of the central metal atom (character tables); (iii) do not consider the ligand orbitals individually, but use linear combinations of these orbitals adapted to the symmetry of the complex: symmetry-adapted orbitals [or symmetry-adapted linear combinations (SALCs)]; (iv) allow metal and ligand orbitals to interact. Only orbitals of the same symmetry can interact, since their overlap is not zero.

Tables 1.2 and 1.3 show the symmetry labels of the s, p and d orbitals for O_h, T_d and other geometries. All that is required, therefore, is to construct the ligand group orbitals (by means of their projection operators) and combine them adequately with the orbitals of the central ion. Obviously, this approach will be valid for both σ and π bonds. Two conditions must be met with these ligand orbitals: they must be close in energy to those of the metal, and their overlap must also be substantial.

In order to calculate the energies it is necessary to turn to quantum methods, which may be very simple or extraordinarily complicated. The simplest method is what is known as the Angular Overlap Model; this will be discussed below and offers solutions to the energy calculations which are accurate enough for the objectives set within this book. Another more sophisticated method is the extended-Hückel model, which can be applied using a PC and the now widely available software program CACAO (**c**omputer **a**ided **c**omposition of **a**tomic **o**rbitals) [7]. Other increasingly sophisticated methods require significant calculation times. These methods will not be discussed in this chapter.

1.3.1
Molecular Orbitals of an Octahedral Complex

1.3.1.1 σ Molecular Orbitals

The symmetry labels of the s, p, and d orbitals are known (Table 1.2), so all that is required is to calculate the group orbitals on the basis of the ligand orbitals which can be adequately combined. There are several ways of representing the ligand orbital that is involved in the σ interaction: as an s orbital, as a p orbital, or as a hybrid (s–p) orbital directed toward the metallic center (Figure 1.7). It should be borne in mind that the order of the six orbitals in Figure 1.7 is totally arbitrary, as is the choice of the coordinate axes.

In the O_h point group the six σ orbitals of the ligands can be grouped according to the symmetry group orbitals $a_{1g} + e_g + t_{1u}$ (readers can easily verify this). By applying the projection operators we obtain the group orbitals as combinations of these six orbitals, and their adaptation to the symmetry of the central ion orbitals can be correlated. This calculation is rarely made in any book on Group Theory. Due to the high symmetry of an octahedral complex the group orbitals can be derived by simple intuition, and represented as in Figure 1.8, without needing to resort to projection operators. These results are given in Table 1.4 and Figure 1.8.

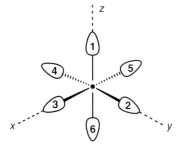

Figure 1.7 Ligand orbitals (σ character) in an octahedral complex.

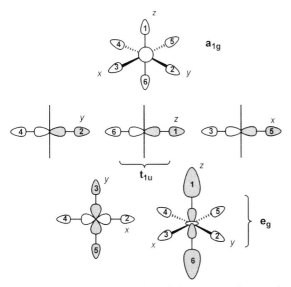

Figure 1.8 Ligand group orbitals and the corresponding metal ion orbitals, with their symmetry label (σ-bond; O_h symmetry).

Table 1.4 Correlation between ligand group orbitals and those of the central ion (O_h).

Metal orbital	Symmetry	Ligand group orbitals
s	a_{1g}	$\sigma_1 + \sigma_2 + \sigma_3 + \sigma_4 + \sigma_5 + \sigma_6$
p_z	t_{1u}	$\sigma_1 - \sigma_6$
p_x		$\sigma_3 - \sigma_5$
p_y		$\sigma_2 - \sigma_4$
d_{z^2}	e_g	$2\sigma_1 + 2\sigma_6 - \sigma_2 - \sigma_3 - \sigma_4 - \sigma_5$
$d_{x^2-y^2}$		$\sigma_2 - \sigma_3 + \sigma_4 - \sigma_5$
$d_{xy,\ xz,\ yz}$	t_{2g}	–

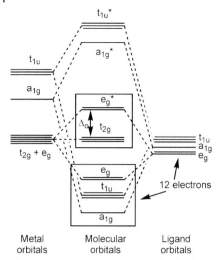

Figure 1.9 MO diagram (only σ orbitals) for an O_h complex.

Remark: for the e_g SALCs we require a combination that matches the z^2 and $x^2 - y^2$ metal orbitals. Remembering that z^2 is, actually, $2z^2 - x^2 - y^2$, the former has positive lobes along the z axis that have twice the amplitude of the toroidal negative region in the xy plane. For this reason σ_1 and σ_6 are multiplied by 2.

Therefore, six group orbitals have been obtained from six ligand σ orbitals. The combination with the six central ion orbitals of the same symmetry will give rise to 12 MO: six bonding and six antibonding. The t_{2g} orbitals of the central ion (d_{xy}, d_{xz}, d_{yz}) do not have adequate symmetry and do not play a role in this type of bond. They remain, therefore, as nonbonding orbitals.

Given the above premises, the relative energy diagram of the MOs can be drawn from the relative energies of the central metal orbitals (s, p, d). Figure 1.9 shows this diagram, without taking into account any quantitative scale. The main characteristic of this diagram is that there are *always* twelve electrons in the deep bonding orbitals, as they come from the two electrons provided by each of the six ligands. Therefore, the 'important' orbitals from the point of view of the complex are the t_{2g} (nonbonding) and e_g^* (anti-bonding) orbitals. The electrons from the central ion will be placed one by one in these orbitals. We have thus arrived, albeit from a completely different angle, at the same splitting achieved with crystal field theory. The energy separation between the t_{2g} and e_g orbitals will thus be termed Δ_o. The placement of the electrons for configurations d^{4-7} will depend on the value of Δ_o. In crystal field theory this separation was a measure of the field strength; in MO theory the separation will depend upon the degree of overlap between the metal's orbitals and those of the ligand.

1.3.1.2 π Molecular Orbitals

Figure 1.10 shows schematically the ligand orbitals capable of forming π bonds. They must be perpendicular to the σ bonds. There is no unique specification of

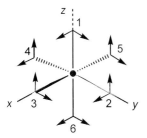

Figure 1.10 A representation of the π-ligand group orbitals (O_h symmetry).

Table 1.5 π group orbitals with their symmetry labels (O_h).

Symmetry	Group orbitals	Symmetry	Group orbitals
t_{1u}	$p_{2z} + p_{3z} + p_{4z} + p_{5z}$	t_{2g}	$p_{2x} + p_{3y} - p_{4x} - p_{5y}$
	$p_{1x} + p_{2x} + p_{6x} + p_{4x}$		$p_{1y} + p_{3z} - p_{6x} - p_{5x}$
	$p_{1y} + p_{3y} + p_{6y} + p_{5y}$		$p_{1x} + p_{2z} - p_{6y} - p_{4z}$
t_{1g}	$p_{2x} - p_{3y} - p_{4x} + p_{5y}$	t_{2u}	$p_{2z} - p_{3z} + p_{4z} - p_{5z}$
	$p_{1x} - p_{2z} - p_{6y} + p_{4z}$		$p_{1x} - p_{2x} + p_{6x} - p_{4x}$
	$p_{1y} - p_{3z} - p_{6x} + p_{5z}$		$p_{1y} - p_{3y} + p_{6y} - p_{5y}$

the direction of the local x and y axes, but the choice and notation in Figure 1.10 prove to be convenient in practice. It has internal consistency in that all p_x are oriented in the same x direction, all p_y in the same y direction and all p_z in the same z direction [8]. The reducible representation of these twelve ligands is t_{2g} + $t_{1u} + t_{2u} + t_{1g}$. By applying the projection operators (or simply by the same matching procedure indicated for σ orbitals) we obtain the ligand group orbitals that are able to form π bonds. The group orbitals with their symmetry labels are shown in Table 1.5.

As there are no central ion orbitals with adequate symmetry for the t_{1g} and t_{2u} group orbitals, we are left with the t_{2g} and t_{1u} group orbitals. The t_{1u} orbitals of the central ion have interacted with the other σ group orbitals of the ligands, those of the same symmetry, and thus they are not considered in this section. We are left, therefore, with the t_{2g} group orbitals, which can interact with the t_{2g} (nonbonding) orbitals of the metal ion. This π interaction is shown schematically in Figure 1.11.

The influence of these π interactions will vary according to the energy of the group orbitals (t_{2g}); these may be full p orbitals (for example, a Cl^- ion) (Figure 1.11A). As these orbitals are highly stable they will be situated further down the MO energy scale shown in Figure 1.9. In contrast, they may be high-energy empty orbitals, such as the d orbitals of a PR_3, or a π anti-bonding molecular orbital of a carbonyl group (Figure 1.11B). In this case they will be situated at the upper end of the energy scale (Figure 1.9). Figure 1.12 illustrates the energy diagram for the

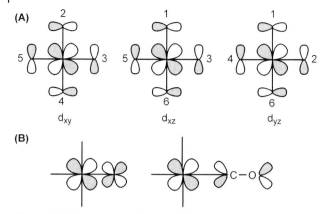

Figure 1.11 Ligand group orbitals and the corresponding metal ion orbitals, with their symmetry label (π-bond; O_h symmetry).

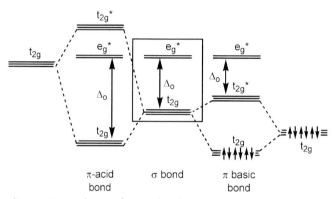

Figure 1.12 MO diagram for π-acid and π-basic ligands, in O_h geometry.

two cases. The stable ligands which are full of electrons are referred to as π-basic because, in addition to the $\sigma(L \rightarrow M)$ bond, they generate a π bond with the same direction: $\pi(L \rightarrow M)$. In contrast, the high-energy, empty ligands are known as π-acidic, as they receive electron density from the metal $\pi(L \leftarrow M)$, rather than vice versa. This phenomenon is called *back donation*. It is logical to assume that for a π-basic bond to occur the metal must be highly positive, that is, have a high formal oxidation state. In contrast, π-acidic ligands serve to stabilize metal ions with high electron density, that is, with a low – or even negative – oxidation state. This is what occurs with PR_3, CO, CN^-, etc. Many organic ligands with double, triple, or delocalized bonds also fall into this category.

One of the important effects arising from the contribution of the two types of π bond is the variation produced in the Δ_o value (Figure 1.12). The π-basic ligands reduce the value of this parameter. Therefore, the first ligands of the spectrochemical series are those which produce this type of interaction: the halides. In contrast, PR_3, CO, CN^-, etc. come at the end of the spectrochemical series as they yield a very high value of Δ_o, due to their being π-acidic ligands.

1.3.2
Molecular Orbitals of a Tetrahedral Complex

The steps to be followed here are the same as in the case of octahedral complexes. They must first be carried out for σ interactions and then for π interactions. In the T_d point group the four σ orbitals of the ligands can be grouped according to the group orbitals $a_1 + t_2$ (Figure 1.13). This can be verified by means of group theory. By applying the corresponding projection operators we obtain the group orbitals as linear combinations of these four orbitals, and their adaptation to the symmetry of the central ion orbitals can be correlated. This result is shown in Table 1.6 (the numbering refers to that used in Figure 1.13).

In Figure 1.13 only the s and p orbitals of the central ion are taken into account, but it should be remembered that the d_{xy}, d_{xz} and d_{yz} orbitals have the same symmetry (t_2) as the three p orbitals. Thus, there will always be a mixture of both types of orbital. Bearing this important point in mind, the corresponding energy diagram for the molecular orbitals obtained is shown schematically in Figure 1.14.

For π-group orbitals, although this is not immediately obvious due to their orientation, these orbitals belong to the symmetry species $t_1 + t_2 + e$. The t_1 group orbitals do not correspond to any central ion orbital and the t_2 group form σ bonds. The orientation and the MOs with e symmetry, as well the energy diagram including the σ and π bonds, are difficult to visualize due to the presence of mixed orbitals. For a complete study see Ref. [2].

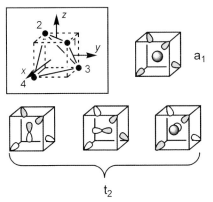

Figure 1.13 Ligand group orbitals and the corresponding metal ion orbitals, with their symmetry label (σ-bond; T_d symmetry).

Table 1.6 σ group orbitals for T_d symmetry.

Symmetry	Metal orbitals	Group orbitals
a_1	s	$\sigma_1 + \sigma_2 + \sigma_3 + \sigma_4$
t_2	(p_x, p_y, p_z)	$\sigma_1 + \sigma_2 - \sigma_3 - \sigma_4$
	(d_{xy}, d_{xz}, d_{yz})	$\sigma_1 - \sigma_2 - \sigma_3 + \sigma_4$
		$\sigma_1 - \sigma_2 + \sigma_3 - \sigma_4$

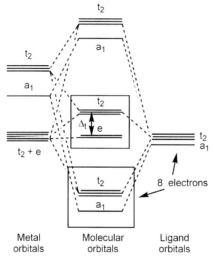

Figure 1.14 MO diagram (only σ orbitals) for a T_d complex.

1.3.3
Molecular Orbitals of a Square-planar Complex

In the D_{4h} point group the four σ orbitals of the ligands can be grouped according to the group orbitals $a_{1g} + e_u + b_{1g}$. This can be verified by means of group theory. By applying the corresponding projection operators we obtain the group orbitals as linear combinations of these four orbitals, and their adaptation to the symmetry of the central ion orbitals can be correlated. This result is shown in Table 1.7 and Figure 1.15.

There are two kinds of π group orbitals: those in the molecular plane and those which are perpendicular to it (Figure 1.16). The first four belong to the symmetry species $a_{2g} + e_u + b_{2g}$, while the four perpendicular ones belong to the species $a_{2u} + e_g + b_{2u}$. By applying the corresponding projection operators we obtain the group orbitals as linear combinations of these eight orbitals, and their adaptation to the symmetry of the central ion orbitals can be correlated. This result is shown in Table 1.8 and Figure 1.16A for b_{2g} and e_u π orbitals. The corresponding π_z MO orbitals can be easily deduced by the reader following an analogous procedure.

Table 1.7 σ group orbitals for D_{4h} orbitals.

Symmetry	Metal orbitals	Group orbitals
a_{1g}	s, d_{z^2}	$\sigma_1 + \sigma_2 + \sigma_3 + \sigma_4$
e_u	(p_x, p_y)	$\sigma_1 + \sigma_2 - \sigma_3 - \sigma_4$
		$-\sigma_1 + \sigma_2 + \sigma_3 - \sigma_4$
b_{1g}	$d_{x^2-y^2}$	$\sigma_1 - \sigma_2 + \sigma_3 - \sigma_4$

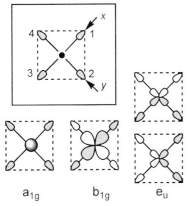

a_{1g} b_{1g} e_u

Figure 1.15 Ligand group orbitals and the corresponding metal ion orbitals, with their symmetry label (σ-bond; square planar geometry).

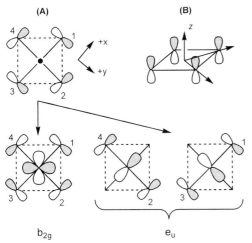

b_{2g} e_u

Figure 1.16 Ligand group orbitals and the corresponding metal ion orbitals, with their symmetry label (π-bond; square planar geometry).

Table 1.8 π group orbitals for D_{4h} geometry (the numbering is that used in Figure 1.16).

Symmetry	Metal orbitals	Group orbitals
a_{2u}	p_z	$p_{1z} + p_{2z} + p_{3z} + p_{4z}$
e_g	(d_{xy}, d_{xz})	$p_{1z} - p_{3z}; p_{2z} - p_{4z}$
b_{2u}	—	$p_{1z} - p_{2z} + p_{3z} - p_{4z}$
a_{2g}	—	$p_{1y} - p_{2x} - p_{3y} + p_{4x}$
e_u	(p_x, p_y)	$p_{2x} + p_{4x}; p_{1y} + p_{3y}$
b_{2g}	d_{xy}	$p_{1y} + p_{2x} - p_{3y} - p_{4y}$

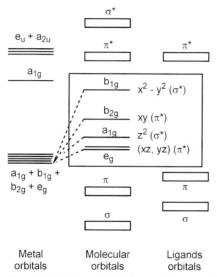

Figure 1.17 Schematic MO diagram for a complex with square-planar geometry.

The corresponding – and simplified – energy diagram for the MO obtained is shown schematically in Figure 1.17. The ordering of energies will be discussed in Section 1.4.4.

1.3.4
Mixed Ligands and Other Geometries

The reader interested in octahedral complexes with one carbonyl (π-acidic) ligand and one π-basic ligand as well as in more complicated complexes, such as [MCl$_2$L$_4$], [MCl$_3$L$_3$], [M(CO)$_2$L$_4$], [M(CO)$_3$L$_3$] may consult the excellent treatment given by Y. Jean in his book [6].

For other geometries, such as square-pyramidal, trigonal bipyramidal (ML$_5$), trigonal-planar (ML$_3$), "butterfly" (ML$_4$), linear or angular ML$_2$ complexes, and their relationships, the reader may also consult the same book [6].

1.3.5
Nobel Prizewinning Discoveries of Complexes

1.3.5.1 Metallocenes

The discovery of the remarkably stable organometallic compound ferrocene [$Fe(\eta^5-C_5H_5)_2$] occurred in 1951. Two research workers, Ernst-Otto Fisher in Munich and Geoffrey Wilkinson in London, were awarded the Nobel Prize in 1973 for their contributions.

The discussion of the bonding does not depend critically on whether the preferred rotational orientation of the two rings is staggered (D_{5d}) or eclipsed (D_{5h}); in any event, the barriers to ring rotation in all types of arene-metal complexes are very low, ca. 10–20 kJ mol^{-1}. Applying group theory and looking at the number of nodal planes, it is easy to demonstrate that the shapes and relative energies of the five π-orbitals in an isolated $C_5H_5^-$ are as shown in Figure 1.18 ($C_5H_5^-$ has 6 electrons). Considering, now, the two rings together and assuming D_{5d} symmetry, the representation of the 10 π-orbitals is given in Table 1.9. The reduction for the representation Γ_π gives: $a_{1g} + a_{2u} + e_{1g} + e_{1u} + e_{2g} + e_{2u}$.

Considering the symmetry labels (and energies) of the iron(II) orbitals, and combining them with the group orbitals of the two Cp ligands, an approximate MO diagram for ferrocene is shown in Figure 1.19. The principal bonding interac-

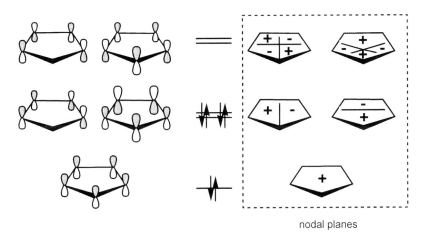

nodal planes

Figure 1.18 MOs for a $C_5H_5^-$ ring.

Table 1.9 Reducible representation of the 10 π-orbitals.

D_{5d}	E	$2C_5$	$2C_5^2$	$5C_2$	i	$2S_{10}$	$2S_{10}^3$	$5\sigma_d$
Γ_π	10	0	0	0	0	0	0	2

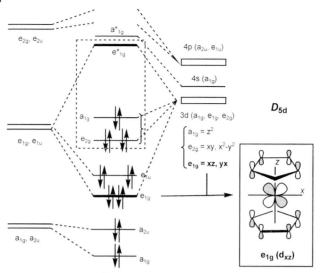

Figure 1.19 Scheme of the MOs in the ferrocene structure.

tion is that giving rise to the strongly bonding e_{1g} and strongly antibonding e_{1g}^* orbitals. To give one concrete example of how ring and metal orbitals overlap, the nature of this particular important interaction is illustrated in Figure 1.19 (inset). The other d orbitals (z^2, $x^2 - y^2$, and xy) make up a block of three nonbonding or nearly nonbonding orbitals. In fact, $x^2 - y^2$ and xy (e_{2g}) are stabilized by bonding interactions with the π^* orbitals of appropriate symmetry on the Cp rings. The order given in this figure may be different to that found in other books, mainly for the antibonding MOs. The metal orbitals e_{2g} and a_{1g} can be reversed, depending on the metal and the type of calculation.

1.3.5.2 Carbenes

Three chemists were awarded the Nobel Prize in 2005, *"for the development of the metathesis method in organic synthesis"*: Y. Chauvin, R. H. Grubbs and R. R. Schrock. Two of them (Grubbs and Schrock), have developed catalysts that improve considerably the metathesis effect. These catalysts are metal-carbene derivatives [9, 10]. Carbene complexes, whose general formula is $[L_nM=CR_2]$ formally contain an M=C double bond. Two group orbitals can be constructed for the bond with the metal center: a hybrid, sp and a pure orbital, p (Figure 1.20A).

There are two limiting cases: *Fisher carbenes* and *Schrock carbenes*. If we consider the carbene as an L-type ligand (with two electrons in the s type orbital) it therefore acts as a σ donor, which interacts with an empty orbital of the metal (e.g. z^2). In this model the p orbital is empty, so the carbene acquires a π-acceptor character (Figure 1.20A). This orbital can be either higher or lower in energy than the d orbitals on the metal.

When the p orbital is higher in energy than the d orbital, we obtain a typical back-donation scheme with the formation of a bonding MO mainly located on the

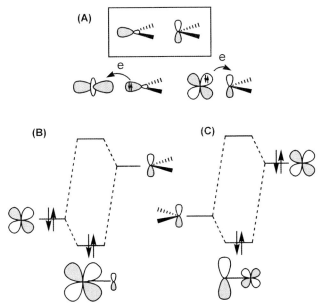

Figure 1.20 HOMO and LUMO in carbene complexes (see text for explanation).

metal (Figure 1.20B). However, if the p orbital is lower in energy than the d orbital, the occupied bonding MO is mainly concentrated on the carbene so that, in a formal sense, two electrons have been transferred from the metal to this ligand (Figure 1.20C), which leads to an 'increase' in the metal's oxidation state by two units.

The first situation (Figure 1.20B) is found for metals with d orbitals low in energy (right-hand side of the periodic table). In addition the presence of π-acceptor ligands also leads to a lowering of the level of the d orbitals. As far as the carbene is concerned, the energy of the p orbital is raised if the substituents are π-donors, that is, they have lone pairs (halogens, OR, NR_2, etc.). Carbene complexes that possess these characteristics are called *Fischer carbenes.* $[(CO)_5W=C(Ph)(OMe)]$ and $[Cp(CO)(PPh_3)Fe=CF_2]^+$ are paradigmatic examples. These carbenes are $[M(\delta-)=C(\delta+)]$, with an electrophilic character for the carbon center.

The second situation (Figure 1.20C) is found for metals on the left-hand side of the periodic table, and must have π-donor ligands to destabilize the d orbital. Moreover, to ensure that the p orbital on the carbene is as low in energy as possible, its substituents cannot be π donor: CH_2 is itself a good candidate and, more generally, alkyl substituents are suitable (the term 'alkylidene' is often used for a carbene substituted by alkyl groups). Two examples are $[Cp_2(CH_3)Ta=CH_2]$ and $[CpCl_2Ta=C(H)(CMe_3)]$. The metal in this group of complexes is usually considered to be oxidized by two units by the alkylidene ligand, so the two examples mentioned above contain Ta(V). These carbenes are called *Schrock carbenes.* Now the electron polarization is $[M(\delta+)=C(\delta-)]$.

1.4
Angular Overlap Model [11]

The angular overlap model is a simple and quantitative method that enables the energy of the molecular orbitals to be calculated from certain pre-established simplifications and tables. It is based on the quantum consideration that an orbital, described in polar coordinates, has a radial part and an angular part: $\Phi = R(r) \cdot Y(\theta, \phi)$ (Figure 1.1). The expressions of Φ for the d orbitals are shown in Table 1.1.

1.4.1
Overlap Integral

From the two orbitals μ and v we can derive what is known as the overlap integral, given by: $S_{\mu v} = <\phi_\mu | \phi_v> = S'_{\mu v}(\lambda, r) \cdot S''_{\mu v}(\lambda, \theta, \phi)$. λ depends on the type of atomic orbital, which may be σ, π or δ molecular orbitals.

A simple example to study is the typical case of angular dependence in the overlap between two orbitals: an s orbital that rotates around a d_{z2} orbital, modifying the overlap angle (Figure 1.21). The overlap will be proportional to $3\cos^2\theta - 1$ (Table 1.1), and is shown graphically in Figure 1.21. It can be seen that for an angle θ of 54.73° the overlap S is zero.

1.4.2
Energy of the Molecular Orbitals

The interaction energy associated with two atomic orbitals that overlap to form a molecular orbital is: $E_{\mu v} = <\phi_\mu | \hat{H} | \phi_v>$. The energy is negative (stabilization of the MO formed) when the overlap is positive, and positive (destabilization of the MO formed) when the overlap is negative. The *angular overlap model* is the simplest way of calculating the energies of the MO as it only considers the angular part of the wavefunction. The value of the stabilization (or destabilization) energy is $E \approx S_{ij}^2$. The model is only useful for comparisons and relative calculations, and not for absolute calculations.

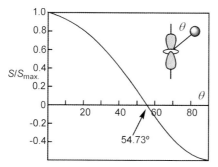

Figure 1.21 Variation of the relative overlap (S/S_{max}) varying the θ angle (see text).

1.4.3
The Additive Character Rule

When an atomic orbital of a metal overlaps with several atomic orbitals of ligands, the relative calculation of the stabilization (or destabilization) energy is given by what is termed the *additivity rule:* $E \approx \Sigma_n S_\lambda^2$ $(n$ = ligand; λ = σ, π, δ).

As we are interested in coordination complexes the sign given to the energy value must be taken into account. In general, the sign of the angular overlap parameter is given by the donor or acceptor nature of the ligands with respect to the central atom (transition ion). Although the σ bonds are always donors, this is not the case for the π bonds. For donor ligands the sign will be positive (destabilization of the MO, as we are considering anti-bonding orbitals). When the ligands are acceptors (π-acids) the sign is negative (stabilization of the anti-bonding molecular orbital), but *always* after considering the σ contribution, which is anti-bonding.

1.4.4
Tables for Calculating the Angular Overlap Parameters

Table 1.10 gives the general expressions for the relative values of S, with respect to θ and ϕ, for various types of overlap involving a central transition ion and the s and p orbitals of the ligands.

This general table can be used to construct other simplified tables which give the values of E_σ and E_π for the complexes with the commonest geometry directly. For example, Figure 1.22 shows the positions of various ligands, which readily enables the parameters E_σ and E_π to be calculated for the most usual geometries. These parameters are given directly in Table 1.11, as energy values proportional to S^2.

Table 1.10 Angular dependence of overlap integrals (for the s and p orbitals of the ligands with regard to the d orbital of the central ion).

S	Expression	S	Expression
s, z^2	$(3H^2 - 1)S_\sigma/2$	x, z^2	$\sqrt{3}FH^2 S_\pi + F(3H^2 - 1)S_\sigma/2$
s, xz	$\sqrt{3}\ FHS_\sigma$	x, xz	$-H(1 - 2F^2)S_\pi + \sqrt{3}F^2 HS_\sigma$
s, yz	$\sqrt{3}\ GHS_\sigma$	x, yz	$FGH(\sqrt{3}S_\sigma + 2S_\pi)$
s, xy	$\sqrt{3}\ FGS_\sigma$	x, xy	$-G(1 - 2F^2)S_\pi + \sqrt{3}F^2 GS_\sigma$
s, $x^2 - y^2$	$\sqrt{3}\ (F^2 - G^2)S_\sigma/2$	x, $x^2 - y^2$	$-F(1 - F^2 + G^2)S_\pi + \sqrt{3}F(F^2 - G^2)S_\sigma/2$
z, z^2	$-\sqrt{3}H(1 - H^2)S_\pi + H(3H^2 - 1)S_\sigma/2$	y, z^2	$\sqrt{3}\ GH^2 S_\pi + G(3H^2 - 1)S_\sigma/2$
z, xz	$-F(1 - 2H^2)S_\pi + \sqrt{3}\ FH^2 S_\sigma$	y, xz	$FGH(\sqrt{3}S_\sigma + 2S_\pi)$
z, yz	$-G(1 - 2H^2)S_\pi + \sqrt{3}\ GH^2 S_\sigma$	y, yz	$-H(1 - 2G^2)S_\pi + \sqrt{3}G^2 HS_\sigma$
z, xy	$FGH(\sqrt{3}S_\sigma + 2S_\pi)$	y, xy	$-F(1 - 2G^2)S_\pi + \sqrt{3}\ FG^2 S_\sigma$
z, $x^2 - y^2$	$-H(G^2 - F^2)S_\pi + \sqrt{3}H\ (F^2 - G^2)S_\sigma/2$	y, $x^2 - y^2$	$-G(G^2 - F^2 - 1)S_\pi + \sqrt{3}G(F^2 - G^2)S_\sigma/2$

$F = \sin\theta \cos\phi;\ G = \sin\theta \sin\phi;\ H = \cos\theta.$

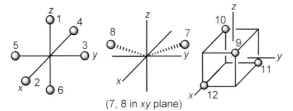

(7, 8 in *xy* plane)

Figure 1.22 Fixed position for ligands in several common geometries.

Table 1.11 Angular proportionality factors for E_σ and E_π. The numbering (1–12) corresponds to that used in Figure 1.22.

Position of the ligand	z^2	$x^2 - y^2$	xz	yz	xy
1 σ	1	0	0	0	0
π	0	0	1	1	0
2 σ	1/4	3/4	0	0	0
π	0	0	1	0	1
3 σ	1/4	3/4	0	0	0
π	0	0	0	1	1
4 σ	1/4	3/4	0	0	0
π	0	0	1	0	1
5 σ	1/4	3/4	0	0	0
π	0	0	0	1	1
6 σ	1	0	0	0	0
π	0	0	1	1	0
7 σ	1/4	3/16	0	0	9/16
π	0	3/4	1/4	3/4	1/4
8 σ	1/4	3/16	0	0	9/16
π	0	3/4	1/4	3/4	1/4
9 σ	0	0	1/3	1/3	1/3
π	2/3	2/3	2/9	2/9	2/9
10 σ	0	0	1/3	1/3	1/3
π	2/3	2/3	2/9	2/9	2/9
11 σ	0	0	1/3	1/3	1/3
π	2/3	2/3	2/9	2/9	2/9
12 σ	0	0	1/3	1/3	1/3
π	2/3	2/3	2/9	2/9	2/9

Another way of presenting these tables is to consider the various geometries with the sum of the contribution of each ligand, calculated previously. This offers a much quicker way of comparing the stability of the molecular orbitals formed for each geometry. The most noteworthy cases are shown in Table 1.12.

Table 1.12 σ and π interaction energy for various geometries.

Geometry	z^2		$x^2 - y^2$		xy		xz		yz	
	σ	π	σ	π	σ	π	σ	π	σ	π
MY linear, $C_{\infty v}$	1	0	0	0	0	0	0	1	0	1
MY$_2$ linear, $D_{\infty h}$	2	0	0	0	0	0	0	2	0	2
angular, C_{2v}	1/2	0	3/2	0	0	2	0	1	0	1
MY$_3$ facial trivacant, C_{3v}	3/2	0	3/2	0	0	2	0	2	0	2
triangular, D_{3h}	3/4	0	9/8	3/2	9/8	3/2	0	3/2	0	3/2
T form, C_{2v}	3/2	0	3/2	0	0	2	0	1	0	3
MY$_4$ tetrahedron, T_d	0	8/3	0	8/3	4/3	8/9	4/3	8/9	4/3	8/9
square-planar, D_{4h}	1	0	3	0	0	4	0	2	0	2
trigonal pyramid, C_{3v}	7/4	0	9/8	3/2	9/8	3/2	0	5/2	0	5/2
cis-divacant, C_{2v}	5/2	0	3/2	0	0	2	0	3	0	3
MY$_5$ trigonal bipyramid, D_{3h}	11/4	0	9/8	3/2	9/8	3/2	0	7/2	0	7/2
square pyramid, C_{4v}	2	0	3	0	0	4	0	3	0	3
MY$_6$ octahedron, O_h	3	0	3	0	0	4	0	4	0	4
trigonal prism, D_{3h}	3/8	0	9/16	3/2	9/16	3/2	2.25	1.5	2.25	1.5
MY$_7$ pentagonal bipyramid, D_{5h}	13/4	0	15/8	5/2	15/8	5/2	0	9/2	0	9/2
MY$_8$ cube, O_h	0	16/3	0	16/3	8/3	16/9	8/3	16/9	8/3	16/9
square antiprism, D_{4d}	0	2/3	4/3	32/9	4/3	32/9	8/3	37/9	8/3	37/9
MY$_{12}$ icosahedron, I_h	12/5	24/5	12/5	24/5	12/5	24/5	12/5	24/5	12/5	24/5

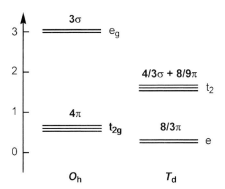

Figure 1.23 Energy calculation from the overlap angular model for O_h and T_d complexes.

1.4.5
Examples of Use of the Above Tables

The relative energies of the MO of octahedral and tetrahedral complexes can be easily deduced by using Tables 1.11 and 1.12, and the results of this are shown schematically in Figure 1.23. The quotient $\Delta S_{\text{tet}}/\Delta S_{\text{oct}}$ gives the value 4/9, as pointed out above.

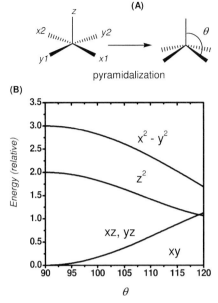

(A)

pyramidalization

(B)

Figure 1.24 Walsh diagram for the pyramidalization of a square-pyramidal geometry.

Table 1.13 $E(\sigma) = \Sigma S_\sigma^2$ of the four equatorial orbitals in a square-base pyramid complex.

	S/S_σ	S^2/S_σ^2	S^2/S_σ^2	ϕ	S^2/S_σ^2
s, $x^2 - y^2$	$\sqrt{3}/(F^2 - G^2)/2$	$3/4(F^2 - G^2)^2$	$3/4(\sin^2\theta \cos^2\phi - \sin^2\theta \sin^2\phi)^2$	0	$3/4(\sin^2\theta)^2$
				90	id
				180	id
				270	id
s, z^2	$(3H^2 - 1)/2$	$1/4(3H^2 - 1)^2$	$1/4(3\cos^2\theta - 1)^2$	–	$1/4(3\cos^2\theta - 1)^2$
s, xz	$\sqrt{3}FH$	$3F^2H^2$	$3\sin^2\theta \cos^2\phi \cos^2\theta$	0	$3\sin^2\theta \cos^2\theta$
				90	0
				180	$3\sin^2\theta \cos^2\theta$
				270	0
s, yz	$\sqrt{3}GH$	$3G^2H^2$	$3\sin^2\theta \sin^2\phi \cos^2\theta$	0	0
				90	$3\sin^2\theta \cos^2\theta$
				180	0
				270	$3\sin^2\theta \cos^2\theta$
s, xy	$\sqrt{3}FG$	$3F^2G^2$	$3\sin^2\theta \cos^2\phi \sin^2\theta \sin^2\phi$	0	0
				90	0
				180	0
				270	0

Starting from Table 1.10 it is easy to construct the Walsh diagram for the pyramidalization of a square-base pyramid complex, which is a common occurrence in this type of complex. The phenomenon of pyramidalization is shown schematically in Figure 1.24A. If we restrict our attention to the σ bonds of the four equatorial orbitals the functions that must be represented are $E(\sigma) \approx \Sigma S_\sigma^2$ (Table 1.13).

Therefore, the final values, summing the four values for the four ϕ angles, are: $x^2 - y^2 = 3(\sin^2\theta)^2$; xz, $yz = 6\sin^2\theta \cos^2\theta$; $xy = 0$; $z^2 = (3\cos^2\theta - 1)^2$. To these values must be added the overlap σ with the ligand in the apical position, which is invariant when pyramidalized and has a maximum value of $S/S_\sigma = 1$ for the z^2 orbital and 0 for the remaining cases. The representation of these trigonometric functions for angles from 90° (non-pyramidalization) to 120° gives the Walsh diagram shown in Figure 1.24B.

References

1 Figgis, B.N., Hitchman, M.A. (2000) *Ligand Field Theory and its Applications*, Wiley-VCH, Weinheim.

2 Cotton, F.A. (1990) *Chemical Applications of Group Theory*, 3rd edn., John Wiley & Sons, New York.

3 Jorgenssen, C.K. (1971) *Modern Aspects of Ligand Field Theory*, Elsevier, New York.

4 Huheey, J.A., Keiter, E.A., Keiter, R.L. (1993) *Inorganic Chemistry: Principles of Structure and Reactivity*, 4th edn., Harper Collins, New York.

5 Tudela, D., A Common Inorganic Chemistry Textbook Mistake: Incorrect Use of Pairing Energy in Crystal Field Stabilization Energy Expressions, *J. Chem. Educ.* 1999, *76*, 134.

6 Jean, Y. (2005) *Molecular Orbitals of Transition Metal Complexes*, Oxford University Press, Oxford.

7 C.A.C.A.O: Package of Programs for Molecular Orbital Analysis, Mealli, C., Proserpio, D.M. MO Theory made visible (CS), *J. Chem. Educ.* 1990, *67*, 399.

8 Kettle, S.F.A. (1992) *Symmetry and Structure*, John Wiley & Sons, Chichester.

9 (a) Schrock, R.R., Murdzek, J.S., Bazan, G.C., Robbins, J. , DiMare, M., O'Regan, M., Synthesis of molybdenum imido alkylidene complexes and some reactions involving acyclic olefins, *J. Am. Chem. Soc.* 1990, *112*, 3875–3886; (b) Bazan, G.C., Oskam, J.H., Cho, H.-N., Park, L.Y., Schrock, R.R., Living ring-opening metathesis polymerization of 2,3-difunctionalized 7-oxanorbornenes and 7-oxanorbornadienes by Mo(CHCMe₂R)(NC₆H₃-iso-Pr₂-2,6)(O-*tert*-Bu)₂ and Mo(CHCMe₂R)(NC6H₃-iso-Pr₂-2,6)(OCMe₂CF₃)₂, *J. Am. Chem. Soc.* 1991, *113*, 6899–6907.

10 (a) Nguyen, S.T., Johnson, L.K., Grubbs, R.H., Ring-opening metathesis polymerization (ROMP) of norbornene by a Group VIII carbene complex in protic media, *J. Am. Chem. Soc.* 1992, *114*, 3974; (b) Wu, W., Nguyen, S.T., Grubbs, R.H., Ziller, J.W., Reactions of Ruthenium Carbenes of the Type (PPh₃)₂(X)₂Ru=CH–CH=CPh₂ (X=Cl and CF₃COO) with Strained Acyclic Olefins and Functionalized Olefins, *J. Am. Chem. Soc.* 1995, *117*, 5503–5511.

11 (a) Burdett, J.K., A new look at structure and bonding in transition metal complexes. *Adv. Inorg. Chem.* 1978, *21*, 113; (b) Hoggard, P.E., Angular overlap model parameters, *Struct. Bond.*, 2004, *106*, 37. See the explanation given in, Purcell, K.F., Kotz, J.C. (1979) *Inorganic Chemistry*, Saunders Company, Philadelphia.

2
Classification of Ligands and Design of Coordination

2.1
Introduction

From the *electronic* point of view, ligands are molecules or ions with atoms from the non-metallic elements, which act as electron donors. The most common atoms are N, P, O, S, X (halogens) and C (in what are termed organometallic compounds).

Whatever the classification, a given ligand may belong to more than one group and, therefore, they should not be divided into *mutually exclusive* groups or classes. Nevertheless, there is an aspect that should be highlighted: the *structural aspect*. This refers to the nature and number of the ligand's donor atoms and how they bond to the metal ion. For example, the NH_3 ligand must bond through the nitrogen atom, whereas the ethylenediamine (en) ligand $NH_2CH_2CH_2NH_2$ has two active centers and can therefore form complexes using only one nitrogen (fairly uncommon) or two. In the latter case it can bond to a single metal ion acting as what is known as a *chelating* ligand, or to two centers acting as a *bridging* ligand. A ligand as simple as the O^{2-} ion can bond to one, two, three or even four metal centers.

Structurally, ligands can be classified according to the *dentate* nature of their donor groups, as well as in terms of the *polynucleating charac-teristics* of the ligand considered as a whole. In this chapter we focus mainly on the first aspect, polynucleating ligands will be considered in Chapter 5.

In general, we should not say that ligands ARE. . . . but that they ACT AS ...

Coordination Chemistry. Joan Ribas Gispert
Copyright © 2008 WILEY-VCH Verlag GmbH & Co. KGaA, Weinheim
ISBN: 978-3-527-31802-5

2.2
Classification of Ligands According to Their Dentate Nature

2.2.1
Monodentate Ligands

Let us consider the water molecule, H_2O. As it only has one electron donor center, the oxygen atom, it can be regarded as a monodentate ligand. However, it can act in several ways: if it bonds to a single metal center we would say it is a monodentate ligand that "acts" as a monodentate ligand, but if it bonds to two metal centers we would say that it is a ligand that "acts" as a bridging ligand. In contrast, if we consider the previous example of ethylenediamine, this appears to be a typical bidentate ligand (it has two donor centers capable of bonding with one or two metal ions). However, if it bonds through a single nitrogen atom it will then *act* as a monodentate ligand (extremely rare).

Monoatomic ligands are either monodentate or bridging. Halido (X^-) and oxido (O^{2-}) have been most investigated. The bridging properties of these ligands will be extensively discussed in Chapter 5. In recent years, unexpected monoatomic species acting as monodentate or bridging ligands have been reported. For example, both bridging and terminal borylene (B–R) coordination modes have now been realized. Braunschweig et al. reported the synthesis and structural characterization of the first compound to contain a boron center bonded solely to transition metals [1].

2.2.2
Polydentate (Non-encapsulating) Ligands

When a ligand with more than one donor atom acts as a polydentate ligand, it can do so in several ways, bonding to various metals and giving rise to a polynuclear complex or to a single metal and forming what is known as a chelate (from the Greek word for "claw" or "pincer"). In this section we will focus solely on this second aspect, while the formation of polynuclear complexes by bridging ligands will be studied later in Chapter 5.

A chelating ligand (or simply chelate) is a ligand in which the bonding atoms and the metal center form a closed ring, known as the *chelate ring*. Therefore, a chelating ligand must have at least two donor atoms to enable formation of this ring. The ring is characterized by how many members it has, the lowest possible number being three. The number of bonding atoms that can bond simultaneously to a metal center determines the dentate nature of the ligand. Thus, we refer to bidentate, tridentate, tetradentate and, in general, to polydentate ligands. A very important aspect to take into account in a bidentate ligand, or in each part of a polydentate ligand, is what is known as *bite* (Figure 2.1). Reference is usually made to bite, bite angle (θ) and normalized bite, which is defined as $b = d/r$.

The value of the parameter b gives a general idea of the extent to which a ligand will act as a chelating or a bridging ligand. Very small b values (three-membered

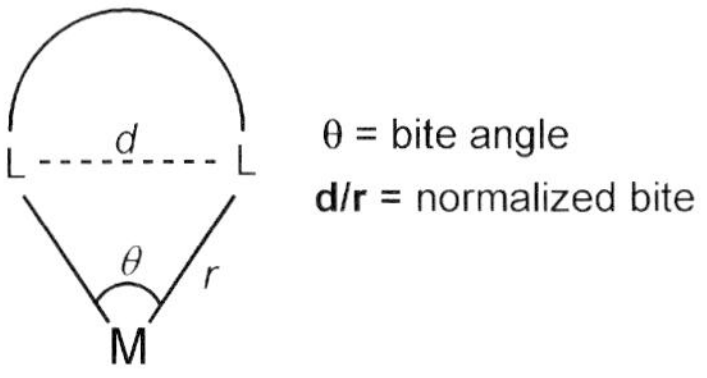

Figure 2.1 Concept of bite.

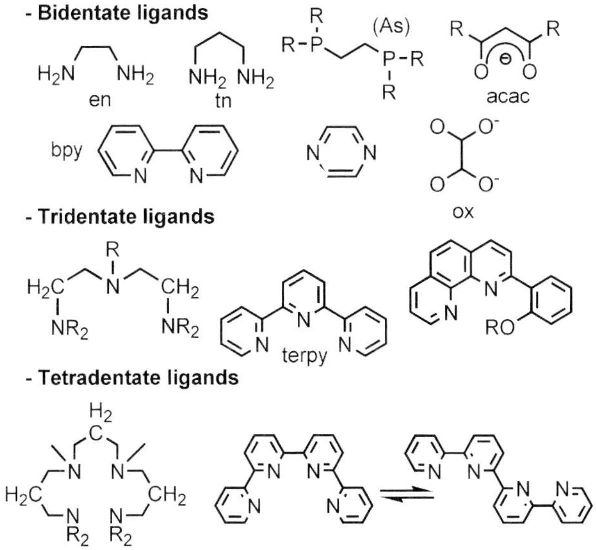

Figure 2.2 Selected bi- tri- and tetradentate ligands.

ring) force the ligand to act as a bridge, four-membered rings allow the ligand to act as a chelate or a bridge, rings of five or six members favor chelate formation, while still larger rings favor the formation of bridging or chelating ligands in the *trans*-form (not very common).

Polydentate ligands can be further sub-divided into open-chain and closed-chain. Open-chain ligands that branch from one atom (or group of atoms) are called *polypodal* ligands; the others are termed *non-polypodal* ligands. Closed chains will be studied later.

2.2.2.1 Non-polypodal Open-chain Ligands

These can be classified according to their dentate nature. Figure 2.2 shows some of the most important bi-, tri- and tetradentate ligands.

Bipyridine (bpy) and ethylenediamine (en) have a marked tendency to act as chelating ligands; 1,3-diaminopropane (tn) is likewise a good chelating ligand, although it can also adopt a conformation suitable for bridging (some examples are reported). 1,3-bis(diphenylphosphane)propane acts as a bridging and chelating ligand in [{Ni(CN)$_2$(dppp)}(μ-dppp)]. Pyrazine cannot act as a chelate but shows a marked tendency to act as a bridge, while the oxalate ligand acts as both a chelate and a bridge.

The stability of five- and six-membered chelate rings in coordination complexes is usually explained in terms of the *chelate effect*, in which a positive entropy term renders chelation thermodynamically favorable (see Chapter 7). The process of chelation becomes less entropically favorable for systems in which chelate rings containing more than six atoms can be formed. Blackman [2] reports a search on the Cambridge Structural Database, revealing that in 2004 there were 9891 metal complexes containing the −N(CH$_2$)$_2$N− fragment and 2762 containing the chelated −N(CH$_2$)$_3$N− fragment, but only 140, 59 and 6 containing the chelated −N(CH$_2$)$_4$N−, −N(CH$_2$)$_5$N− and −N(CH$_2$)$_6$N− fragments, respectively. In the majority of these latter complexes, the large chelate rings are incorporated in macrocyclic or cryptand ligands, and this preorganization undoubtedly aids chelation.

Of the chelating polypyridines, 2,2′-bipyridine (bpy) is the bidentate prototype and 2,2′:6′,2″-terpyridine (terpy) is the tridentate. Tridentate ligands can be divided into *meridionally coordinating* pincers, such as terpy, and tridentate *facially coordinating* such as tris(pyrazol-1-yl)borate ("scorpionates") (see Section 2.3.1).

The next higher homolog, quaterpyridine (Figure 2.2) commonly acts as a bridging ligand, twisting about the central bond to adopt a helical conformation, which binds two M^{2+} centers in a bidentate fashion. Rigidifying the central bond of this ligand by incorporating a phenanthroline subunit might prohibit bridging and induce mononuclear coordination [3].

Figure 2.3 shows penta- and hexadentate ligands. Due to their highly polydentate nature, many of these ligands do not always tend to bond with a single metal, but rather form polynuclear species. However, some pentadentate ligands form 1:1

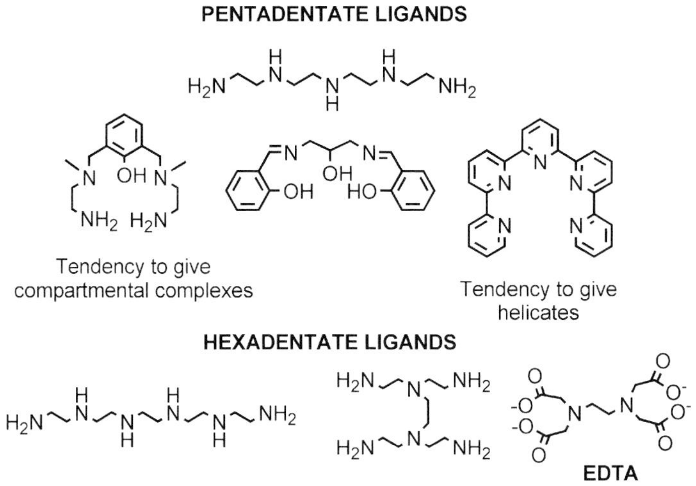

PENTADENTATE LIGANDS

Tendency to give compartmental complexes

Tendency to give helicates

HEXADENTATE LIGANDS

EDTA

Figure 2.3 Selected penta- and hexadentate ligands.

coordination with Ln(III) salts, with additional co-ligands bound to the metal centers [4]. Noteworthy among the hexadentate ligands is ethylenediamine tetraacetate (EDTA), a highly versatile ligand that can bond to a single metal ion through the two nitrogen atoms and four of the oxygen atoms of the carboxylate groups.

Ligands whose dentate nature is above six can yield a high coordination number with lanthanide ions or give rise to more complex polynuclear species, some of which will be studied in Chapter 5.

Summary Most of the open polydentate ligands can give mononuclear complexes and/or polynuclear ones. It is interesting to emphasize the tendency of certain polynucleating ligands to give mononuclear helical [5], or polynuclear helicate complexes. Helicate complexes will be extensively studied in Chapter 14.

2.2.2.2 Polypodal Open-chain Ligands

As their name suggests these are open-chain polydentate ligands in which the donor groups are all bound to a central atom (which may be a donor such as N or a non-donor like C). The central atom generally has three or more arms. Some typical examples of these polypodal (tripodal and tetrapodal) ligands are shown in Figure 2.4. They can be further sub-divided according to their central atom.

Nitrogen Derivatives These ligands contain a tertiary N atom bonded to three arms, each of which contains an N-donor atom, via at least one methylene group on each arm. The C_{3v} symmetric ligands tren, (tris(2-aminoethyl)amine), (Figure 2.4) and trpn, (tris(3-aminopropyl)amine), in which the three arms are the same length, have been known for many years.

Coordination to a metal ion generally results in a complex having a lower symmetry than the free ligand, depending on the coordination number of the metal ion. Four-coordinate complexes of tripodal amine ligands, in which all four N atoms are coordinated to a single metal ion, are relatively rare, as the ligands are often too constrained to be able to accommodate the metal ion. These rare examples display distorted tetrahedral geometries. The steric constraints imposed by such ligands often result in trigonal bipyramidal (tbp) geometries (with a new L

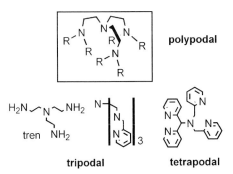

Figure 2.4 Selected polypodal ligands.

ligand) [6]. Examples of hypodentate coordination, in which the ligand binds using less than its full complement of donor atoms have also been reported [2, 7].

Boron and Carbon Derivatives The most important ligands are the poly(pyrazolyl)borates, $[H(R)B(pz)_3]$ and the carbon analogs. Due to their importance, they will be treated in depth later in this chapter.

Molecule Derivatives In order to form a polypodal ligand it is not necessary for the central group to be monoatomic, as in the previous cases. Indeed, it can be a molecule (with or without donor groups) to which the binding groups are attached via pendant arms. This class of ligands, currently of great interest due to their extraordinary coordinating potential and versatility, are generally referred to as *podands*.

2.2.2.3 Closed Polydentate Ligands: Macrocycles

For nearly 40 years, since the pioneering work of Curtis et al., there has been an interest in the kinetic and thermodynamic stability of macrocyclic complexes. Currently a great deal of attention is being focused on macrocyclic ligands because they play an important role in many aspects of chemistry, medicine and the chemical industry.

To date, the majority of these synthetic cyclic ligands contain oxygen, nitrogen, phosphorus and sulfur donor atoms. Macrocycles can be subdivided, thus, according to the ligand's donor atom, the two most important groups being the *crown ethers* and the *nitrogenated macrocycles*.

Crown Ethers The first synthetic crown ethers were synthesized by Pedersen in the late 1960s. The donor atoms are oxygen (ether-type) and have a great capacity to coordinate alkali and alkaline-earth metal ions. He showed that crowns, ranging in size from 9 to 60 atoms, could form complexes with alkali, alkaline-earth and alkylammonium cations. At the time of the publication (1967) synthesis of a ring containing 60 atoms was nearly inconceivable and Pedersen recognized that some unusual mechanism must be at work. This mechanism is now known as the *"template"* effect. Some of the most important crown ethers, together with the system for naming them, are shown in Figure 2.5A. The non-systematic names consist of, in order, (i) the number and kind of hydrocarbon rings, (ii) the number of atoms of the polyether ring, (iii) the word "crown" (or just the letter "C") and (iv) the number of oxygen atoms in the polyether ring.

Oxygenated macrocycles in which one (or more) of the oxygen atoms has been substituted by sulfur or nitrogen, which modifies the hard nature of oxygen atoms, are also regarded as crown ethers. Replacing two oxygen atoms with sulfur in the 18-crown-6 reduces the stability of the complex with K^+ but increases it considerably with Ag^+ (soft nature of S and the Ag^+ ion).

In 1971, Pedersen himself published the synthesis of *only*-sulfur-containing benzo-crown compounds. Thioethers are hypothetically capable of σ-bonding, as well as acting as π-donors and π-acceptors. Much work has been done on the

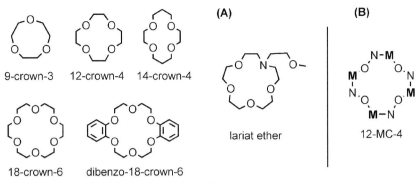

Figure 2.5 Crown-ethers; lariat ethers and metallacrowns.

coordination chemistry of thioethers in the past few years. They stabilize, for example, the very unusual Ag^{2+} and Au^{2+} oxidation states [8].

The possibility of adding functional groups to crown ethers has shown that the 'host–guest' capacity for stabilization is greater when the complexing cavity is well defined in advance. In the early 1980s, crown ethers modified by attaching one or more flexible side arms containing donor groups were synthesized. This class of compound has come to be known as the "*lariat ethers*" (Figure 2.5).

The selective character of crown ethers for binding alkali metals will be discussed in Chapter 14. Other crown-ether analogs, such as spherands etc., will also be discussed in Chapter 14.

Metallacrowns   In metallacrowns, a metal ion forms part of the macrocycle ring. Although often labile, these are sometimes kinetically inert so the integrity of the ring can be comparable with organic macrocycles. The most common forms involve the 'substitution' of ethylene groups from the sing with a metal ion adjacent to a nitrogen atom. (Figure 2.5B). A detailed discussion of this issue can be found in a review article by V. L. Pecoraro et al. [9].

Nitrogenated Non-aromatic Macrocycles   In this type of ligand the donor atoms are three or more nitrogens. Just as crown ethers are good ligands for alkali and alkaline-earth ions, nitrogenated macrocycles are good ligands for almost all transition elements. They may contain only single bonds, which gives them great flexibility, or double bonds which makes them more rigid. Macrocyclic ligands are easily obtained through the condensation of R—NH$_2$ groups with R—CO groups, often by means of template synthesis (in the presence of a cation). Macrocycles with single amine bonds can then be obtained through reduction of the C=N (imine) bonds.

Noteworthy among the macrocycles containing only simple C—N bonds are 9-ane-N3 (1,4,7-triazacyclononane) and 14-ane-N4 (1,5,8,12-tetraazacyclotetradecane, *cyclam*) (Figure 2.6A). The first of these is very effective, acting as a tridentate ligand with the cation outside the cycle cavity, due to its small size. With four or

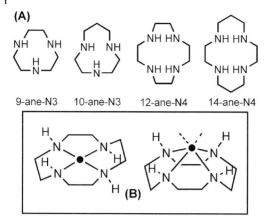

(A)

9-ane-N3   10-ane-N3   12-ane-N4   14-ane-N4

(B)

Figure 2.6 (A) Nitrogenated macrocyclic ligands and (B) their flexibility.

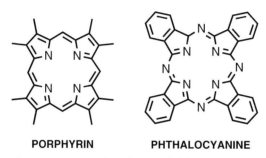

PORPHYRIN   **PHTHALOCYANINE**

Figure 2.7 Drawing of porphyrin and phthalocyanine.

more nitrogen atoms the cation is able to fit into the macrocycle cavity, depending on the relationship between the size of the cation and the methylene chain length in the rings. Other nonplanar coordination is also possible (Figure 2.6B).

Nitrogenated Aromatic Macrocycles: Porphyrins and Phthalocyanines   Porphyrins normally consist of four pyrrole rings linked by *meso*-carbons at the α-pyrrole carbons. Figure 2.7 shows the *porphyrin* ring as an example of a rigid nitrogenated ligand that appears in the heme groups in biochemistry, and the *phthalocyanine* (tetrabenzotetraazaporphyrin) ring, used as a pigment when coordinated and known for a variety of practical applications over the past 70 years.

The metal ion can fit either in the center of the macrocycle or outside it. It is worth recalling the paradigmatic example of iron–haemoglobin which will be discussed in Chapter 17. Large cations, such as the lanthanides, do not fit in the central ring and thus tend to form double- or triple-layer sandwich molecules.

There has been vast research interest in the preparation of periphery-substituted porphyrins in view of their utility as models to mimic cytochrome activity, photo-

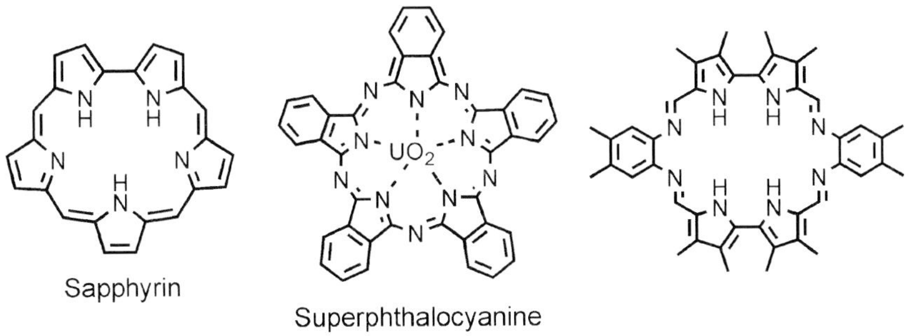

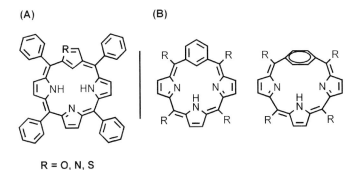

Sapphyrin

Superphthalocyanine

Figure 2.8 Selected expanded porphyrins.

(A)   (B)

R = O, N, S

Figure 2.9 Selected confused porphyrins.

chemical electron transport and ion channels [10]. Extension of the aromatic core has also been investigated, such as in perylenophthalocyanines [11]. Furthermore, many types of porphyrin analogs have been reported, which deviate from the basic tetrapyrrole structure [12]:

- *Expanded porphyrins.* Expanded porphyrins and phthalocyanines differ in terms of their larger central core, while they retain an extended electron configuration (Figure 2.8). The size and functionality of expanded porphyrins can potentially be varied to allow access to a large number of coordination modes. Furthermore, the preorganization of two or more metal centers at short separations (2.5–6 Å), would pave the way to studies involving cooperativity in catalysis and enzyme mimicking. An interesting review of transition-metal complexes of expanded porphyrins has been published recently [13].

- *Confused porphyrins.* The construction of carbaporphyrinoids (named *confused* porphyrins, Figure 2.9A), opens the way for a potential accommodation of a large variety of metal ions that allow exploration of organometallic complexes in a porphyrin-like environment. Atypical oxidation states of metal ions trapped in organometallic environments have been detected [14].

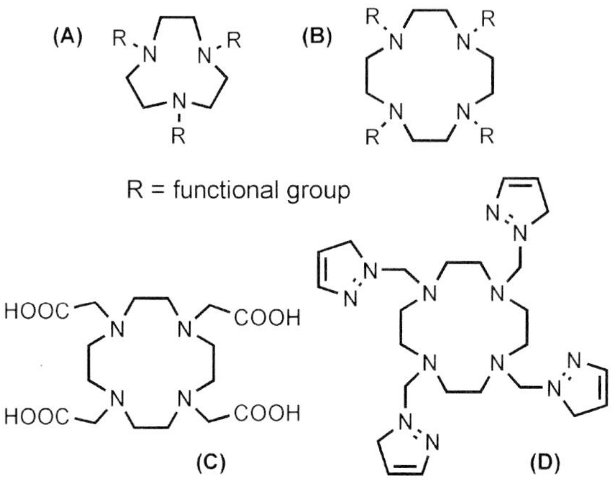

Figure 2.10 Nitrogenated macrocycles with pendant donor groups.

- *Benziporphyrins.* Similar to confused porphyrins, benziporphyrins are a class of porphyrin-like ligands in which the macrocyclic is formally constructed from three pyrrole rings and one benzenoid ring that are connected by meso bridges (Figure 2.9B). These ligands form organometallic complexes with transition metals and stabilize weak metal–arene agostic interactions [15].

Nitrogenated Macrocycles with Pendant Donor Groups [16]   Many macrocycles with one or more pendant chains containing a potentially coordinating group such as carboxylate, pyridine, amine, alcohol(phenol) or thiol have been prepared (Figure 2.10). Their importance lies in the fact that they are able to encapsulate a wide range of divalent and trivalent metal ions of various sizes. The corresponding complexes usually give rise to high stability constants, and these ligands have become useful in medicinal chemistry, e.g. in the formulation of diagnostic agents or in the design of drugs with antitumor properties, as well as in analytical applications.

1,4,7-triazacyclononane has served as the base for the synthesis of numerous derivatives with functional groups on the three nitrogen atoms, thus enabling their coordinating capacity to be increased (Figure 2.10A) and readily giving rise to octahedral complexes. In contrast, substituted tetraazamacrocycles may favor eight coordination (Figure 2.10B). This is what happens with alcohol groups as functional groups coordinating to ions such as Li^+, Na^+, K^+, Cd^{2+}, Hg^{2+}, Pb^{2+}, La^{3+} and Eu^{3+}, giving rise to square antiprism geometry. One of the most widely studied ligands of this type is that shown in Figure 2.10C. Its Zn^{2+} complex is octahedral, with two uncoordinated carboxylate groups; with Na^+ it forms a complex showing square antiprism geometry (coordination number 8), while with several lanthanides it gives nine coordination as it coordinates an additional H_2O molecule. It is used in medicine as it is a good complexing agent of ^{90}Y, which is employed in radioimmunotherapy. This complex is both thermodynamically stable and kinetically inert. The ligand shown in Figure 2.10D has been used to synthesize

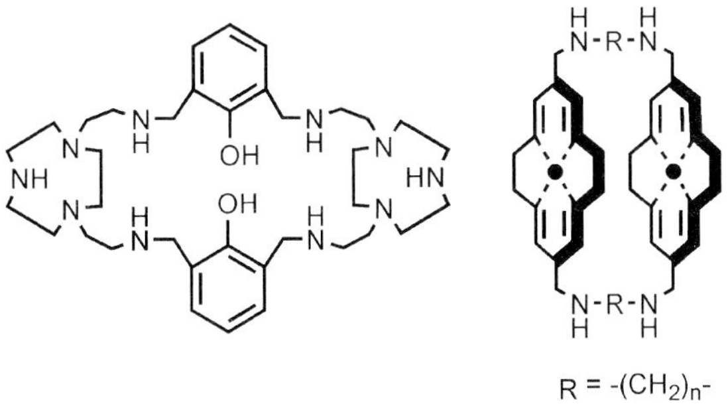

Figure 2.11 Macropolycyclic ligands.

a Na$^+$ complex that is unique in terms of its eight coordination with nitrogen atoms only.

Macropolycycles   The rational design and synthesis of macropolycycles capable of accommodating two or more metal centers in close proximity and in predetermined spatial arrangements is an active area of research. Two of them, with different features, are shown in Figure 2.11. Face-to-face bis-macrocyclic complexes are very interesting. By subtle changes in the structures of the bis-macrocyclic units, we can modify the strength of interaction between the metallic centers. The electrostatic interactions in the binuclear complexes become stronger with shortening of the alkyl linker. The increase in intermetallic interactions leads to the increased stability of the intermediate mixed-valence states. This aspect will be discussed in Chapter 13.

2.2.3
Cryptands and Encapsulating Ligands

In 1969 Dietrich, Lehn and Sauvage synthesized the first macrobicyclic ligands with donor oxygen groups and called them *cryptands*. The aim of this synthesis was to place (*hide*) a cation inside a three-dimensional cage, hence the name cryptand (cryptate), from the Greek word *kryptos* (hidden) (Figure 2.12).

In 1977, aiming more specifically at encapsulation of transition ions, Sargeson et al. synthesized the octaazacryptand complex (sepulchrate, sep), followed in 1984 by the ligand sarcophagine (sar) (Figure 2.13A) [17]. The sep and sar ligands were termed *encapsulating ligands*, giving rise to what are known as *cage complexes*. Subsequently, the terms cryptand and encapsulating ligand have been used almost interchangeably. Some authors, however, argue that the two groups should be distinguished: cryptand ligands or cryptate complexes are those in which there is a single atom (C, N) that closes the cage at each of the vertices of the three-dimensional macrocycle; in contrast, if the atom is substituted by a molecule the resulting ligand should be called an encapsulating ligand and the complex a cage complex (Figure 2.13B). The main groups of encapsulating ligands contain catechol or bipyridine functional groups. The tris(catechol) complexes are models of

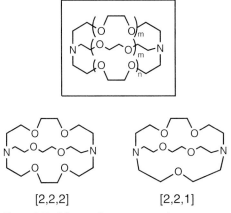

[2,2,2]   [2,2,1]

Figure 2.12 Scheme of some cryptands.

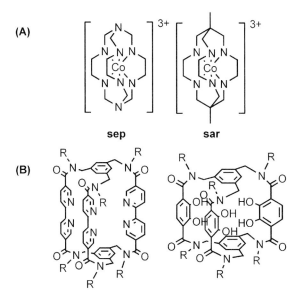

Figure 2.13 Cage complexes and encapsulating ligands.

the siderophore groups produced by certain microorganisms, capable of "protecting" small amounts (even traces) of Fe^{3+}. Therefore, these types of complex are known as siderands. Their stability constant may reach $10^{60} M^{-1}$, and they are stable up to pH 3.

This distinction is rather artificial and many authors consider other differences. Cotton [18], for example, considers that cryptates are synthesized apart from metal ions and then used to form complexes whilst encapsulating ligands are synthesized around the metal ion and cannot release it (sepulchrates, for example), allowing studies under extreme conditions. Such ligands can enforce unusual coordination geometries, such as trigonal prismatic instead of octahedral.

2.2.3.1 Cryptands with Ether-type Oxygen Atoms

Also known as Lehn cryptates, the most important examples are shown in Figure 2.12. These ligands are termed m,m,n-cryptand. In an analogous manner to crown ethers, they coordinate alkali metals. These new species (cryptates), such as [K^+(2,2,2-cryptand)], may stabilize and give crystals with radical anions, not easy to isolate [19], and with Zintl phases (Chapter 6). However, their main interest is that they have been of great importance in developing new species that could not be remotely imagined within the chemistry of alkali metal ions; indeed, they have allowed the stabilization of *alkalide* anions as well as of what are known as *electrides* (electron trapped in the cage (1984)).

The study of crystalline *alkalides* began in 1974 with the synthesis and structure determination of {Na^+(cryptand-[2,2,2])}Na^-. After more than a decade's search the first crystalline *electride*, the [Cs^+(18C6)$_2 \cdot e^-$] (18C6 = 18-crown-6), was synthesized in 1983, and structurally characterized in 1986 [20]. The electron is trapped in a nearly spherical cavity of radius ~2.4 Å, surrounded by eight sandwich-complexed cesium cations. Almost all alkalides and electrides synthesized contain alkali metal cations complexed by crown ethers, cryptands or their aza-analogs. The optical, electrical and magnetic properties range from those of weakly interacting localized electrons, to antiferromagnetism and spin-paired states, to electron delocalisation.

A *thermally stable electride* has been recently reported by Dye et al. [21a] with a cryptand[2,2,2] in which each of the linking arms contains a piperazine ring. The isostructural sodide, with Na^- anions in the cavities is also stable at and above room temperature. More recently, stable alkalides have also been prepared with calix[4]pyrrole, such as Li^+(calix[4]pyrrole)M^- (M^- = electron, Li^-, Na^-, and K^-) [21b].

Finally, cryptands have enabled the properties of crown ethers to be improved with respect to molecular recognition, ion transport and catalysis (see Chapter 14).

2.2.3.2 Cryptands with Nitrogen Atoms (Azacryptands) [22]

The simplest type here is that obtained from the tripod ligand tris(amino)ethylamine, *tren*, with a dialdehyde. The synthesis is valid for any tris(amino)alkylamine (Figure 2.14A), and involves a Schiff reaction. In all these cases the aldolic condensation initially gives rise to double bonds that can easily be reduced to single bonds with BH_4^-. The two groups differ in terms of their flexibility and coordinating capacity. In general, imine ligands (with soft nitrogen sp^2) coordinate well to Cu^+ but not to Cu^{2+}. In contrast, aminocryptands coordinate readily to Cu^{2+}, due to the (harder) sp^3 nature of the donor nitrogen atom. Amino derivatives are more flexible, thus enabling the cation to adopt a more suitable geometry. Obviously, long chains are more flexible and better able to adapt to a given cation. When they coordinate to one or more metals the configuration is often preorganized for coordination. This leads to the stabilisation effect known as the *cryptate effect*.

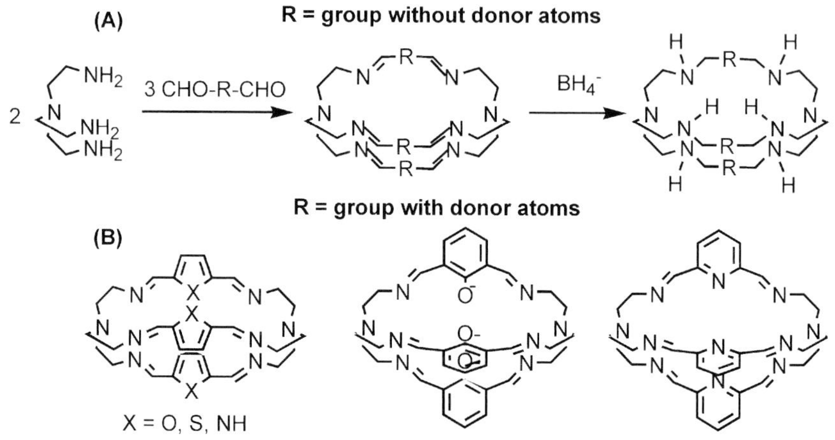

Figure 2.14 Azacryptands.

The most influential factor in the stability and lability of complexes is the R group of the bridging groups (Figure 2.14B). The presence of an anionic functional group (such as the phenoxy group) makes the complex formed much more stable. When kinetic lability is high these ligands are good ion transporters; when it is low they may be used as sequestering agents (toxicity studies in cations such as Cu^+, Tl^+ and Hg^{2+}). If the organic chain with the R group is very short (two carbons, for example), mono- as well as di-nuclear complexes can be obtained. In the latter case the metal–metal distance is very small, of the order of 2.4–2.5 Å which is indicative of a strong metal–metal interaction. For example, the mixed valence complexes of Cu^+–Cu^{2+} formed with these ligands are class III (see Chapter 13).

2.2.3.3 Cascade Effect

The term "cascade complex" or "cascade synthesis" was coined by Lehn and coworkers who noted that azacryptand ligands often incorporated two metal ions that are linked via a bridging anion (Figure 2.15). A number of different bicyclic ligands have been shown to be capable of forming cascade complexes with a variety of metal ions, along with "cascade" anions, such as halides and pseudohalides, hydroxide, carbonate, perchlorate and sulfate [23]. Often these ligands are held in unfamiliar geometries by the constraints of the cryptand skeleton.

Despite the scarcity of authenticated examples of *cascade reactivity*, the main challenge of these complexes is their capability for promoting unusual chemical reactivity. Nelson and coworkers, for example, have recently studied the tendency of different azacryptates to catalyze CO_2 reactions (in methanol) within these sterically-protected host cavities, generating bridging methylcarbonates [24].

2.2.3.4 Other Cage-like Complexes

Tetraaza-based Macrobicyclic and Macrotricyclic Complexes A recent theme in the synthesis of macrocyclic metal ions has been the extension of the ligand

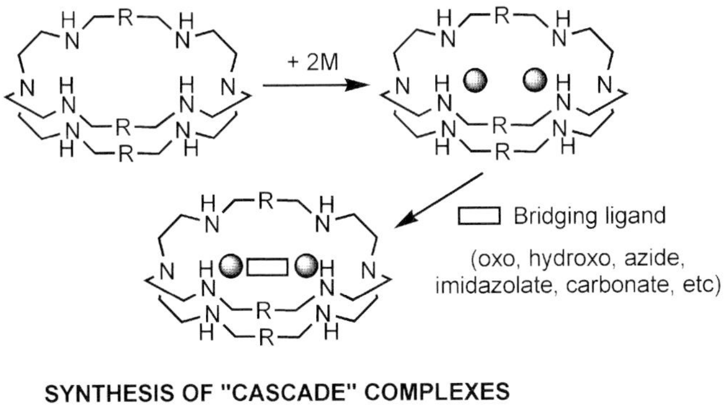

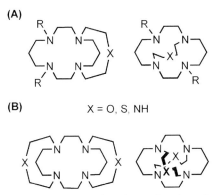

SYNTHESIS OF "CASCADE" COMPLEXES

Figure 2.15 Cascade effect.

(A)

(B)   X = O, S, NH

Figure 2.16 Tetraaza-based macrobicyclic and macrotricyclic ligands.

framework by use of a pendant smaller macrocyclic arm, capable of establishing additional coordination bonds. (Figure 2.16A). Such species result in both mono- and poly-nuclear complexes with constrained donor environments, but with the potential for specific site substitution [25]. Complexes formed by these ligands have a "hemicryptate" structure. A similar area of interest is the coordination of metal ions by macrotricyclic ligands, wherein the possibility exists of complete encapsulation of the metal center, resulting in complexes that are incapable of undergoing simple substitution reactions. Some examples are shown in Figure 2.16B.

Adamanzanes (Bowl and Cages) [26] A new class of macrobyciclic ligands is the so-called *bowl-adamanzanes* (Figure 2.17A). They stabilize low oxidation states and their metal ion complexes are extremely inert. Weisman in 1990 reported the synthesis of the first small bowl-adamanzane (Figure 2.17A(i)) and described its coordination compound with Li^+ and Na^+. It was found that Li^+ binds about 200 times better than Na^+. A large number of coordination compounds with

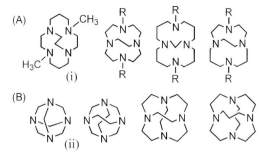

Figure 2.17 Adamanzanes (bowl and cages).

bowl-adamanzanes has been reported, including ones with mono- di- and tri-valent metal ions. Octahedral, trigonal bipyramidal, square pyramidal and tetrahedral geometries have been observed [26].

Strapping of the two secondary amine groups in the cyclic bowl-adamanzane leads to tricyclic tetramine cages, the so-called *cage-adamanzanes* (Figure 2.17B). The smallest member in this family, hexamethylentetramine (Figure 2.17B(ii)) was prepared more than a century ago and its crystal structure, reported in 1923, was the first structure of an organic compound obtained by X-ray diffraction.

These cages possess extreme acid–base properties. They are obtained as inside monoprotonated forms and it seems certain that the proton is encapsulated during the synthesis and not taken up by the cage after its formation. In fact, several observations indicate that these syntheses are templated-assisted with a coordinated proton serving as the template. In the solid state the proton is located at one of the four identical N atoms. However, in solution, the NMR spectra reflect a tetrahedral symmetry meaning the proton must shuttle rapidly. Thus, the inside coordinated proton is completely encapsulated by the organic framework of the macrotricycle.

Cage-adamanzanes are obvious candidates for the encapsulation of metal ions. This fascinating goal, however, has not yet been achieved. Dye and coworkers published the synthesis of the crystalline salt $AdzH^+Na^-$ [27].

2.2.3.5 Molecular Capsules

The potential of certain molecules that combine the properties of a conical cavity with those of a functionalized center may be highlighted through cyclodextrins, calix[*n*]arenes, resorcin[*n*]arenes, pyrogallol[*n*]arenes, calix[*n*]pyrroles, etc. (i.e. "*cavitand*" ligands) [28]. A variety of metallocavitands have been designed in which the metal atom is loosely tethered at the entrance to a receptor cavity.

If the cavity adopts the shape of a truncated cone, i.e. if it possesses two inequivalent entrances, then, ideally, the whole complex could function as a funnel able to select substrates according to their size and shape. Indeed, metallocavitands with coordination sites directed towards the interior of the generic cavity provide interesting systems for studying host–guest complexation processes and stabilization of coordination compounds of unusual forms, by adjusting the size and form of the pocket. It is also possible to activate and transform small molecules or to stabilize reactive intermediates. These pockets, furthermore, have the remarkably ability to fix and transform small molecules such as H_2O and CO_2 [29]. Calixarene derivatives, owing to their importance and versatility will be discussed later in this chapter.

2.3
Versatile Ligands

In general, it can be stated that *all ligands are versatile*, although some are characterized by great versatility when coordinating. Simple ligands such as nitrito, carbonato, oxalato or azido are very versatile and they will be discussed in Chapter 5. Some more complicated ligands merit special attention here, due to their current relevance and their ability to give rise to complexes with very peculiar characteristics.

2.3.1
Scorpionates

Scorpionates are a type of tridentate ligand with the general structure $[RR'B(pz)_2]^-$, where pz is either an unsubstituted or C-substituted pyrazolyl group that can coordinate metals to give complexes $[RR'B(\mu\text{-}pz)_2ML_n]$. A six-membered ring is formed and, in most instances, this has a boat conformation where the pseudoaxial R' group is directed towards the metal and may form a full or partial bond to the metal (Figure 2.18A). This is the reason for the name "scorpionate". Since poly(pyrazolyl)borates were first synthesized by Trofimenko in 1967 over 150 different scorpionate ligands have been synthesized. In 2004 Trofimenko published an interesting review article [30].

If R' in $[RR'B(pz)_2]^-$ is another pyrazolyl group and is identical to the bridging pz groups, then the ligand has C_{3v} symmetry $[RB(pz)_3]^-$. Such ligands are referred

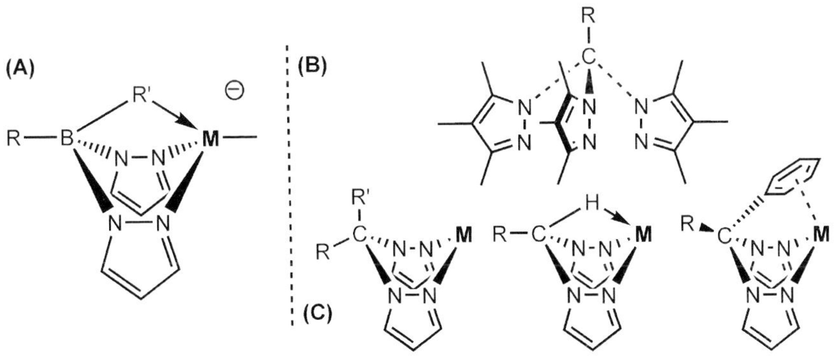

Figure 2.18 Several scorpionates.

to as "*homoscorpionates*". If not, the ligand is described as "*heteroscorpionate*", a description that also includes ligands where R′ is a pyrazolyl group different from the two other bridging pyrazolyl units.

2.3.1.1 Boron-Derivatives: 'Typical' Scorpionates

Tris(pyrazolyl)borates (Tp=HB(pz)$_3$] generally coordinate as tridentate ligands through three nitrogen atoms of the pyrazole rings (κ^3-$N,N′,N″$), thereby providing effective steric shielding of the metal center. By introducing suitable substituents (e.g. Me, CF$_3$, ^{t}Bu, Ph), particularly in the 3-position of the pyrazolyl rings, this effect can be tuned to a large extent. The tris(pyrazolyl)borate ligands are flexible and act like 'molecular vices', opening their tripodal structure for larger metals and closing around smaller metal ions. Placing sterically hindered groups on the central boron does influence the flexibility of the ligands, particularly when they have to open far from their ideal geometry in order to coordinate particular cations.

Several recent contributions have demonstrated that the coordination mode of Tp ligands can be far more versatile that hitherto anticipated. Besides the common κ^3-$N,N′,N″$ coordination mode the bidentate κ^2-$N,N′$ as well as κ^1 have been reported [31]. Naked Tp$^-$ ("κ^0") has been reported for the first time in [Rh(H)$_2$(PMe$_3$)$_4$][Tp] [32].

More recently there has been interest in the chemistry of scorpionates that incorporate softer Lewis donors such as phosphorus or sulfur in place of nitrogen, especially since the greater π-acidic character of these donors might be exploited for the binding of unusual low-valent electron-rich, transition metal centers. The so-called Janus scorpionate ligands (with N and S donor centers) have also been reported [33].

2.3.1.2 Carbon-Derivatives

The central boron atom can be replaced by other elements such as aluminum, indium, gallium, carbon, silicon, etc. Such a change can either preserve or alter the charge of the ligand. Poly(pyrazolyl)alkanes, [R$_n$C(pz)$_{4-n}$] ($n=0$, 1, 2) (Figure 2.18B) are stable and flexible polydentate ligands isoelectronic with poly(pyrazolyl)borates. Despite the fact that tris(pyrazolyl)methane was first reported in 1937, until recently its chemistry and that of its homologs, has been relatively unexplored, mainly due to difficulties in the preparation of large quantities of these ligands. Several reviews of this class of heteroscorpionates have been recently reported [34]. Figure 2.18C shows selected examples of coordination modes.

2.3.2
Calixarenes [35]

Calix[*n*]arenes belong to a group of macrocyclic ligands, prepared by condensation reactions of *para*-substituted phenols and formaldehyde. Examples with between four and eight phenol groups have been obtained. The number of phenol groups

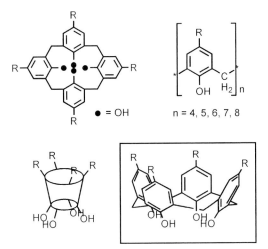

Figure 2.19 Different drawings of calixarenes.

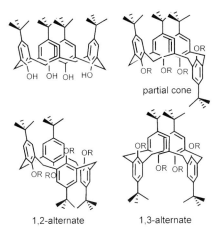

partial cone

1,2-alternate   1,3-alternate

Figure 2.20 Four possible conformations of the calix[4]arene.

is designated by *n* and the general term is calix[*n*]arene. Some of their representations found in the literature are shown in Figure 2.19.

The simplest and most common family has four phenolic residues and thus they are known as calix[4]arenes. In general, these have C_{4v} symmetry and resemble a torus or chalice; *calix* is the Greek word *chalice*, hence the origin of their name. The *p-tert*-butylcalix[4]arenes are the most studied. These molecules and their derivatives have been extensively studied for their many interesting properties and reactivity, and their chemistry has been applied in areas as diverse as catalysis, enzyme mimics, host–guest chemistry (for cations, anions and molecules), selective ion transport, and sensors. It has been found that, as expected, the affinity of a calixarene towards a metal ion depends on the conformation of the calix[4]arene (Figure 2.20).

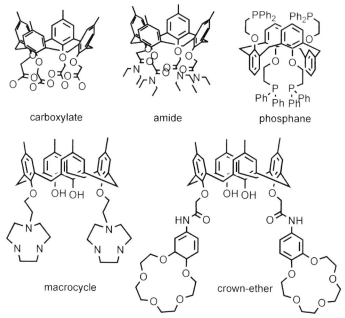

Figure 2.21 Functionalization of calixarenes.

2.3.2.1 Functionalization and Complexation

Calixarenes are very similar to crown ethers and cryptands in their complexing capacity with respect to alkali and alkaline-earth metal ions, but are much more versatile. Indeed, calixarenes have been widely regarded as important macrocyclic host molecules in host–guest chemistry (Chapter 14). Furthermore, the functionalization of calix[4]arenes (upper and/or lower rim) has led to the isolation of cavity-shaped podands displaying highly selective complexation properties (Figure 2.21).

Selected examples of metal complexes are shown in Figure 2.22. A review of these multivalent ligands has been published recently [36].

In 2005 Haino and coworkers developed calix[5]arene-based receptors for dumbbell-shaped C_{120} (two linked fullerenes) [37]. Several complexes have recently illustrated the ability of calix[6]arene H_6L to form large cluster complexes in a rational manner: Co_5, Ni_5, $CoNi_4$, $MnNi_4$, $CuNi_4$, Cu_{10} complexes [38].

An interesting characteristic of certain metal calixarenes is that they may form liquid crystals; this is due to the different radius of the two rings, which can form a *head–tail* arrangement. For example, the calixarene coordinated to the oxotungsten group is synthesized in dimethylformamide (DMF) with solvent molecules. At room temperature these molecules are occluded inside the calixarene, giving a normal *isotropic* solution. When heated the DMF molecules are eliminated and the liquid is reordered, giving a *mesophase* (Figure 2.23).

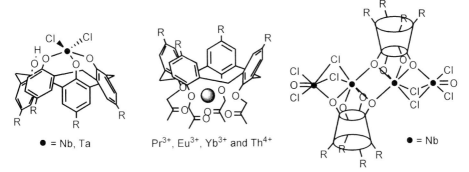

Figure 2.22 Several complexes formed with calixarenes.

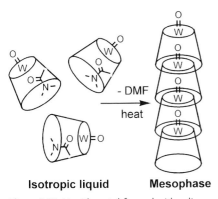

Isotropic liquid   **Mesophase**

Figure 2.23 Liquid crystal formed with calixarenes.

2.3.2.2 Deprotonated Calix[n]arene, Calixanions [39]

The discovery that calixarenes transport alkali metal ions in water/organic membrane systems has led to particular interest in deprotonated calixanions. The base M_2CO_3 (M = alkali metal ion) is a good choice for making calix[4]arene monoanions, while the same strategy with calix[6]arenes and calix[8]arenes leads to the formation of dianions, although the respective monoanions can also be obtained.

Calixanions can vary in terms of ring size, conformation, the nature of the cation, and the degree of deprotonation. The properties and reactivity of calixanions are very sensitive to each variable. The calix[6]arene dianion, for example, is very selective for K^+ from a K/Na solution, but there is no selectivity when crystallizing the same dianion from a K/Rb solution. Indeed, the first example of a mixed alkali metal complex, C6·KRb, has been obtained [39].

2.3.3
Fullerenes

Fullerenes are spherical carbon molecules with the formula C_{60} (in general C_n, where n is very large) (Figure 2.24A). Fullerenes that obey the isolated pentagon rule (IPR) avoid pentagon–pentagon contacts by having each or their 12 pentagons surrounded by 5 hexagons. This arrangement minimizes the steric strain that results for misalignment of the p-orbitals on the fullerene surface. For C_{60} and C_{70} there is only one way of arranging the pentagons and hexagons to produce an IPR structure, but for higher fullerenes several isomers can obey the IPR rule.

In recent years various metal complexes have been synthesized in which fullerenes act as ligands. Two types of complexes can be distinguished: *exohedral* and *endohedral* (depending on where the metal ion is situated, that is, outside or inside the organic fullerene).

2.3.3.1 Exohedral Metallofullerenes

The exohedral group is by far the most numerous. The most common modes of coordination are shown in Figure 2.24: π and σ organometallic modes and non-organometallic modes. The preferential reactivity of a C–C bond has been observed

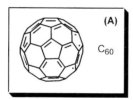

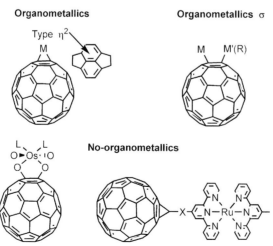

Figure 2.24 Different kinds of metallofullerenes.

in many reactions with transition metal complexes. Recall that, in C_{60} double bond character is associated with the C—C (hexagonal) edges. C—C displaces, thus, ethene from $[Pt(\eta^2\text{-}C_2H_4)(PPh_3)_2]$. Both C_{60} and C_{70} exhibit alkene-like character in addition reactions. In these products, the two carbon atoms that are directly bonded to the transition metal atom are pulled out from the fullerene cluster towards the metal atom.

2.3.3.2 Endohedral Metallofullerenes

It was first proposed in 1985 that fullerenes could confine atoms in their interior because of their closed-cage structure. Indeed, endohedral fullerenes, the closed-cage carbon molecules that incorporate atoms or a molecule inside the cage, are not only of scientific interest but are also expected to be important for their potential use in various fields such as molecular electronics, catalysis and biomedical applications. However, development of their applications has been hampered by severe limitations in their production, which has relied only on physical methods, such a co-vaporization of carbon and metal atoms. These metallofullerenes are generally produced by the arc-discharge method.

M $\subset$ C_{60} (M = lithium, alkaline-earth and lanthanide) are well known. Other cages are made of higher fullerenes such as Ca $\subset$ C_{72} and Ca $\subset$ C_{74}, with the peculiar feature that the free fullerenes have not been isolated. M $\subset$ C_{82} (M = La, Y, Ce, Pr, etc.) is one of the most abundantly produced endohedral metallofullerenes. The metal atoms tend to be located not in the center of the interior space but close to the carbon cage [40]. In most metallofullerenes there is a significant charge-transfer interaction between the encapsulated metal and the cage. Not only metallic atoms can be placed inside the fullerene: N $\subset$ C_{60}, P $\subset$ C_{60}, Kr $\subset$ C_{60}, N $\subset$ C_{70}, have been characterized. The atoms are freely suspended in the fullerene cage and exhibit properties resembling those of ions in electromagnetic traps (Figure 2.25A).

In some larger fullerenes, two or more metal atoms can be inserted, such as in La$_2$ $\subset$ C_{80}, Y$_2$ $\subset$ C_{84}, Sc$_3$ $\subset$ C_{82}. Some of the endohedral fullerenes have both non-metallic and metallic atoms, such as in Sc$_2$C$_2$ $\subset$ C$_n$ (n = 80, 82, 84) [41], M$_3$N $\subset$ C$_n$ (n = 80, 84, 86, 88; M = Ln^{3+}) (Figure 2.25B). M$_3$N $\subset$ C$_n$ have been recently

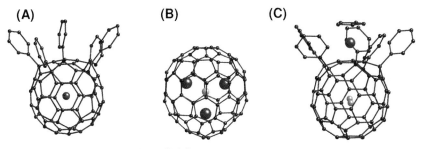

(A)   **(B)**   **(C)**

Figure 2.25 Several endohedral metallofullerenes.

synthesized in high yields by the so-called trimetallic nitride template (TNT) process (inclusion of N_2/He in the synthetic process) [42]. The isolation of pure isomers has been recently reported [43].

An alternative approach to synthesizing endohedral fullerenes is *"molecular surgery"*, in which the cage is opened and then closed in a series of organic reactions. Murata and coworkers succeeded in fully incorporating a H_2 molecule into a derivative on an open-cage fullerene, allowing the direct observation of the encapsulated molecule by X-ray (synchrotron) diffraction analysis (Figure 2.25C). This compound can be regarded as a nanosized container for a single H_2 molecule, in which hydrogen storage can be controlled by pressure and temperature. According to the authors, at present this $H_2 \subset C_{60}$ is the only molecule which can allow us to discuss the electronic state of a H_2 molecule that is entirely isolated from the atmosphere [44].

Most of the ligands studied in this chapter, such as crown ethers, cryptands, calixarenes, etc. exhibit "molecular-recognition" characteristics. These features will be extensively discussed in Chapter 14.

References

1 Braunschweig, H., Kollann, C., Rais, D. Transition-Metal Complexes of Boron – New Insights and Novel Coordination Modes, *Angew. Chem. Int. Ed. Eng.* 2006, *45*, 5254–5274.

2 Lundin, N.J., Hamilton, I.G., Blackman, A.G. Synthesis and hypodentate Cu(II) complexes of new tripodal tetraamine ligands incorporating long pendant arms, *Polyhedron*, 2004, *23*, 97–102.

3 Zong, R., Thummel, R.P. 2,9-Di-(2′-pyridyl)-1,10-phenanthroline: a Tetradentate Ligand for Ru(II), *J. Am. Chem. Soc.* 2004, *126*, 10800.

4 Albrecht, M., Mirtschin, S., Osetska, O., Dehn, S., Enders, D., Fröhlich, R., Pape, T., Hahn, E.F. Pentadentate Ligands for the 1:1 Coordination of Lanthanide(III) Salts, *Eur. J. Inorg. Chem.* 2007, 3276.

5 Drew, M.G.B., Parui, D., De, S., Naskar, J.P., Datta, D. An Eight-Coordinate Mononuclear Double Helical Complex, *Eur. J. Inorg. Chem.* 2006, 4026.

6 Blackman, A.G. The coordination chemistry of tripodal tetraamine ligands, *Polyhedron*, 2005, *24*, 1–39.

7 Constable, E.C. Higher Oligopyridines as a Structural Motif in Metallosupra-molecular Chemistry, in: Karlin, K.D. (Ed.), *Progress in Inorganic Chemistry*, vol. *42*, 1994, Wiley, New York, pp. 95–100.

8 Shaw, J.L., Wolowska, J., Collison, D., Howard, J.A.K., McInnes, E.J.L., McMaster, J., Blake, A.J., Wilson, C., Schröder, M. Redox Non-innocence of Thioether Macrocycles: Elucidation of the Electronic Structures of Mononuclear Complexes of Gold(II) and Silver(II), *J. Am. Chem. Soc.* 2006, *128*, 13827–13839.

9 Pecoraro, V.L., Stemmler, A.J., Gibney, B.R., Bowdin, J.J., Wang, H., Kampf, J.W., Barwinski, A. Metallacrowns: A New Class of Molecular Recognition Agents, *Prog. Inorg. Chem.* 1997, *45*, 83.

10 Even, P., Boitrel, B. Crown porphyrins, *Coord. Chem. Rev.* 2006, *250*, 519–541.

11 Cammidge, A.N., Gopee, H. Perylenophthalocyanines, *Chem. Eur. J.* 2006, *12*, 8609–8613.

12 Takeuchi, Y., Matsuda, A., Kobayashi, N. Synthesis and Characterization of meso-Triarylsubporphyrins, *J. Am. Chem. Soc.* 2007, *129*, 8271.

13 Sessler, J.L., Tomat, E. Transition-Metal Complexes of Expanded Porphyrins, *Acc. Chem. Res.* 2007, *40*, 371–379.

14 Xie, Y., Morimoto, T., Faruta, H. Sn(IV) Complexes of N-Confused Porphyrins and Oxoporphyrins – Unique Fluorescence "Switch-On" Halide Receptors, *Angew. Chem. Int. Ed.* 2006, *45*, 6907–6910.

15 (a) Stepien, M., Latos-Grazynski, L., Szterenberg, L., Panek, J., Latajka, Z. Cadmium(II) and Nickel(II) Complexes of Benziporphyrins. A Study of Weak Intramolecular Metal–Arene Interactions, *J. Am. Chem. Soc.* 2004, *126*, 4566; (b) Stepien, M., Latos-Grazynski, L., Szterenberg, L. Conformational Flexibility of Nickel(II) Benziporphyrins, *Inorg. Chem.* 2004, *43*, 6654–6662.

16 Wainwright, K.P. Synthetic and structural aspects of the chemistry of saturated polyaza macrocyclic ligands bearing pendant coordinating groups attached to nitrogen, *Coord. Chem. Rev.* 1997, *166*, 35.

17 (a) Sargeson, A.M.The potential for the cage complexes in biology, *Coord. Chem. Rev.* 1996, *151*, 89; (b) Harrowfield, J.M., Koutsantonis, G.A., Kraatz, H-B., Nealon, G.L., Orlowski, G.A., Skelton, B.W., White, A.H. Cages on Surfaces: Thiol Functionalisation of Co^III Sarcophagine Complexes, *Eur. J. Inorg. Chem.* 2007, 263.

18 Cotton, F.A., Wilkinson, G., Murillo, C.A., Bochmann, M. (1999), *Advanced Inorganic Chemistry*, 6th edn. John Wiley, New York.

19 Nakamoto, M., Yamasaki, T., Sekiguchi, A. Stable Mononuclear Radical Anions of Heavier Group 13 Elements: [(^tBu_2MeSi)_3E^•−]·[K^+(2.2.2-Cryptand)] (E = Al, Ga), *J. Am. Chem. Soc.* 2005, *127*, 6954.

20 (a) Ellaboudy, A., Dye, J.L. Cesium 18-Crown-6 Compounds. A Crystalline Ceside and a Crystalline Electride, *J. Am. Chem. Soc.* 1983, *105*, 6490; (b) Dawes, S.B., Ward, D.L., Huang, R.H., Dye, J.L. First Electride Crystal Structure, *J. Am. Chem. Soc.* 1986, *108*, 3534.

21 (a) Redko, M.Y., Jackson, J.E., Huang, R.H., Dye, J.L. Design and Synthesis of a Thermally Stable Organic Electride, *J. Am. Chem. Soc.* 2005, *127*, 12416; (b) Chen, W., Li, Z-R., Wu, D., Li, Y., Sun, C-C., Gu, F.L., Aoki, Y. Nonlinear Optical Properties of Alkalides Li^+(calix[4]pyrrole)M^− (M = Li, Na, and K): Alkali Anion Atomic Number Dependence, *J. Am. Chem. Soc.* 2006, *128*, 1072.

22 Nelson, J., McKee, V., Morgan, G. Coordination Chemistry of Azacryptands, *Prog. Inorg. Chem.* 1998, *47*, 167.

23 (a) Bond, A.D., Derossi, S., Harding, C.J., McInnes, E.J.L., McKee, V., McKenzie, C. J., Nelson, J., Wolowska, J. Cascade complexation: a single cyano bridge links a pair of Cu(II) cations, *Dalton Trans.* 2005, 2403; (b) Ravikumar, I., Suresh, E., Ghosh, P. A Perfect Linear Cu-NNN-Cu Unit Inside the Cryptand Cavity and Perchlorate Entrapment within the Channel Formed by the Cascade Complex, *Inorg. Chem.* 2006, *45*, 10046.

24 Dussa, Y., Harding, C., Dalgaard, P., McKenzie, C., Kadirvelraj, R., McKee, V., Nelson, J. Cascade chemistry in azacryptand cages: bridging carbonates and methylcarbonates, *J. Chem. Soc. Dalton Trans.* 2002, 1704.

25 Ingham, A., Rodopoulos, M., Coulter, K., Rodopoulos, T., Subramanian, S., McAuley, A. Synthesis, characterization and reactivity of some macrobicyclic and macrotricyclic hetero-clathrochelate complexes, *Coord. Chem. Rev.* 2002, *233–234*, 255.

26 Springborg, J. Adamanzanes – bi- and tricyclic tetraamines and their coordination compounds, *Dalton Trans.* 2003, 1653.

27 Redko, M.Y., Vlassa, M., Jackson, J.E., Misiolek, A.M., Huang, R.H., Dye, J.L. "Inverse Sodium Hydride": A Crystalline Salt that Contains H^+ and Na^−, *J. Am. Chem. Soc.* 2002, *124*, 5928.

28 (a) Canary, J.W., Gibb, B.C. Selective Recognition of Organic Molecules by Metallohosts, *Prog. Inorg. Chem.* 1997, *45*, 1; (b) Jeunesse, C., Armspach, D., Matt, D. Playing with podands based on cone-shaped cavities. How can a cavity influence the properties of an appended metal centre?, *Chem. Commun.* 2005, 5603; (c) Power, N.P., Dalgarno, S.J., Atwood, J.L. Robust and stable pyrogallol[4]arene molecular capsules facilitated *via* an octanuclear zinc coordination belt, *New J. Chem.* 2007, *31*, 17.

29 Kersting, B. Carbon Dioxide Fixation by Binuclear Complexes with Hydrophobic

Binding Pockets, *Angew. Chem. Int. Ed.* 2001, *40*, 3987.

30 Trofimenko, S. Scorpionates: genesis, milestones, prognosis, *Polyhedron*, 2004, *23*, 197.

31 Edelman, F.T. Versatile Scorpionates – New Developments in the Coordination Chemistry of Pyrazolylborate Ligands, *Angew. Chem. Int. Ed.* 2001, *40*, 1656.

32 Paneque, M., Sirol, S., Trujillo, M., Gutierrez-Puebla, E., Monge, M.A., Carmona, E. Denticity Changes of Hydro tris(pyrazolyl)borate Ligands in Rh^I and Rh^{III} Compounds: From κ^3- to Ionic "κ^0"-Tp', *Angew. Chem. Int. Ed.* 2000, *39*, 218.

33 Gardinier, J.R., Silva, R.M., Gwengo, C., Lindeman, S.V. A metallic tape stabilized by an unprecedented (μ_5-κ^2,κ^2,κ^2,κ^1,κ^1-) scorpionate binding mode, *Chem. Commun.* 2007, 1524.

34 (a) Bigmore, H.R., Lawrence, S.C., Mountford, P., Tredget, C.S. Coordination, organometallic and related chemistry of tris(pyrazolyl)methane ligands, *Dalton Trans*, 2005, 635; (b) Pettinari, C., Pettinari, R. Metal derivatives of poly(pyrazolyl)alkanes: I. Tris(pyrazolyl)alkanes and related systems, Metal derivatives of poly(pyrazolyl)alkanes: II. Bis(pyrazolyl)alkanes and related systems, *Coord. Chem. Rev.* 2005, *249*, 525 and 663.

35 Mandolini L. and Ungaro R. (Eds.), 2000, "*Calixarenes in Action*". Imperial College Press, London.

36 Baldini, L., Casnati, A., Sansone, F., Ungaro, R. Calixarene-based multivalent ligands, *Chem. Soc. Rev.* 2007, *36*, 254.

37 Haino, T., Seyama, J., Fukunaga, C., Murata, Y., Komatsu, K., Fukazawa, Y. Calix[5]arene-Based Receptor for Dumb-Bell-Shaped C_{120}, *Bull. Chem. Soc. Jpn.* 2005, *78*, 768.

38 Kajiwara, T., Shinagawa, R., Ito, T., Kon, N., Iki, N., Miyano, S. *p-tert*-Butylthiacalix[6]arene as a Clustering Ligand. Syntheses and Structures of Co_5^{II}, Ni_4^{II}, and Mixed-Metal $M^{II}Ni_4^{II}$ (M = Mn, Co, and Cu) Cluster Complexes, and a Novel Metal-Induced

Cluster Core Rearrangement, *Bull. Chem. Soc. Jpn.* 2003, *76*, 2267.

39 Hanna, T.A., Liu, L., Angeles-Boza, A.M., Kou, X., Gutsche, C.D., Ejsmont, K., Watson, W.H., Zakharov, L.N., Incarvito, C.D., Rheingold, A.L. Synthesis, Structures, and Conformational Characteristics of Calixarene Monoanions and Dianions, *J. Am. Chem. Soc.* 2003, *125*, 6228.

40 (a) Beavers, C.M., Zuo, T., Duchamp, J.C., Harich, K., Dorn, H.C., Olmstead, M.M., Balch, A.L. $Tb_3N@C_{84}$: An Improbable, Egg-Shaped Endohedral Fullerene that Violates the Isolated Pentagon Rule, *J. Am. Chem. Soc.* 2006, *128*, 11352; (b) Yang, S., Troyanov, S.I., Popov, A.A., Krause, M., Dunsch, L. Deviation from the Planarity – a Large Dy_3N Cluster Encapsulated in an I_h-C_{80} Cage: An X-ray Crystallographic and Vibrational Spectroscopic Study, *J. Am. Chem. Soc.* 2006, *128*, 16733.

41 Iiduka, Y., Wakahara, T., Nakajima, K., Nakahodo, T., Tsuchiya, T., Maeda, Y., Akasaka, T., Yoza, K., Liu, M.T.H., Mizorogi, N., Nagase, S. *Angew. Chem. Int. Ed.* 2007, *46*, 5562.

42 Zuo, T., Beavers, C.M., Duchamp, J., Campbell, A., Dorn, H.C., Olmstead, M. M., Balch, A.L. Isolation and Structural Characterization of a Family of Endohedral Fullerenes Including the Large, Chiral Cage Fullerenes $Tb_3N@C_{88}$ and $Tb_3N@C_{86}$ as well as the I_h and D_{5h} Isomers of $Tb_3N@C_{80}$, *J. Am. Chem. Soc.* 2007, *129*, 2035.

43 Stevenson, S., Mackey, M.A., Coumbe, C.E., Phillips, J.P., Elliot, B., Echegoyen, L. Rapid Removal of D_{5h} Isomer Using the "Stir and Filter Approach" and Isolation of Large Quantities of Isomerically Pure $Sc_3N@C_{80}$ Metallic Nitride Fullerenes, *J. Am. Chem. Soc.* 2007, *129*, 6072.

44 (a) Komatsu, K., Murata, M., Murata, Y. Encapsulation of Molecular Hydrogen in Fullerene C_{60} by Organic Synthesis, *Science*, 2005, *307*, 238; (b) Murata, M., Murata, Y., Komatsu, K. Synthesis and Properties of Endohedral C_{60} Encapsulating Molecular Hydrogen, *J. Am. Chem. Soc.* 2006, *128*, 8024.

Bibliography

Lindoy, L.F. (1989) *The Chemistry of Macrocyclic Ligand Complexes*, Cambridge University Press, Cambridge.

Constable, E.C. (1999) *Coordination Chemistry of Macrocyclic Compounds*, Oxford Chemistry Primers, Oxford University Press, Oxford.

Cram, D.J., Cram, M.J. (J.F. Stoddart Ed.) (1994) *Container Molecules and their Guests*, The Royal Society of Chemistry, Cambridge.

Trofimenko, S. (1999), *Scorpionates. The Coordination Chemistry of Polypyrazolylborate Ligands*, Imperial College Press, London.

3
Stereochemistry and Distortions in Coordination Compounds

3.1
Stereochemistry

The stereochemistry of coordination compounds is governed by *electrostatic and electronic* factors. The former concern the ligand repulsions that tend to stabilize the most energetically favorable geometry. The latter refer to the stability of each possible geometry for a given electron configuration, d^n.

3.1.1
Ligand–Ligand Repulsion: Kepert's Model [1]

This model is based on the hypothesis that ligands are point charges. The electrostatic repulsions between ligands favor the geometry in which the system's potential energy is a minimum. Thus, for example, the theory of repulsions predicts that three-coordinated complexes tend to adopt trigonal planar geometry, while for coordination numbers 4–12, assuming equal point ligands, the most stable geometries are the polyhedra shown in Figure 3.1. For some coordination numbers, such as 6 and 12, there is one polyhedron that clearly takes preference over the others: the octahedron and the icosahedron, respectively. However, for other coordination numbers the situation is more complex, as there is more than one polyhedron with similar energy.

3.1.2
Coordination Numbers: Various Factors

Although coordination numbers range from 1 to 12, the literature contains very few examples with coordination numbers at one extreme or the other (1, 11, 12). For example, for coordination number 1 only a few cases have been described of organometallic compounds with very sterically hindered ligands, such as 1,3,5-triphenylbenzene: $[Cu\{C_6H_2(C_6H_5)_3\}]$ and $[Ag\{C_6H_2(C_6H_5)_3\}]$.

Some factors favor low coordination (up to 6) while others favor high coordination (from 7 to 12). In general, the factors that favor low coordination are:

Coordination Chemistry. Joan Ribas Gispert
Copyright © 2008 WILEY-VCH Verlag GmbH & Co. KGaA, Weinheim
ISBN: 978-3-527-31802-5

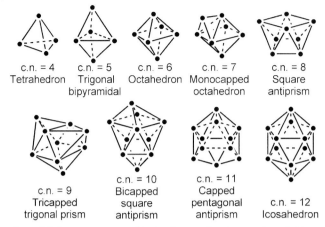

c.n. = 4 — Tetrahedron c.n. = 5 — Trigonal bipyramidal c.n. = 6 — Octahedron c.n. = 7 — Monocapped octahedron c.n. = 8 — Square antiprism

c.n. = 9 — Tricapped trigonal prism c.n. = 10 — Bicapped square antiprism c.n. = 11 — Capped pentagonal antiprism c.n. = 12 — Icosahedron

Figure 3.1 Representative polyhedra for coordination numbers 4–12.

(i) cations that are not too voluminous (first transition row); (ii) large, voluminous ligands. Steric considerations hinder high coordination; (iii) soft ligands and low oxidation state metals. The π-acid bond compensates for the loss of stability due to the few ligands present. Low oxidation state metals are electron rich and do not need the additional electron density provided by extra ligands. In contrast, the following factors favor high coordination numbers: (i) large cations, such as those of the second and third transition rows, and f^n elements; (ii) low steric hindrance of ligands; (iii) high oxidation state and hard ligands (such as fluoride, oxo, etc.).

3.1.3
Coordination Number 2

Most of the known examples are given with d^{10} ions (Group 11 and Hg^{2+}): $[M(NH_3)_2]^+$ (M = Cu, Ag); $[MCl_2]^-$ (M = Cu, Ag, Au); $[M(CN)_2]^-$ (M = Ag, Au) and $[Hg(CN)_2]$. These are linear complexes that readily react with their own ligands to give rise to greater coordination: $[Hg(CN)_2] + 2CN^- \rightarrow [Hg(CN)_4]^{2-}$. The recently reported copper(I) silanechalcogenates $[^tBuPCu^IESiPh_3]$ (E = O,S,Se) are also linear [2]. In some cases, great stability is achieved due to steric factors, such as in $[Cu(IPr)(R)]$ (IPr = 1,3-bis(2,6-diisopropylphenyl)imidazol-2-ylidene; R = Me, Et, anilide, ethoxide, phenoxide, etc) (Figure 3.2) [3]. The C—Cu—R linkage is nearly linear. Few cases have been described with other d^n ions, and all concern sterically hindered ligands.

The low stability of this coordination number is illustrated in Group 11 cyanocomplexes: whereas the derivatives of Ag^+ and Au^+ are two-coordinated, that of Cu^+, $K[Cu(CN)_2]$, has a chain structure in which the coordination number of Cu^+ is 3.

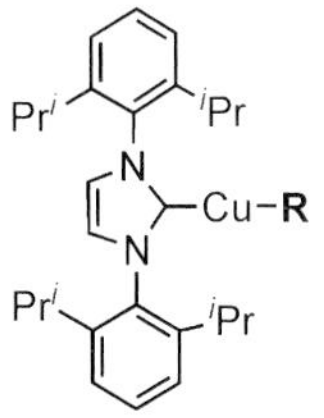

Figure 3.2 Ligand with strong tendency to give coordination number 2 (see text).

3.1.4
Coordination Number 3 [4]

Three-coordinated complexes are most common among the d^{10} ions of Groups 10, $[Pt^0(PPh_3)_3]$; 11, $K[Cu^I(CN)_2]$ (see above); $[Cu^I(SPMe_3)_3](ClO_4)$; and 12, $[Hg^{II}I_3]^-$. Although less common, for non-d^{10} configurations some important examples are $[Rh^I(PPh_3)_3]^+$ and a few organometallic complexes $[MR_3]$ or $[MR_2(thf)]$ (M = Mn, Rh, Ir). Some amide complexes also show this coordination mode, for example $[M(NR_2)_3]$, $[M(NR_2)_2L]$ and $[M(NR_2)L_2]$, with bulky substituents such as SiMe$_3$. Surprisingly, a few examples of rare earth complexes $[M\{N(SiMe_3)_2\}_3]$ have been reported, where M = Nd, Eu, Dy, Er, Yb) [4].

3.1.4.1 **Stereochemistry**
Discrete complexes with three-coordination usually have trigonal planar geometry. The main exception to this generalization is with low-spin d^8 systems, which clearly favour a T- or Y-shape geometry. MO calculations indicate that the potential energy surface for $[Au^{III}(CH_3)_3]$ is determined by the *HOMO orbital degeneracy* of the symmetrical D_{3h} point group and favor distortion to T- and Y- shaped geometries of lower energies, the former representing the minimum [5a].

For other d^n configurations, steric effects are predominant. For example, the two complexes in Figure 3.3A show clear distortion with respect to regular triangular geometry. In the first case the tridentate nitrogenated ligand forces the Cu to be coordinated in a T-shape, while in the second example the bidentate dimethyl-phenanthroline ligand forces the Cu to adopt Y-shaped geometry. In $[Cu^IL\{N\{(C_3F_7)C(Dipp)N\}_2]$ (Figure 3.3B), where L = CH$_3$CN, CNBu^t, CO; Dipp = 2,6-Diiso-propylphenyl), the copper center adopts trigonal planar geometry, with a Y-shape [5b]. A singular T-shaped platinum(II) boryl complex (Figure 3.3C) has been obtained from a square-planar precursor with a very long Pt–Br distance due to the high *trans*-influence of the B(Fc)Br group [5c]. The bulky tricyclohexylphosphane ligands cooperate to inhibit coordination of nucleophiles, such as THF or CH$_3$CN.

In general, three-coordinated complexes currently show coordinative unsaturation, thus explaining the low frequency of coordination number 3 among the structural data. A unique dinuclear three-coordinated Cr(I) complex, with μ-η^2-N$_2$

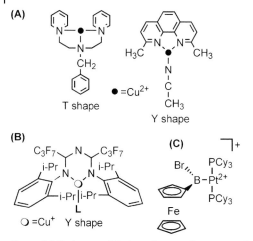

Figure 3.3 T-shape and Y-shape for coordination number 3.

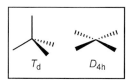

M.O. (Overlap Angular Model)

Figure 3.4 MO diagrams for the two most important geometries for coordination number 4.

ligand has been reported recently, showing marked reactivity with a variety of molecules, such as π-acids or potential oxidants, increasing the coordination number and the oxidation state of the metal [6].

Most trihalides of metals belonging to Groups 4–7 have been characterized in the gas phase as trigonal molecules, but they present solid state structures based on MX_6 octahedra such as ZrI_3 (1D), or BiI_3 (2D) structures [7].

3.1.5
Coordination Number 4

Essentially, four-coordinated complexes may be tetrahedral or square-planar (Figure 3.4). For electrostatic reasons four-coordinated complexes should be tetra-

Table 3.1 Potential energy values for various configurations (only σ interaction).

d^n	$E\ (T_d)$	$E\ (D_{4h})$		$\Delta(D_{4h} - T_d)$	
		(low spin)	(high spin)	(low spin)	(high spin)
d^1	0	0	0	0	0
d^2	0	0	0	0	0
d^3	4/3 σ	0	0	−4/3 σ	−4/3 σ
d^4	8/3 σ	σ	σ	−5/3 σ	−5/3 σ
d^5	4 σ	σ	4σ	−3 σ	0
d^6	4 σ	σ	4σ	−3 σ	0
d^7	4 σ	σ	4σ	−3 σ	0
d^8	16/3 σ	2 σ	4σ	−10/3 σ	−4/3 σ
d^9	20/3 σ	5 σ	5σ	−5/3 σ	−5/3 σ
d^{10}	8 σ	8 σ	8σ	0	0

hedral; square-planar coordination should be very rare. However, as we will see below, there are many square-planar complexes due to electronic reasons. A mapping of the stereochemistry and symmetry of four-coordinated transition-metal complexes has been reported [8].

3.1.5.1 Energy of the MOs (Angular Overlap Model)

Figure 3.4 shows the relative potential energy values of the MO for the two geometries. Table 3.1 gives the values of the corresponding energies and the difference between the two geometries. It should be borne in mind that tetrahedral complexes are *always* high spin; for square-planar complexes the high and low spin complexes are distinguished as follows: (i) in the case of low spin the four orbitals of lowest energy are filled first, prior to beginning the occupation of the $x^2 - y^2$; (ii) in the case of high spin, 3σ is lower than P (pairing energy) and the molecular orbitals are occupied as in the tetrahedral case.

This table is very important for complexes with *only monodentate ligands*. It can be deduced that for configurations d^0–d^2, d^5, d^6, d^7 (high spin) and d^{10} there is no energy difference from the electronic point of view; therefore, electrostatic considerations will take precedence and most complexes will be tetrahedral. For other configurations there may be equilibrium between the two energies (electrostatic and electronic), and thus either of the two geometries may arise. For example, d^4 complexes show no clear structural preference and have been found with perfect tetrahedral, square or distorted structures, such as $[CrCl_4]^{2-}$ with a scissor distorted geometry [8]. Four-coordinated d^8 complexes are mostly square-planar, especially when the crystal field created by the ligands is large. Many structural reasons explain why square-planar copper(II) complexes, d^9, are commonly found.

3.1.5.2 Tetrahedral Complexes

Tetrahedral complexes are mainly produced with large ligands (Cl^-, Br^-, I^-, etc.) and small transition ions, as in this way the ligand–ligand repulsions are minimized. There are three types of these ions: (i) noble gas configuration (d^0), such as Be^{2+}; (ii) pseudo-noble gas configuration (d^{10}), such as Zn^{2+}. For Zn complexes most structures (84%) are tetrahedral, but there is a significant 11% of square-planar structures [8]; (iii) any configuration that is not favored by electronic aspects. For example, complexes of Co^{2+} (d^7) with weak crystal field ligands (halides, for example) show no electronic difference between T_d and D_{4h}. Therefore, it is not surprising that more tetrahedral complexes have been identified for Co^{2+} than for any other ion. Furthermore, d^4 and high-spin d^5 complexes also include tetrahedral geometry.

3.1.5.3 Square-planar Complexes

Square-planar complexes are sterically less favored than tetrahedral complexes. Thus, very large ligands tend not to yield this geometry, and neither do very small ligands favor it as they tend to increase the coordination number, giving rise to octahedral coordination. The d^8 configuration is the most favored for this type of geometry. In general, it usually arises from ligands with low steric hindrance, which create a strong field, and with π-acid characteristics, which compensate for the loss of energy by not shifting from five- to six-coordination. Cyanide ligand forms highly stable square-planar complexes with the Ni^{2+} ion (although an excess of CN^- will favor five-coordination); ammonia ligands, water, etc. form octahedral complexes while more voluminous ligands, such as halides, form tetrahedral complexes. In contrast, Pd^{2+}, Pt^{2+} and Au^{3+} ions form square-planar complexes, even with halides, due to their greater crystal field energy.

3.1.5.4 *cis*-Divacant Octahedron

Although *cis*-divacant O_h geometry (also known as *sawhorse*) is not very common it is of great interest for its reactivity on the open side of the metal atom, the ability to form agostic interactions and the possibility of fluxional behavior through an intermediate square-planar geometry. The symmetry point group is C_{2v}. These structures include $[CoCl_4]^{2-}$ (d^7) and $[RuH_4]^{4-}$ (d^8) [8].

3.1.5.5 Polydentate Ligands: Steric Effects

Polydentate ligands can modify the predicted electronic structure. The ideal normalized bite is 1.63 for tetrahedral bond angles and 1.41 for square geometry [8]. Ligands with larger bites are well adapted to all structural situations (square or tetrahedral limits).

The geometry given by tridentate ligands depends on the nature of these ligands. For example, a potentially meridional ligand, such as terpy, does not give tetrahedral geometry without great distortion, while potentially facial ligands, such as cyclic triazacyclononane, give nearly tetrahedral structures. Nickel(II) scorpionate complexes [Ni(Tp)X] (X = Cl, Br, I) (see Chapter 2) show four-coordinate C_{3v} geometry [9].

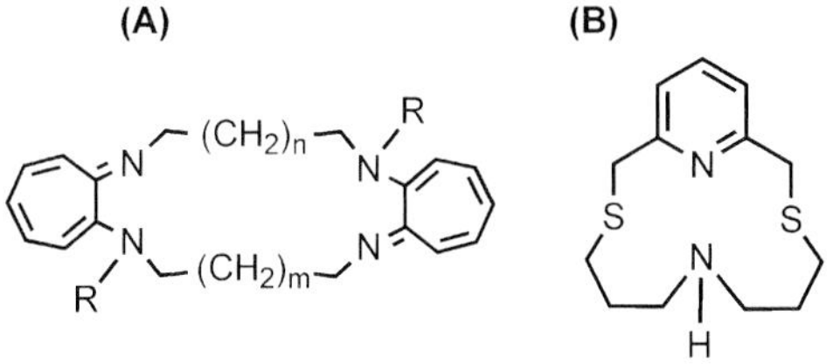

(A)   **(B)**

Figure 3.5 Two ligands that can give different kind of geometries (see text).

Tetradentate tripod ligands give distorted tetrahedral complexes, while macrocyclic aromatic ligands favor square-planar geometry. The flexibility of the nonaromatic macrocycles, like cyclam, allows major distortions. A good example of a constrained geometry imposed on d^n ions by macrocyclic ligands such as those shown in Figure 3.5A has been reported: the coordination sphere only approaches the tetrahedron when relaxing the rigidity of the ligands by increasing the size of the aliphatic rings ($m + n$) [10].

3.1.5.6 $D_{4h} \leftrightarrow O_h$ Equilibrium

Many square-planar complexes of first-row ions present a $D_{4h} \leftrightarrow O_h$ equilibrium. This is achieved by adding molecules from the solvent. Some of these equilibria are reversible (through heating they lose the solvent, and in its presence acquire it anew). For example, Lipschitz salts, $[Ni(bidentate\text{-}amine)_2]^{2+}$, are easily converted into six-coordinated complexes. Many other Ni^{2+} complexes have been reported with octahedral geometry triggered by the coordination of two solvent molecules to the nickel ion, which depends on the temperature and the solvent [11a]. Such an equilibrium gives rise to the phenomena known as *thermochroism* and *solvatochroism*, respectively. The color of a dry nitromethane solution of the square-planar complex $[Ni(L)](ClO_4)_2$ (**L** = Figure 3.5B) turns from red to cyan, purple, blue, yellow-green and pink, following addition of halides, acetonitrile, water, pyridine, and bpy, respectively [11b]. The geometry is changed from square-planar to octahedral. These solvent molecules can be easily removed by heating the resulting complexes, which indicates the reversibility of the process.

The equilibrium is so delicate that certain dinuclear complexes of Ni^{2+} may have one ion square-planar and the other octahedral. $[Ni(acac)_2]$ trimerizes to give $[Ni_3(acac)_6]$. In contrast, with the trimethylacetylacetonate ligand only the square-planar complex is obtained, while with ligands with groups intermediate between CH_3 and $C(CH_3)_3$ there is an equilibrium between several coordinations.

3.1.6
Coordination Number 5

There is no regular polyhedron able to give five-coordination. The two non-uniform polyhedra that give rise to this coordination are the *trigonal bipyramid* and the *square-base pyramid* (Figure 3.6A). Figure 3.6B shows the approach given by Addison et al. for intermediate geometries [12]. tbp and sp geometries are very

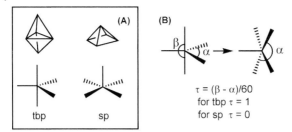

$$\tau = (\beta - \alpha)/60$$
for tbp $\tau = 1$
for sp $\tau = 0$

Figure 3.6 The two most stable geometries for c.n. = 5. Definition of the τ parameter (see text).

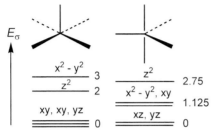

Figure 3.7 MO diagram for square-pyramidal and trigonal bipyramidal geometries (coordination number 5).

close in energy, as demonstrated by both molecular orbital arguments [13] and electrostatic ligand repulsion calculations [1]. The transition from a trigonal bipyramid to a square-base pyramid does not require a large amount of energy and, therefore, this isomerization occurs readily. From the square-base pyramid it is possible to return to the original tbp or its isomer. The movement along the reaction coordinate that connects the two bipyramids is generally known as Berry pseudorotation.

3.1.6.1 Electronic Aspects

Figure 3.7 shows the energy values of the MOs for trigonal bipyramid and square-base pyramid geometries based on the angular overlap model and taking into account only σ interactions. For any configuration, sp complexes are more stable from the electronic point of view when considering only σ interactions. Introducing π interactions produces a variation in the values, although in general the difference is not very great. This calculation will be left to the reader to carry out.

In the case of the Cu^{2+} ion, with a strong tendency towards five-coordination, subtle variations in the ligands, or even in temperature, can change the coordination. For example, $[CuX_5]^{3-}$ (X = Cl, Br) species are usually tbp at high temperatures and sp at low ones.

One of the spontaneous tendencies of complexes with square-planar geometry is what is known as *pyramidalization*. This tendency is the result of both steric (repulsion between the apical ligand and the four equatorial ligands) and electronic

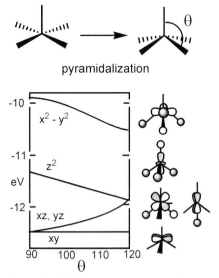

pyramidalization

Figure 3.8 MO diagram for the pyramidalization effect in square-pyramidal geometry.

effects. Pyramidalization is quantified by means of the value of the angle θ between the apical and equatorial ligands (Figure 3.8). This phenomenon of pyramidalization can be studied semi-quantitatively by calculating the MOs using the angular overlap method, as performed in Chapter 1. Figure 3.8 shows the Walsh diagram when varying the value of the angle θ from 90° to 120°. It is easily demonstrated that d^6 ions with low spin will tend toward $\theta = 90°$, while those with high spin will produce pyramidalization. For the remaining d^n there may be equilibrium between the two configurations. In any event, most complexes show a certain degree of pyramidalization due to steric effects and ligand–ligand repulsions.

3.1.6.2 [ML$_5$] Complexes

ML$_5$ molecules (assuming the five ligands to be the same) may present both types of structure and/or any intermediate one ($0 < \tau < 1$). A given complex often presents one structure or another depending exclusively on the packing forces. A special case is that of the $[Ni(CN)_5]^{3-}$ anion in the double complex salt $[Cr(en)_3][Ni(CN)_5]$. In the same crystal the $[Ni(CN)_5]^{3-}$ coexists as a square-base pyramid and a distorted trigonal bipyramid.

3.1.6.3 [M(bidentate)$_2$L] Complexes

Rules cannot be established with respect to geometry as both structures (tpb and sp) continue to have very similar energies, and which one of them arises depends on other factors. Hathaway and coworkers have shown, for example, that $[Cu(AA)_2L]$ complexes (AA = bpy or phen) give a variety of stereochemistries depending on the ligand X and the counter ion [14]. In the series $[Cu(hfacac)_2L]$ (hfacac = hexafluoroacetylacetonate, L = NH$_3$, H$_2$O), when L = NH$_3$ the geometry is tbp whereas

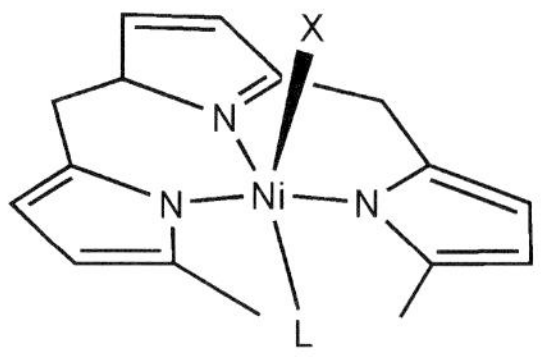

tbp tendency

Figure 3.9 Trigonal bipyramid tendency in tripyrrin ligand: effect of the other ligands (see text).

with L = H_2O it is sp [15]. Some complexes present both geometries, for example, chloro-bis(diphenylphosphaneethane)cobalt(II): the red isomer has square-planar geometry, while that of the green one is trigonal bipyramid. Certain species, such as VO in $[VO(L–L)_2]$ complexes, favor sp geometry.

3.1.6.4 [M(tridentate)L₂] Complexes

Two types of tridentate ligands can be distinguished. In the first group the ligand is based on a flexible aliphatic chain such as $RA(CH_2CH_2AR_2)_2$ (A = N, P) or $X(CH_2CH_2X)_2$ (X = O, S). The second group contains more rigid tridentate ligands, such as terpyridine (terpy). From the stereochemical point of view, the former tends to adopt square-planar geometry, while the latter prefers trigonal bipyramid geometry. In general, the voluminous substituents (R) in a tridentate ligand with an aliphatic chain exert a force such that the two L ligands tend to separate as much as possible, favouring trigonal bipyramid geometry.

Tridentate *tripyrrin* ligands (Figure 3.9) with Ni^{2+} give distorted tbp complexes in which the geometry is tuned by the other ligands [16]. The behavior is mainly associated with the fact that the terminal methyl groups of the tripyrrolic ligand shield the fourth equatorial position and therefore force the anionic co-ligand to bind in a different location, i.e. above or below the NiN_3 plane. With Cl and H_2O the geometry is clearly a distorted trigonal bipyramid with two of the tripyrrin nitrogens occupying the apical and one occupying an equatorial position. With chelated oxalate (X and L sites) the N donors of the tripyrrin ligand occupy one apical and two equatorial positions of the central Ni^{2+} ion. With chelated nitrate (bite angle of the nitrate ligand of 59°), a trigonal bipyramidal coordination of Ni^{2+} similar to that of oxalate is found, with a greater distortion interpreted as an intermediate geometry between a pentacoordinated square pyramidal and a tetracoordinated pseudotetrahedral one (one Ni– length is 2.3 Å).

3.1.6.5 [M(tetradentate)L] Complexes

Tetradentate ligands favor a given geometry in accordance with their characteristics. For example, tripod ligands favor tbp geometry (Figure 3.10A). Certain flexible tripod ligands may also give square-planar geometry. For example, the tripod ligand $[N(CH_2CH_2PR_2)_3]$ gives very distorted tbp geometry with Co^{2+} and Cl^- ions, while with I^- it gives square-planar geometry (Figure 3.10B). In general, macrocy-

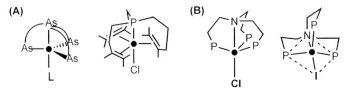

Figure 3.10 Effect of tripodal ligands in certain complexes with coordination number 5.

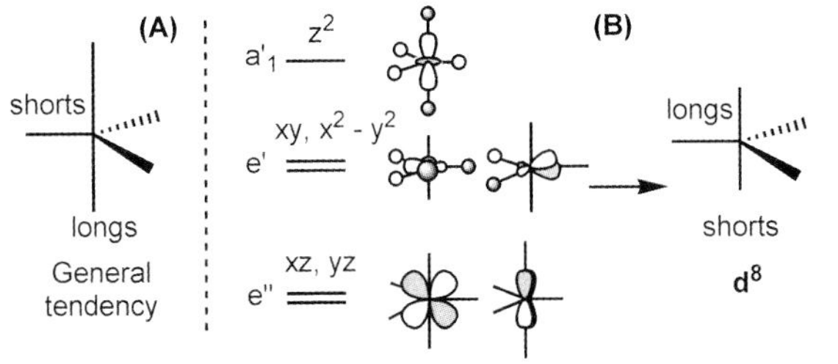

Figure 3.11 Metal–ligand distances in d^n (mainly d^8) configuration, in trigonal bipyramid geometry.

clic ligands with four donor atoms tend to give square-pyramidal coordination, if a new monodentate ligand is added. Two ligands in *trans*-positions can also be added, giving rise to octahedral geometry.

3.1.6.6 Bond Lengths in tbp and sp Geometry

An important characteristic of five-coordination is the difference between the positions of the ligands (non-equivalent bonds due to symmetry). The reasons are both electrostatic (ligand–ligand repulsions) and electronic in nature.

The trigonal bipyramid has two types of non-equivalent position: the three equatorial positions are equivalent to one another but not with respect to the axial positions. The axial ligands experience greater repulsion than do the equatorial ones [1]. This results in d(axial) > d(equatorial) (Figure 3.11A).

A study of the electronic influence for each d^n configuration confirms this tendency in most, but not all, cases. For certain electronic configurations, the tendency of the bond lengths is inverted. This can be checked by using the corresponding MOs energy diagram (Figure 3.11B). The only orbital that has electronic density from the ligands in the axial position is the $a'_1(z^2)$ orbital; the orbitals of e' symmetry $(xy, x^2 - y^2)$ have electron density from the equatorial ligands and the e'' symmetry orbitals (xz, yz) do not have electronic density from the ligands at all. Therefore, when only the e'' $(d^0–d^4)$ orbitals are occupied there is no electronic reason why the electrostatic tendency should be modified: the equatorial distances will be shorter. When one or several electrons occupy the e' orbitals (antibonding) the equatorial distance is weakened and becomes longer. This occurs especially in d^8 $(e')^4$ complexes. In d^9 complexes there is the opposite position of the electron in a'_1, which weakens the axial distance.

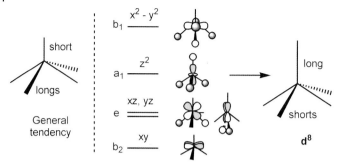

Figure 3.12 Metal–ligand distances in d^n (mainly d^8) configuration, in square-pyramidal geometry.

The square-base pyramid has non-equivalent positions: the four basal positions are equivalent to one another, but not with respect to the apical position. Taking into account pyramidalization, a ligand in the basal position experiences more repulsions than does the apical ligand. This results in d(basal) > d(apical) (Figure 3.12A). A study of the electronic influence for each d^n configuration confirms this tendency in most, but not all, cases. For certain electron configurations, the tendency of the bond lengths is inverted. This can be checked by using the corresponding MOs energy diagram (Figure 3.12B). The only orbital that has electronic density from the ligands in the axial position is the $a_1'(z^2)$ orbital; there is electronic density from the basal ligands in the orbitals with b_1 (x^2-y^2) and e (xz, yz) symmetry; there is no electronic density from the ligands in the b_2 (xy) orbital. Therefore, for d^0–d^2 configurations only electrostatic reasons will take effect; for d^3–d^6 the anti-bonding nature of the e orbitals will make the basal distances longer (increased electrostatic effect). Upon reaching d^7 (low spin) the apical σ orbital that overlaps perfectly with the z^2 will destabilize this bond, but does not always compensate the destabilization of the four e electrons. In contrast, in the transition to d^8 (low spin) (with two electrons in z^2) the destabilization is very strong and weakens the apical bond, thus making it longer. These results are shown schematically in Figure 3.12B. In accordance with theoretical predictions, low spin d^7 and d^8 complexes have shorter basal distances.

Readers interested in the electronic aspects of transition metal pentacoordinated complexes should refer to the article by Hoffmann et al. from early 1975 [17].

3.1.7
Coordination Number 6

The most stable geometry for six-coordination is octahedral (trigonal antiprism) (Figure 3.13A). The trigonal prism (D_{3h}) is unstable for both electrostatic and electronic reasons. Other geometries are only found in the presence of polydentate ligands that force the stereochemistry.

(A)

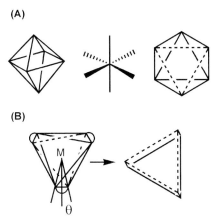

(B)

Figure 3.13 Octahedron vs. trigonal prism geometry (coordination number 6).

3.1.7.1 O_h with Bidentate Ligands

[M(A–A)$_2$(L)(L')] complexes When *b* is very small and there is a great difference in the nature of the two monodentate ligands the extraordinarily rare pentagonal (deformed) pyramid geometry may arise. Known examples are: [CrO(O$_2$)$_2$(py)] and (NH$_4$)$_4$[VO(O$_2$)$_2$(NH$_3$)] (*b* of the peroxide = 0.77). When *b* is large, as in (NH$_4$)$_4$[VO(ox)$_2$(H$_2$O)] (*b* = 1.24) the structure is the conventional octahedral one.

[M(A-A)$_3$] complexes From Figure 3.13B it can be deduced that the regular octahedron has $\theta = 30°$ and $b = \sqrt{2}$. As *b* becomes $< \sqrt{2}$ the two triangular faces tend to be eclipsed, and an intermediate geometry between O_h and D_{3h} is increasingly stabilized. The ideal trigonal prism corresponds to $\theta = 0°$. If we restrict ourselves to ions in the first transition row the ligands can be divided into three large groups. In the first the parameter *b* varies between 1.50 and 1.40 and comprises ligands that form six-membered chelate rings, such as 1,3-propanediamine, acetylacetonate and derivatives (θ close to 30°). In the second group *b* varies between 1.40 and 1.25 and contains ligands with five-membered chelate rings, such as phenanthroline and ethylenediamine. In the third group *b* varies between 1.25 and 1.05 and comprises ligands that form four-membered chelate rings, such as nitrate and dithiocarbamates (θ close to 15–20°) [1].

3.1.7.2 Trigonal-prismatic Geometry

Calculations made by Hoffmann et al. give a lower energy for D_{3h} over O_h for certain d^n configurations, mainly for d^0–d^2 [18]. Furthermore, other d^7–d^{10} structures may produce a stabilization of the trigonal prismatic geometry. The size of the metal ion, the bite angle of chelated ligands, and the steric hindrances, etc. will influence the equilibrium between different geometries. The following cases are particularly noteworthy:

Complexes with Monodentate Ligands Although rare, even some examples with only monodentate ligands have been reported [19]. In 1986 it was predicted that the neutral hexacoordinated permethyl complexes, $[M(CH_3)_6]$ (and MH_6) should be non-octahedral [20]. Indeed, d^1-$[Re(CH_3)_6]$ has a regular prismatic structure. Neutral d^0 complexes such as $[W(CH_3)_6]$ and $[Mo(CH_3)_6]$ have a C_{3v} distorted trigonal-prismatic structure. Given that all compounds such as WF_6, WCl_6, $[W(OR)_6]$, $[W(NR_2)_6]$ and their molybdenum analogs are octahedral, Kaupp calculated that for the system $[W(CH_3)_nCl_{6-n}]$ the trigonal-prismatic structure will prevail for $n = 3$–6, while for $n \leq 2$ the octahedral structure predominates [21]. Seppelt and coworkers reported several structures for $[M(CH_3)_nX_{6-n}]$ complexes in which they found this predicted tendency for trigonal-prismatic vs. octahedral geometry [22]. According to these authors, the trigonal-prismatic structure is retained for hexacoordinated neutral complexes when only one (or two) of the ligands has π-donor abilities (X, OR, NR).

Complexes with Dithiolate (or Dithiolene) $S_2C_2R_2^{2-}$ Ligands The first reported example of a trigonal prismatic complex $[Re(1,2\text{-}S_2C_2Ph_2)_3]$ dates back to 1965 [23]. Several tris(dithiolate) complexes with trigonal-prismatic or very distorted octahedral geometry have been reported (Table 3.2). The study of factors which contribute to the stabilization of the trigonal-prismatic geometry was reported in 1966 [24].

With respect to the X-ray structural determination of these complexes, the S–S distance on the triangular faces always takes a value close to 3.0 Å (0.6 Å shorter than the van der Waals distance). This relatively short distance, nominally non-bonded, indicates that there are interligand bonding forces present in these complexes, which are considerably stronger than in classical octahedral complexes.

It can be concluded that for these (and similar tris-chelated) complexes the trigonal prismatic structure is only possible for d^0 to d^2 configurations if the normalized bite is smaller than 1.34. If either of the two conditions is not met, a twisted octahedron is preferred with the twist angle related to the normalized bite [19].

Table 3.2 Angular values of dithiolate complexes ($\theta = 0$, trigonal prism; $\theta = 30$, octahedron).

	Group 4	Group 5	Group 6	Group 7	Group 8
d^0		VL_3; $\theta = 0$			
d^0	ZrL_3^{2-}; $\theta = 19.6$	NbL_3^-; $\theta = 0.6$	MoL_3; $\theta = 0$		
		TaL_3^-; $\theta = 16$			
d^1		VL_3^-; $\theta = 17$		ReL_3; $\theta = 0$	
				TcL_3; $\theta = 4.6$	
d^2			MoL_3^{2-}; $\theta = 14$		
			WL_3^{2-}; $\theta = 14$		
d^4					FeL_3^{2-}; $\theta = 24$

The geometry of the MS_6 polyhedron in these complexes has been discussed at various levels of sophistication to understand the trigonal-prismatic arrangement in neutral species. The mono- and di-anions display a more distorted geometry, which is intermediate between an octahedral and a trigonal-prismatic geometry. Recent studies of Wieghardt et al. have proved that the neutral complexes contain a Mo(v) ion, a ligand π-radical monoanion, and two closed-shell ligand dianions, $[M^V(L)_2(L\bullet)]^0$, whereas the monoanions $[M(L)_3]^-$ contains a central Mo(V) or W(V) ion (d^1) and three closed-shell, dianionic ligands [25].

It is worth noting that certain sulfides and selenides of Nb, Ta, Mo, W (MS_2, MSe_2) that have d^1 and d^2, respectively, present the trigonal prism structure in the solid crystalline state [7].

Complexes with Bis(bidentate), Expanded Hexadentate or Cryptand Ligands The rigid pyridazine-based ligands shown in Figure 3.14A, display a tendency to give, in a few cases, dinuclear $[M_2L_3]$ complexes with trigonal prism geometry for each M. However, the outcomes of the self-assembly are usually binuclear species with only two side-by-side ligands (instead of three), with the typical octahedral coordination being completed by axial ligands [26]. It is worth noting that by eliminating the rigidity of the ligands (i.e. through the possibility of rotation around the N–N moiety) the so-called helicates are synthesized (see Chapter 14).

A well-known strategy for obtaining trigonal-prismatic complexes is using rigid, penta- or hexa dentate ligands that force the geometry upon a complex by means of steric hindrances. Certain rigid cryptand ligands, such as those shown in Figure 3.14B, tend to encapsulate cations with d^5 configuration, favoring the *tendency* toward trigonal prism geometry. This tendency is preferred for d^5 metal ions, in which there is no crystal field effect. It has not been possible to synthesize complexes with other d^n ions that, for electronic reasons, do not "allow" this distortion. For example, starting from a Mn^{3+} salt yields a Mn^{2+} complex, while starting from a Fe^{2+} salt leads to a Fe^{3+} complex. It is worth noting that with less rigid ligands (Figure 3.14C) it is possible to obtain complexes with any d^n configuration.

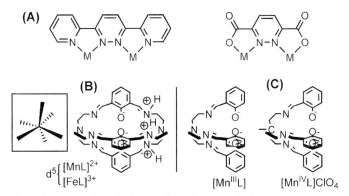

Figure 3.14 Some ligands that can favor the trigonal prism geometry.

Crystal Lattice Effects Very recently, Reedijk and coworkers [27] reported the structure of [Mn^{II}(acac)$_2$(bpy)], this being the first example of a mixed-ligand complex with 'innocent' bidentate ligands that present trigonal-prismatic coordination geometry. The related phenanthroline complex, [Mn^{II}(acac)$_2$(phen)] is octahedral. Since the rigidity of the dinitrogen diimine ligand does not seem to play a role, and the energy difference between different ligand environments for the two complexes is quite small the packing effect in the crystal lattice must play an important role in the final solid-state structure.

3.1.8
Coordination Number 7

Coordination number seven is not common in transition metal chemistry: an estimate based on the number of transition metal σ-bonded complexes found in the Cambridge Structural Database reveals that heptacoordinated complexes represent only 1.8% of the total number of structures reported [28]. With seven-coordination, no regular polyhedron can describe the coordination sphere. Among the various irregular (non-uniform) polyhedra only three of them are important from the chemical point of view: the pentagonal bipyramid (pbt), the capped octahedron (co) and the capped trigonal prism (ctp) (Figure 3.15), the pbt being by far the most abundant. It should be noted that the three polyhedra have comparable stability and there is practically no potential energy barrier between them. The electron configuration of the metal ions is seen to be preferred from d^0 to d^4, with this latter configuration, d^4, being the most common [28, 29]. The heptacoordi-

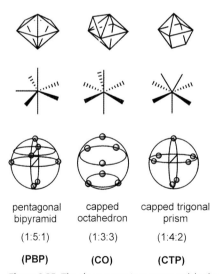

pentagonal bipyramid	capped octahedron	capped trigonal prism
(1:5:1)	(1:3:3)	(1:4:2)
(PBP)	**(CO)**	**(CTP)**

Figure 3.15 The three most important polyhedra for coordination number 7.

nated species exhibit extremely interesting chemical properties and catalytic activity, and therefore have become a challenging research area.

3.1.8.1 Hepta(monodentate) Complexes

Complexes with only monodentate ligands are, mainly, $[MF_7]^{n-}$, $[M(H_2O)_7]^{n+}$, $[M(CNR)_7]^{n+}$ and $[M(CN)_7]^{n-}$, along with other $[MA_6B]$, $[MA_5B_2]$ and $[MA_4B_3]$, $[MA_3B_3C]$, $[MA_2B_2C_3]$ complexes.

No clear structural preference is found for ML_7 complexes according to their electron configuration or type of ligands. Different structures are even known for the same ligand. In $[WF_7]^-$ different polyhedra are obtained depending on the counter cation. Small changes in the steric aspects of the ligands produce a variation in the geometry: with Mo the ligand CNMe adopts capped octahedral geometry, whereas the ligand CNBu adopts capped trigonal prism geometry. $[MoF_7]^-$ and $[WF_7]^-$ exhibit co structures, while $[TaF_7]^{2-}$ and $[NbF_7]^{2-}$ display ctp structures. Very recently, $[M^V(N_3)_7]^{2-}$ (M = Nb, Ta) have been reported. The heptaazido dianions possess monocapped triangular prismatic structures [30]. With uranium(IV) the structure is monocapped octahedron or pentagonal bipyramid. It should be noted that these structures exhibit fluxionality and generally are deformed from the idealized arrangements.

MA_5B_2 compounds prefer pbp geometry, as they have two short M–B distances (UO_2, for example) in the axial position of the pentagonal bipyramid.

3.1.8.2 Complexes with Polydentate Ligands

Several general families of complexes with polydentate ligands have been identified. No general trends can be found. Complexes with pentadentate crown ethers are practically all pbp with the crown ether occupying the equatorial plane.

3.1.9
Coordination Number 8

There are various polyhedra with eight vertices, as shown in Figure 3.16. The two most stable polyhedra in terms of coordination chemistry are the square antiprism and the triangular dodecahedron [31]. In general, octacoordination is most common for early transition ions (Groups 3–7) of the second and third series. Due to the principle of electroneutrality, this coordination (as well as the remaining higher coordinations) will be favored with high oxidation states.

3.1.9.1 Complexes with Monodentate Ligands

Square Antiprismatic Geometry The square antiprism is shown in Figure 3.17A. All the ligands are identical and the stereochemistry is fixed by the angle α. The point symmetry is D_{4d}. Assuming the hard sphere model and equal bond lengths throughout, $a = b$ and $\alpha = 59.26°$. Polyhedron stereochemistry calculated by minimizing the repulsion energies lengthens the polyhedron along the z axis (minimum

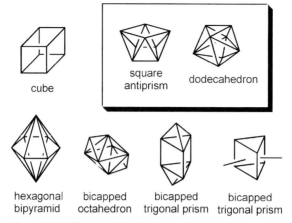

Figure 3.16 Different polyhedra for coordination number 8.

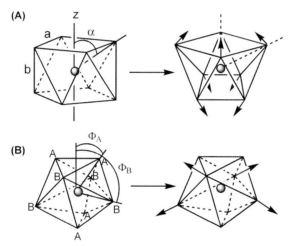

Figure 3.17 Distortions in the two most important polyhedra for coordination number 8 (square-antiprism and dodecahedron).

energy for $\alpha = 57.1°$) (Figure 3.17A). This reduction of the angle is observed in practically all the known structures. For example in $H_4[W(CN)_8] \cdot 8H_2O$ $\alpha = 57.6°$ and in $Cs_4[U(NCS)_8]$ $\alpha = 56.7°$, whereas in $Na_3[W(CN)_8] \cdot 4H_2O$ $\alpha = 59.1°$.

Triangular Dodecahedron Geometry The triangular dodecahedron is shown in Figure 3.17B. In contrast to the square antiprism, the eight ligand positions are not equivalent. There are two groups of four ligands (A and B) that are equivalent among themselves. The point symmetry is D_{2d}. In the rigid sphere model Φ_A and Φ_B are 36.85° and 110.54°, respectively. As in the case of the square antiprism, the calculations that minimize the potential energy show that there is a slight deviation from the theoretical values of the angles: 37.3° and 108.6°, in line with the distortion shown in Figure 3.17B. Fewer examples are known here than in the previous case, and these include $K_4[Mo(CN)_8] \cdot 2H_2O$ ($\Phi_A = 36.0°$ and $\Phi_B = 107.1°$) and $(Bu_4N)_3[Mo(CN)_8]$ ($\Phi_A = 37.2°$ and $\Phi_B = 107.5°$).

3.1.9.2 Relationship Between Structures with Eight-coordination

Potential energy calculations for both structures show that there is practically no potential barrier between them and therefore many types of intermediate structure would be expected. For example, the complex $(Et_3NH)_2(H_3O)_2[Mo(CN)_8]$ is an almost perfect intermediate structure between the two geometries. It should be noted that in solution, and regardless of the geometry, NMR studies of ^{13}C for the octacyanomolybdates show that the eight positions are identical, which implies a fluxional mechanism between the two stereochemistries.

Many structures have been shown to undergo C_{2v} distortion (which is equivalent, at the limit, to the bicapped trigonal prism geometry, see Figure 3.16). Examples of complexes that undergo this distortion are $Na_3[W(CN)_8] \cdot 4H_2O$, $[ZrF_8]^{4-}$, and $Li_4[UF_8]$. Although cubic geometry is not at all favored for steric reasons this geometry is given with monodentate ligands, mainly because of packing effects. For example, the complex $Li_6(BeF_4)[ZrF_8]$ is a dodecahedron distorted toward cubic geometry due to the position of the lithium ions in the crystal lattice; $Cs_4[M(NCS)_8]$ is an ideal square antiprism, while the tetraethylammonium salt, $(Et_4N)_4[M(NCS)_8]$, is an almost perfect cube ($M = U^{IV}, Th^{IV}$).

Given the corresponding energies it is not surprising that $[ML_8]$ complexes show a preference for square antiprism geometry. In contrast, the situation could be different in $[ML_4L'_4]$ complexes as the dodecahedron has two groups of four non-equivalent ligands. Therefore, when L and L' are very different, the triangular dodecahedron will have structural preferences. Well-characterized examples include $[ThCl_4(Me_2SO)_4]$, $[MoH_4(PPh_2Me)_4]$ and $[Mo(CN)_4(CNMe)_4]$.

3.1.9.3 Complexes with Bidentate-chelated Ligands

The most widely studied and abundant complexes are of the type $[M(A–A)_4]$ (A–A = nitrate, oxalate, dithiocarbamate, β-diketonate). In general, the geometry depends on the bite (b) parameter of the ligand. Small values of b stabilize the geometry of the regular dodecahedron such as $K_3[Cr(O_2)_4]$, $(Ph_4As)[Fe(NO_3)_4]$, for $b = 1.12–1.20$ the square antiprism becomes more stable with the four bidentate ligands on the square edges, such as $Na_4[Zr(OX)_4]$. When $b \geq 1.3$, two well-defined minima appear in the potential energy curves, and these correspond to the two enantiomers of a square antiprism with D_4 symmetry in which the four bidentate ligands occupy the edges that join the squares, such as $[Nb(BuCOCHCOBu)_n]$ (see Chapter 4).

3.1.9.4 Complexes with Polydentate Ligands

Tetradentate macrocyclic ligands, mainly phthalocyanines, porphyrins and O_4-crown ethers, usually form complexes in which their four donor atoms remain coplanar, leading to sandwich structures that can, in principle, be cubic, square antiprismatic or intermediate.

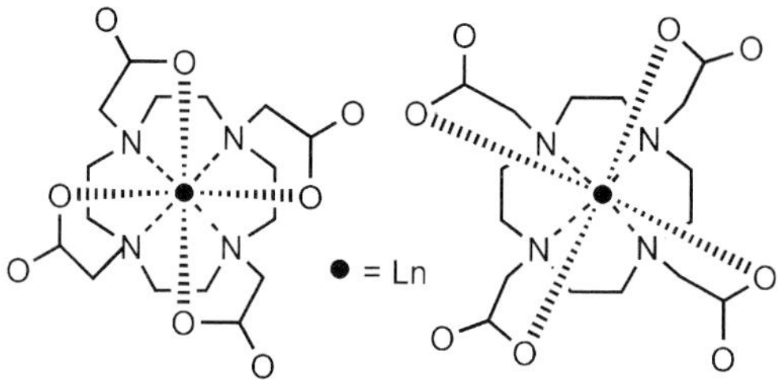

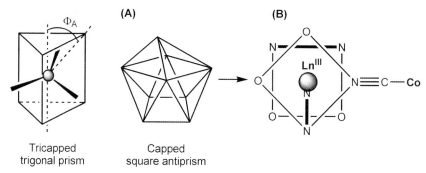

Figure 3.18 DOTA ligand and its tendency to give square antiprism geometry. Distortion towards the cube.

(A)   (B)

Tricapped
trigonal prism

Capped
square antiprism

Figure 3.19 Two polyhedra for coordination number 9.

One of the polydentate ligands most widely studied for its tendency to give eight-coordination is the anion 1,4,7,10-tetraazacyclododecane-N,N',N'',N'''-tetraacetate (DOTA). Its complexes with lanthanide ions have been studied in detail for their high stability. The eight donor groups of the DOTA ligand tend to give square antiprism geometry, although there are also examples of structures intermediate between the square antiprism and the cube (Figure 3.18).

3.1.10
Coordination Number 9

The two polyhedra with nine vertices that, even when distorted, offer the best adaptation to a spherical surface are the tricapped trigonal prism and the mono-capped square antiprism (Figure 3.19A). The known [ML$_9$] complexes show the tricapped trigonal prismatic geometry. The angle Φ_A for the rigid sphere model is 41.81°, but upon minimising the total repulsion energy the optimum angle becomes 44.4°. Some examples are [Pr(H$_2$O)$_9$](BrO$_3$)$_3$ (47.4°), K$_2$[ReH$_9$] (43.2°).

Few [Ln(A–A)L$_7$] or [Ln(A–A)$_2$L$_5$] have been described. In 2006 Matoga et al. reported the [W(CN)$_7$(η^2-O$_2$)]$^{3-}$ anion (isolable as its [PPh$_4$]$^+$ salt), with η^2 coordination mode of the peroxido ligand with very small bite angle (41.0°). The complex ion exhibits the rare trapezoidal tridecahedral geometry and represents the new class of nine-coordinate complexes with one bidentate and seven monodentate ligands [32a]. In 2005 Diaz and coworkers reported a series of trinuclear cyanido-bridged Ln_2M complexes (M = Co^{3+} and Fe^{3+} from $[M(CN)_6]^{3-}$ and Ln^{3+} = La, Ce, Pr, Nd and Sm) with two bpy as blocking ligands (thus $[Ln(A-A)_2L_5]$) [32b]. In all cases the coordination polyhedra can be described as slightly distorted capped square antiprisms. The apical position of such polyhedra is occupied in all cases by an N atom of the bpy ligand (Figure 3.19B).

$[M(A-A-A)_3]$ complexes have the tricapped trigonal prism structure, regardless of the value of b, such as in $Na_6[Ln\{O(CH_2COO)_2\}_3](ClO_4)_3$ [33], and $[Ln\{HB(N_2C_3H_3)_3\}_3]$ (homoscorpionate) [34]. Polydentate ligands can stabilize any geometry.

3.1.11
Coordination Numbers 10, 11 and 12

From the geometric point of view the four most stable polyhedra for coordination number 10 are: bicapped square antiprism, tetracapped trigonal prism, sphenoco-rona and pentagonal antiprism (Figure 3.20). Known complexes with ten-coordination are of two types: (i) $[M(A-A)_5]$. Once again its stereochemistry depends on the parameter b of the bidentate ligands. For a small b (<1) the most stable geometry is the bicapped square antiprism, such as in $K_2[Er(NO_3)_5]$ and $Na_6[Th(CO_3)_5]\cdot12H_2O$. As the value of b increases other structures stabilize, although very few examples are known. $K_4[Th(ox)_4]\cdot nH_2O$ (polymer) presents the two structures (antiprism and sphenocorona), depending on n. (ii) Polydentate ligands. There is no systematization. As an example, the Nd^{3+} ion in $Na_3[Nd(TTH A)]\cdot2.5NaClO_4\cdot7H_2O$ (H_6TTHA = triethylenetetraminehexacetic acid) is a 4,4-bicapped square antiprism [35].

The coordination number 11 is a rare coordination and very few well-characterized examples are known. An icosahedron with a missing vertex is found in several nitrate complexes of lanthanide with other monodentate ligands, $[Ln(NO_3)_5L]$, $[Ln(NO_3)_4L_3]$, $[Ln(NO_3)_3L_5]$, L = H_2O, methanol. An extensive review of these complexes has been published [36].

From the geometric point of view many polyhedra could give twelve-coordination. The most stable geometry is, without doubt, the icosahedron, which is one of the five regular polyhedra (Figure 3.1). Known compounds are of the

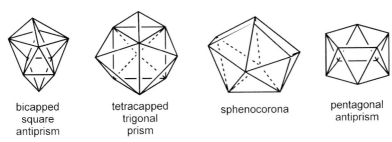

| bicapped square antiprism | tetracapped trigonal prism | sphenocorona | pentagonal antiprism |

Figure 3.20 Polyhedra for coordination number 10.

[M(A–A)$_6$] type, with varying degrees of distorted icosahedral geometry. A total of five isomers, with T_h, D_{3d}, D_3, D_2, and C_2 symmetry are possible for an icosahedral [M(L–L)$_6$] structure. Two of them, namely the T_h and the D_3 isomers, have been observed in complexes with bidentate ligands having a sufficiently small bite angle (for example, with NO_3^-). In 1995 Bandurkin reported an extensive review of these complexes [36]. In recent years new complexes have been obtained with polydentate ligands, one highly unusual example being [Ln(NO$_3$)$_3$(18-crown-6)].

3.1.12
More Than One Coordination Type in the Same Crystal

It is worth mentioning some cases in which the same crystal presents more than one coordination number and geometry for a given metal ion, showing that different geometries can have very similar energy. We have already described in this chapter the case of [Ni(CN)$_5$]$^{3-}$. Other paradigmatic cases are: (i) Bi(NO$_3$)$_3$ reacts with cucurbit[8]uril, (**Q8**), to give [{Bi(NO$_3$)(H$_2$O)$_5$}$_2$(**Q8**)] [Bi(NO$_3$)$_3$(H$_2$O)$_4$]$_2$ [Bi(NO$_3$)$_5$]$_2$·**Q8**·19H$_2$O, whose structure includes three discrete Bi complexes, with nine-coordination in the moiety with the QB unit and ten-coordination in the other two moieties (Figure 3.21) [37]. (ii) In the complex [Pb$_4$(L)$_3$(NO$_3$)$_2$(H$_2$O)], (L = oxydiacetate) there are three crystallographically independent lead atoms which are seven- eight- and nine-coordinated, respectively [38].

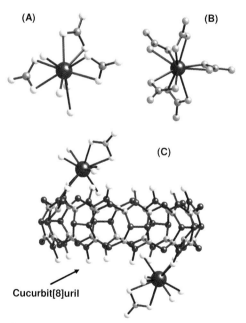

(A)　　　　　　　　　　　　　**(B)**

(C)

Cucurbit[8]uril

Figure 3.21 Bi(NO$_3$)$_3$ and cucurbit[8]uril, (**Q8**), giving three different geometries. Large black spheres = Bi (see text).

3.2
Distortions in Coordination Compounds Due to Electronic Factors: Jahn–Teller Effect

3.2.1
Introduction

In the previous sections we have seen how most complexes present distortions with respect to hypothetical regular geometry. Many of these distortions are due to steric factors. However, some of them are the result of electronic characteristics and must be studied separately. The most important distortion due to electronic effects is known as the Jahn–Teller distortion or effect [39, 40]. If it is assumed that the displacement of atoms with respect to their equilibrium state can be represented by means of a displacement coordinate, Q_i (normal vibration mode), the corresponding Hamiltonian can be written as follows (theory of perturbations):

$$\hat{H}(Q_i) = H^0 + H_i Q_i + \frac{1}{2} H_{ii} Q_i^2$$

where $H_i = (\delta H / \delta Q_i)_0$ and $H_{ii} = (\delta^2 H / \delta Q_i^2)_0$ (H_{ii} is the force constant)

Using only the first- and second-order perturbation gives rise to the following energy expression:

$$E(Q_i) = E_0 + Q_i \langle \Phi_0 | H_i | \Phi_0 \rangle + \frac{1}{2} Q_i^2 \left\{ \langle \Phi_0 | H_{ii} | \Phi_0 \rangle - 2 \sum_n \frac{|\langle \Phi_0 | H_i | \Phi_n \rangle|^2}{E_n - E_0} \right\}$$

E_0 is the initial energy prior to the perturbation; the first integral corresponds to what is known as the *first-order* Jahn–Teller effect, while the second integral plus the sum corresponds to the *second-order* Jahn–Teller effect. The sum over n extends to all the excited energy terms. For the molecule to be more stable the first- and second-order terms must be negative. As an initial approach each unoccupied orbital above the HOMO can be made to correspond with an excited term.

3.2.2
First-order Jahn–Teller Effect

The first-order Jahn–Teller effect occurs if the integral $\langle \Phi_0 | H_i | \Phi_0 \rangle$ is different from 0. As H is a totally symmetric energy Hamiltonian, $H_i = (\delta H / \delta Q_i)_0$ will be represented in the same way as Q_i (normal vibration mode). Therefore, the symmetry condition required so that the integral is not nil is $\Gamma(\Phi_0) \otimes \Gamma(\Phi_0) = \Gamma(Q_i)$, where $\Gamma(\Phi_0)$ and $\Gamma(Q_i)$ are the representations of the molecule's ground state (without distortion) and the distortion coordinate, respectively. Two cases can be described:

1. Φ_0 corresponds to a non-degenerate species. The direct product $\Gamma(\Phi_0) \otimes \Gamma(\Phi_0)$ must always be the totally symmetrical representation, and therefore the normal

mode that can stabilize the molecule by distortion is totally symmetrical. Consequently, $<\Phi_0|H_i|\Phi_0> \neq 0$ only when Q_j corresponds to the totally symmetrical normal vibration mode. *However, this results in a totally symmetrical distortion, which is tantamount to saying that there is no distortion.*

2. Φ_0 corresponds to a degenerate species. In a non-linear symmetry point group the direct product of a degenerate species by itself *always* contains species with symmetry other than the totally symmetrical form. Therefore, $<\Phi_0|H_i|\Phi_0> \neq 0$ for *a normal mode that is not totally symmetrical (Q_j)* of the molecule under study.

> The first-order Jahn–Teller theorem states that a non-linear molecule with a degenerate ground electronic state is stabilized by means of a distortion, in accordance with a normal vibration mode Q_j of the molecule.

The Jahn–Teller theorem can also be expressed as follows: "*A non-linear molecule with partially occupied degenerate orbitals is unstable with respect to distortion*".

3.2.2.1 Distortions of Octahedral Geometry

Let us consider a d^9 or a d^7 (O_h) complex. The electron configuration will be $(t_{2g})^6(e_g)^3$ and $(t_{2g})^6(e_g)^1$, respectively, and the corresponding energy term is ^2E_g. $E_g \otimes E_g = A_{1g} + A_{2g} + E_g$. The normal vibration modes of an octahedral complex are: $A_{1g} + E_g + T_{2g} + T_{1u} + T_{2u}$. The direct product contains the normal vibration mode E_g. Therefore, there will be distortion in accordance with this normal mode, as shown in Figure 3.22A. This implies that the complex will become distorted through elongation or compression, and thus the unpaired electron will move to an orbital of lower energy, thereby stabilising the structure, as shown in Figure 3.22B.

The Jahn–Teller effect will also occur if Φ_0 is degenerate in the t_{2g} orbitals: $T_{2g} \otimes T_{2g} = A_{1g} + E_g + T_{1g} + T_{2g}$. In this case there are two symmetry species that

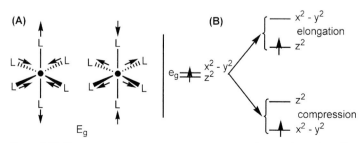

Figure 3.22 Elongation or compression in O_h geometry due to Jahn–Teller effect.

correspond to two normal vibration modes in O_h symmetry: E_g and T_{2g}. If the distortion occurs according to the normal vibration mode E_g, the effect of this distortion in a $(t_{2g})^1$ system will also be elongation or compression.

It is not possible to predict the magnitude of the distortion. The only predictable event is that distortion e_g > distortion t_{2g}, given the non-bonding nature of the t_{2g} orbitals and the antibonding nature of e_g ones. Therefore, the cations that show the most marked Jahn–Teller effect are those derived from e_g^1 and e_g^3 configurations. Noteworthy examples are the complexes of Cu^{2+} ($d^9 : e_g^3$), Mn^{3+} ($d^4 : e_g^1$) and Cr^{2+} ($d^4 : e_g^1$).

Although it is not possible to predict the sign of the distortion (elongation or compression) most Cu^{2+} complexes with Jahn–Teller distortion undergo elongation, owing to $4s$–$3d_{z2}$ orbital mixing in D_{4h} symmetry, which slightly lowers the energy of the d_{z2} compared to d_{x2-y2}. In this point group, the $4s$ orbital of the metal belongs to the symmetry species a_{1g}. Given that the orbital z^2 also belongs to the symmetry species a_{1g}, there may be configuration interaction between the two. This additional stabilization is greater in a Jahn–Teller-elongated Cu^{2+} ion (where d_{z2} contains two electrons) than in a Jahn–Teller-compressed one (where it only contains one). This interaction is shown in Figure 3.23. Although the configuration interaction is more intense in the compression, the presence of two electrons in the a_{1g} orbital favors elongation over compression.

The Jahn–Teller effect is also revealed when studying the formation constants of $[M(en)_3]^{2+}$ complexes, where $M = Cu^{2+}$ or other similar cations that do not show the Jahn–Teller effect. The stability constants K_1 ($[M(H_2O)_6]^{2+} + en \leftrightarrow [M(H_2O)_4(en)]^{2+}$) and K_2 ($[M(H_2O)_4(en)]^{2+} + en \leftrightarrow [M(H_2O)_2(en)_2]^{2+}$) follow the general trend when shifting from Mn^{2+} to Zn^{2+}; in contrast, K_3 ($[M(H_2O)_2(en)_2]^{2+} + en \leftrightarrow [M(en)_3]^{2+}$) is greatly reduced in the case of Cu^{2+}. This is because the Jahn–Teller effect "forces" the lengthening of two bonds in the *trans position*. Therefore, it will only be stable if the bite of the chelate ligand allows this lengthening. This is not the case with the bidentate ethylenediamine ligand, whose bite is small, but it may occur in complexes such as $[Cu(hfacac)_2(en)]$, in which the hexafluoroacetylacetonate (hfacac) ligand does permit lengthening due to its greater bite.

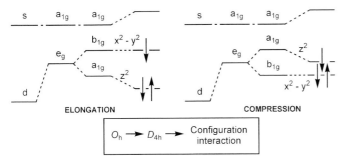

Figure 3.23 MO diagram for elongation vs. compression in Jahn–Teller effect.

3.2.2.2 Distortions in Tetrahedral Geometry

The molecular orbitals centered on the metal ion belong to the symmetry species e and t_2. Assuming unpaired electrons in either of the two groups of orbitals and applying the direct product rule, we get: $E \otimes E = A_1 + A_2 + E$ and $T_2 \otimes T_2 = A_1 + E + T_1 + T_2$.

The normal vibration modes of a T_d complex are: A_1 (totally symmetrical), E, $2T_2$. Therefore, there may be a Jahn–Teller effect through the normal modes E and T_2. In tetrahedral complexes Jahn–Teller distortion will be more effective when the unpaired electrons occupy the t_2 orbitals, this being due to the disposition of these orbitals with respect to the ligands. Consequently, the effect will be heightened in complexes with d^3, d^4, d^8 and d^9 configuration. d^8 complexes show very slight distortion. The effect can be seen more clearly in complexes with d^9 configuration.

3.2.2.3 Jahn–Teller effect in Other Non-O_h (T_d) Geometries

The D_{3h} trigonal-planar structure of the d^8 Au^{3+} halides is Jahn–Teller distorted to T-, Y-, and L-shaped geometries. This distortion was highlighted computationally early in 1992 [41]. The D_{3h} structure has two electrons in a doubly degenerate e' orbital, which splits by bending into a doubly occupied b_2 (HOMO) and an empty a_1 (LUMO) orbital in the distorted C_{2v} symmetry. The T-shaped structure of AuF_3 was confirmed by electron diffraction studies [42a]. In 2006 was reported the crystal structure of $[NEt_4][AuCl_2][AuCl_3]$, which contains Y-shaped $AuCl_3$, thus affording the first solid-state evidence for first-order Jahn–Teller distortion in $AuCl_3$ [42b].

3.2.2.4 Dynamic Jahn–Teller Effect and Crystal Stereochemistry

In a first-order deformation it is not possible to predict the axis along which this deformation will take place. Assuming octahedral symmetry, if the three axes of the octahedron are equivalent (types of ligand, etc.) the distortion will not be produced in a way that is predetermined by the Jahn–Teller effect but rather, statistically, three distortions will take place.

Such disorder may be *dynamic* (essentially temperature dependent) or *static* (temperature independent). In fact, however, the *static* Jahn–Teller effect in solids can also be *dynamic* when the energy barrier between the three possible elongations is very low and they are all given simultaneously. In other words, if kT is greater than the energy difference between the minima the octahedron "appears" regular. This means that during the time of the structural measurement the distances appear to be the same. *In many cases the static or dynamic nature of the distortion depends solely on the time scale of the technique used, and therefore the terminology is mistaken.*

As indicated, the static/dynamic nature may also depend on temperature. For example, salts of the $[Cu(NO_2)_6]^{4-}$ anion present highly singular characteristics: the salts of K_2Ba, K_2Ca and K_2Sr are elongated anions at room temperature (four distances at 2.05 Å and two distances at 2.3 Å); the salt of K_2Pb is a regular octahedron at room temperature but at $0\,°C$ it presents compression; the salts of Cs_2Ba and Rb_2Ba show dynamic compression, even at room temperature.

The fact that a given distortion (in other words, a privileged direction) exists in a crystal is due more to *cooperative effects* of the crystal itself than to the Jahn–Teller effect. The crystal, for certain reasons related to the stability of the lattice, may block the dynamic nature of the distortions and render them static: *among all the possible distortions predicted by vibrational effects the only one that occurs in the solid is the distortion favored by the symmetry of the environment and, to a certain extent, permitted by it.*

The copper(II) Tutton salts, of general formula $M_2[Cu(H_2O)_6](SO_4)_2$, provide good examples of the way in which the interplay of Jahn–Teller vibronic coupling and lattice forces decide the structure adopted by the $[Cu(H_2O)_6]^{2+}$ ion. The ammonium Tutton salt is especially interesting in this respect because it exists in two distinct forms. When deuterated it adopts a structure (form A) in which the long axis of the distorted $[Cu(D_2O)_6]^{2+}$ cation is different than in the hydrogenous salt (form B). This change is accompanied by slight alterations in the hydrogen bonding interactions involving the ammonium and sulfate groups. Application of pressure causes the deuterated compound to switch to form B, the change exhibiting hysteresis when the pressure is decreased. The $K_2[Cu(H_2O)_6](SO_4)_2$ adopts form A; the $Rb_2[Cu(H_2O)_6](SeO_4)_2$ compound also adopts this form, but the corresponding potassium salt exists in form B.

The dynamic Jahn–Teller coupling in axially distorted vanadium(III) complexes has been reported. In general, the dynamic nature is especially observable in solution. Thus, for example, the complex $[V(CO)_6]$ $(t_{2g})^5$ presents wider bands in the absorption spectrum than does $[Cr(CO)_6]$ $(t_{2g})^6$.

3.2.3
Second-order Jahn-Teller Effect or Pseudo-Jahn-Teller Effect (PJTE)

3.2.3.1 Concept and Conditions
If the Hamiltonian of first-order distortion is nil, the second-order term may be operative. As a way of better understanding the phenomenon, let us consider the following question: what will happen if in a given molecule there are two frontier orbitals of very similar energy?

> There may be a reduction in molecular symmetry (through distortion) and in the new symmetry sub-group the two molecular orbitals already belong to the same symmetry species; therefore, they can interact, with one of them stabilizing and the other destabilizing.

Figure 3.24 shows this ideal situation. Without entering into quantum discussions (beyond the scope of this book), from the second term of the Jahn–Teller

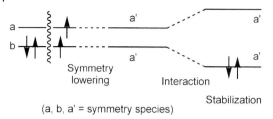

(a, b, a' = symmetry species)

Figure 3.24 Restricted MO diagram showing the second order Jahn–Teller effect.

Hamiltonian (Section 3.2.1), we only have to see whether $<\Phi_n|H_i|\Phi_0> \neq 0$. For this to occur $\Gamma\Phi_n \otimes \Gamma\Phi_0 = \Gamma Q_j$. Note that if the ground state of the non-distorted molecule is non-degenerate, then Φ_0 corresponds to the HOMO and Φ_n to the LUMO, and thus the symmetry rule becomes: Γ (HOMO) $\otimes \Gamma$ (LUMO) = Γ (Q_j). Furthermore, if the difference in energies (denominator; see Section 3.2.1) is large the term on the right is negligible and there is no distortion. Consequently, a requisite for second-order Jahn–Teller distortion is a small HOMO–LUMO gap.

3.2.3.2 Examples

Mononuclear Complexes   All five-coordinated copper(II) complexes involve a non-degenerate ground state. Thus, the PJTE is applicable. One of the most widely studied examples is the stereochemistry of the $[CuX_5]^{3-}$ anion, as compared with the distortion of other diamagnetic analogs: $[CuCl_5]^{3-}$ vs. $[ZnCl_5]^{3-}$. Both complexes can be considered derivatives of a trigonal bipyramid (D_{3h}) but with two very different types of distortion: according to the vibrational coordinate E′ in the case of $[CuCl_5]^{3-}$ (tendency toward a square-base pyramid) and according to the vibrational coordinate A_2'' in the case of $[ZnCl_5]^{3-}$ (tendency toward tetrahedral coordination with an additional fifth ligand). Figure 3.25A depicts the two normal vibration modes, E′ and A_2''. In the case of $[CuCl_5]^{3-}$, the MO diagram (frontier orbitals only) is shown in Figure 3.25B. The direct product $e′ \otimes a_1′ = e′$. The distortion reduces the D_{3h} symmetry to C_{2v}. The two new a_1 (C_{2v}) orbitals are able to mix, thus stabilising the molecule toward the square-base pyramid structure [43].

If we now consider the $[ZnCl_5]^{3-}$ ion, the two frontier orbitals in Figure 3.25B are full (d^{10}). Therefore, the distortion cannot originate from these orbitals, and it is necessary to take into account one of the anti-bonding orbitals of higher energy. Calculations performed for these systems indicate that the orbital a_2'' (mixture of the p_z orbital of Zn and ligand group orbitals of the type π) is at a higher energy. The direct product of HOMO (a_1') by LUMO (a_2'') is A_2''. Therefore, it is not surprising that the zinc complex is distorted toward the formation of a "tetrahedral" complex with a fifth additional chloride ligand.

The distortion in tris(dithiolate) metal complexes (d^0) is another typical example. In addition to their tendency toward trigonal prismatic geometry, some of the tris(dithiolate) complexes (mainly d^0) possess another interesting structural feature:

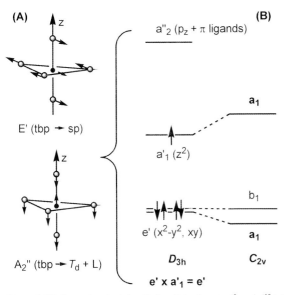

Figure 3.25 Second-order Jahn–Teller distortion in d^9 and d^{10} trigonal bipyramidal geometries.

a bending of the S–C–C–S plane away from the S–M–S plane. The two planes present a certain angle α that ranges from 0° to 30° (Figure 3.26).

The distortion can be attributed to a second-order Jahn–Teller effect [44]. Applying MO calculations it is possible to demonstrate that the energy diagram of the molecular orbitals, with their symmetry labels, is as shown in Figure 3.26. In the case of a d^0 ion (such as Mo^{6+}) all the metal d orbitals are empty. The energy difference between the HOMO (a'_2) and the LUMO (a'_1) is very small. Therefore, if $\langle a'_2 | Q_i | a'_1 \rangle \neq 0$ a second-order Jahn–Teller effect may occur. The direct product of $a'_2 \otimes a'_1$ (in the D_{3h} point group) is a'_2, which corresponds precisely to the normal vibration mode of the molecule in which the dithiolate ligands turn a certain angle (Figure 3.26). It should be noted that the atoms of S remain fixed, if this were not the case we would be dealing with a rotation, rather than vibration, of the molecule. The D_{3h} symmetry has been reduced to C_{3h}. The change in the symmetry point group means that the two molecular orbitals a'_1 and a'_2 come to have the same symmetry species, a' (Table of correlations $D_{3h} \rightarrow C_{3h}$). There will be an interaction between them, with one of them stabilizing and the other destabilizing.

Dinuclear Complexes The anionic dinuclear complex of the salt $(AsPh_4)_2$ $[(C_3OS_4)Cu(\mu\text{-}C_2S_4)Cu(C_3OS_4)]$, in which $C_2S_4^{2-}$ is the tetrathiooxalate anion and $C_3OS_4^{2-}$ corresponds to the anion 1,3-dithiol-4,5-dithiolate, presents a marked second-order Jahn–Teller distortion [45]. A peculiar characteristic of this complex is that the two Cu^{2+} ions are not square-planar, but rather show a tetrahedral distortion of 28.3° (Figure 3.27).

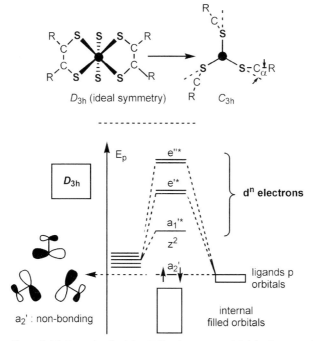

D_{3h} (ideal symmetry) C_{3h}

a_2' : non-bonding

Figure 3.26 Second-order Jahn–Teller distortion in M-dithiolate complexes (see text).

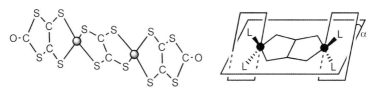

Figure 3.27 Second-order Jahn–Teller distortion in a dinuclear copper(II) tetrathiooxalate complex (see text).

The molecular orbitals diagram (frontier orbitals only) is shown in Figure 3.28. The HOMO is practically an orbital of the tetrathioxalate bridge ligand. Above this, at a fairly similar energy, lie the two combinations (symmetric and antisymmetric) of the magnetic orbitals of the two Cu^{2+} ions (d_{xy}, assumed square-planar geometry). Calculations performed using the extended Hückel method show that the energy gap between the HOMO and the LUMO is very small and thus second-order Jahn–Teller interaction may occur provided there is a normal vibration mode of suitable symmetry in order to mix the HOMO and LUMO orbitals. The direct product of the symmetry species of HOMO and LUMO is: $b_{1u} \otimes b_{2u} = b_{3g}$ and $b_{1u} \otimes b_{1g} = a_u$ (assumed planar symmetry: D_{2h}).

The normal vibration modes of the "core" of the molecule are: $\mathbf{b_{3g}} + \mathbf{a_u} + b_{2g} + b_{1u}$. Therefore, *there are two normal vibration modes* with the same symmetry as the direct product of the species HOMO and LUMO. Thus, the complex will distort

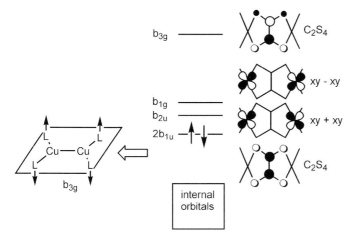

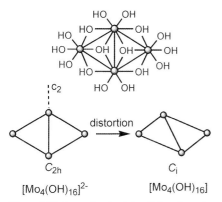

planar structure (D_{2h})

Figure 3.28 MO diagram corresponding to the complex of Figure 3.27 (see text).

Figure 3.29 Second-order Jahn–Teller distortion in $[Mo_4(OH)_{16}]^{2-}/[Mo_4(OH)_{16}]$

in accordance with one of the two normal modes. Semi-quantitative calculations demonstrate that the deformation via the normal vibration mode b_{3g} produces a more favorable distortion. The symmetry of the complex shifts from D_{2h} to C_{2h}. The orbitals b_{1u} and b_{2u} become b_u (same symmetry species, which enables them to interact and stabilize).

Another interesting example can be found in reference [46].

Tetranuclear Complexes   When the anionic complex $[Mo_4(OH)_{16}]^{2-}$ is oxidized to two electrons to give rise to the neutral complex $[Mo_4(OH)_{16}]$ it undergoes a severe distortion, as depicted in Figure 3.29. In both cases the four metal atoms are joined by means of Mo–Mo bonds and Mo–OH–Mo bonds. The anion has C_{2h} symmetry with four, equal peripheral Mo–Mo distances [47].

—— 15b$_u$

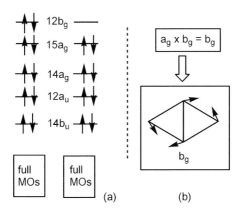

(a)   (b)

[Mo$_4$(OH)$_{16}$]$^{2-}$ [Mo$_4$(OH)$_{16}$]

Figure 3.30 MO diagram corresponding to the complex of Figure 3.29 (see text).

Molecular orbital calculations demonstrate that the electron configuration corresponding to the most external layers is ... $(14b_u)^2 (12a_u)^2 (14a_g)^2 (15a_g)^2 (12b_g)^2 (15b_u)^0$ (Figure 3.30). The energy calculation shows that there is a large energy difference between the 12b$_g$ (HOMO) and the 15b$_u$, (LUMO). In contrast, the orbitals 15a$_g$ and 12b$_g$ have approximately the same energy [47]. Therefore, when the complex is oxidized to two electrons, the HOMO and LUMO change; they will now be 15a$_g$ and 12b$_g$, respectively. As there is approximately the same energy the conditions are ideal for a second-order Jahn–Teller distortion. The direct product of the representations A$_g$ ⊗ B$_g$ is B$_g$ (point group C$_{2h}$), and B$_g$ is one of the normal vibration modes of the cluster (Figure 3.30). Therefore, it is to be expected that the deformation occurs in accordance with this normal vibration mode.

References

1 Kepert, L. (1982) *Inorganic Stereochemistry*, Springer-Verlag, Berlin.

2 Medina, I., Jacobsen, H., Mague, J.T., Fink, M.J. Dicoordinate Copper(I) Silanechalcogenolates, *Inorg. Chem.* 2006, *45*, 8844.

3 Goj, L.A., Blue, E.D., Munro-Leighton, C., Gunnoe, T.B., Petersen, J.L. Cleavage of X–H Bonds (X = N, O, or C) by Copper(I) Alkyl Complexes to Form Monomeric Two-coordinate Copper(I) Systems, *Inorg. Chem.* 2005, *44*, 8647.

4 Alvarez, S. Bonding and stereochemistry of three-coordinated transition metal compounds, *Coord. Chem. Rev.* 1999, *193–195*, 13.

5 (a) Komiya, S., Albright, T.A., Hoffmann, R., Kochi, J.K. Reductive Elimination and Isomerization of Organogold Complexes. Theoretical Studies of Trialkylgold Species as Reactive Intermediates, *J. Am. Chem. Soc.* 1976, *98*, 7255; (b) Dias, H.V.R., Singh, S. Copper(I) Complexes of Fluorinated Triazapentadienyl Ligands:

Synthesis and Characterization of [N{(C₃F₇)C(Dipp)N}₂]CuL (Where L = NCCH₃, CNBuᵗ, CO; Dipp = 2,6-Diisopropylphenyl), *Inorg. Chem.* 2004, *43*, 5786; (c) Braunschweig, H., Radacki, K., Rais, D., Scheschkewitz, D. A T-Shaped Platinum(II) Boryl Complex as the Precursor to a Platinum Compound with a Base-stabilized Borylene Ligand, *Angew. Chem. Int. Ed.* 2005, *44*, 5651.

6 Monillas, W.H., Yap, G.P.A., MacAdams, L.A., Theopold, K.H. Binding and Activation of Small Molecules by Three-Coordinate Cr(I), *J. Am. Chem. Soc.* 2007, *129*, 8090.

7 Wells, A.F. (1984) *Structural Inorganic Chemistry*, 5th edn. Clarendon Press, Oxford.

8 Cirera, J., Alemany, P., Alvarez, S. Mapping the Stereochemistry and Symmetry of Tetracoordinate Transition-metal Complexes, *Chem. Eur. J.* 2004, *10*, 190.

9 Desrochers, P.J., Tesler, J., Zvyagin, S.A., Ozarowski, A., Krzystek, J., Vicic, D.A. Electronic Structure of Four-Coordinate C_{3v} Nickel(II) Scorpionate Complexes: Investigation by High-Frequency and -Field Electron Paramagnetic Resonance and Electronic Absorption Spectroscopies, *Inorg. Chem.* 2006, *45*, 8930.

10 Jaynes, B.S., Doerrer, L.H., Liu, S., Lippard, S.J. Synthesis, Tuning of the Stereochemistry, and Physical Properties of Cobalt(II) Tropocoronand Complexes, *Inorg. Chem.* 1995, *34*, 5735.

11 (a) Boiocchi, M., Fabrizzi, L., Foti, F., Vázquez, M. Further insights on the high–low spin interconversion in nickel(II) tetramine complexes. Solvent and temperature effects, *Dalton Trans.* 2004, 2616; (b) Tamayo, A., Casabó, J., Escriche, L., Lodeiro, C., Covelo, B., Brondino, C.D., Kivekas, R., Sillampaa, R. Color Tuning of a Nickel Complex with a Novel N₂S₂ Pyridine-containing Macrocyclic Ligand, *Inorg. Chem.* 2006, *45*, 1140.

12 Addison, A.W., Rao, T.N., Reedijk, J., van Rijn, J., Verschoor, G.C. Synthesis, structure, and spectroscopic properties of copper(II) compounds containing nitrogen–sulphur donor ligands; the crystal and molecular structure of aqua[1,7-bis(*N*-methylbenzimidazol-2′-yl)-2,6-dithiaheptane]copper(II) perchlorate, *J. Chem. Soc. Dalton Trans.* 1984, 1349.

13 Burdett, J.K. (1990), *Molecular Shapes. Theoretical Models of Inorganic Stereochemistry*. John Wiley, New York.

14 Murphy, G., O'Sullivan, C., Murphy, B., Hathaway, B. Comparative Crystallography. 5. Crystal Structures, Electronic Properties, and Structural Pathways of Five [Cu(phen)₂Br][Y] Complexes, Y = [Br]⁻·H₂O, [ClO₄]⁻, [NO₃]⁻·H₂O, [PF₆]⁻, and [BPh₄]⁻, *Inorg. Chem.* 1998, *37*, 240.

15 Pinkas, J., Huffman, J.C., Chisholm, M.H., Caulton, K.G. Origin of Different Coordination Polyhedra for Cu[CF₃C(O)CHC(O)CF₃]₂L (L = H₂O, NH₃), *Inorg. Chem.* 1995, *34*, 5314.

16 Bröring, M., Prikhodovski, S., Brandt, C.D. Structural diversity in five-coordinate nickel(II) complexes of the tripyrrin ligand, *Inorg. Chim. Acta*, 2004, *357*, 1733.

17 Rossi, A.R., Hoffmann, R. Transition metal pentacoordination, *Inorg. Chem.* 1975, *14*, 365.

18 Hoffmann, R., Howell, J.M., Rossi, A.R. Bicapped Tetrahedral Trigonal Prismatic, and Octahedral Alternatives in Main and Transition Group Six-coordination, *J. Am. Chem. Soc.* 1976, *98*, 2484.

19 Alvarez, A., Avnir, D., Llunell, M., Pinsky, M. Continuous symmetry maps and shape classification. The case of six-coordinated metal compounds, *New J. Chem.* 2002, *26*, 996.

20 Demolliens, A., Jean, Y., Eisenstein, O. Deviation from the Ideal Octahedral Field vs. Alkyl Distortion in d⁰ Metal-Alkyl Complexes: A MO Study, *Organometallics*, 1986, *5*, 1457.

21 Kaupp, M. Charting No-Man's Land in d⁰ Transition Metal Six-Coordination: Structure Predictions for the Complexes [WCl₅CH₃], [WCl₄(CH₃)₂], and [WCl₃(CH₃)₃], *Angew. Chem. Int. Ed.* 1999, *38*, 3034.

22 Roessler, B., Pfennig, V., Seppelt, K. Trigonal-Prismatic versus Octahedral Molecular Structures in [(CH₃)ₙMX₆₋ₙ] Compounds, *Chem. Eur. J.* 2001, *7*, 3652.

23 Eisenberg, R., Ibers, J.A. Trigonal Prismatic Coordination. The Molecular Structure of Tris(*cis*-1,2-diphenylethene-1,2-dithiolato)rhenium, *J. Am. Chem. Soc.* 1965, *87*, 3776.

24 Stiefel, E.I., Eisenberg, R., Rosenberg, R.C., Gray, H.B. Characterization and Electronic Structures of Six-coordinate Trigonal Prismatic Complexes, *J. Am. Chem. Soc.* 1966, *88*, 2956.

25 Kapre, R.R., Bothe, E., Weyhermuller, T., George, S.D., Wieghardt, K. Electronic Structure of Neutral and Monoanionic Tris(benzene-1,2-dithiolato)metal Complexes of Molybdenum and Tungsten, *Inorg. Chem.* 2007, *46*, 5642.

26 Sun, W.-W., Cheng, A.-L., Jia, Q.-X., Gao, E.-Q. Metal-coordination-directed Assembly of Binuclear Trigonal Prisms and Three-dimensional Hydrogen-bonded Networks, *Inorg. Chem.* 2007, *46*, 5471.

27 van Gorkum, R., Buda, F., Kooijman, H., Spek, A.L., Bouwman, E, Reedijk, J. Trigonal-Prismatic vs. Octahedral Geometry for Mn(II) Complexes with Innocent Didentate Ligands: A Subtle Difference as Shown by XRD and DFT on [Mn(acac)$_2$(bpy)], *Eur. J. Inorg. Chem.* 2005, 2255.

28 Casanova, D., Alemany, P., Bofill, J.M., Alvarez, S. Shape and Symmetry Of Heptacoordinate Transition-Metal Complexes: Structural Trends, *Chem. Eur. J.* 2003, *9*, 1281.

29 Hoffman, R., Beier, B.F., Muetterties, E.L., Rossi, A.R. Seven-coordination. A molecular orbital exploration of structure, stereochemistry, and reaction dynamics, *Inorg. Chem.* 1977, *16*, 511.

30 Haiges, R., Boatz, J.A., Yousufuddin, M.K., Christe, O. Monocapped Trigonal-Prismatic Transition-Metal Heptaazides: Syntheses, Properties, and Structures of [Nb(N$_3$)$_7$]$^{2-}$ and [Ta(N$_3$)$_7$]$^{2-}$, *Angew. Chem. Int. Ed.* 2007, *46*, 2869.

31 Casanova, D., Llunell, M., Alemany, P., Alvarez, S. The Rich Stereochemistry of Eight-Vertex Polyhedra: A Continuous Shape Measures Study, *Chem. Eur. J.* 2005, *11*, 1479.

32 (a) Matoga, D., Szklarzewicz, J., Mikuriya, M. [PPh$_4$]$_3$[W(CN)$_7$(O$_2$)]·4H$_2$O

as the Representative of the [M(L)$_7$(LL)] Class for Nine-coordinate Complexes, *Inorg. Chem.* 2006, *45*, 7100; (b) Figuerola, A., Ribas, J., Llunell, M., Casanova, D., Maestro, M., Alvarez, S., Diaz, C. Magnetic Properties of Cyano-Bridged Ln^{3+}–M^{3+} Complexes. Part I: Trinuclear Complexes (Ln^{3+} = La, Ce, Pr, Nd, Sm; M^{3+} = Fe$_{LS}$, Co) with bpy as Blocking Ligand, *Inorg. Chem.* 2005, *44*, 6939.

33 Baggio, R., Garland, M.T., Perec, M. Synthesis and X-ray Structure of the Mononuclear Nine-coordinate Gadolinium(III) Hydrogen Oxydiacetate [Gd(C$_4$H$_5$O$_5$)$_3$]·C$_4$H$_6$O$_5$·H$_2$O, *Inorg. Chem.* 1997, *36*, 950.

34 Apostolidis, C., Rebizant, J., Kanellakopulos, B., von Ammon, R., Dornberger, E., Müller, J., Powietzka, B., Nüber, B. Homoscorpionates (hydridotris(1-pyrazolyl)borato complexes) of the trivalent 4f ions. The crystal and molecular structure of [(HB(N$_2$C$_3$H$_3$)$_3$]$_3$Ln^{III}, (Ln = Pr, Nd), *Polyhedron*, 1997, *16*, 1057.

35 Mondry, A., Starynowicz, P. Crystal Structure and Absorption Spectroscopy of a Neodymium(III) Complex with Triethylenetetraaminehexaacetic Acid, Na$_3$[Nd(TTHA)]·2.5NaClO$_4$·7.617H$_2$O, *Inorg. Chem.* 1997, *36*, 1176.

36 Bandurkin, G.A., Dzhurinskii, B.F. *Russ. J. Inorg. Chem.* 1995, *40*, 1399.

37 Sokolov, M.N., Mitkina, T.V., Gerasko, O.A., Fedin, V.P., Virovets, A.V., Llusar, R. Coordination of Bimuth(III) to Cucurbit[8]uril. Preparation and X-ray Structure of [{Bi(NO$_3$)(H$_2$O)$_5$}$_2$(Q8)][Bi (NO$_3$)$_3$(H$_2$O)$_4$]$_2$[Bi(NO$_3$)$_5$]$_2$·Q8·19H$_2$O, *Z. Anorg. Allg. Chem.* 2003, *629*, 2440.

38 Svensson, G., Olson, S., Albertsson, J. Coordination between lead and oxydiacetic acid, *Acta Chem. Scand.* 1998, *52*, 868.

39 (a) Murphy, B., Hathaway, B. The stereochemistry of the copper(II) ion in the solid state – some recent perspectives linking the Jahn–Teller effect, vibronic coupling, structure correlation analysis, structural pathways and comparative X-ray crystallography, *Coord. Chem. Rev.* 2003, *243*, 237; (b) Bersuker, I.B. Modern aspects of the Jahn–Teller effect theory and applications to molecular problems, *Chem. Rev.* 2001, *101*, 1067; (c)

Ceulemans, A., Lijnen, E. The Jahn–Teller Effect in Chemistry, *Bull. Chem. Soc. Jpn*, 2007, *80*, 1229.

40 Bersuker, B. (1996), *Electronic Structure and Properties of Transition Metal Compounds.* John Wiley, New York.

41 (a) Réffy, B., Kolonits, M., Schulz, A., Klapotke, T.M., Hargittai, M. Intriguing Gold Trifluoride – Molecular Structure of Monomers and Dimers: An Electron Diffraction and Quantum Chemical Study, *J. Am. Chem. Soc.* 2000, *122*, 3127; (b) Flower, K.R., Pritchard, R.G., McGown, A.T. [NEt$_4$][AuCl$_2$][AuCl$_3$]: Solid-State Evidence of Essentially Y-Shaped Jahn–Teller-Distorted AuCl$_3$, *Angew. Chem. Int. Ed.* 2006, *45*, 6535.

42 Schwerdtfeger, P., Boyd, P.D.W., Brienne, S., Burrell, A.K. Relativistic effects in gold chemistry. 4. Gold(III) and gold(V) compounds, *Inorg. Chem.* 1992, *31*, 3411.

43 Reinen, D., Friebel, C. Copper(2+) in 5-coordination: a case of a second-order Jahn–Teller effect. 2.

Pentachlorocuprate(3−) and other Cu(II)L$_5$ complexes: trigonal bipyramid or square pyramid? *Inorg. Chem.* 1984, *23*, 791.

44 Campbell, S., Harris, S. New Explanation for Ligand Bending in Transition Metal Tris(dithiolate) Complexes, *Inorg. Chem.* 1996, *35*, 3285.

45 Vicente, R., Ribas, J., Alvarez, S., Seguí, A., Solans, X., Verdaguer, M. Synthesis, X-ray diffraction structure, magnetic properties, and MO analysis of a binuclear (μ-tetrathiooxalato)copper(II) complex, (AsPh$_4$)$_2$[(C$_3$OS$_4$)CuC$_2$S$_4$Cu(C$_3$OS$_4$)], *Inorg. Chem.*, 1987, *26*, 4004.

46 Escuer, A., Vicente, R., Ribas, J., Solans, X. Magnetic Transition and Structural Asymmetrization in the Ferromagnetic Compound [{Ni$_2$(Medpt)$_2$(N$_3$)$_2$}(μ-(1,1-N$_3$)$_2$)], an Example of a Dynamic Second-Order Jahn–Teller Effect, *Inorg. Chem.* 1995, *34*, 1793.

47 Cotton, F.A., Fang, A. A Second-Order Jahn–Teller Effect in a Tetranuclear Metal Atom Cluster Compound, *J. Am. Chem. Soc.* 1982, *104*, 113.

4
Isomerism in Coordination Compounds

4.1
Introduction

Isomers are *different* compounds, each having the same chemical formula but with different physical and chemical properties. Isomerism is an important aspect of the chemistry of coordination compounds because ligand interconversion often occurs readily in solution. When a pure complex is dissolved, the solution may contain a variety of compounds, including *isomers* of the original solid-state structure. There are many categories of isomers, not mutually exclusive. A schematic representation of the different kinds of isomers is shown in Figure 4.1.

4.2
Constitutional or Structural Isomers

Structural isomers mean two or more molecules with the same kind and amount of elements. However, these elements are connected to one another in different ways. One must *break* σ-bonds to convert one structural isomer into another. The most important types of structural isomerism are described below.

4.2.1
Ionization Isomerism

These isomers simply differ in the distribution of ions between those directly coordinated and counterions present in the crystal structure. They have been known since Werner's time. Some examples are: $[CoCl_2(en)_2]NO_2$ and $[CoCl(en)_2(NO_2)]Cl$; $[CoCl(en)_2(H_2O)]SO_4$ and $[Co(en)_2(SO_4)]Cl \cdot H_2O$.

Analytical tests, electronic and IR spectroscopy can differentiate between the two isomers.

Coordination Chemistry. Joan Ribas Gispert
Copyright © 2008 WILEY-VCH Verlag GmbH & Co. KGaA, Weinheim
ISBN: 978-3-527-31802-5

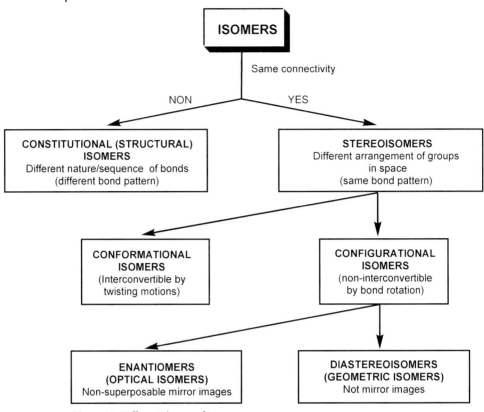

Figure 4.1 Different classes of isomers.

4.2.2
Hydrate (Solvate) Isomerism

Hydrate (solvate) isomerism is similar to ionization isomerism considering the water (or solvent) to be acting as ligands or crystallization molecules. An example is $[CrCl_2(H_2O)_4]Cl \cdot 2H_2O$ and $[CrCl(H_2O)_5]Cl_2 \cdot H_2O$. The electronic spectra are different.

4.2.3
Coordination Isomerism

This may occur when the cation and anion of a salt are both complexes, the two isomers differing in the distribution of the ligands between the cation and anion. A typical case is $[Cr(NH_3)_6][Co(CN)_6]$ and $[Co(NH_3)_6][Cr(CN)_6]$. Coordination isomerism can also occur in polynuclear complexes with a bridging ligand. For example: $[(NH_3)_4Co\text{-}(\mu\text{-}OH)_2\text{-}CoCl_2(NH_3)_2]^{2+}$ and $[Cl(NH_3)_3Co\text{-}(\mu\text{-}OH)_2\text{-}CoCl(NH_3)_3]^{2+}$.

Pseudo coordination isomerism is different. With 2-(aminomethyl)pyridine, **1**, and pyridine-2,6-dicarboxylate ion, **2**, nickel(II) gives a neutral (mixed ligand) complex [Ni·**1**·**2**·(H$_2$O)], whereas with copper(II) an ionic complex [Cu·**1**$_2$·(CH$_3$OH)][Cu·**2**$_2$] forms wherein each complex ion contains exclusively one type of ligand [1].

4.2.4
Linkage Isomerism

Ambidentate ligands (Chapter 2) such as NO$_2^-$, NCS$^-$, CN$^-$, etc. may coordinate in two or more ways. Thiocyanate ion, for example, is usually bound through the N-atom with hard transition metal ions, while it is bound through the S-atom with soft transition metal ions. The nitrito-κN and nitrito-κO linkage isomerism has been fully investigated, since Jorgensen in 1894. Linkage isomerism in polynuclear cyanide derivatives is rare. It has been reported, for example, in the products obtained from the reaction of [Cr(CN)$_3$(Tp)]$^-$ with [Cu(H$_2$O)$_6$]$^{2+}$, (Tp$^-$ = hydrotris (pyrazoe-1-yl)borate) [2].

Coordination compounds featuring induced linkage isomerism have been developed as interesting candidates for molecular switches. Indeed, photoinduced linkage isomerizations have been observed in certain late transition metal complexes containing NO$_2^-$, N$_2$, SO$_2$, and dmso (dimethylsulfoxide). In these compounds, the reversible rearrangement of the ambidentate ligand triggered by a metal-centered redox (or photoinduced) process results in a hysteresis-like bistable behavior. [Ru(dmso)(NH$_3$)$_n$]$^{3+/2+}$ derivatives, where the ambidentate sulfoxide ligand is S-bound to Ru^{2+} but isomerizes to O-bound after oxidation, is a paradigmatic example [3a]. In [Ru(dmso)(L)(terpy)]$^{2+}$ (L = bpy, tetramethylethylenediamine, 2-pyridine-carboxylate) the dmso ligand can be isomerized either electrochemically or photochemically [3b,c].

A reversible N- to C-bonded rearrangement of tris(2-pyridylmethyl)-1,4,7-triaza cyclononane (tmptacn) in [Co(tmptacn)]$^{3+}$ has been reported [4]. The complex undergoes a novel base-catalyzed N- to C- bonded rearrangement in which the tacn nitrogen is displaced by the α-carbon which deprotonates and binds to the metal as a carbanion. The reaction is completely but very slowly reversed in acid (Figure 4.2) [4].

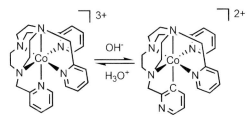

Figure 4.2 Reversible N- to C-bonded rearrangement of tris (2-pyridylmethyl)-1,4,7-triazacyclononane (tmptacn) in [Co(tmptacn)]$^{3+}$.

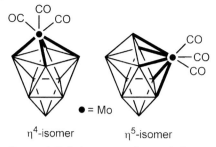

η^4-isomer η^5-isomer

Figure 4.3 $[(\eta^5\text{-Pb}_9)\text{Mo(CO)}_3]^{4-}$ vs. $[(\eta^4\text{-Pb}_9)\text{Mo(CO)}_3]^{4-}$ isomers.

Hapto-coordinated arenes also give linkage isomerization [5]. The linkage isomerization of $[\text{Os}^{II}(\eta^2\text{-benzene})(\text{NH}_3)_5]^{2+}$ has been studied by ^{1}H NMR. At 20 °C there is a single broad peak, corresponding to the six ring protons. Lowering the temperature to −87 °C causes this broad feature to split into three well resolved peaks. The mechanism responsible for averaging the ring proton signals was thought to be an intrafacial linkage isomerism (*ring walk*), in which the metal moves from one pair of carbons to the next by either an η^3 or η^1 intermediate. In comparison to the benzene analog, the naphtalene complex shows much slower rates of isomerization.

Haptotropisomerism or *haptotropic migrations*, occurs with π complexes in which the π-bound ligand features multiple coordination possibilities. Most reports deal with haptotropic migration across naphthalene derivatives or more extended aromatic π systems.

A particular case has been recently reported by Fässler et al. who isolated the novel isomers $[(\eta^5\text{-Pb}_9)\text{Mo(CO)}_3]^{4-}$ and $[(\eta^4\text{-Pb}_9)\text{Mo(CO)}_3]^{4-}$ with the countercation $[\text{K(2.2.2-crypt)}]^+$ (Figure 4.3) [6]. For the structure of the anionic cage Pb_9^{4-}, see Chapter 6.

4.2.5
Polymerization Isomerism

This represents an additional way in which an empirical formula may give incomplete information about the nature of a complex. Some examples are $[\text{PtCl}_2(\text{NH}_3)_2]$ ($n = 1$); $[\text{Pt(NH}_3)_4][\text{PtCl}_4]$ ($n = 2$); $[\text{PtCl(NH}_3)_3][\text{PtCl}_3(\text{NH}_3)]$ ($n = 2$); $[\text{Pt(NH}_3)_4][\text{PtCl}_3(\text{NH}_3)]_2$ ($n = 3$); $[\text{PtCl(NH}_3)_3]_2[\text{PtCl}_4]$ ($n = 3$). ($n =$ number of units)

4.2.6
Ligand Isomerism

4.2.6.1 With Different Connectivity
If two ligands are isomers, the corresponding complexes (with the same metal) are also isomers, in which the ligand connectivity changes. For example, 1,2-diaminopropane and 1,3-diaminopropane, *n*-halobenzoates ($n = 1$, 2 or 3), etc.

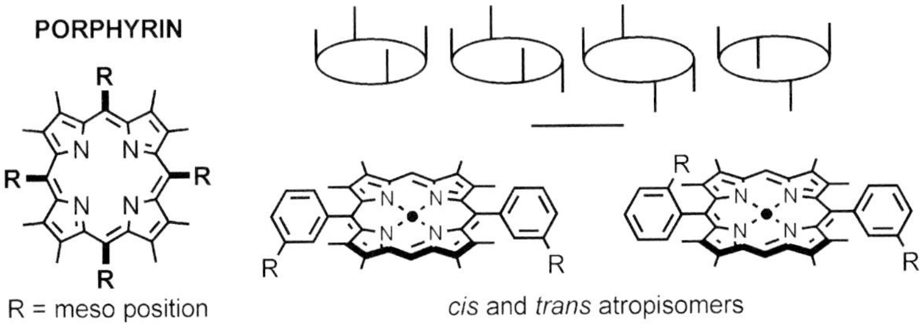

Figure 4.4 Ligand atropisomerism: metallo-porphyrin derivatives.

4.2.6.2 Ligand Conformational Isomerism (*Atropisomer* Ligands)

In some cases is not the connectivity of the ligands that changes, but only their conformational character, giving *ligand conformational isomerism (atropisomerism)*. In a restricted organic meaning, *conformers* are equivalent structures that arise as a result of the rotation of carbon atom about a σ bond. Usually, at normal temperature, the different conformers are not distinguishable because the energy required to convert from one conformer to another due to a rotation is small enough to allow the conversion to occur very rapidly.

Atropisomers are a subclass of conformers which can be isolated as separate chemical species and which arise from restricted rotation about a single bond. The boundary between atropisomers and conformers (and therefore between *configuration* and *conformation*) is delineated by Oki's arbitrary definition that atropisomers are conformers which interconvert with a half-life of more than 1000 seconds at a given temperature, the time considered to be the minimum lifetime for a molecule to be isolable, although the term is used more flexibly in general usage. Indeed, an atropisomer is a special case where the molecule can be equally well described as a 'conformational isomer' and a 'configurational isomer'

The most reported cases of atropisomerism in ligands concern the metallo-porphyrin derivatives (Figure 4.4). Porphyrins with four identical substituted encumbered groups in the *meso* position are derivatives with restricted rotation about the R-porphyrin bond. There are four possible atropisomers, αβαβ, $α_2β_2$, αβ₃, $β_4$ (Figure 4.4). Playing with the nature of the ligands it is possible to obtain only one of the atropisomers. Furthermore, the presence of different axial ligands on the metal of the metallo-porphyrins is 'non-innocent' due to (i) a minimization of the steric repulsion between the substituents and the axial ligand and (ii) a maximization of their attractive van der Waals interactions by favorable packing.

4.3
Stereoisomers

As shown in Figure 4.1, stereoisomers can be divided into *conformational isomers* and *configurational isomers*. The main difference lies in the energy barrier between the isomers: if this energy barrier is small the isomers are conformational; if the energy barrier is high, the isomers have separate identity and they are configuration isomers.

4.3.1
Atropisomers: Conformational Isomers

In this section we are going to treat the molecule as a whole, not with the ligands (as in the previous section). Conformational isomers have the same connection between the atoms but differ in dihedral angles. One does not need to break any bond to convert one conformational isomer into another because the σ-bonds rotate freely.

One of the best studied cases of atropisomers are ruthenium complexes coordinated to two monodentate lopsided ligands, for example, 1-methylimidazole (MeIm), 1,2-dimethylimidazole (Me$_2$Im), 1-methylbenzimidazole (MeBim) and 1,5,6-trimethylbenzimidazole (Me$_3$Bim) (Figure 4.5). Two different examples are *cis-cis-cis*-[RuCl$_2$(DMSO)$_2$(L)$_2$] [7] or [Ru(bpy)$_2$(L)$_2$] [8], L being one of these lopsided ligands. For two *cis*-L lopsided ligands, the corresponding atoms of each ligand can be on the same side or opposite sides of the N—M—N plane, giving head-to-head (HH) and the head-to-tail (HT) orientations, respectively.

For *cis,cis,cis*-[RuCl$_2$(DMSO)$_2$(L)$_2$] complexes [7], four atropisomers (2HH + 2HT) are possible since each L ligand can flip between two orientations but, depending on L, all four atropisomers have not always been isolated. In [Ru(bpy)$_2$(L)$_2$] when L = Me$_2$Im and MeBim, the two L ligands can be oriented in restricted ways on the ruthenium giving, theoretically, four atropisomers. However, symmetry requirements lead to two of the HH conformers being identical and thus only three isomers can be detected (Figure 4.6) [8]. The best way to get an exact idea of these four (three) atropisomers is by using ball and stick models. With these models, all the orientations of the benzimidazole ligand may be easily seen.

A very singular case is the conformational dynamics of Cu$^+$/Cu^{2+} complexes with tripodal ligands, giving steric control of molecular motions [9]. Cu$^+$ is ligated by all three arms of the ligand MeTQA [(R)-1-(quinolin-2-yl)-N,N-bis(quinolin-2-ylmethyl)ethanamine] (Figure 4.7). However, in the presence of SCN$^-$, a strongly coordinating anion for copper(I), one of the arms is forced out of binding to the metal (rotation on a σ bond). In this ligand environment, Cu$^+$ displays four-coordination with a tetrahedral geometry. Oxidation of Cu$^+$ to Cu^{2+} triggers the

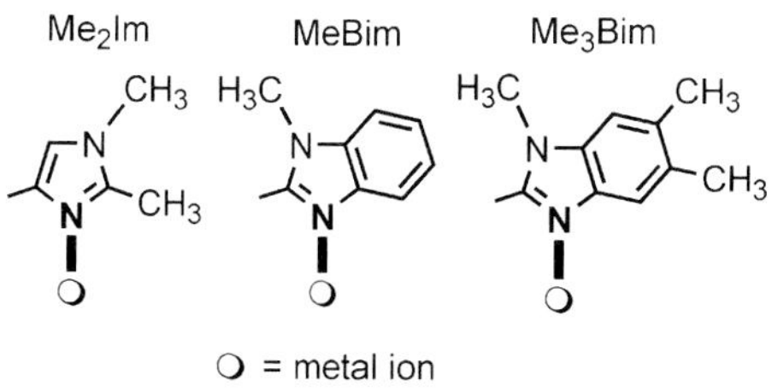

Figure 4.5 Monodentate lopsided ligands able to give atropisomer complexes (see text).

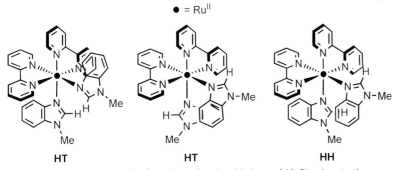

Figure 4.6 Atropisomers in [Ru(bpy)$_2$(L)$_2$] when L = Me$_2$Im and MeBim (see text).

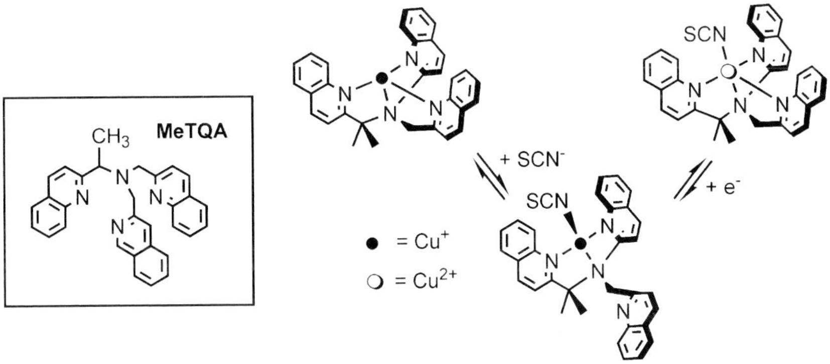

HT   HT   HH

Figure 4.7 Conformational dynamics of Cu$^+$/Cu^{2+} complexes with tripodal ligands, giving steric control of molecular motions (see text).

dissociated arm back into metal penta-coordination. This process could be cycled repeatedly by the reduction and oxidation between Cu$^+$ and Cu^{2+}. This is a new example of materials that perform specific tasks triggered by an external input.

4.3.2
Geometrical Isomers: Diastereoisomers

The most typical cases are the known *cis* and *trans* isomers in square-planar or octahedral complexes (Figure 4.8A). It is possible, but not always easy, to tune the synthesis of one or other isomer: In *cis/trans*-[Pt(aryl)$_2$(dmso)$_2$] complexes, smaller aryl groups such as phenyl and tolyl tend to give *cis*-coordination while bulkier ligands prefer the *trans*-configuration [10]. The co-crystallization of the *cis*- and *trans*-isomers in the same crystal is rare: the pyridine adduct of bis(benzoylacetonato-*O*,*O*′)copper(II) crystallizes in both its *cis*- and *trans*-isomers within the same crystal. It seems that the presence of π–π pyridine interactions assists the crystallization of both isomers [11].

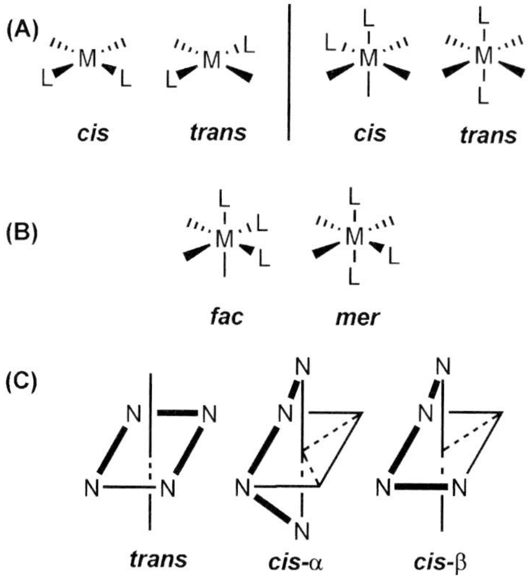

Figure 4.8 Geometrical isomers.

With three identical monodentate ligands or a tridentate ligand, two main isomers are possible: *fac* (facial) and *mer* (meridional) (Figure 4.8B). The existence of one isomer does not necessarily imply the existence of the other. The number of possible geometric isomers may be high, if one considers at the same time the conformational isomers of the ligand(s). In a few cases all these possible isomers can be separated. Jackson et al., for example, reported the synthesis, separation and assignment of the seven isomers (four *mer* and three *fac*) of $[CoCl(dien)(ibn)]^{2+}$ (dien = diethylenetriamine; ibn = 1,2-diamino-2-methylpropane). Their structures in DMSO solution have been uniquely determined by two-dimensional NMR spectroscopy [12].

Triethylenetetramine (a tetradentate ligand) can wrap around a cobalt(III) ion in three different ways: the *trans* and two *cis* (α and β) isomers, as shown in Figure 4.8C. The isomeric distribution and catalyzed isomerization of $[Co^{III}LX]^{2+,3+}$ complexes with pentadentate macrocyclic ligands (L) has been recently reported [13].

4.3.3
Enantiomers

A chiral molecule (from the Greek *cheir* = hand) exists in two forms, known as enantiomers. Enantiomers are non-superimposable mirror images of one another. The word 'enantiomer' also comes from the Greek, *enantios* meaning 'opposite'. The two enantiomers can exist together in any composition. If they are present in the same proportion the resulting compound is termed *racemic*. If a *chiral environment* is defined as a system that contains chiral constituents (generally, chiral molecules), the behavior of the enantiomers is unique; several cases can be distinguished: (i) the enantiomers behave in exactly the same way as in an achiral environment; (ii) the enantiomers behave differently in a non-racemic chiral environment; (iii) the enantiomers behave in opposing ways for certain phenomena in a racemic chiral environment.

> In a chiral environment, the two enantiomers may and must be considered as chemically different compounds.

The most important chiral environment we can consider is the biosphere. Enantiomers behave differently in terms of their biological aspects and, thus, for example, in certain pharmaceutical applications enantiomer mixtures are not considered to be pure substances.

The most widely studied phenomenon of interaction between an enantiomer and a racemic environment is the interaction between an enantiomer and linearly polarized light *(that can be considered as an environment)*. Linearly polarized light is a racemate of light polarized in a clockwise and anticlockwise direction. Each enantiomer exhibits what is called optical activity. Each isomer of the pair is capable of rotating the plane of the polarized light. One isomer rotates this light to the right through '*x*' number of degrees, while the other isomer of the pair rotates the light to the left through the same number of degrees. Therefore, the enantiomers are also called *optical isomers* due to their ability to divert the plane of polarized light. In fact this is the only difference between the two isomers; all their other physical properties are exactly the same. This makes it extremely difficult to separate the two isomers should they be mixed, as they often are. The experimental study of these isomers is the same as that reported for organic chiral compounds *(optical rotatory dispersion*, ORD and *circular dichroism*, CD) [14].

4.4
Chirality in Coordination Compounds: Nomenclature of Chirality [15]

There are various systems for naming chiral complexes. The two most widely used are the one based on the Cahn, Ingold and Prelog (CIP) rules and the system known as the *skew-line convention*.

4.4.1
R, S Symbols (CIP System)

R and *S* are the letters used to assign absolute stereochemistry based on the Cahn–Ingold–Prelog 'Sequence Rules'. After assigning priorities to the substituents around an asymmetric center the molecule is viewed such that the bond from the asymmetric center to the substituents of lower priority is going away from the viewer, or into the page. The priority for naming the ligands is given by the donor atoms of highest atomic number. Although this system was originally

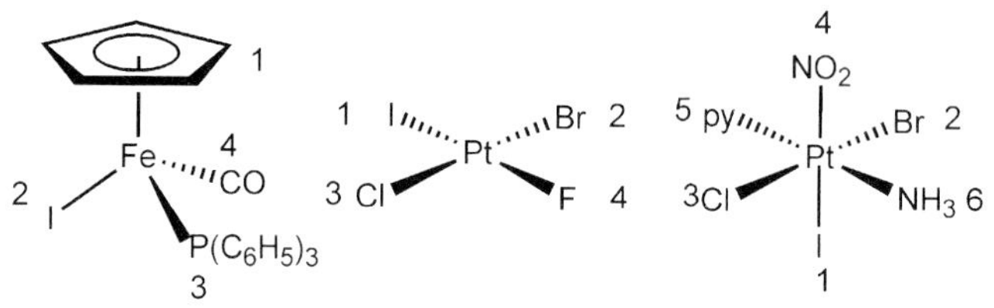

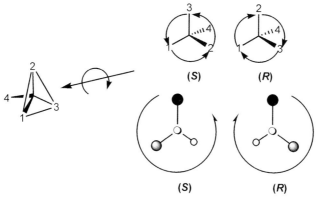

Figure 4.9 The priority for naming the ligands (see text).

Figure 4.10 *R* and *S* definition in T_d symmetry.

formulated for asymmetric carbon it can, in principal, be extended to any element and coordination number (Figure 4.9).

In the case of the tetrahedral complex (Figure 4.9) the cyclopentadienyl ligand (or similar) is regarded as a high priority monodentate ligand. In the case of the octahedral complex there are three ligands with the same atom (nitrogen) in the first coordination sphere. The priority is given by the second sphere, and therefore NO_2 > pyridine > NH_3. Thus, finally, it is necessary to follow these three steps: (i) assign priorities to all groups attached to the chiral center; (ii) align the smallest group away from you; (iii) determine in what direction priorities decrease, clockwise or anticlockwise (Figure 4.10).

Looking at the chiral center with the substituent of lower priority (4) going away from the viewer, if the groups descend in priority clockwise then the complex is named *R* (from the Latin word *rectus*, right); if the groups descend in priority anticlockwise the complex is named *S* (from the Latin word *sinister*, left). The symbol *R* or *S* is added in brackets as a prefix to the name of the chiral compound. A racemic mixture is designated as (*R,S*).

The CIP system is very useful for assigning the *conformation* of chiral organic ligands. Although it was devised for tetrahedral carbon it is usually extrapolated to chiral aromatic ligands. For example, the derivatives of bis-naphthalene in Figure 4.11 are denominated *R* and *S*. One only has to consider the direction in which the *upper ring* (bold line) turns with respect to the hypothetical plane that the molecule would have were it planar.

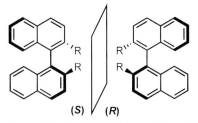

Figure 4.11 *R* and *S* definition in bis-naphthalene derivatives.

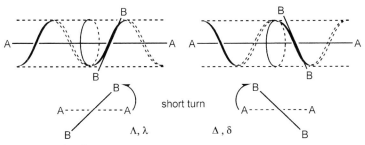

Figure 4.12 Skew-line convention.

Remarks (i) The symbols *R* and *S* do not show any correlation with the sign of the optical rotation; (ii) The *R/S* convention is useful for the determination of chirality at tetrahedral centers. The same principles are readily extendable to geometries other than tetrahedral. However, according to the IUPAC recommendations [15] *R* and *S* are replaced by *C* (clockwise) and *A* (anticlockwise) when applied to other polyhedra.

4.4.2
Skew-line Convention

This is mainly used for octahedral complexes. Two skew-lines define a helical system (Figure 4.12). One of the skew-lines AA determines the axis of a helix upon a cylinder whose radius is equal to the length of the common normal to the two skew-lines AA and BB. The other of the skew-lines, BB, is tangent to the helix and determines the pitch of the helix. In Figure 4.12 the two skew-lines AA and BB are seen in projection onto a plane orthogonal to their common normal [15].

 To define the chirality the principal line AA must be situated behind the line BB. If, in order to superimpose AA with BB, the line AA must be turned anticlockwise (to the left) then AA and BB describe a helix of left chirality (Λ, λ). If, for AA to coincide with BB, it must be turned to the right, AA and BB are said to describe a helix of right chirality (Δ, δ) (Figure 4.12). The capital letters indicate the configuration of the complex; the lower case letters refer to the conformation of the ligand.

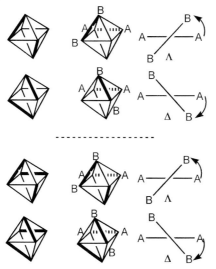

Figure 4.13 Δ or Λ nature of the two enantiomers *cis*-[M(chelate)$_2$(H$_2$O)$_2$] and [M(chelate)$_3$].

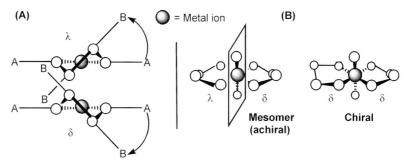

Figure 4.14 (A) The skew-line system for denoting the conformation of the chelate rings; (B) mesomer (achiral) and chiral species (see text).

Figure 4.13 shows the Δ or Λ nature of the two enantiomers *cis*-[M(chelate)$_2$(H$_2$O)$_2$] and [M(chelate)$_3$]. The simplest and most intuitive way of formulating the [M(aa)$_3$] complexes consists in comparing them with three-blade helices. In the first case the helix, as it advances, turns anticlockwise (toward the left): it is known as helix Λ. In the second case the helix, as it advances, turns clockwise (toward the right): it is termed Δ.

The skew-line system is very useful for denoting the conformation of the chelate rings (Figure 4.14A). Finally, we can describe the interaction between the *configuration (molecular chirality)* and the *conformation (chirality of the ring)* for tris(chelate) complexes. If we imagine that the ligands are planar there will only be two possible enantiomers, Δ and Λ. However, as the ligands can have δ or λ conformations, there will be four pairs of enantiomers. In other words, the configuration of the molecule may be Δ or Λ, and for a given configuration the ring conformations may be $\delta\delta\delta$, $\delta\delta\lambda$, $\delta\lambda\lambda$ or $\lambda\lambda\lambda$. Since a ring conformation δ is obviously the enantiomer of the λ conformation, the four pairs of isomers are $\Delta\delta\delta\delta$, $\Delta\delta\delta\lambda$, $\Delta\delta\lambda\lambda$, $\Delta\lambda\lambda\lambda$ and $\Lambda\lambda\lambda\lambda$, $\Lambda\lambda\lambda\delta$, $\Lambda\lambda\delta\delta$, $\Lambda\delta\delta\delta$.

Remark   A complex may contain two chiral ligands and be achiral. Consider two chelate ligands in the *trans* position in an octahedral complex (Figure 4.14B). If one has a δ conformation and the other λ the molecule has a symmetry plane and is therefore achiral. In this case it is a *mesomeric* complex.

4.5
Chirality in Coordination Compounds: Origin and Examples in Mononuclear Complexes

It should be remembered that not only asymmetric molecules (C_1, lack of any symmetry element) are chiral: *disymmetric* molecules (lack of any improper symmetry axis S_n) are also chiral. From a mathematical point of view $S_1 = \sigma$ and $S_2 = i$. Therefore, in most complexes one only has to check if they contain symmetry planes or an inversion centre to determine whether they are chiral or not. Only in a very few cases is it necessary to investigate the presence or absence of other elements of S_n symmetry. The chirality of a coordination compound may have several causes which derive from the study of the different parts of the complex. Let us study each of these in turn.

4.5.1
Different Nature of the Ligands

The different nature of the ligands, as occurs with asymmetric carbon, is the simplest case to imagine. It is difficult to synthesize complexes with a high coordination number and where all the substituents are different.

4.5.2
Distortion from Achiral to Chiral Geometries

Two examples could illustrate this case: in hexacoordinated ML_6 complexes with monodentate ligands the two extremes (octahedron, O_h and trigonal prism, D_{3h}) are achiral structures. In between, the molecular symmetry is lowered to D_3 and is therefore chiral. A bis-chelate complex, $[M(chelate)_2]$, can be distorted between square-planar and tetrahedral geometry becoming, therewith, a chiral species. The two end structures are of D_{2h} and D_{2d} achiral symmetries, while the intermediate geometries are of D_2 symmetry and therefore chiral. The problem is the barrier to enantioisomerization.

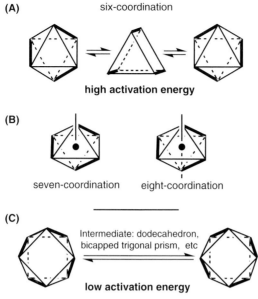

Figure 4.15 Spatial arrangement of the achiral chelated rings
in six, seven and eight coordination complexes.

4.5.3
Spatial Arrangement of the Achiral Chelated Rings

This is the most typical situation in coordination complexes. For example, this situation arises in $[M(L–L)_3]$ (point group D_3) (L–L = chelate ligand) or *cis*-$[M(L–L)_2L'L']$ (point group D_2) octahedral complexes (Figure 4.15A). In general, complexes with larger normalized bite values tend to have larger twist angles and slower rates of isomerization. Mechanistically, either bond-rupture or non-bond-rupture mechanisms are implicated. d-orbital effects on this stereochemical non-rigidity have been studied. For example with the same triscatecholate ligands, racemization of the Ti^{4+} complex is clearly faster than that of the Ga^{3+} complex. The Ge^{4+} analog is extremely inert. The isomerization of the d^{10} complexes is slower than that of the d^0 Ti^{4+} analogs.

Together with these classical six-coordinated complexes, seven [16a] and eight-coordinated [16b] enantiomers (with three bidentate ligands) form a series of chiral propeller molecules, with monodentate ligands on the threefold axis (Figure 4.15B).

Figure 4.15C represents schematically a $[M(L-L)_4]$ (square anti-prism, D_4) complex. In octahedral geometry the two chiral complexes can be isolated. In contrast, this is very difficult in octacoordinated complexes due to the low activation energy that must be overcome to move from one enantiomer to another.

Assuming octahedral geometry, chirality may occur with any class of polydentate ligands. There are two ways of placing a single tridentate ligand (Figure 4.16A)

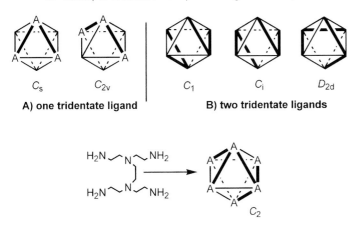

A) one tridentate ligand   **B) two tridentate ligands**

C) one hexadentate ligand

Figure 4.16 Chirality (or not) in O_h symmetry, with some polydentate ligands.

and three ways of placing two tridentate ligands (Figure 4.16B). The symmetry point groups indicate that only the configuration of two tridentate ligands giving C_1 symmetry can present two enantiomers. Flexible aliphatic triamines, such as diethylenetriamine (dien), usually form this type of structure. More rigid amines, such as terpy, tend to form the non-chiral configuration D_{2d}. However, in the D_{2d} configuration the non-planarity of the ligands breaks the symmetry (losing the symmetry plane) and it becomes C_2 (chiral).

A similar study can be done with other polydentate ligands, but such a study is beyond the scope of this chapter. An interesting review of tetradentate ligands has been reported by Scott et al. [17]. By way of example, we could highlight the hexadentate ligand shown in Figure 4.16C which, with the Co^{2+} cation forms two enantiomers of symmetry C_2, both of which have been isolated.

4.5.4
The Coordination of Chiral Ligands

The coordination of ligands that are themselves chiral leads to the formation of chiral complexes. It is necessary to distinguish between chiral ligands with a *stereogenic* center (chiral) and those without a *stereogenic* center.

4.5.4.1 The Coordination of Ligands With a Stereogenic Center
A typical example is shown in Figure 4.17A, in which the chiral ligand is the alaninate anion. Figure 4.17B shows the four possible complexes that can be obtained with three *S*-alaninate chiral ligands. All of these have been synthesized and separated with Co^{3+} and Rh^{3+}. Four further isomers are obtained with *R*-alaninate.

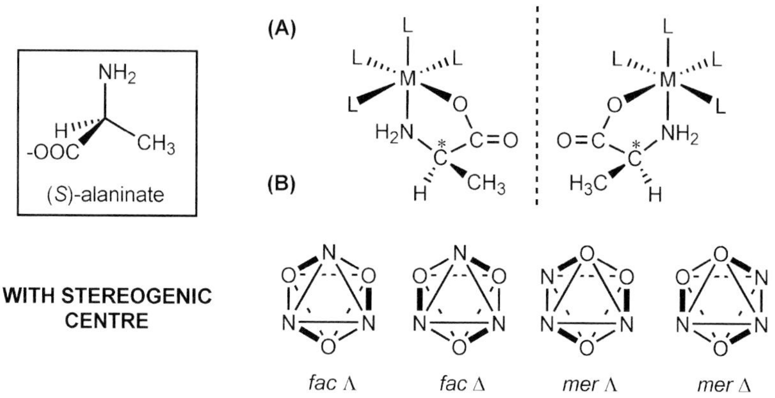

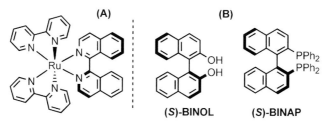

(S)-alaninate

WITH STEREOGENIC CENTRE

fac Λ   *fac* Δ   *mer* Λ   *mer* Δ

Figure 4.17 Chirality due to the presence of a chiral ligand (alaninate anion).

WITHOUT STEREOGENIC CENTRE

(A)   **(B)**

(S)-BINOL   **(S)-BINAP**

Figure 4.18 Chirality due to the presence of ligands without a stereogenic center.

4.5.4.2 The Coordination of Ligands Without a Stereogenic Center

Figure 4.18A shows an octahedral chiral complex resulting from a chiral ligand without a stereogenic center. The non-planarity of the 1,1′-bis(isoquinoline) molecule is due to the repulsions between the hydrogen atoms of the benzene rings. Two other ligands without stereogenic centers are depicted in Figure 4.18B.

4.5.5
The Coordination of Pro-chiral Ligands

Certain non-chiral ligands become chiral when they are coordinated and lose improper symmetry elements. This situation is usually quite frequent when planar ligands are distorted upon becoming coordinated and lose their planarity as a result of ligand–ligand interactions. Square-planar complexes are, in most cases, achiral species. Distortion of the planarity leads to a helical arrangement of two bidentate ligands (Figure 4.19), with C_2 symmetry.

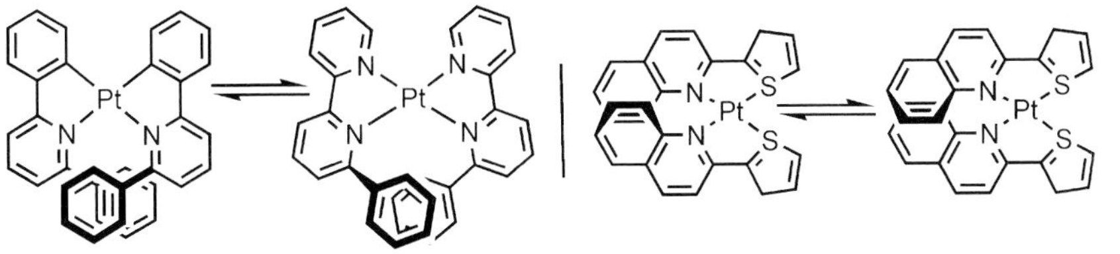

Figure 4.19 Chirality due to the presence of pro-chiral ligands.

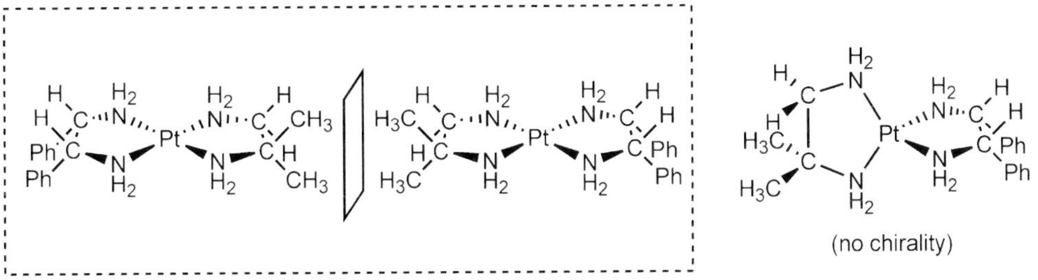

Figure 4.20 Chirality due to the presence of two coordinated ligands that mutually break improper rotation elements.

4.5.6
The Presence of Two Coordinated Ligands That Mutually Break Improper Rotation Elements Present in Each Ligand Individually

A few cases are known. The example shown in Figure 4.20 was first identified in 1939. This is the square-planar compound [Pt(isobutylenediamine)(*meso*-stilbene-diamine)] that presents two enantiomers. Assuming tetrahedral geometry there would be a symmetry plane that would prevent the existence of chirality.

4.5.7
Planar Chirality

Arene ligands bearing two non-equivalent substituents in 1,2 or 1,3 positions relative to each other, coordinated in a η^6-fashion to a transition metal, generate *planar chirality* at the metal with respect to the face of the coordinated arene [18].

4.6
Chirality in Polynuclear Coordination Compounds: Examples

Here it is worth remembering an important aspect: *a molecule with a single chiral center (stereocenter) will always be chiral. If there is more than one asymmetric center then it does not necessarily have to be chiral.*

4.6.1
Chiral Polynuclear Complexes

Double- or triple-helical polynuclear complexes that are formed spontaneously by self-assembly of two or three oligodentate ligands and several metal ions are called *helicates* (see Chapter 14). *Due to their helicity, such molecules are chiral.* The helicates described in the literature are usually racemic mixtures. Other non-helicate examples can be found in Ref. [19].

4.6.2
Chiral Polygons and Polyhedra

It is possible to synthesize polygons and polyhedra that are both highly symmetrical and chiral. Focusing only on polyhedra, there are only three point groups *T*, *O* and *I* that have multiple higher order (three- four- five-fold) axes but no improper ones. Any molecule belonging to one of the three higher-order pure rotation groups is necessarily chiral. However, as indicated by Cotton: *"So far as we are aware, no molecules of O or I symmetry have been reported in the literature"* [20].

In 2001 a *T*-symmetrical icosahedron was reported with a set of four equal tridentate ligands (substituted cyclohexane-1,3,5-triols) placed around a central cation large enough to support 12-coordination, namely Ba^{2+}, in such a way that they cover four non-touching triangular faces of an icosahedron (Figure 4.21) [21a]. The resulting icosahedral geometry retains the full rotational symmetry of the tetrahedron. This arrangement can occur in two enantiomorphs and, in fact, a racemic mixture is obtained. The use of Sr^{2+} results in the formation of a cage with coordination number 10. The resulting SrO_{10} polyhedron has chiral C_3 symmetry.

Raymond et al. [21b] synthesized an enantiostable molecule of *T* symmetry, $[Ga_4L_6]$ (L = catecholate; Figure 4.22). The synthesis, however, is not enantiospecific and thus a racemic mixture that forms racemic crystals was obtained (ΛΛΛΛ or ΔΔΔΔ). However, with NEt_4^+ as encapsulated cation, enantiopures ΔΔΔΔ- and ΛΛΛΛ-$[Et_4N]_{11}[(Et_4N)\subset Ga_4L_6]$ with complete retention of chirality of the metal centers, are easily formed. Many similar complexes have been obtained to date.

Note   An interesting 'microreview' of chiral molecules containing a metal–metal bond has been published recently by Cotton and Murillo [22].

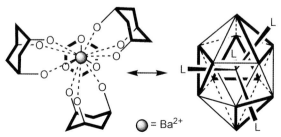

Figure 4.21 A chiral *T*-symmetrical icosahedron (see text).

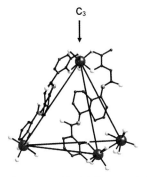

Figure 4.22 [Ga$_4$L$_6$] (L = catecholate): an enantiostable molecule of *T* symmetry. Large black spheres = Ga.

4.7
Separation of Chiral Isomers

Optically pure compounds are of importance in many domains of chemistry. In spite of recent progress in this area, the control of the chirality at the metal center during the synthesis remains an attractive challenge.

4.7.1
Resolution of Racemic Complexes (Stereospecific Synthesis)

In most cases, the preparation of optically pure complexes involves an initial racemic synthesis followed by a resolution process. The separation of the optically active compound from the racemic product mixtures is achieved by various methods. The formation of tightly associated contact ion pairs between *prostereogenic* or *chiral* cations and *chiral* anions can therefore occur and lead, as a result, to large chemical and physical differences among the diastereomeric salts. Lacour and Hebbe-Viton have published a review of the main chiral anions which mediate in asymmetric chemistry [23]. They distinguish:

- Pure anions, such as the natural compounds available like tartaric, mandelic or 10-camphorsulfonic acids. These rather hydrophilic anions are usually available in large quantities in one enantiomeric form. These anions, however, possess a rather large number of potential conformers. To reduce them it is possible to coordinate the moieties to metal ions and the derived, more rigid, anionic complexes have proven their efficiency over the years in a large number of applications. The most characteristic of these compound is the antimonyl (*L* or *D*)-tartrate anion which is commercially available as its sodium or potassium salt.

- Intrinsically chiral-metallo-organic complexes, such as tris(oxalato)metallate(III) anionic complexes.

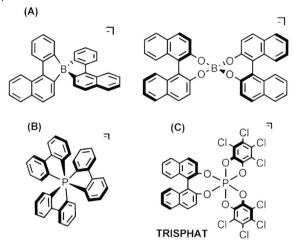

Figure 4.23 Chiral tetrahedral borate anions and hexacoordinated phosphate anions.

- Tetrahedral borate anions, such as those shown in Figure 4.23A.

- Hexacoordinated phosphate anions like those shown in Figure 4.23B and C.

4.7.1.1 Resolution by Solubility Differences of Solids

A simpler method involves precipitation with suitable chiral counteranions. For example, $[M(aa)_3]^{n+}$-type complexes can be separated with enantiomerically pure tartrate or similar anions. This discrimination is due to the formation of ionic pairs. For example, an anion such as $[Sb_2(d\text{-tartrate})_2]^{2-}$ has been observed to have a skeleton that adapts well to the L-shaped channels of a Λ complex, and this makes precipitation or co-crystallization easier. Another aspect to be borne in mind is the ease of formation of hydrogen bonds with one complex or another; these offer stability to the final structure.

The enantiopure Λ form of the tributylammonium salt of the chiral TRISPHAT (Figure 4.23C) is able to induce an efficient resolution of a racemic mixture of *cis*-$[Ru(dmp)_2(NCCH_3)_2](PF_6)_2$ (dmp = 2,9-dimethyl-1,10-phenantroline) due to the spontaneous and selective precipitation of the heterochiral pair [Δ-$[Ru(dmp)_2(NCCH_3)_2][\Lambda\text{-Trisphat}]_2$ [24].

4.7.1.2 Resolution by Chromatography

Most ion-pair chromatographic resolution involves the addition of a non-racemic counterion to the mobile phase. Selected examples are: the resolution of a dinuclear cobalt(III) helicate by ion-pair chromatography with a concentrated solution of enantiopure sodium(+)-antimonyltartrate as the eluent; the resolution of an iron(II) dinuclear double helicate, (racemic sulfate salt), absorbed onto a cation exchanger column and eluted with an aqueous solution of the sodium salt of O,O'-di-4-toluoyl-*L*-tartrate [25]. The $[Ru(bpy)_3]^{2+}$ ion can be separated into the diastereomeric homochiral, [Δ-RuL₃][Δ-TRISPHAT] salt by column chromatography over silica gel (eluent CH_2Cl_2) [26].

4.7.1.3 Resolution by Asymmetric Extraction

The lipophilicity of the TRISPHAT confers on its salts an affinity for organic solvents and, once dissolved, the ion pair does not have a tendency to pass to aqueous layers. This rather uncommon property was used to develop a simple and practical resolution procedure for chiral cationic coordination compounds by asymmetric extraction. A solution of [Bu₃NH][Λ-TRISPHAT] in $CHCl_3$ was added to a racemic mixture of orange-colored $[Ru(Me_2bpy)_3]^{2+}$ and $[Ru(Me_2phen)_3]^{2+}$salts. Upon vigorous stirring of the biphasic mixtures, a partial transfer of coloration occurred from the aqueous layer to the organic one. The selectivity ratio for the enantiomers of the cations was 49:1 and 35:1 in the organic and aqueous layer, respectively [26].

4.7.2
Enantioselective Synthesis

4.7.2.1 Spontaneous Selectivity or Total Spontaneous Resolution

In a very few cases the synthesis of solid complexes give a pure enantiomer complex. For example, $[Ln^{III}(dbm)_3(H_2O)]$ (Ln = Pr, Sm and Er; dmb = dibenzoyl-methane) display seven-coordinate propeller-shaped molecules (capped-octahe-dron with the aqua ligand capping one of the surfaces) which are chiral and crystallize as conglomerates [16a]. The Λ and Δ-enantiomers crystallize separately. Analysis of the crystal structures reveals 'supramolecular' interactions, which explains how stereochemical information can be transferred between stacks of molecules. Because the crystallization starts without seeding, the overall prepara-tion may be regarded as *absolute asymmetric synthesis* [16a]. These complexes race-mize rapidly in solution. The stereoselectivity is, thus, a feature derived from the nucleation.

Another significant example is the absolute asymmetric synthesis of *helical* complexes formed with $[Al(OC_6H_3Ph_2)_3]$ ($OC_6H_3Ph_2$ = 2,6-diphenylphenoxide) and different aldehydes, $[Al(OC_6H_3Ph_2)_3(R-CHO)]$ [27]. These crystals are stereochemi-cally labile and thus enantiomerize rapidly in solution. Preferential crystallization was achieved without stirring the solution (this feature seems to be very general). Once again, the stereoselectivity derives from the nucleation, being transferred from 'crystal-to-crystal'.

4.7.2.2 Predetermined Chirality

By definition, the theoretical maximum yield in a resolution cannot exceed 50% for each of the two pure enantiomers. To overcome this limitation, stereoselective synthetic methods that allow the preparation of optically active complexes have been developed. The best method is the *induction of helicity* by chiral ligands. When chiral nonracemic ligands are used, one isomer may be formed preferentially. This has been termed predetermination of *chirality-at-metal*.

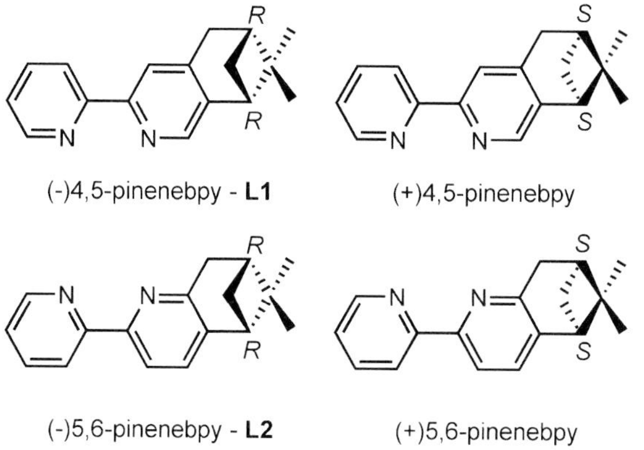

(-)4,5-pinenebpy - **L1**　　　　　(+)4,5-pinenebpy

(-)5,6-pinenebpy - **L2**　　　　　(+)5,6-pinenebpy

Figure 4.24 Two chiragen ligands (see text).

Smirnoff, one of the last collaborators of Alfred Werner, described clearly the diastereoselectivity synthesis of chiral octahedral platinum(IV) complexes. By carefully measuring the optical rotation values, Smirnoff deduced that $[PtCl_2(pn)_2]^{2+}$ and $[Pt(pn)_3]^{4+}$ are formed stereoselectively, if non-racemic 1,2-diaminopropane (pn) is used as a ligand.

Scott et al. published in 2003 a review on the predetermination of chirality at octahedral centers with tetradentate ligads [17]. An important class of chiral pyridine ligands form part of *chiragens* developed by von Zelewsky et al. [28] and as other groups. The key step of this development was the synthesis of the two molecules **L1** and **L2** (Figure 4.24). Both enantiomers of these ligands are easily accessible.

These ligands were called CHIRAGEN (from *chira*lity *gen*erator). Von Zelewsky synthesized more than 100 different ligands following the same principles [28]. Some interesting examples of chiral complexes with some chiragens are:

- Platinum(II) forms square planar complexes. With chiragens, chiral configurations have been reported, although the configuration at the metal is fixed only in the solid state and is floppy in solution [28].

- An inherently chiral family of complexes is the *helicates* (see Chapter 14) which are obtained as racemates if no chiral induction is present. In this respect, the 5,6-chiragens consisting of two 5,6-pinene bpy moieties connected through a bridge represent a very good choice (Figure 4.25A) [28].

- The diastereoselective formation of $[M_4L_6]$ tetrahedral cage complexes has been reported [29]. For example, a single diastereoisomer of the cage complexes $[M_4L_6](BF_4)_8$ [M = Co²⁺, Zn²⁺; L = chiragen derivative, (Figure 4.25B) can be easily prepared in a good yield [30].

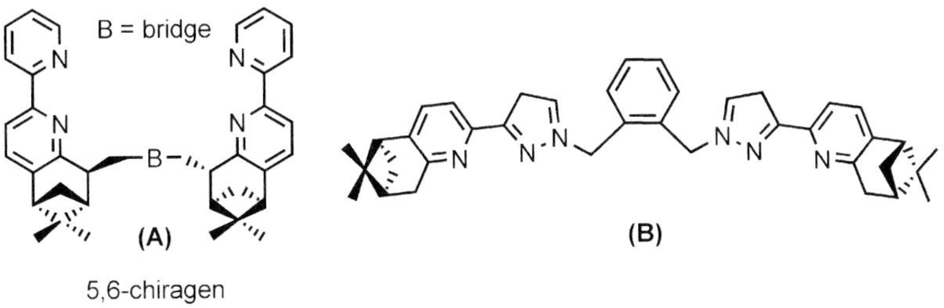

5,6-chiragen

Figure 4.25 Two chiragen ligands that give chiral helicates (A) and chiral [M₄L₆] tetrahedral cages (B) respectively.

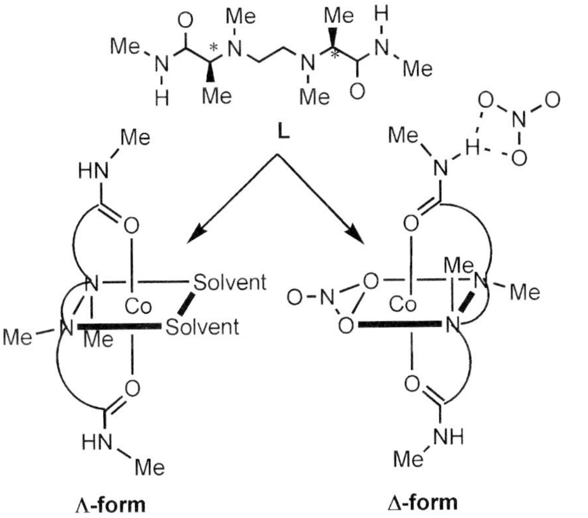

Figure 4.26 An example of dynamic helicity inversion (see text).

4.7.3
Dynamic Helicity Inversion

Dynamic helicity inversion often plays an important role in biological DNA and protein systems. However, labile transition metal complexes with helicity inversion are still very limited. Miyake et al. recently demonstrated that the helicity of an octahedral Co^{2+} complex with the chiral tetradentate ligand L (Figure 4.26) was dynamically inverted in CH_3CN/CH_2Cl_2 by simply adding NO_3^- anion. When two NO_3^- anions interact, respectively, with the Co^{2+} center and the amide hydrogen atom, the complex helicity changes from Λ to Δ (Figure 4.26). The present L-Co^{2+} exhibits, thus, 'solvato-diastereomerism', in which the nature of the solvent employed dynamically induces helicity inversion of the labile metal complex. Similar helicity inversion has been observed with Ni^{2+} and Cu^{2+} complexes [31].

4.8
From Chirality (Coordination Compounds) to Chirality (Application to the Enantioselective Synthesis of Organic Molecules) [32]

From 1965–1970 there are many examples showing that tailor-made chiral metal complexes have an interesting catalytic activity for the production of a broad range of enantiomerically pure compounds. The demand for enantiomerically pure compounds with a desired biological activity is growing in fine-chemical synthesis. An obvious reason for the development is that the *opposite* enantiomer of a chiral pharmaceutical or chemical has at best no activity, or worse, causes side effects.

Such chemicals are obtained through catalytic processes in which the intimate mechanism is well or not-well known. The catalytic species is often formed *in situ* from a given cation and a chiral ligand. We will first give a rapid overview of the main chiral ligands that have been used either in catalysis research or, mainly, in the chemical and pharmaceutical industries.

4.8.1
Overview of Chiral Ligands

Chiral bidentate ligands have dominated the field of asymmetric transition metal catalysis for more than 30 years. However, for a few years it has been known that metal complexes based on monodentate ligands provide highly enantioselective catalysts. The main reason for the preference for monodentate over bidentate ligands is their relative ease of preparation.

Numerous chiral phosphanes have been designed and synthesized over the past four decades [33a]. The most important chiral phosphanes are shown in Figure 4.27. Optically active trialkylphosphanes are known to hardly racemize, even at quite high temperature. On this basis, Imamoto et al. [33b] designed a new class of P-chiral phosphane ligands, 1,2-bis(alkylmethylphosphane)ethanes (abbreviated as BisP*) (alkyl = *tert*-butyl, 1,1-diethylpropyl, 1-adamantyl, ciclopentyl, ciclohexyl) (Figure 4.27, inset). UREAphos, i.e. urea appended chiral phosphite has been recently reported [34].

Optically active *non-phosphane* molecules, BINOL (Figure 4.18) and their derivatives have also played an important role in asymmetric catalysis. Very recently two reviews on the design and applications of linked-BINOL chiral ligands in asymmetric catalysis have been published [35].

4.8.2
Overview of Main Catalytic Processes

4.8.2.1 Asymmetric Hydrogenation
(R,R)-1,2-bis[o-methoxyphenyl)phenylphosphane]ethane, (DIPAMP, Figure 4.27) was a landmark discovery at an early stage in the history of asymmetric hydrogenation reactions. The synthesis of L-Dopa (Figure 4.28A), a drug for the treatment of Parkinson's disease, has been developed and is applied on an industrial scale.

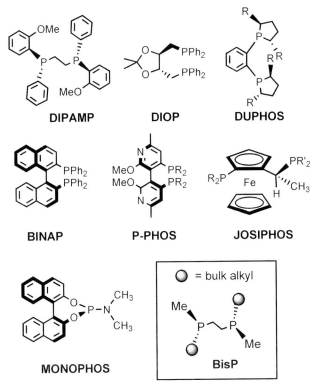

Figure 4.27 Some important P-derivatives chiral ligands.

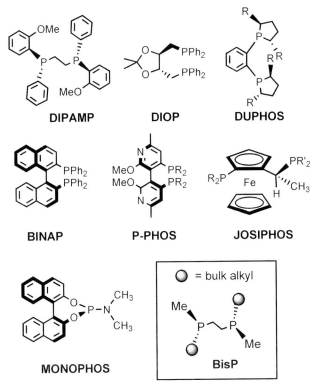

Figure 4.28 Synthesis of L-Dopa and S-Naproxen (see text).

The reaction is not very sensitive to the type of diphosphane used, although DIPAMP is the most widely used. This synthesis proceeds from the asymmetric hydrogenation of cinnamic acid derivatives developed by Knowles [32], who received the Nobel Prize for Chemistry in 2001 together with Noyori (BINAP) and Sharpless (asymmetric epoxidation).

BINAP belongs to the second generation of chiral ligands for asymmetric hydrogenation. It was introduced by Noyori [32] and has been particularly explored for reduction of prochiral alkenes with ruthenium catalysts. An impressive list of rather complex organic molecules has been hydrogenated with high stereoselectivity. An important case is the synthesis of Naproxen (and analogs) which are important anti-inflammatory for rheumatic diseases (Figure 4.28B). With this system a high enantioselectivity can be achieved (97%) albeit at rather high pressures (135 bar).

UREAphos (see above) has been recently proved successful for rhodium catalysed asymmetric hydrogenation of various substrates [34].

4.8.2.2 Asymmetric Hydroformylation

Asymmetric hydroformylation of alkenes (Figure 4.29A) is potentially an important reaction for the synthesis of chiral aldehydes as intermediates in drug synthesis. In 1992, an important breakthrough appeared in the patent literature when Babin and Whiteker reported the asymmetric hydroformylation of various alkenes with *ees* up to 90%, using bulky chiral diphosphites. However, the most interesting ligand discovered for asymmetric hydroformylation is undoubtedly BINA-PHOS (and related ligands), introduced by Takaya and Nozaki (Figure 4.29B) [36]. The Rh⁺ complex of (*R,S*-BINAPHOS) provides very high enantioselectivity.

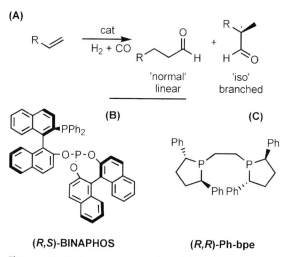

(R,S)-BINAPHOS   **(R,R)-Ph-bpe**

Figure 4.29 (A) Asymmetric hydroformylation of alkenes;
(B) some important ligands used for this process (see text).

Very recently, Klosin et al. reported an excellent new ligand, a disubstituted phospholane, (*R,R*)-Ph-bpe, (Figure 4.29C) for the rhodium catalyzed hydroformylation of styrene, allyl cyanide and vinyl acetate at high temperatures (80–100 °C) [37]. This new Rh-ligand catalyst displays the best regio- and enantioselectivity reported to date for the first two reactions.

4.8.2.3 Asymmetric Epoxidation (and Similar Reactions)

Epoxidation of alkenes is a powerful reaction for the introduction of oxygen into hydrocarbons. Epoxides are widely used as epoxy resins, paints, surfactants, and intermediates in various organic syntheses. Let us emphasize some processes:

- The Katsuki–Sharpless epoxidation of allylic alcohols [32]. The reaction uses a chiral Ti^{4+} catalyst, *tert*-butylhydroperoxide as the oxidant and it works only for allylic alcohols as the substrate.

- The Jacobsen asymmetric epoxidation of alkenes. As oxygen donors H$_2$O$_2$, PhIO, NaClO (bleach), peracids, pyridine-*N*-oxides, etc., have been used. [Mn^{III}(R-salen)] complexes have been developed as highly enantioselective catalysts for the epoxidation of aryl-conjugated, *cis*-disubstituted olefins, etc (Figure 4.30) [38]. Highly active unsymmetrical macrocyclic oligomeric Co-Salen catalysts have been recently reported [39].

A recently developed Pt(II)-catalyst displays high activity and complete substrate selectivity in the epoxidation of terminal alkenes, with H$_2$O$_2$ as oxidant [40]. Various lacunary polyoxometalates have been reported to be active for the catalytic epoxidation. For example, [γ-H$_2$SiV$_2$W$_{10}$O$_{40}$]$^{4-}$ catalyzes the epoxidation of alkenes with H$_2$O$_2$. In this process the role of vanadium has been proven to be essential [41].

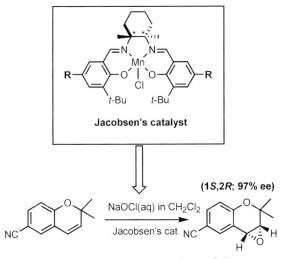

Figure 4.30 Jacobsen asymmetric epoxidation of alkenes.

4.8.2.4 **Other Important Asymmetric Reactions**

Cyclopropane molecules can be synthesized by the addition of carbenes to alkenes. The carbene can be generated via the transition metal catalyzed decomposition of diazo compounds. In recent years, most of the synthetic efforts have focused on the enantioselective synthesis of cyclopropanes [42]. Derivatives of salycilaldehyde and chiral aminoalcohols were amongst the first effective ligands for copper catalysts [32]. Chiral iodomethyl zinc phosphates (BINOL derivatives) have been tested as good cyclopropanating ligands with many representative substrates (unfunctionalized alkenes) [43]. Cyclopropanation of 2,5-dimethyl-2,4-hexadiene provides chrisantemic acid, a natural product used as an insecticide [32].

Palladium-L-catalyzed asymmetric cycloaddition reactions have been reported [44]. Cationic iron salts have been found to be good catalysts for intra- and intermolecular addition of carboxylic acids to olefins, affording the corresponding esther in good yields. See Ref. [45] for more details.

4.8.2.5 **Heterogenized Asymmetric Catalysts**

Immobilization on basic metal oxides leads to increased styrene conversion (epoxidation reaction) with PhIO catalyzed by a manganese(III) porphrin [46].

Nanoparticles can be used as novel supports to prepare '*heterogenized*' asymmetric catalysts that are more accessible to the reactants. Very recently, Neumann et al. reported a direct aerobic epoxidation of alkenes catalyzed by metal (Ag and Ru) nanoparticles stabilized by the $H_5PV_2Mo_{10}O_{40}$ polyoxometalate and supported on α-alumina [47].

The difficulty in recovering nanoparticle-supported asymmetric catalysts by settling or filtration, however, precludes their widespread application. Superparamagnetic materials are intrinsically nonmagnetic but can be readily magnetized in the presence of an external magnetic field. Recent advances in the synthesis of superparamagnetic materials facilitate their exploitation in many technological and biomedical applications. Lin et al. recently reported the design of novel magnetically recoverable heterogenized chiral catalysts and their application in highly enantioselective asymmetric hydrogenation of aromatic ketones [48]. They used a chiral ruthenium catalyst (BINAP derivative) immobilized on magnetite nanoparticles. The heterogenized catalyst can be readily recycled by magnetic decantation and used for asymmetric hydrogenation up to 14 times without loss of activity and enatioselectivity.

4.9
Valence Tautomerism

The valence tautomerism interconversion is an analogous process to low-spin to high-spin crossover phenomena (Chapter 10). Valence tautomeric complexes combine *redox-active ligands* and transition metal ions with two or more accessible oxidation states, exhibiting two nearly degenerate electronic states with localized electronic structures. Charge distribution in such electronic isomers has an appreciable sensitivity to the environment so an external perturbation, like photons, temperature and/or pressure, may lead to an intramolecular electron transfer between both redox active units and therefore to a reversible interconversion between the two degenerate electronic states.

In 2005 Ruiz-Molina et al. published a review on valence tautomerism, with the title '*New challenges for electroactive ligands*' [49]. Indeed, these molecules are actively investigated as possible candidates for information storage and photoswitching effects.

4.9.1
Conditions and Types

Even though the number of metal complexes based on redox-active ligands is considerable, those exhibiting valence tautomerism are rather limited because they must simultaneously satisfy two conditions: (i) the degree of covalence in the interaction between metal ions and electroactive ligands must be low; (ii) the energy of their frontier orbitals must be similar [49].

Most of the valence tautomeric complexes reported so far are based on quinone or quinone-type ligands with a series of transition metal ions such as Co, Cu, Ni and Mn (metal dioxolene complexes). They are 'non-innocent' electroactive ligands that may exist as neutral quinones (Q), radical semiquinones (SQ^-) or dianionic catecholates (Cat^{2-}) (Figure 4.31A), although in valence tautomeric complexes they are only found in the SQ^- and Cat^{2-} forms due to the limiting binding ability of the quinone ligand. Valence tautomers are characterized by displacement of electron density from one redox center to the other, for example in these cases,

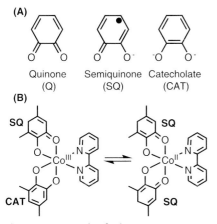

Figure 4.31 Example of valence tautomerism.

M^{II}–SQ and M^{III}–Cat complexes, which interconvert by intramolecular electron transfer between the two redox centers. Thermally induced valence tautomerism in Co complexes was first reported in 1980 in relation to the equilibrium $[Co^{III\text{-}LS}(Cat)(SQ)(N\text{-}N)] \leftrightarrow [Co^{II\text{-}HS}(SQ)_2(N\text{-}N)]$ (Figure 4.31B) [49]. Many other Co(III)/Co(II) couples have been reported to date [49].

Some copper complexes containing dioxolene ligands may yield, under the right conditions, the valence tautomeric interconversion. One of the first examples involving the Cu^{2+}/Cu^+ pair was reported in a copper-containing amine oxidase enzyme that catalyzes the oxidation of amines to aldehydes, an important process involved in relevant biological functions such as growth regulation and tissue maturation [50]. These results and other biological processes have motivated the synthesis and characterization of different copper-quinone synthetic models.

$[Mn^{III}(macro)(dtbcat)]^+$ (macro = tetraazamacrocycle; dtbcat = 3,5-di-*tert*-butylcatecholate) show valence tautomerism, tunable by varying the counterion [51]. Nickel [52], and iron-semiquinone [53] complexes exhibiting valence tautomerism have been reported. Polynuclear complexes can also exhibit this behavior: dinuclear cobalt bis(dioxolene) or tetraoxolene valence tautomers are well known examples [54], as well as 1-D Mn(III) systems with quinone subunits [55].

Within non-quinone ligands, it is worth noting the metalloporphyrins. The cation radicals of metalloporphyrins have been extensively studied because they serve as model systems for biological-redox intermediates. A well-recognized case is the heme system, in which the porphyzinoid macrocyclic ligand can participate directly in the electron-transfer process [56].

4.9.2
Photo-induced Valence Tautomerism

When the charge-transfer bands of some Co valence tautomeric compounds are excited at low temperature, metastable redox isomers can be created after irradiation. The lifetimes of the metastable states can be more than several hours. These transformations can involve changes in the magnetic properties of the compounds as well as their color. Hence, these materials can be regarded as novel photomagnetic materials. The photoresponsive behavior of these valence tautomeric complexes is similar to those of spin-crossover complexes (light-induced excited spin-state trapping – LIESST- effects). The reader can consult a recent review on this topic [57].

References

1 Alcok, N.W., Clarkson, G.J., Lawrance, G.A., Moore, P. Metal-Mediated Pseudo Coordination Isomerism in Complexes of Mixed Neutral Didentate and Dianionic Tridentate Pyridine-Containing Ligands, *Aust. J. Chem.* 2004, *57*, 565.

2 Harris, T.D., Long, J.R. Linkage isomerism in a face-centered cubic $Cu_6Cr_8(CN)_{24}$

cluster with an $S = 15$ ground state, *Chem. Comm.* 2007, 1360 and references therein.

3 (a) Tomita, A., Sano, M. Preparations and Electrochemical Properties of Pyrazine-Bridged Ruthenium-Binuclear Complexes Exhibiting Molecular Hysteresis, *Inorg. Chem.* 2000, *39*, 200; (b) Rachford, A.A., Petersen, J.L., Rack, J.J. Designing Molecular Bistability in Ruthenium Dimethyl Sulfoxide Complexes, *Inorg., Chem.* 2005, *44*, 8065; (c) Rachford, A.A., Rack, J.J. Picosecond Isomerization in Photochromic Ruthenium-Dimethyl Sulfoxide Complexes, *J. Am. Chem. Soc.* 2006, *128*, 14318.

4 Jackson, W.G., McKeon, J.A., Hockless, D.C.R., Willis, A.C. The Reversible and Stereoselective N- to C-bonded Rearrangement of Tris(2-pyridylmethyl)-1,4,7-triazacyclononanecobalt(III), *Inorg. Chem.* 2006, *45*, 4119.

5 Harman, W.D. Conformational and linkage isomerizations for dihapto-coordinated arenes and aromatic heterocycles: controlling the stereochemistry of ligand transformations, *Coord. Chem. Rev.*, 2004, *248*, 853.

6 Yong, L., Hoffmann, S.D., Fässler, T.F. Crystal Structures of [K(2.2.2-crypt)]$_4$[Pb$_9$Mo(CO)$_3$]–Isolation of the Novel Isomers [(η5-Pb$_9$)Mo(CO)$_3$]$^{4-}$ beside [(η4-Pb$_9$)Mo(CO)3]$^{4-}$, *Eur. J. Inorg. Chem.* 2005, 3663.

7 (a) Marzilli, L.G., Iwamoto, M., Alessio, E., Hansen, L., Calligaris, M. The Rare Head-to-Head Conformation of Untethered Lopsided Ligands Discovered in Both Solution and Solid States of 1,5,6-Trimethylbenzimidazole Re(v) and Ru(II) Complexes, *J. Am. Chem. Soc,* 1994, *116*, 815; (b) Iwamoto, M., Alessio, E., Marzilli, L.G. Observation of an Unusual Molecular Switching Device. The Position of One 1,2-Dimethylimidazole Switched "On" or "Off" the Rotation of the Other 1,2-Dimethylimidazole in cis,cis,cis-Ru^{II}Cl$_2$(Me$_2$SO)$_2$(1,2-dimethylimidazole)$_2$, *Inorg. Chem.* 1996, *35*, 2384; (c) Alessio, E., Calligaris, M., Iwamoto, M., Marzilli,

L.G. Orientation and Restricted Rotation of Lopsided Aromatic Ligands. Octahedral Complexes Derived from *cis*-RuCl$_2$(Me$_2$SO)$_4$, *Inorg. Chem.* 1996, *35*, 2538.

8 (a) Velders, A.H., Hotze, A.C.G., Haasnoot, J.G., Reedijk, J. The First Observation and Full Characterization of All Atropisomers and Their Allowed Interconversions in an Octahedral Bis(bipyridine)ruthenium(II) Complex with Two Lopsided Bicyclic Ligands, as Studied by 2D NMR Techniques at Variable Temperature, *Inorg. Chem.* 1999, *38*, 2762; (b) Verders, A.H., Hotze, A.C.G., van Albada, G.A., Haasnoot, J.G., Reedijk, J. Tuning the Rotational Behavior of Lopsided Heterocyclic Nitrogen Ligands (L) in Octahedral *cis*-[Ru(bpy)$_2$(L)$_2$](PF$_6$)$_2$ Complexes. A Variable-Temperature ^{1}H NMR Study, *Inorg. Chem.* 2000, *39*, 4073; (c) Velders, A.H., Massera, C., Ugozzoli, F., Biagini-Cingi, M., Manotti-Lanfredi, A.M., Haasnoot, J.G., Reedijk, J. A Simple Example of the Fluxional Behaviour of Ruthenium-Coordinated C$_2$-Symmetric Monodentate Ligands – Synthesis, ^{1}H NMR Spectroscopic Study, and Crystal Structure of cis-[Ru(bpy)$_2$(4Pic)$_2$](PF$_6$)$_2$, *Eur. J. Inorg. Chem.* 2002, 193.

9 Zhang, J., Siu, K., Lin, C.H., Canary, J.W. Conformational dynamics of Cu(I) complexes of tripodal ligands: steric control of molecular motion, *New J. Chem.* 2005, *29*, 1147.

10 Klein, A., Schurr, T., Knödler, A., Gudat, D., Klinkhammer, K-W., Jain, V.K., Zalis, S., Kaim, W. Multiple isomerism (cis/trans; syn/anti) in [(dmso)$_2$Pt(aryl)$_2$] complexes: A combined structural, spectroscopic, and theoretical investigation, *Organometallics*, 2005, *24*, 4125.

11 Lennartson, A., Hakansson, M., Jagner, S. cis- and trans-Bis(benzoylacetonato)pyridinecopper(II): co-crystallisation of isomers and reversible pyridine loss with retention of crystallinity, *New. J. Chem.* 2007, *31*, 344.

12 Zhu, T., Jackson, W.G. Synthesis, Separation, and Assignment of the Seven Geometric Isomers of [Co(dien)(ibn)Cl]$^{2+}$, *Inorg. Chem.* 2003, *42*, 88.

13 Aullón, G., Bernhardt, P.V., Bozoglián, F., Font-Bardía, M., Macpherson, B.P., Martínez, M., Rodríguez, C., Solans, X. Isomeric Distribution and Catalyzed Isomerization of Cobalt(III) Complexes with Pentadentate Macrocyclic Ligands. Importance of Hydrogen Bonding, *Inorg. Chem.* 2006, *45*, 8551.

14 Drago, R.S. (1992), *Physical Methods for Chemists* (2nd Edn.), Saunders College Publishing. Orlando (USA).

15 Nomenclature of Inorganic Chemistry (2005) *IUPAC Recommendations 2005*, Royal Society of Chemistry, Cambridge.

16 (a) Lennartson, A., Vestergren, M., Hakansson, M. Resolution of Seven-Coordinate Complexes, *Chem. Eur. J.* 2005, *11*, 1757; (b) Hakansson, M., Vestergren, M., Gustafsson, B., Hilmersson, G. Isolation and Spontaneous Resolution of Eight-Coordinate Stereoisomers, *Angew. Chem. Int. Ed.* 1999, *38*, 2199.

17 Knight, P.D., Scott, P. Predetermination of chirality at octahedral centres with tetradentate ligands: prospects for enantioselective catalysis, *Coord. Chem. Rev.* 2003, *242*, 125.

18 Therrien, B., Süss-Fink, G. New arene ruthenium complexes with planar chirality, *Inorg. Chim. Acta*, 2004, *357*, 219.

19 (a) Patterson, B.T., Keene, F.R. Isolation of Geometric Isomers within Diastereoisomers of Dinuclear Ligand-Bridged Complexes of Ruthenium(II), *Inorg. Chem.* 1998, *37*, 645; (b) Soloshonok, V.A., Ueki, H. Design, Synthesis and Characterization of Binuclear Ni(II) Complexes with Inherent Helical Chirality, *J. Am. Chem. Soc.* 2007, *129*, 2426.

20 Cotton, F.A., Murillo, C.A., Yu, R. Deliberate synthesis of the preselected enantiomer of an enantiorigid molecule with pure rotational symmetry *T*, *Dalton Trans.* 2005, 3161.

21 (a) Sander, J., Hegetschweiler, K., Morgenstern, B., Keller, A., Amrein, W., Weyhermüller, T., Müller, I. T-Symmetrical Icosahedra: A New Type of Chirality in Metal Complexes, *Angew. Chem. Int. Ed.* 2001, *40*, 4179; (b) Terpin,

A.J., Ziegler, M., Johnson, D.W., Raymond, K.N. Resolution and Kinetic Stability of a Chiral Supramolecular Assembly Made of Labile Components, *Angew. Chem. Int. Ed.* 2001, *40*, 157.

22 Cotton, F.A., Murillo, C.A. Chiral Molecules Containing Metal-Metal Bonds, *Eur. J. Inorg. Chem.* 2006, 4209.

23 Lacour, J., Hebbe-Viton, V. Recent developments in chiral anion mediated asymmetric chemistry, *Chem. Soc. Rev.* 2003, *32*, 373.

24 Hamelin, O., Pécaut, J., Fontecave, M. Crystallization-Induced Asymmetric Transformation of Chiral-at-metal Ruthenium(II) Complexes Bearing Achiral Ligands, *Chem. Eur. J.* 2004, *10*, 2548.

25 Rapenne, G., Patterson, B.T., Sauvage, J.P., Keene, F.R. Resolution, X-ray structure and absolute configuration of a double-stranded helical diiron(II) bis(terpyridine) complex, *Chem. Commun.* 1999, 1853.

26 Lacour, J., Torche-Haldimann, S., Jodry, J.J., Ginglinger, C., Favarger, F. Ion pair chromatographic resolution of tris (diimine)ruthenium(II) complexes using TRISPHAT anions as resolving agents, *Chem. Commun.* 1998, 1733.

27 Johansson, A., Hakansson, M. Absolute Asymmetric Synthesis of Stereochemically Labile Aldehyde Helicates and Subsequent Chirality Transfer Reactions, *Chem. Eur. J.* 2005, *11*, 5238.

28 (a) Knof, U., von Zelewski, A. Predetermined Chirality at Metal Centers, *Angew. Chem. Inter. Ed.* 1999, *38*, 302; (b) von Zelewski, A. Stereoselective synthesis of coordination compounds, *Coord. Chem. Rev.* 1999, *190–192*, 811; (c) von Zelewski, A., Mamula, O. The bright future of stereoselective synthesis of co-ordination compounds, *J. Chem. Soc., Dalton Trans.* 2000, 219.

29 Saalfrank, R.W., Demleitner, B., Glaser, H., Maid, H., Bathelt, D., Hampel, F., Bauer, W., Teichert, M. Enantiomerisation of Tetrahedral Homochiral [M_4L_6] Clusters: Synchronized Four Bailar Twists and Six Atropenantiomerisation Processes Monitored by Temperature-Dependent Dynamic ^{1}H NMR Spectroscopy, *Chem. Eur. J.* 2002, *8*, 2679.

30 Argent, S.P., Riis-Johannessen, T., Jeffery, J.C., Harding, L.P., Ward, M.D. Diastereoselective formation and optical activity of an M$_4$L$_6$ cage complex, *Chem. Commun.* 2005, 4647.

31 (a) Miyake, H., Yoshida, K., Sugimoto, H., Tsukube, H. Dynamic Helicity Inversion by Achiral Anion Stimulus in Synthetic Labile Cobalt(II) Complex, *J. Am. Chem. Soc.* 2004, *126*, 6524; (b) Miyake, H., Sugimoto, H., Tamiaki, H., Tsukube, H. Dynamic helicity inversion in an octahedral cobalt(II) complex system *via* solvato-diastereomerism, *Chem. Commun.* 2005, 4291.

32 van Leeuwen, V.N.M. (2004), *Homogeneous Catalysis: Understanding the Art*. Kluwer Academic Publishers, Dordrecht.

33 (a) Wu, J., Chan, A.S.C. P-Phos: A Family of Versatile and Effective Atropisomeric Dipyridylphosphine Ligands in Asymmetric Catalysis, *Acc. Chem. Res.* 2006, *39*, 711; (b) Imamoto, T., Watanabe, J., Wada, Y., Masuda, H., Yamada, H., Tsuruta, H., Matsukawa, S., Yamaguchi, K. P-Chiral Bis(trialkylphosphine) Ligands and Their Use in Highly Enantioselective Hydrogenation Reactions, *J. Am. Chem. Soc.* 1998, *120*, 1635.

34 Sandee, A.J., van der Burg, A.M., Reek, J.N.H. UREAphos: supramolecular bidentate ligands for asymmetric hydrogenation, *Chem. Comm.* 2007, 864.

35 (a) Shibasaki, M., Matsunaga, S. Design and application of linked-BINOL chiral ligands in bifunctional asymmetric catalysis, *Chem. Soc. Rev.* 2006, *35*, 269. (b) Brunel, J.M. Update 1 of BINOL: A Versatile Chiral Reagent. *Chem. Rev.* 2007, *107*, PR1.

36 Sakai, N., Mano, S., Nozaki, K., Takaya, H. Highly enantioselective hydroformylation of olefins catalyzed by new phosphine phosphite-rhodium (I) complexes, *J. Am. Chem. Soc.* 1993, *115*, 7033.

37 Axtell, A.T., Cobley, C.J., Klosin, J., Whiteker, G.T., Zanotti-Gerosa, A., Abboud, K.A. Highly Regio- and Enantioselective Asymmetric Hydroformylation of Olefins Mediated by 2,5-Disubstituted Phospholane Ligands, *Angew. Chem. Int. Ed.* 2005, *44*, 5834.

38 (a) Adam, W., Fell, R.T., Stegmann, V.R., Saha-Möller, C.R. Synthesis of Optically Active α-Hydroxy Carbonyl Compounds by the Catalytic, Enantioselective Oxidation of Silyl Enol Ethers and Ketene Acetals with (Salen)manganese(III) Complexes, *J. Am. Chem. Soc.* 1998, *120*, 708; (b) Egami, H., Irie, R., Sakai, K., Katsuki, T. Enantioselective Epoxidation of Conjugated Z-Olefins with Newly Modified Mn(salen) Complex, *Chem. Lett.* 2007, *36*, 46.

39 Zheng, X., Jones, C.W., Weck, M. Ring-Expanding Olefin Metathesis: A Route to Highly Active Unsymmetrical Macrocyclic Oligomeric Co-Salen Catalysts for the Hydrolytic Kinetic Resolution of Epoxides, *J. Am. Chem. Soc.* 2007, *129*, 1105.

40 Colladon, M., Scarso, A., Sgarbossa, P., Michelin, R.A., Strukul, G. Regioselectivity and Diasteroselectivity in Pt(II)-Mediated "Green" Catalytic Epoxidation of Terminal Alkenes with Hydrogen Peroxide: Mechanistic Insight into a Peculiar Substrate Selectivity, *J. Am. Chem. Soc.* 2007, *129*, 7680.

41 Nakagawa, Y., Mizuno, N. Mechanism of [γ-H$_2$SiV$_2$W$_{10}$O$_{40}$]$^{4-}$-Catalyzed Epoxidation of Alkenes with Hydrogen Peroxide, *Inorg. Chem.* 2007, *46*, 1727.

42 Lebel, H., Marcoux, J-F., Molinaro, C., Charette, A.B. Stereoselective Cyclopropanation Reactions, *Chem. Rev.* 2003, *103*, 977.

43 Lacasse, M.C., Poulard, C., Charette, A.B. Iodomethylzinc Phosphates: Powerful Reagents for the Cyclopropanation of Alkenes, *J. Am. Chem. Soc.* 2005, *127*, 12440.

44 Trost, B.M., Stambuli, J.P., Silverman, S.M., Schwörer, U. Palladium-Catalyzed Asymmetric [3 + 2] Trimethylenemethane Cycloaddition Reactions, *J. Am. Chem. Soc.* 2006, *128*, 13328.

45 Komeyama, K., Mieno, Y., Yukawa, S., Morimoto, T., Takaki, K. Cationic Iron-catalyzed Addition of Carboxylic Acids to Olefins, *Chem. Lett.*, 2007, *36*, 752.

46 Hitomi, Y., Mukai, H., Ohyama, J., Shinagawa, M., Shishido, T., Tanaka, T. Adsorption of Manganese Porphyrin on

Metal Oxides and Its Enhanced Catalytic Activity on Epoxidation Reaction, *Chem. Lett.*, 2007, 660.

47 Maayan, G., Neumann, R. Direct aerobic epoxidation of alkenes catalyzed by metal nanoparticles stabilized by the $H_5PV_2Mo_{10}O_{40}$ polyoxometalate, *Chem. Comm.* 2005, 4595.

48 Hu, A., Yee, G.T., Lin, W. Magnetically Recoverable Chiral Catalysts Immobilized on Magnetite Nanoparticles for Asymmetric Hydrogenation of Aromatic Ketones, *J. Am. Chem. Soc.* 2005, *127*, 12486.

49 Evangelio, E., Ruiz-Molina, D. Valence Tautomerism: New Challenges for Electroactive Ligands, *Eur. J. Inorg. Chem.* 2005, 2957.

50 Dooley, D.M., McGuirl, M.A., Brown, D.E., Turowski, P.N., McIntire, W.S., Knowles, P.F. A Cu(I)-semiquinone state in substrate-reduced amine oxidases, *Nature*, 1991, *349*, 262.

51 Caneschi, A., Dei, A. Valence Tautomerism in a o-Benzoquinone Adduct of a Tetraazamacrocycle Complex of Manganese, *Angew. Chem. Int. Ed.* 1998, *37*, 3005.

52 Ohtsu, H., Tanaka, K. Chemical Control of Valence Tautomerism Of Nickel(II) Semiquinone and Nickel(III) Catecholate States, *Angew. Chem. Int. Ed.* 2004, *43*, 6301.

53 Shaikh, N., Goswami, S., Panja, A., Wang, X.-Y., Gao, S., Butcher, R.J., Banerjee, P. New Route to the Mixed Valence Semiquinone-Catecholate Based Mononuclear Fe[III] and Catecholate Based Dinuclear Mn[III] Complexes: First Experimental Evidence of Valence Tautomerism in an Iron Complex, *Inorg. Chem.* 2004, *43*, 5908.

54 (a) Hearns, N.G.R., Korcok, J.L., Paquette, M.M., Preuss, K.E. Dinuclear Cobalt Bis(dioxolene) Complex Exhibiting Two Sequential Thermally Induced Valence Tautomeric Transitions, *Inorg. Chem.* 2006, *45*, 8817; (b) Tao, J., Maruyama, H., Sato, O. Valence Tautomeric Transitions With Thermal Hysteresis around Room Temperature and Photoinduced Effects Observed in a Cobalt-Tetraoxolene Complex, *J. Am. Chem. Soc.* 2006, *128*, 1790.

55 Attia, A.S., Pierpont, C.G. Valence Tautomerism within a Linear Polymer Consisting of Pyrazine-Bridged Manganese-Quinone Subunits. Synthesis and Characterization of [Mn[III](μ-pyz)(3,6-DBSQ)(3,6-DBCat)]$_n$, *Inorg. Chem.* 1997, *36*, 6184.

56 Weiss, R., Bulach, V., Gold, A., Terner, J., Trautwein, A.X. Valence-tautomerism in high-valent iron and manganese porphyrins, *J. Biol. Inorg. Chem.* 2001, *6*, 831.

57 Sato, O., Cui, A., Matsuda, R., Tao, J., Hayami, S. Photo-induced Valence Tautomerism in Co Complexes, *Acc. Chem. Res.* 2007, *40*, 361.

5
Polynucleating Ligands: From Di- and Polynuclear Complexes to Nanomolecules

5.1
Introduction

Polynuclear transition metal chemistry is an area of modern science whose interfaces with many disciplines have provided invaluable opportunities for crossing boundaries both within and between the sciences of chemistry, physics and biology. Indeed, molecular multimetallic compounds have unusual and useful catalytic, optical, photochemical, electronic and magnetic properties.

This class of compounds is found in the literature with several names such as polynuclear complexes, multinuclear coordination compounds, polymetallic compounds, cages, aggregates, clusters, etc. In the current language of inorganic chemistry a "cluster" may take on many meanings (Chapter 6). In this chapter the word cluster is used equally with the term "polynuclear complex" for molecules or ions with no metal–metal bonding.

Inorganic chemists have made little progress in discovering general approaches to preparing compounds containing large numbers of metal centers. However, over the last 15 years several groups have been developing excellent routes to such compounds, trying to introduce an element of *design* into the assembly process. Thus, we can loosely say for *strategies*: (i) The simplest strategy is the use of polynucleating ligands in a predetermined (or controlled) manner; (ii) the named "*controlled stepwise synthesis*" is a more complex strategy which may give new polynuclear complexes in a more controlled manner. In general, it is the "*aggregation of relatively small building-block complexes*". This strategy can be called: the "*complexes as metals and complexes as ligands*" strategy. We shall deal with all these strategies in this chapter. We first deal with selected examples of polynucleating ligands.

5.2
Polynucleating Ligands

Ligands that bridge two or more metal ions are called *bridging ligands*. The resulting complex is often called a *bridged complex*. A bridged complex can be a dimer,

Coordination Chemistry. Joan Ribas Gispert
Copyright © 2008 WILEY-VCH Verlag GmbH & Co. KGaA, Weinheim
ISBN: 978-3-527-31802-5

a trimer, a polynuclear complex or a coordination polymer. Polynucleating ligands can be "arbitrarily" divided into three broad groups: (i) simple monoatomic bridging ligands; (ii) polyatomic bridging ligands; (iii) polytopic (compartmental or not) ligands.

5.2.1
Monoatomic Bridging Ligands

The simplest bridging ligands are monoatomic ligands (X^-, O^{2-}, S^{2-}, etc.). An oxido (O^{2-}) or halido (X^-) ligand can form bonds to two, three or even four metal ions, giving rise to di-, tri- or tetranuclear fragments, respectively. This is because these ligands have four pairs of free electrons directed toward the vertices of a tetrahedron. Although not *intrinsically* monoatomic groups, OH^-, OR^-, etc. can be considered as monoatomic, because the bridging part is *only* the oxygen atom.

Just as coordinated water can lose a proton to give hydroxido/oxido complexes, so alcohols can lose a proton to give alkoxido/aryloxido (RO^-) complexes. Thus, these ligands are good candidates for the preparation of clusters. Some of the most important clusters derived from hydroxido or alkoxido bridges will be studied later in this chapter.

5.2.2
Polyatomic Bridging Ligands

The list could be infinite. We shall deal now with some simple polyatomic bridging ligands which often yield discrete polynuclear complexes.

5.2.2.1 Diatomic Ligands

Diatomic molecules or ions can act as bridging ligands. Typical examples are the dioxygen molecule, the superoxide ion (O_2^-) and the peroxide ion (O_2^{2-}). Synthetic copper-dioxygen complexes are of great interest as structural and/or functional models for O_2 transport proteins such as hemocyanin and dioxygen activating copper proteins such as tyrosinase (see Chapter 17).

5.2.2.2 Small Versatile Ligands

Representative examples are the *nitrite, carbonate, oxalate, carboxylate, and azide* anions. Let us summarize the most important coordination modes of these ligands (Figure 5.1).

The nitrite ion (NO_2^-) can act as an N-ligand and as an O-ligand (linkage isomerism, Chapter 4). The carbonate dianion (CO_3^{2-}) is known as an extremely versatile ligand, able to generate coordination compounds with manifold nuclearity [1]. Since each oxygen atom of the carbonato ion may act as a terminal donor atom or as a monoatomic bridge, the ligand has been found to lead to di-, tri-, tetra-, penta- and even hexanuclear complexes. More than 15 coordination modes have been structurally characterized. The oxalate anion is a very versatile ligand that can adopt a variety of coordination modes, both terminal (monodentate, bidentate

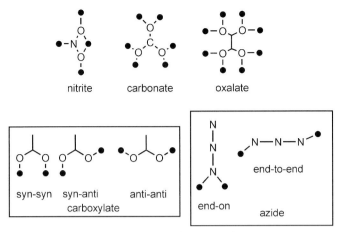

Figure 5.1 Some coordination modes of nitrito, carbonato, oxalato, carboxylato and azido ligands (see text).

chelating) and bridging (μ–μ_8) [2]. The carboxylate ions are a very important class of ligands for the prepararion of polynuclear metal complexes. The most common bonding modes are monodentate, chelating and bidentate bridging. It is not uncommon to have more than one coordination type in the same complex. The azide ion presents also great versatility, giving end-to-end (1,3) or end-on (1,1) coordination modes, together with many mixed possibilities. However, azide ion is best known as a ligand which tends to give 1-D, 2-D or 3-D systems rather than discrete polynuclear ones [3].

5.2.3
Polytopic Ligands

Di(poly)topic ligands are ligands with two (or more) *different coordination sites*. Polytopic ligands fall into two major classes: compartmental and non-compartmental. Compartmental ligands are polynucleating ligands in which the metal ions share at least one donor atom, which acts as a bridge. In other words, a compartmental ligand can hold metal ions side by side. Non-compartmental ligands contain two separate coordination sites linked by one or more bridging units without sharing any atom. Furthermore, the polytopic (compartmental or not) ligands can be symmetric or asymmetric. Owing to their intrinsic importance we will treat the compartmental ligands separately in the next section.

5.2.4
Compartmental Bridging Ligands

An "old" approach to the preparation of polynuclear transition metal complexes is the use of compartmental multinucleating ligands. Five types of compartmental ligands can be distinguished: (i) polyketones; (ii) macrocycles (2+2 condensation); (iii) acyclic ligands open at one end (2+1 condensation); (iv) acyclic ligands open at the base (1+2 condensation) and (v) polypodal ligands. The latter four types usually correspond to Schiff base formation reactions.

5.2.4.1 Polyketones

1,3,5-triketones are easily deprotonated and give rise to highly stable dianionic species (Figure 5.2A). These ligands form dinuclear complexes; mononuclear species can only be obtained under extreme conditions. Ligands based on 2,6-diacetylphenol (Figure 5.2B) easily undergo deprotonation at the –OH group and form dinuclear complexes with the phenoxide oxygen bridging two metal ions.

β-Diketones and their derivatives have been extensively studied in order to prepare polynuclear complexes within *"rational design"* schemes. H_3L (Figure 5.2C) for example, possesses three ionizable protons that can be removed gradually, and this has allowed the preparation of a variety of complexes with different metals featuring various nuclearities [4]. These complexes contain the ligand in its di- or tri-anionic form, depending on the amount of base used in the reaction system. A typical result is $[Cu_8(OMe)_8L_2X_2]$ (X = Cl, Br, NO_3, ClO_4) that consists of two rows of four Cu^{2+} ions held together by two L^{3-}. With ClO_4^-, the $[Cu_8]$ moiety exists as a discrete octanuclear cluster in the solid state, but with NO_3^-, Cl^- and Br^- these $[Cu_8]$ entities are linked by the anions to give one-dimensional systems (Figure 5.2D).

5.2.4.2 Schiff Base Derivatives

The synthesis of compartmental macrocyclic ligands is normally achieved by a template Schiff-base condensation between 2,6-diformyl-4-substituted phenols and α,ω-diamines. The compartmental ligands of Figure 5.3 can be synthesized from the same reagents with only the reaction ratios being varied. Therefore, they are described as (2+2), (2+1) and (1+2) ligands, the first number indicating the

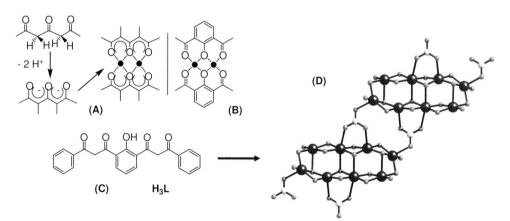

Figure 5.2 Coordination modes of polyketones (see text).

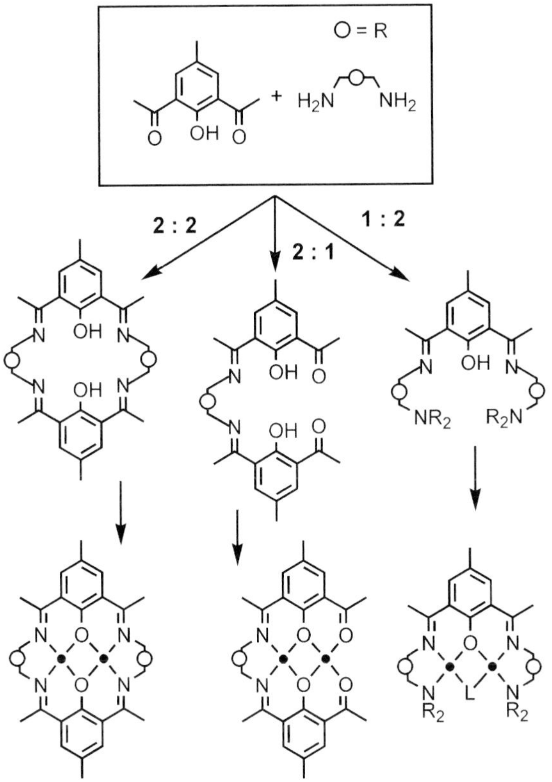

Figure 5.3 Schiff-base derivatives acting as ligands (see text).

proportion of ligands with the C=O group that are involved in the synthesis and the second the proportion of ligands with the NH$_2$ group.

Asymmetric modifications of these symmetric systems were designed in order to achieve a different recognition process at the two adjacent chambers and consequently to favor the formation of heterodinuclear complexes that are not otherwise accessible. The preferential binding of a different metal to each separate subunit has been termed *"haptoselectivity"*.

5.3
Building Block Strategy or Controlled Synthesis: Complexes as Metals/ Complexes as Ligands (Complexes as Tectons)

The synthesis of novel clusters is achieved through the reaction of a complex that acts as a "metal" with another complex which acts as a "ligand". Thus, *"building-blocks"* are used instead of the typical isolated metal ions and the "classical" ligands studied in previous chapters. The great advantage of this strategy lies in the facility to use a given building block in each step of the process and to incorporate specific blocks in selected places. This approach provides an efficient means of controling both the nuclearity and the dimensionality of a polymetallic edifice.

5.3.1
The "Complexes as Metals" Approach

Diaz et al. have reported several polynuclear systems linking together heterodinuclear (Cu–Ni) or trinuclear copper(II) entities through hydrogen bonds or thiocyanato/selenocyanato bridges [5]. In this case, the trinuclear copper(II) entity is the building block (the *"metal"*) and the water (hydrogen bonds) or the thiocyanato bridging ligand links these units to give more complicated (1D, for example) entities. The heterotrinuclear [NiCuNi] (*complex as metal*) reacts with N_3^-, giving a new hexanuclear derivative through *end-to-end* azido bridging ligands [6].

5.3.2
The "Complexes as Ligands" Approach

The most successful strategic approach for preparing heteronuclear complexes is that of using metal complexes as "ligands", i.e. metal complexes containing potential donor groups for another metal ion with empty coordination sites. For brevity, let us mention only the polycyanometalate derivatives. Cyanometalates, $[M(CN)_n]^{m-}$, are characteristic examples of metal complexes that can be used as "ligands". It is well known that when $[Fe(CN)_6]^{3-}$ reacts with Fe^{2+} salts, the so-called Prussian Blue is formed. However, if most of the free sites of the metallic ion are blocked with convenient ligands, the synthesis can be directed towards new discrete or low-dimensional species (see next section).

5.3.3
The "Complexes as Ligands and Complexes as Metals" Approach

This strategy has allowed the systematic synthesis of a series of polycyanometalate derivatives $[M-(CN)_n-M'L_m]$. The structure of one of them, $[Cr\{CN-Cu(tren)\}_6]^{9+}$, is shown in Figure 5.4A [7]. With this strategy, a unique case has been recently reported by Dunbar et al. [8]: the reaction of $[Fe(CN)_6]^{3-}$ (*complex-ligand*) and $[Co(tmphen)_2]^{2+}$ (tmphen = 3,4,7,8-tetramethyl-1,10-phenantroline) (*complex-metal*) leads to the compound $\{[Co(tmphen)_2]_3[Fe(CN)_6]_2\}$ whose structure is shown in Figure 5.4B. A particular feature of this compound is the presence of three terminal cyanides per Fe^{3+} that can be used for subsequent coordination of additional metal centers. Thus, the pentanuclear complex can be again considered as a *"complex-ligand"*, which is able to link six $[Ni(H_2O)_6]^{2+}$ ions to give the undecanuclear cluster $\{[Ni(H_2O)_5]_6[Co(tmphen)_2]_3[Fe(CN)_6]_2\}^{13+}$. As indicated by the authors, this is a *"step-wise assembly"* [8].

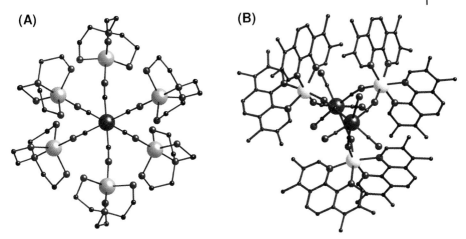

Figure 5.4 Two cases of the "complexes as ligands and complexes as metals" approach (see text) (spheres in (A), Cr = large black, Cu = large gray; in (B) Co = large gray, Fe = large black).

5.4
High Nuclearity Clusters: Generalities

Loosely this field can be regarded as dominated by a few types of ligands: (i) simple oxido and/or alkoxido ligands that easily produce high nuclearity assemblies, (ii) carboxylato ligands; (iii) several organic ligands such as di-2-pyridyl ketone, 2-pyridone (2-hydroxypyridine), etc. Finally, in many cases, the use of more than one of these ligands (*heteroleptic systems*) produces extraordinary high-nuclearity clusters (oxido + carboxylato ligands, for example).

5.5
Oxido-hydroxido High-nuclearity Clusters (Excluding Polyoxometalates)

5.5.1
Non-protected Compounds

Polynuclear derivatives from trivalent cations (Al, Ga, Cr, Fe, Ln) can be obtained by controlled hydrolysis of their salts. However, with this procedure it is not easy to obtain clusters of different nuclearity because of the great tendency to give insoluble oxides such as $M(OH)_3$, M_2O_3 or $MO(OH)$. However, in certain limited conditions, "non-protected" clusters can be obtained. Let us mention some examples:

In $[Al_8(\mu_3\text{-}OH)_2(\mu_2\text{-}OH)_{12}(H_2O)_{18}]^{10+}$ (Figure 5.5A) there is a core of edge-shared AlO_6 octahedra that are organized into cubane-like moieties with μ_3-OH units in the central core, an outer ring of μ_2-OH units and an outermost ring of bound water molecules [9]. $[Al_{13}(\mu_3\text{-}OH)_6(\mu_2\text{-}OH)_{18}(H_2O)_6]^{15+}$ and the analogous $[Ga_{13}(\mu_3\text{-}OH)_6(\mu_2\text{-}OH)_{18}(H_2O)_{24}](NO_3)_{15}$ (the first hydroxido-cluster with Ga isolated to date)

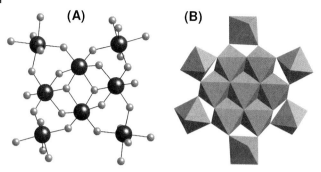

Figure 5.5 $[Al_8(\mu_3\text{-}OH)_2(\mu_2\text{-}OH)_{12}(H_2O)_{18}]^{10+}$ and $[Ga_{13}(\mu_3\text{-}OH)_6(\mu_2\text{-}OH)_{18}(H_2O)_{24}]^{15+}$ (see text).

are shown in Figure 5.5B [10]. The peripheral gallium atoms are each coordinated to four water ligands.

A unique example of a germanium oxido-hydroxido cluster is the $[Ge_{10}O_{24}(OH)_3]$ cluster built from six GeO_4 tetrahedra and four GeO_6 octahedra [11]. The importance of this cluster is its packing in the lattice creating a mesoporous entity with crystalline pore walls. In such materials, an important factor influencing the pore size is the number of M atoms in the rings delineating the channels. This germanium cluster has the lowest framework density of any inorganic material and channels that are defined by 30-rings. As a typical and known example, the faujasite structure exhibits 12-ring pores.

5.5.2
Protected Compounds

The best method to avoid the complete hydrolysis of $[M(H_2O)_6]^{3+}$ is the so-called *"protection"* of the terminal groups. This protection can achieved in several ways, most commonly by organic ligands with steric hindrance. Two representative general examples are given below.

5.5.2.1 Alkoxido Derivatives
Such species are usually prepared in solution through the controlled hydrolysis of the metallic alkoxides, $[M(OR)_n]$ or the corresponding complexed alkoxides, $[M(OR)_{n-x}L_x]$, where L is a complexing ligand. The stability of these clusters increases with their nuclearity and is enhanced by the presence of complexing ligands that chelate or bridge the metallic centers.

Some examples are: $[Ti_{16}O_{16}(OEt)_{32}]$, $[Ti_{16}O_{16}(OEt)_{32\text{-}x}(OR)_x]$ (Figure 5.6A) [12], and $[Mn_{19}O_{12}(OC_2H_4OCH_3)_{14}(HOC_2H_4OCH_3)_{10}]$ $(OC_2H_4OCH_3$ = methoxiethoxide), (Figure 5.6B). The latter is flat disk-shaped and all Mn atoms are divalent and octahedrally coordinated by oxygen atoms in a CdI_2-type layer structure. There is a central Mn^{2+} atom, six in the interior of the ring and twelve peripheral ones [13]. The unique $[Fe_{13}F_{24}O_4(OMe)_{12}]^{5-}$ has an ideal α-Keggin structure $([XM_{12}O_{40}]^{3-}$ (see Section 5.6.2.3) with a central tetrahedral $\{FeO_4\}$ core (Figure 5.6C) [14].

(A) **(B)** **(C)**

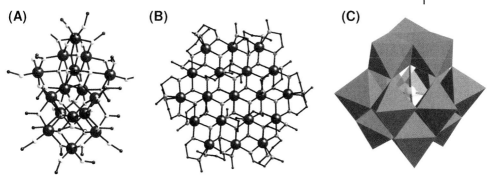

Figure 5.6 Three alkoxido derivatives (see text for their formulae); (in (A), Ti = large black spheres in (B), Mn = large black spheres).

(A) **(B)**

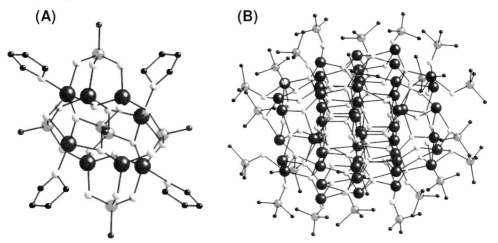

Figure 5.7 Two polymetal-siloxane derivatives, coming from silanols (see text) (spheres in (A), Fe = large black, Si = large gray; in (B), Bi = large black).

At this point it is worth mentioning the polymetal-siloxane derivatives, based on the *silanols* (like alcohols but with Si). Metalasiloxanes are important because many industrially important processes are catalyzed by transition metal complexes immobilized on silica surfaces and, furthermore, they have also been envisaged as single-source precursors for modified zeolites. $[Fe_8(thf)_4(RSiO_3)_2(RSi(OH)O_2)_4(\mu_3\text{-}OH)_2]$ is a paradigmatic example (Figure 5.7A). Large bismuth-oxido clusters (such as Bi_4, Bi_9, Bi_{18}, Bi_{22}, Bi_{33} and Bi_{50}) have been recently reported, starting from bismuth silanolates $[Bi(OSiR_2R')_3]$. All of them contain μ-silanolates in the framework (Bi_{50} in Figure 5.7B) [15].

5.5.2.2 Assembly of Metal-oxido(alkoxido)-bridged Clusters with Blocking Ligands

A typical example is the supertetrahedral decanuclear Ni^{2+} cluster $[Ni_{10}(O)(thme)_4(dbm)_4(O_2CPh)_2(EtOH)_6]$ (H_3thme = 1,1,1-tris(hydroxymethyl)ethane; Hdbm = dibenzoylmethane). The triangular Ni_6 faces are near planar and are each held together by a fully deprotonated $\mu_6\text{-}thme^{3-}$ (Figure 5.8A) [16].

(A)   **(B)**

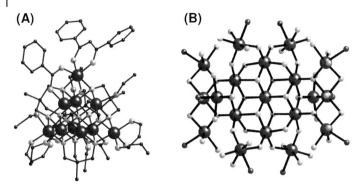

Figure 5.8 $[Ni_{10}(O)(thme)_4(dbm)_4(O_2CPh)_2(EtOH)_6]$ and $[Fe_{19}(\mu_3\text{-}O)_6(\mu_3\text{-}OH)_6(\mu\text{-}OH)_8(heidi)_{10}(H_2O)_{12}]$ (see text). (large black spheres, Ni in (A); Fe in (B)).

Addition of tripodal chelating ligands of the general form $[N(RCOOH)_2R']$, in which R' can be an organic residue, to solutions of $[Fe(H_2O)_6]^{3+}$ can halt the hydrolysis process through the stabilization of intermediate phases composed of close-packed cores, which are a portion of the typical brucite structure, $Mg(OH)_2$. For example, with the tripodal H_3heidi, $N(CH_2COO)_2(CH_2CH_2OH)$ the $[Fe_{19}(\mu_3\text{-}O)_6$ $(\mu_3\text{-}OH)_6(\mu\text{-}OH)_8(heidi)_{10}(H_2O)_{12}]$ cluster is obtained [17] (Figure 5.8B). It contains trapped iron hydroxide mineral portions produced by a modified hydrolysis of iron(III) in the presence of the ligand.

5.6
Polyoxometalates (POMs)

Polyoxometalates are a large family of metal–oxygen clusters of the early transition metals in high oxidation states, most commonly V^V, Mo^{VI} and W^{VI}. The "formal" oxidation state 5+, 6+ favors the presence of the terminal M=O moiety which impedes the total polymerization of the final product. POMs form typically in an acidic aqueous solution and can be isolated as solids with an appropriate counter-cation, for example alkali metal cations, NH_4^+ etc. Their diversity in structure and composition allows a wide versatility in terms of shape, polarity and redox potentials.

In almost all General Inorganic Chemistry books [18], the typical V, Mo and W polyoxometalates are covered in more or less detail. Let us remember that a simplified conventional nomenclature is sufficient. The heteroatom, if present, is considered as the central atom of a complex, and the addenda as the ligands. The polyanion is placed in square brackets and thus separated from the countercations, as illustrated by the following examples: $[SiW_{12}O_{40}]^{4-}$,12-tungstosilicate or dodecatungstosilicate; $H_3[PMo_{12}O_{40}]$,12-molybdophosphoric acid; $Na_5[PMo_{10}V_2O_{40}]$, sodium decamolybdodivanadophosphate.

5.6.1
Nanosized Polyoxometalates (NS-POMs)

The aggregation process of polyoxometalates allows us to enter into the "*nano-world*". This can be achieved, for example, by replacing hexavalent metal centers with one or more lower valent metals, such as V^{5+}, Nb^{5+} etc. This substitution results in highly charged nucleophilic compounds. Polyoxometalates activated in this manner can then react with cationic complexes to produce new polyoxometalates with extended metal–oxygen frameworks. Furthermore, the occupation of their surface by oxygen atoms offers the possibility of coordination to other transition metals and hence their use as pillars to link transition (and non-transition) ions.

5.6.2
Selected Examples of POMs and NS-POMs

5.6.2.1 Vanadates
Polyoxovanadates show the occurrence of different basic types of polyhedra, VO_4, VO_5 and VO_6. V_2O_5 dissolves in NaOH to give colorless solutions. At pH > 13 the main ion is the mononuclear VO_4^{3-}. As the basicity is reduced, a series of complicated reactions occurs, yielding, in the pH range 2 to 6, the orange decavanate ion which can exists in several protonated forms: $[V_{10}O_{27}(OH)]^{5-}$, $[V_{10}O_{26}(OH)_2]^{4-}$, $[V_{10}O_{25}(OH)_3]^{3-}$ and $[V_{10}O_{24}(OH)_4]^{2-}$. Many salts of the decavanadate ion, for example $Ca_3V_{10}O_{28} \cdot 18H_2O$, have been crystallized and their structures well established [18]. They consist of ten octahedra fused together (Figure 5.9A). Spectroscopic studies (Raman, ^{17}O and ^{51}V NMR) indicate that this structure persists in solution.

Polyoxovanadates may also form flexible ring structures, $[V_3O_9]^{3-}$ and $[V_4O_{12}]^{4-}$ through tetrahedral arrays. Hayashi and coworkers have recently reported the crystal structures of larger cyclic host molecules (with V_6, V_8, and V_{10}) that consist of VO_4 units, using metal cationic species as effective templates in the oligomerization of the vanadates [19].

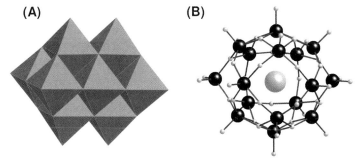

(A)   **(B)**

Figure 5.9 Decavanadate ion, $[V^V_{10}O_{28}]^{6-}$ and $[V^V_{13}V^{IV}_3O_{42}]^{7-}$ cluster, hosting a chloride ion (Cl^- = large gray sphere).

5.6.2.2 Nano-sized Vanadates

The best example is given by the $\{V_{19}\}$ clusters, showing a range of different structures: they are formally generated by linking twelve $\{V^{IV}O_6\}$ octahedral, six $\{V^VO_4\}$ tetrahedral and one central $\{V^VO_4\}$ unit. $\{V_{22}\}$ and $\{V_{34}\}$ clusters are built up from $\{O=VO_4\}$ pyramids. In many cases, the structures depend to a large extent on the size, the shape and the charge of a template (in most cases an anion) incorporated as guest in the final structure. The $[V^V_{13}V^{IV}_3O_{42}]^{7-}$ cluster, for example, hosts a chloride ion (Figure 5.9B) [20]. The encapsulation of other anions (X) into anionic vanadate cages has also been reported (X = ClO_4^-, SCN^-, CH_3COO^-).

5.6.2.3 Molybdates and Tungstates

The "classic" polyanions of molybdenum and tungsten are of two types: the *isopolyanions (homopolyanions)* and the *heteropolyanions*.

Homopolyanions The cluster species $[M_6O_{19}]^{2-}$ (*Lindqvist* anions) is common to both metals (and with Nb, Ta and mixed systems) (Figure 5.10A). The two main molybdenum species in solution are $[Mo_7O_{24}]^{6-}$ and $[Mo_8O_{26}]^{4-}$ (Figure 5.10B,C), depending on the pH [18]. For tungsten in the pH range 5 to 7.8, the equilibria involve WO_4^{2-}, $[W_6O_{20}(OH)_2]^{6-}$, $[W_7O_{24}]^{6-}$, $[HW_7O_{24}]^{5-}$, and $[H_2W_{12}O_{42}]^{10-}$. These polyanions are built primarily of MO_6 octahedra through sharing of vertices or edges, but not faces. It should be noted that the octahedra are actually not regular as a result of M–O π bonding [18].

Other structures do not consist entirely of MoO_6 octahedra, for example, the α-$[Mo_8O_{26}]^{4-}$ ion consists of a central, crown-shaped, ring of six edge-sharing distorted octahedra capped by two tetrahedral MoO_4 units (Figure 5.10D). A very important feature is that these capping MoO_4^{2-} tetrahedra can be replaced by isostructural units, such as $PhAsO_3^-$, opening a new branch of polyoxoanion chemistry, related to catalysis by metal oxide surfaces.

$[n\text{-}Bu_4N]_8[Mo_6O_{19}]_2[\alpha\text{-}(Mo_8O_{26})]$, a rare example in which both the Lindqvist anion and the crown-shaped ring coexist in the same crystal architecture, has been reported recently [21].

Heteropolyanions The largest and best known group is composed of those with the heteroatom(s) surrounded by a cage of MO_6 octahedra. The structure of the $[TeMo_6O_{24}]^{6-}$ anion (Figure 5.11A) is known as *Anderson*-type structure.

(A)   **(B)**   **(C)**   **(D)**

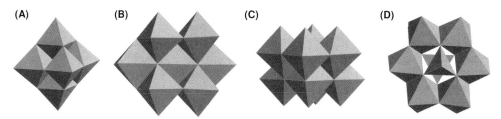

Figure 5.10 Typical homopolyanions (see text).

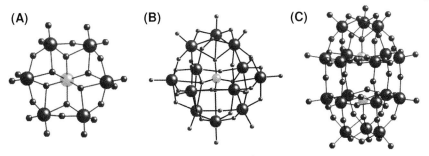

Figure 5.11 Typical heteropolyanions (see text) (gray spheres = heteroatoms).

The $[X^{n+}M_{12}O_{40}]^{(8-n)}$ structure is called the *Keggin* structure (Figure 5.11B). It has full tetrahedral (T_d) symmetry, being very compact but accommodating a great variety of heterocations that differ considerably in size (Si^{4+}, Ge^{4+}, P^{5+}, As^{5+}, Ti^{4+}, Zr^{4+} for Mo; B^{3+}, Al^{3+}, Ga^{3+}, Si^{4+}, Ge^{4+}, P^{5+}, As^{5+}, Fe^{3+}, Co^{2+}, Co^{3+}, Cu^+, Cu^{2+}, Zn^{2+}, Cr^{3+}, Mn^{4+} for W). The $[(X^{n+})_2M_{18}O_{62}]^{(16-2n)-}$ ion is called the *Dawson* structure (Figure 5.11C). It is larger but closely related to that of the Keggin ion. The X in this case is more limited: P and As. Dawson-type heteropolyoxometalates have received increasing attention in the field of heterogeneous catalysis.

5.6.2.4 Nano-sized Molybdates

Most of the very large mixed-valence polyoxomolybdates can be considered as being based on similar building-block units, $\{Mo_n\}$, which can be linked in different ways directly or via other coordinating centers. In general, if the ratio Mo^V: Mo^{VI} is high, structures with Mo^V–Mo^V pairs are preferred, while if it is smaller, highly delocalized systems with the characteristic dark blue color are favored. Larger nanosized polyoxomolybdate (NS-POM) species with greater nanosized cavities are known from the pioneering work of Müller's group [22]. They have: (i) versatile redox chemistry, (ii) the possibility of integration of hetero elements and exchange of ligands, (iii) a tuneable charge/size ratio, (iv) template functionality. Some interesting examples according to their $\{Mo_n\}$ units are:

- $\{Mo_6\}$, formed by basic pentagonal $\{(Mo)Mo_5\}$ units with the tendency to give a very robust fundamental cluster skeleton $\{(pent)_{12}(link)_{30}\}$ (called *keplerates*) where "pent" is the basic pentagonal unit and "link" refers to linker groups. This general formula describes, for example, the $(H_2O,NH_4)_n \subset [\{(Mo^{VI})Mo^{VI}_5O_{21}(H_2O)_6\}_{12}\{Mo^V_2O_4(L)\}_{30}] \cdot \approx 200H_2O$, $\{Mo_{132}\}$-type cluster (L = sulfate and/or acetate), which are easily accessible soluble, spherical, porous nanocapsules with internal surfaces that are tunable with respect to charge and hydrophilicity/phobicity (Figure 5.12) [23]. The same $\{Mo_6\}$ unit allows the synthesis of hetero-giant clusters with other ions such as V^{4+}, Fe^{3+} [24]. With this unit, Müller et al., "*have demonstrated that it is possible to use pentagonal units as building units which play, geometrically speaking, the same role as the pentagonal units in other sphere-based constructions, such as spherical viruses, fullerenes and geodesic domes*" [25].

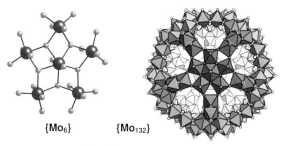

{Mo₆}   **{Mo₁₃₂}**

Figure 5.12 $[\{(Mo^{VI})Mo^{VI}_5O_{21}(H_2O)_6\}_{12}\{Mo^V_2O_4(L)\}_{30}]$ with its {Mo₆} unit.

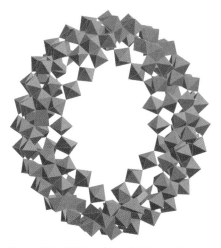

Figure 5.13 $(NH_4)_{25\pm5}[Mo_{154}(NO)_{14}O_{420}(OH)_{28}(H_2O)_{70}]\cdot ca.350H_2O$ (see text).

- {Mo₈}, composed of seven MoO_6 octahedra and one MoO_7 bipyramid, is the building-block unit in $[Mo_{36}O_{112}(H_2O)_{16}]^{8-}$, which was the first reported nanosized polyoxomolybdate anion, known since 1973.

- {Mo₁₁}, composed of three types of building units: (i) the {Mo₈}-type group (see above) + 1{Mo₂} group + 1{Mo₁} group. The {Mo₁₅₄}- and {Mo₁₇₆}-type clusters contain each of these three groups 14 and 16 times, respectively. $(NH_4)_{25\pm5}[Mo_{154}$ $(NO)_{14}O_{420}(OH)_{28}(H_2O)_{70}]\cdot ca.350H_2O$, $[Mo_{176}O_{528}H_{16}(H_2O)_{80}]^{16-}$ are deep-blue, mixed-valence electron-rich species [22, 26]. The approximately ring-shaped structures are extremely beautiful (Figure 5.13). The cavity, like a car tyre, is very broad. In fact, if an acidified aqueous molybdate(VI)-containing solution reacts with an appropriate reducing agent, deep-blue ring-shaped clusters {Mo₁₅₄} and {Mo₁₇₆} are immediately formed in solution, in proportions that depend mainly on the pH. They can be precipitated under certain conditions. If these clusters are maintained in solution, a molecular growth process occurs near its central cavity in the presence of an excess of molybdate and reducing agents, resulting in $\{Mo_{248}\}$-type clusters.

- Larger and highly reduced mixed-valence cluster anions containing 368 Mo atoms can be obtained by reducing an aqueous molybdate solution acidified with sulfuric acid. $Na_{48}[H_xMo_{368}O_{1032}(H_2O)_{240}(SO_4)_{48}] \cdot ca.\ 1000H_2O$ $(x \approx 16)$, has a central ball-shaped fragment $\{Mo_{288}\}$ and two capping units $\{Mo_{40}\}$. This giant molecular container has a huge cavity (diameter $\approx 2.5 \times 4.0\,nm$ at its most extended points) offering space for about 400 water molecules [27]. The high degree of reduction (ca. $112Mo^V$, $256Mo^{VI}$) reflects the fact that reducing conditions favor cluster growth. This cluster has the size of hemoglobin (diameter approximate 6 nm).

When some of these nanoclusters are dissolved in different solvents (water, ethanol and other polar ones) they tend to deprotonate several aqua ligands becoming macroanions, which then slowly self-associate into "blackberry" structures containing >1000 clusters.

5.6.2.5 Nano-sized Tungstates

The structures of very large polytungstate anions ($n > 20$) may practically all be represented in terms of a fragment of the Keggin anion $\{X^{n+}W_{12}O_{40}\}^{(8-n)-}$. This feature may be attributed to a weak, but not negligible, W–W interaction in the $\{W_3\}$ triangles. The largest nano-sized isopolytungstate so far reported, is $[H_{12}W_{36}O_{120}]^{12-}$. The new cluster anion comprises three lacunary $\{W_{11}\}$ cluster subunits linked together by three $\{W_1\}$ bridges and encapsulates a K^+ cation (Figure 5.14A) [28].

Nano-sized heteropolytungstates incorporating different numbers of monovacant lacunary $\{XW_{11}\}$-type building blocks, with $X = B^{3+}$, P^{5+} and As^{5+}, are numerous, as well as those with trivacant, pentavacant and hexavacant lacunary building blocks. Salts of the anion $[P_2W_{21}O_{71}]^{6-}$ (with two trivacant lacunary building blocks) have been known for over a century (since 1892), but a representative structure was only recently solved. The cluster is a sandwich-like arrangement of two $\{PW_9\}$ moieties enclosing three $\{WO_5(H_2O)\}$ octahedral (Figure 5.14B). Other examples (such as a giant tetrapod) can be found in Ref. [29].

5.6.3
Polyoxometalate-supported Transition Metal Complexes

5.6.3.1 From Not-lacunary Polyoxometalates

Conventional polyoxometalates are not basic enough to react, in standard conditions, with d^n cations. However, in recent years, a series of novel hybrid compounds made of Keggin or Anderson polyoxometalates associated with transition metal complex cations [30], as well as with polynuclear complexes [31], have

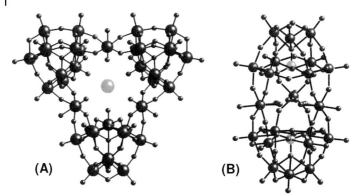

Figure 5.14 Two nanosized tungstates (see text) (in (A), k^+ = large gray sphere; in (B), large gray spheres = heteroatoms).

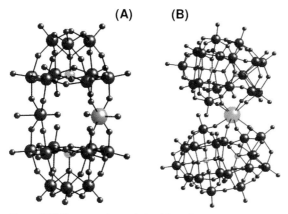

Figure 5.15 Two complexes derived from lacunary polyoxometalates (see text) (in (A), large gray sphere = Pd, Pt; in (B), large gray sphere = lanthanide ion).

been described. The charge density of the surface oxygen atoms can be increased either by reducing some of the metal centers (e.g. Mo^{6+} to Mo^{5+}) or by replacing higher-valent metal centers (e.g. replace Mo^{6+} with V^{4+}).

5.6.3.2 From Lacunary Polyoxometalates

Monovacant (α-$PW_{11}O_{39}$, α-$P_2W_{17}O_{61}$), divacant (γ-$SiW_{10}O_{36}$), and trivacant (α-PW_9O_{34}, α-$P_2W_{15}O_{56}$) species of the Keggin- and Dawson-type polyoxometalates can be easily prepared and all have an extraordinary ability to act as ligands. For example, with $[XW_9O_{34}]^{9-}$ anion it has been possible to stabilize platinum and palladium-oxo complexes, which are difficult to obtain and possible intermediates for oxidative catalysis, such as $K_7Na_9[Pt^{IV}O(H_2O)(PW_9O_{34})_2]$ and $K_{10}Na_3[Pd^{IV}O(OH)(WO)(H_2O)(PW_9O_{34})_2]$ (Figure 5.15A) [32]. These ligands have allowed the preparation of *isolated* magnetic clusters M_3, M_4, M_5, M_6, M_9, Cu_{10}, in the structure [33].

Rare-earth metal ions are known to be highly oxophilic and hence good candidates for these complexation reactions. Each lanthanide ion has several (usually 2–4) additional coordination sites available, which can be used as linkers to one or more other lacunary POM units. This approach should, in theory, result in polymeric or unusually large molecular POM assemblies, such as $[(PEu_2W_{10}O_{38})_4$ $(W_3O_{14})]^{30-}$, $[Ce^{III}_3Sb_4W_2O_8(H_2O)_{10}(SbW_9O_{33})_4]^{19-}$ and, in particular, the cyclic $[As_{12}$ $Ce_{16}(H_2O)_{36}W_{148}O_{524}]^{76-}$, the largest polyoxotungstate reported to date. Double mono-lanthanide homo-polyoxometalates such as $[La(Mo_8O_{26})_2]^{5-}$ or hetero-polyoxometa-lates such as $[Eu^{III}(SiW_{11}O_{39})_2]^{13-}$, $[Ln^{III}(P_2W_{17}O_{61})_2]^{17-}$ (Figure 5.15B) and many others have also been reported [34]. The isolated anions are valuable ligands for multiple lanthanide and actinide coordination due to their open structure and therefore have potential for the separation of nuclear waste. Furthermore, the addition of lanthanides to POMs has recently received renewed interest for the creation of luminescent and electrochromic materials. To date, transition-metal-substituted POMs containing from 1 to 12 lacunary precursors and from 1 to 28 transition metal ions have been obtained.

An interesting review on polyoxometalate clusters has been published recently [35].

5.7
Oxido-carboxylate Clusters (Especially Those Derived from [M₃O(carboxylate)₆] Units)

The stable $(\mu_3$-oxido)hexakis$(\mu$-carboxylato)trimetal(III) moities within $[M_3O$ $(O_2CR)_6L_3]^+$ species have the structure shown in Figure 5.16. These so-called "basic metal carboxylates" have several advantages as polynucleating precursors: (i) they are small clusters themselves; (ii) homovalent M^{III}_3 and heterovalent $M^{III}_2M^{II}$

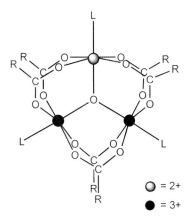

○ = 2+
● = 3+

[M₃O(carboxylate)₆L₃]

Figure 5.16 Scheme of the [M₃O(carboxylate)₆L₃] entity.

triangles are available for V, Cr, Mn and Fe; (iii) they can be prepared with a wide range of carboxylate and terminal ligands and (iv) the central oxide and bridging carboxylates provide sources of ligands with several potential bridging modes. We will give some examples of the iron and manganese derivatives.

5.7.1
Iron

High nuclearity iron(III) carboxylate clusters have been an active research area in recent years owing to their relevance as models for protein active sites and their interesting magnetic properties. The $[Fe_3O]^{7+}$ core can be used as a building block to give rise to larger clusters, mainly the so-called ferric-wheels (see Section 5.8.1). Many spectacular polyhedra have also been reported starting from the same core. As a typical example, oxo-centered iron(III) triangles react with a range of phosphonates to give tetra-, hexa-, hepta, nona-, and tetradecanuclear iron(III) cages [36].

5.7.2
Manganese

Large nuclearities have been encountered in a discrete form and the Mn oxidation states have spanned the range II–IV, including mixed-valency. The metal ions are invariably bridged by combinations of O^{2-}, OH^- and RO^- and usually by one or both oxygen atoms of carboxylate groups. The stimuli for these studies include (i) the search for models of Mn-containing proteins and enzymes, and (ii) the realization that Mn clusters often possess large values of ground state spin, a necessary condition for isolating single molecule magnets (SMM) (see Chapter 10).

In an early review published in 1989, Christou [37] described how the trinuclear, oxido-centered complexes $[Mn_3O(O_2CR)_6L_3]^z$ ($z = 0$, 1+) are an excellent starting point for species with nuclearities up to 12: the complexes with the $[Mn_4(\mu_3\text{-}O)_2]$ core have a "*butterfly*" structure with Mn^{2+}, $3Mn^{3+}$ or $4Mn^{3+}$ ions (Figure 5.17A); the $[Mn_4O_3Cl_4(O_2CR)_3(py)_3]$ species have three Mn^{3+} and one Mn^{4+} ions. The core consists of a Mn_4O_3 partial *cubane* with the vacant apex occupied by a $\mu_3\text{-}Cl^-$ ion; the $[Mn_6O_2(\text{carboxylate})_{10}L_4]$ clusters (Figure 5.17B), can be synthesized starting from $[Mn_3O]$ species or by direct reactions.

The reaction of several $[Mn_aO_b]$ units and dicarboxylates may trigger aggregation into higher nuclearity products. Undoubtedly, the most known and famous carboxylate manganese complexes are the dodecamanganese clusters, {Mn_{12}}, because they were the first reported single molecule magnets (SMMs) (Chapter 10). The archetype is $[Mn_{12}O_{12}(O_2CMe)_{16}(H_2O)_4]$ which can be simply prepared from $Mn(O_2CMe)_2 \cdot 4H_2O$ and $KMnO_4$. All these Mn_{12} complexes are valence-trapped and contain a central $[Mn^{IV}_4O_4]$ cubane around which a non-planar ring of Mn^{3+} ions is held to the central cubane by eight O^{2-} ions (Figure 5.17C).

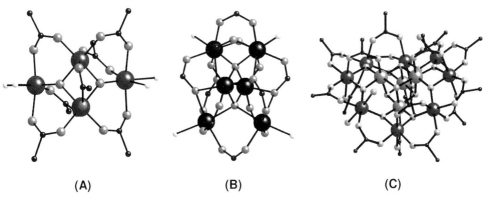

(A)   **(B)**   **(C)**

Figure 5.17 Tetra-, hexa- and dodecamanganese complexes synthesized from [Mn₃O] units (Mn = large spheres).

Figure 5.18 [Mn₈₄O₇₂(O₂CMe)₇₈(OMe)₂₄(MeOH)₁₂(H₂O)₄₂(OH)₆], comprising a {Mn₈₄} torus (Mn = large black spheres).

Finally, although not always directly derived from [Mn₃O] units, we mention some polynuclear carboxylate manganese complexes with nuclearities higher than 12 such as the giant {Mn₈₄} torus with the formula [Mn₈₄O₇₂(O₂CMe)₇₈(OMe)₂₄ (MeOH)₁₂(H₂O)₄₂(OH)₆]. The structure comprises a {Mn₈₄} torus (Figure 5.18) composed of alternating near-linear [Mn₃O₄] and cubic [Mn₄O₂(OMe)₂] subunits.

The torus has a diameter of about 4.2 nm and a thickness of about 12 nm, with a central hole of diameter 1.9 nm. The Mn_{84} molecules order within the crystal in an aesthetically pleasing manner, thus giving nanotubular stacks parallel to the crystal c axis and displaying extensive cylindrical channel formation along one dimension. Molecules in neighboring chains are exactly adjacent, and thus the structure may be alternatively described as consisting of graphite-like $[Mn_{84}]$ sheets lying on top of each other with perfect registry [38].

5.8
Metal Wheels

A particularly symmetric class of polynuclear complexes are the so-called *metal wheels*. In many cases the molecules have an alkali cation in the center of the wheel.

5.8.1
Ferric Wheels

The $[Fe_6]$ unit (Figure 5.19A) is the core of the neutral iron ferric wheels $[Fe_6Cl_6(L)_6]$, H_2L = N-substituted diethanolamines) [39]. The $[Fe_8]$ unit is rarer [40]. By contrast, many $[Fe_{10}]$ ferric wheels have been reported to date, such as $[Fe_{10}(OMe)_{20}(O_2CR)_{10}]$ [40, 41]. They contain ten octahedral Fe^{3+} ions in a planar ring, with each Fe^{3+} bridged to its neighbors by one carboxylate and two methoxide groups. The molecules possess an idealized S_{10} symmetry (Figure 5.19B). The $[Fe_{12}]$, $[Fe_{16}]$ and $[Fe_{18}]$ unit are also known [42].

5.8.2
Wheels of Other Metals

Basic chromium(III) carboxylates, $[Cr_3O(O_2CR)_6(L)_3]$, undergo a complicated rearrangement upon heating giving $[Cr_8(OH)_8(O_2CPh)_{16}]$, and $[Cr_{10}(OR)_{20}(O_2CMe)_{10}]$

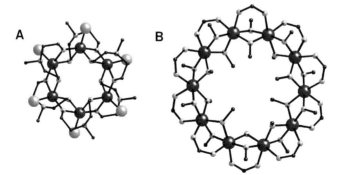

Figure 5.19 Two ferric wheels (Fe_6 and Fe_{10}) (Fe = large black spheres).

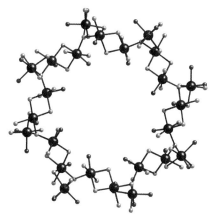

Figure 5.20 The giant Ni$_{24}$ wheel, [Ni$_{24}$(OH)$_8$(mpo)$_{16}$(O$_2$CMe)$_2$ $_4$(Hmpo)$_{16}$] (see text) (Ni = large black spheres).

[43]. These species contain the same metal topology as the classical "ferric wheels" described above. [Ga(OMe)$_2${O$_2$CC(OH)Ph$_2$}]$_{10}$ is another decanuclear wheel (Figure 5.19B) which incorporates a water molecule in the center of the cluster [44].

A novel octanuclear copper(II) nucleoside wheel, which acts as host to the [Cu(H$_2$O)$_6$]$^{2+}$ cation, despite the high positive charge of the wheel (4+) has been recently reported [45].

A giant Ni$_{24}$ wheel [Ni$_{24}$(OH)$_8$(mpo)$_{16}$(O$_2$CMe)$_{24}$(Hmpo)$_{16}$] (Hmpo = 3-methyl-3-pyrazolin-5-one) (Figure 5.20) has been reported [46]. The structure can be described as consisting of an octamer of chemically equivalent trinuclear building blocks, but as a whole it can be compared to the wheels reported in this paragraph.

Winpenny et al. have recently reported an interesting study on the template effect when constructing heterometallic [Cr$_n$Ni] rings (n = 7, 9) [47]. It is possible to choose between octa- nona- and decanuclear wheels by using appropriate secondary ammonium ions as templates. This suggests the balance between the structures can be very fine for specific templates. Finally, it is worth mentioning the novel heterometallic cyanide-bridged molecular wheels, [Fe$_6$Mn$_6$], recently reported [48]. These molecular wheels are SMMs (see Chapter 10).

5.9
Other Significant Organic Polynucleating Ligands

Many organic ligands have been systematically studied over recent years, with very successful results in the synthesis of high-nuclearity clusters. The most important are: phosphonates (O$_3$PR), pyridyl-oximes, di-2-pyridyl-ketones, pyridonates, etc. We will mention, for the sake of brevity, only two of them:

5.9.1
Di-2-pyridylketones

Neutral di-2-pyridyl ketone [49], hereafter abbreviated as $(py)_2CO$, has three potential donor groups, the two 2-pyridyl nitrogens and the carbonyl oxygen. However, there is a chemical characteristic of $(py)_2CO$ that makes this ligand special; that is its carbonyl group. Water and alcohols have been shown to add to the carbonyl group upon coordination of the 2-pyridyl rings to the metal, producing the ligands $(py)_2C(OH)_2$ or $(py)_2C(OR)(OH)$. When these two ligands are deprotonated, interesting (for cluster chemistry) coordination modes are expected and observed. This is due to the ability of the negatively charged oxygen to bridge two or three metal ions. This gives a great coordinative flexibility in the mono- and di-anionic ligands.

Perlepes' group started an interesting approach, linking the coordination potentiality of these $(py)_2CO$ derivatives with the potentiality of the carboxylates. This combination of ligands has been named by the authors "*binary ligand blend*". Carboxylates are employed for two reasons: first, they are able to deprotonate the hydroxy group(s) of $(py)_2C(OR)(OH)$ and $(py)_2C(OH)_2$ under mild conditions and secondly, they are very flexible ligands.

The coordination modes of the dianionic form, $(py)_2CO_2^{2-}$, are shown in Figure 5.21A. With Cu^{2+}, hepta- and dodeca-nuclear complexes have been reported. For example, $[Cu_{12}(O_2CMe)_{12}\{(py)_2CO_2\}_6]$ has a very beautiful structure [50]. It has been named by the authors "*fly-wheel*" and "*chair-within-a-chair*" due to the metal

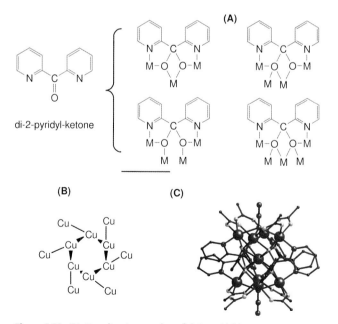

Figure 5.21 (A) Coordination modes of di-2-pyridyl-ketones; (B,C) two representative structures (see text).

topology (Figure 5.21B). With Co^{2+} or Ni^{2+}, the peculiar enneanuclear clusters [M_9 $(OH)_2(O_2CMe)_8\{(py)_2CO_2\}_4$] have been reported, with the salient feature of eight-coordination about the central metal ion [50]. The Co_9 (or Ni_9) skeleton can be considered as a (preformed) host that is able to accommodate small linear anions (N_3^-, OH^-, OCN^-, CN^-) [50]. The products contain a *"ternary ligand blend"* (Figure 5.21C).

The monoanionic forms, [$(py)_2C(OR)(O^-)$] (R = H, Me, Et) form novel polynuclear species such as Ni_4, Co_4 and Mn_4 complexes having a cubane-like structure. The similar copper system has a "double cubane" structure (Cu_8). Surprisingly, in most of these complexes the carboxylate ligands are not crucial for the assembly of the tetranuclear complexes, behaving only as terminal ligands.

5.9.2
2-Pyridonates

The pyridonate group, once deprotonated, shows many coordination modes (Figure 5.22). Pyridonate derivatives are abbreviated as *x*hp, e.g. chp = 6-chloro-2-hydroxypyridonate, bhp = 6-bromo-2-hydroxypyridonate, mhp = 6-methyl-2-hydroxypyridonate, etc. Almost all metal complexes of 2-pyridonate derivatives are *heteroleptic*, i.e. they contain other ligands such as carboxylates, hydroxide or solvent molecules/ions. In general, the final structure is strongly dependent on the solvent used for recrystallization, showing an enormous range of structural types with the "cage-like" structure being predominant [51].

Both nickel and cobalt cubanes of formula [$M_4(OMe)_4(chp)_4(MeOH)_7$] have been reported. The loss of coordinated MeOH molecules creates vacancies on the coordination spheres of the metals, which facilitates the formation of larger polynuclear species, such as the dodecanuclear complex [$Co_{12}(OH)_4(chp)_{18}(Hchp)_2Cl_2$ $(MeOH)_2$], containing two [Co_4O_3Cl] cubanes bridged by a central cluster of four further cobalt ions [52].

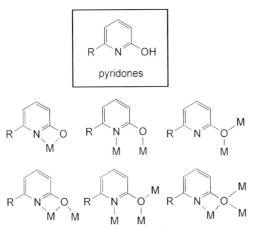

Figure 5.22 Coordination modes of pyridones.

[Ni$_7$(chp)$_{12}$(MeOH)$_6$Cl$_2$] and [Ni$_7$(chp)$_{12}$(OH)$_2$(MeOH)$_6$] contain a pair of vertex-sharing adamantanes [52]. However, working with MeCN as solvent, the nonanuclear cluster [Ni$_9$(chp)$_{16}$(OH)$_2$(MeCN)$_2$] was obtained, in which the sites occupied by external MeOH molecules in the Ni$_7$ complexes are now occupied by two [Ni(chp)$_3$]$^-$ [52]. It is clear, thus, that in all these cases, the less coordinating solvent leads to a higher nuclearity.

5.10
Metallodendrimers [53, 54]

The mixing of complementary ternary components (for example, tripod ligands) may lead to the self-assembly of new polyfunctional species known as *dendrimers*. They are prepared via an iterative, step-wise "cascade synthesis". These types of molecule are referred to by a variety of names including dendrimers, cascade molecules, cascadols, crowned arborols, cauliflower polymers, molecular fractals and "starburst" polymers. They are, in fact, monodisperse macromolecules that expose many end groups at their globular periphery. Synthesis of the dendrimers is iterative and stepwise, the monomeric units being incorporated in each successive iteration (Figure 5.23).

The chemistry of dendritic macromolecules began with the work of Vögtle in 1978, and the first such macromolecules were purely organic. In 1983, De Gennes (Nobel Laureate for Physics in 1991) and Hervet published a mathematical treatment of these dendrimers, and concluded that the limiting aspect is the length of the spacer between the connecting centers. Shortly afterwards, attempts were made to link the idea of dendritic molecules to the mathematical treatment of fractal theory, introduced in 1975–1977 by Mandelbrot, as fractal theory offered a rational description of dendrimers.

Dendrimers may be purely organic or metallodendrimers, and in what follows we shall focus exclusively on the latter. There are two general strategies for synthesising metallodendrimers: the divergent strategy and the convergent strategy.

The *divergent strategy* is the most general alternative and requires bifunctional species. Here, dendrimers are synthesised from the central core to the periphery. The core represents the zero generation, and it has one or more reactive sites at

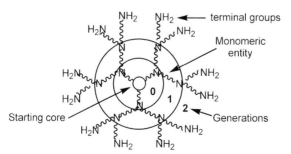

Figure 5.23 Scheme of a dendrimer.

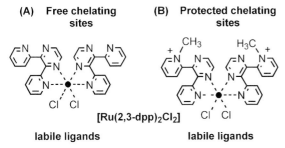

(A) Free chelating sites   **(B) Protected chelating sites**

[Ru(2,3-dpp)₂Cl₂]

labile ligands   labile ligands

Figure 5.24 Functional groups for the synthesis of dendrimers by divergent strategy.

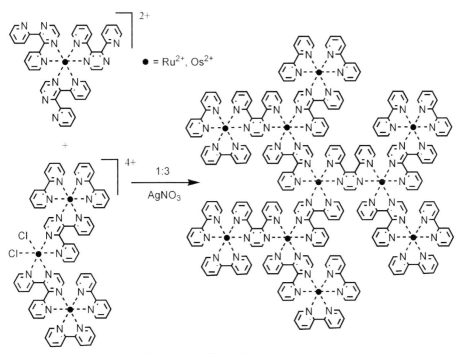

● = Ru²⁺, Os²⁺

1:3

AgNO₃

Figure 5.25 Convergent synthesis of one metallodendrimer (see text).

which a new generation of blocks will bind together covalently, and so on, successively. The repetitive addition of similar (or dissimilar) blocks through the protection/deprotection of functional groups produces new generations of dendrimers (Figure 5.24).

In the *convergent strategy*, dendrimers are built from the periphery to the central core. In other words, the branched arms (dendritic branches) are synthesized first and then, in a final step, linked to a core of varying size. The convergent strategy yields more homogeneity but fewer generations. The convergent synthesis of one metallodendrimer is shown in Figure 5.25.

5.10.1
Classification

Metallodendrimers can be categorized according to where the metal centers are positioned, that is, metals as cores, metals as termini (on the surface), metals as branching centers, metals as connectors, metals at multiple locations and metals as transformation auxiliaries [53].

- *Metals as cores.* Metallated phthalocyanines as the core of dendrimers have been reported since the first one published in 1997. The phthalocyanine core can be either free or, more generally, complexed by a metal (in particular zinc). Phthalocyanine dendrimers from generation 1 to generation 8 have been synthesized [55].

- *Metals as branching centers.* In 1990, Balzani's group synthesized various polypyridine dendritic complexes of Ru and Os. Figure 5.25 shows one of the first complexes of this type, a decanuclear complex of Ru or Os (the chelating ligand bpy can be substituted by similar ligands). These complexes have been found to have interesting luminescent and redox properties, and are used as "antennae" for capturing light (see Chapter 15) [56].

- *Metals as block connectors.* The ligand 2,2′:6′,2″-terpyridine (terpy) has been used as a ligand for binding coordination polymers and oligomers. In 1993, Newkome's group prepared the complex represented in Figure 5.26 (where only the top half is shown) [54].

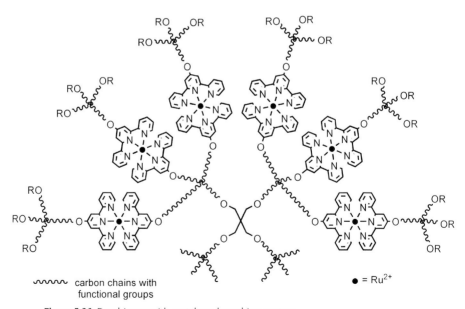

carbon chains with
functional groups

● = Ru^{2+}

Figure 5.26 Dendrimers with metals as branching centers.

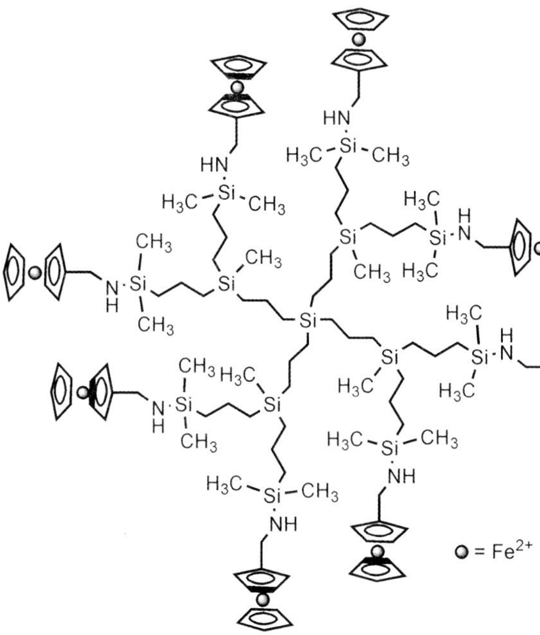

Figure 5.27 Dendrimers with ferrocene-derived terminal groups.

- *Metals as terminal groups: functionalization of the surface.* The grafting of organometallic or transition-metal fragments onto the surface of dendrimers can potentially lead to the resulting species having valuable properties in areas such as catalysis, electrochemistry, photochemistry, and magnetism. For example, dendrimers with ferrocene-derived terminal groups (Figure 5.27) have been studied in detail for their redox properties. Cyclic voltammetry studies indicate eight-electron oxidation, which gives these dendrimers a great electron transfer potential. The ferrocene can be substituted with $[Fe(C_5H_5)(CO)_2]$, $[Co(C_5H_5)(CO)_2]$, $[Cr(C_6H_6)(CO)_3]$ and other similar groups. Dendrimers with other metal complexes on their exterior surface have also been obtained.

It is anticipated that the introduction of two or more different metals into the dendritic system can multiply these effects, so that the synthesis of heterometallic dendrimers has now become a challenge for chemists. Rossell et al. have recently developed an efficient strategy for the synthesis of carbosilane dendrimers containing double metallic layers Ru/Au, Ru/Pd, and Ru/Rh [57].

Cavity-cored with designed building blocks. Recently, cavity-cored dendrimers have received considerable attention because of their elaborate structures and potential applications in delivery and recognition. In 2006, Stang et al. reported [58] the first self-assembly, by a combination of dendritic donor subunits, of nanoscale metallodendrimers that have a hexagonal cavity as a core.

5.10.2
Properties

Metallodendrimers have a great volume and mass: for example, a polypyridine dendrimer with 22 metals has over 1000 atoms and an estimated radius of 5 nm. Such dendrimers are usually soluble in many common solvents (water, acetonitrile, dichloromethane, etc.). One of the most appealing features is the possibility of tuning their properties by changing the number, chemical nature and relative position of functional units within the branched structure. This feature has led to their widespread use in a variety of applications from biology to material sciences, i.e. at the interface of many disciplines.

The intrinsic qualities of dendrimers – monodispersity, multivalency, weak cytotoxicity, and low immunogenicity – make dendrimers ideal molecules for use in medical and biological fields: controlled diffusion of drugs, medical imaging, gene therapy, antiviral and antimicrobial protection, cellular therapy, reparation of tissues, biomaterials, etc. A review of dendrimers and MRI (magnetic resonance imaging) (Chapter 17) has been recently published [59]. These dendrimers have the periphery modified with Gd(III) chelates (see Chapter 17). The monodisperse character of dendrimers creates a unique opportunity to introduce dendritic MRI contrast agents with prolonged vascular retention time into clinical use [59].

Dendrimers "encapsulating" photosensitizers (metal cores) such as metalloporphyrins and metallophthalocyanines have been applied for effective photodynamic therapy (PDT) [60]. PDT involves administration of photosensitizers (PSs) to the body, followed by activation of the PSs at the diseased site using light of specific wavelength. Upon photoirradiation, the PSs undergo photosensitized reaction processes that produce free radicals, superoxide anions or cytotoxic singlet oxygen ($^1O_2$) [60].

In terms of photochemistry, dendrimers have proved especially useful for capturing sunlight. In fact, the photochemical properties of dendrimers are not just the sum of the properties of each unit. Fréchet's group [61] studied the encapsulation of lanthanide ions by poly(benzylether) dendritic ligands in the context of a possible application for optical signal amplification. Indeed, during the course of photophysical studies on these assemblies, it was found that irradiation at wavelengths where the dendrimer backbone absorbed (280–290 nm) resulted in strong luminescence from the lanthanide core. Apparently, energy absorbed by the peripheral dendrimer shell was efficiently transferred to the luminescent Ln^{3+} as the focal point. This channeling of excitation energy from a dendrimer shell to a single core unit was termed the "*antenna effect*" (see Chapter 15). In each of the dendrimer units there is always competition between radiative emission (luminescence) and non-radiative emission. In dendrimers the non-radiative process is improved, to the detriment of the luminescent one. This effect is very important in terms of being able to model the natural aspects of photosynthesis, taking greater advantage of the absorbed energy without losing it in very rapid luminescent processes.

Dendritic-based materials are widely used as soluble supports in homogeneous (and to some extent heterogeneous) catalysis. The dendritic framework can be functionalized with the catalyst either at the core or at the periphery, which determines the catalytic performance to a great extent. Of particular interest is the use of noncovalent (partially ionic) interactions to attach the catalytic entities to a dendritic scaffold. An example of intramolecular cyclocarbonylation reactions with palladium-complexed dendrimers on silica gel as catalysts, allowing the synthesis of oxygen- nitrogen-, or sulfur-containing macrocycles can be found in Ref. [62].

A review on recent progress and applications for metallodendrimers has been published recently [53a].

Finally, I would like to finish with a note of realism, by quoting the words of Helms and Meijer, who recently (2006) stated [63]: "*In academic laboratories, dendrimers have been used as light- and energy-harvesting materials, for drug delivery, as catalysts, and in optoelectronic applications. Yet they have not been widely introduced commercially. This situation is about to change, with several dendrimers now entering the market ... However for large scale applications the focus has switched to hyperbranched polymers, which are easy to manufacture ... For smaller scale applications, especially in the biomedical sector ... dendrimers have also been used to improve existing molecular imaging technologies (a dendrimer functionalized at its periphery with gadolinium chelates) ...; encapsulation of guest molecules in dendrimers can also be used to deliver therapeutic agents throughout the body ... Dendrimers can be used as vehicles to introduce a gene into a cell ...*" [63].

5.11
Future Outlook

The last twenty years have witnessed an explosion in the synthesis of coordination clusters. The vast majority of these, including the most relevant and spectacular, have been made as a result of "serendipitous self-assembly". Thus, aggregates with an extraordinary beauty and/or complexity, such as the giant Keplerate $[\{(Mo^{VI})Mo^{VI}_5O_{21}L_6\}_{12}\{Fe^{III}(H_2O)L\}_{30}]{\sim}150H_2O$ (L = H_2O, CH_3COO^-, etc.) [64], the Keplerate magnetic cluster featuring an icosahedron of Ni^{2+} ions encapsulating a dodecahedron of La^{3+} ions formulated as $[La_{20}Ni_{30}(IDA)_{30}(CO_3)_6(NO_3)_6(OH)_{30}(H_2O)_{12}](CO_3)_6 \cdot 72H_2O$ (IDA = iminodiacetate) [65], and the fullerene-like molecule $[\{\eta^5Cp\text{-}Fe\text{-}\eta^5P_5\}_{12}\{CuCl\}_{10}\{Cu_2Cl_3\}_5\{Cu(CH_3CN)_2\}_5]$ that possesses 90 inorganic core atoms [66], (Figure 5.28) have only been obtained "by accident" and not by design. In this context, coordination chemists have learned the reaction conditions necessary to favor the formation of large polynuclear clusters, however, it currently remains impossible to predict the structure of any new system prepared in this manner. I have no doubt that the accumulated experience in this field, together with more powerful theoretical models and computer capabilities will one day allow chemists to know the structure of the polynuclear coordination complex expected from a reaction system before performing the experiment in the laboratory.

(A)　　　　　　　　　　　**(B)**

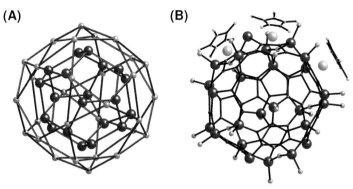

Figure 5.28 (A) Central core of the cluster [La$_{20}$Ni$_{30}$] [Ni (gray, small); La (black, large)]. (B) central core of the fullerene-like molecule described in the text (Cu, black). Only three η^{5}Cp-Fe-η^5-P$_5$ are shown. The other 9 entities are omitted for clarity (Fe, gray).

References

1 Escuer, A., Vicente, R., Kumar, S.B., Mautner, F.A. Spin frustration in the butterfly-like tetrameric [Ni$_4$(μ-CO$_3$)$_2$(aetpy)$_8$][ClO$_4$]$_4$ [aetpy = (2-aminoethyl)pyridine] complex. Structure and magnetic properties, *J. Chem. Soc., Dalton Trans*, 1998, 3473.

2 Yuongme, S., van Albada, G.A., Chaichit, N., Gunnasoot, P., Kongsaeree, P., Mutikainen, I., Roubeau, O., Reedijk, J., Turpeinen, U. Synthesis, spectroscopic characterization, X-ray crystal structure and magnetic properties of oxalato-bridged copper(II) dinuclear complexes with di-2-pyridylamine, *Inorg. Chim. Acta* 2003, *353*, 119.

3 (a) Ribas, J., Escuer, A., Monfort, M., Vicente, R., Cortés, R., Lezama, L., Rojo, T. Polynuclear Ni^{II} and Mn^{II} azido bridging complexes. Structural trends and magnetic behaviour, *Coord. Chem. Rev.* 1999, *193–195*, 1027; (b) Escuer, A., Aromí, G. Azide as a Bridging Ligand and Magnetic Coupler in Transition Metal Clusters, *Eur. J. Inorg. Chem.* 2006, 4721.

4 Aromí, G., Ribas, J., Gamez, P., Roubeau, O., Kooijman, H., Spek, A.L., Teat, S., MacLean, E., Stoeckli-Evans, H., Reedijk, J. Aggregation of [Cu^{II}] Building Blocks into [Cu$_8^{II}$] Clusters or [Cu$_4^{II}$]$_\infty$ Chain through Subtle Chemical Control, *Chem. Eur. J.* 2004, *10*, 6476.

5 Tercero, J., Diaz, C., Ribas, J., Mahía, J., Maestro, M., Solans, X. Synthesis, characterization, and magnetic properties of new complexes based on self-assembled homotrinuclear units Cu^{II}–Cu^{II}–Cu^{II}, *J. Chem. Soc., Dalton Trans.* 2002, 2040.

6 Tercero, J., Diaz, C., Ribas, J., Maestro, M., Mahía, J., Stoeckli-Evans, H. Oxamato-Bridged Trinuclear Ni^{II}Cu^{II}Ni^{II} Complexes: A New (Ni^{II}Cu^{II}Ni^{II})$_2$ Hexanuclear Complex and Supramolecular Structures. Characterization and Magnetic Properties, *Inorg. Chem.* 2003, *42*, 3366.

7 Marvaud, V., Decroix, C., Sciuller, A., Guyard-Duhayon, C., Vaissermann, J., Gonnet, F., Verdaguer, M. Hexacyanometalate Molecular Chemistry: Heptanuclear Heterobimetallic Complexes; Control of the Ground Spin State, *Chem. Eur. J.* 2003, *9*, 1677.

8 Berlinguette, C.P., Dunbar, K.R. The step-wise assembly of an undecanuclear heterotrimetallic cyanide cluster, *Chem. Comm.* 2005, 2451

9 Casey, W.H., Olmstead, M.M., Phillips, B.L. A New Aluminum Hydroxide

Octamer, $[Al_8(OH)_{14}(H_2O)_{18}](SO_4)_5 \cdot 16H_2O$, *Inorg. Chem.* 2005, *44*, 4888.

10 Rather, E., Gatlin, J.T., Nixon, P.G., Tsukamoto, T., Kravtsov, V., Jonhson, D.W. A Simple Organic Reaction Mediates the Crystallization of the Inorganic Nanocluster $[Ga_{13}(\mu_3\text{-}OH)_6 (\mu_2\text{-}OH)_{18}(H_2O)_{24}](NO_3)_{15}$, *J. Am. Chem. Soc.* 2005, *127*, 3242.

11 Zou, X., Conradsson, T., Klingstedt, M., Dadachov, M.S., O'Keeffe, M. A mesoporous germanium oxide with crystalline pore walls and its chiral derivative, *Nature*, 2005, *437*, 716.

12 Fornasieri, G., Rozes, L., Le Calve, S., Alonso, B., Massiot, D., Rager, M.N., Evain, M., Boubekeur, K., Sanchez, C. Reactivity of Titanium Oxo Ethoxo Cluster $[Ti_{16}O_{16}(OEt)_{32}]$. Versatile Precursor of Nanobuilding Block-Based Hybrid Materials, *J. Am. Chem. Soc.* 2005, *127*, 4869.

13 Pohl, I.A.M., Westin, L.G., Kritikos, M. Preparation, Structure, and Properties of a New Giant Manganese Oxo-Alkoxide Wheel, $[Mn_{19}O_{12}(OC_2H_4OCH_3)_{14} (HOC_2H_4OCH_3)_{10}] \cdot HOC_2H_4OCH_3$, *Chem. Eur. J.* 2001, *7*, 3439.

14 Bino, A., Ardon, M., Lee, D., Spingler, B., Lippard, S.J. Synthesis and Structure of $[Fe_{13}O_4F_{24}(OMe)_{12}]^{5-}$: The First Open-Shell Keggin Ion, *J. Am. Chem. Soc.* 2002, *124*, 4578.

15 Mehring, M., Mansfeld, D., Paalasmaa, S., Schürmann, M. Polynuclear Bismuth-Oxo Clusters: Insight into the Formation Process of a Metal Oxide, *Chem. Eur. J.* 2006, *12*, 1767.

16 Shaw, R., Tidmarsh, I.S., Laye, R.H., Breeze, B., Helliwell, M., Brechin, E.K., Heath, S.L., Murrie, M., Ochsenbein, S., Güdel, H-U., McInnes, E.J.L. Supertetrahedral decametallic Ni(II) clusters directed by μ_6-*tris*-alkoxides, *Chem.Comm.* 2004, 1418.

17 Powell, A.K., Heath, S.L., Gatteschi, D., Pardi, L., Sessoli, R., Spina, G., Del Giallo, F., Pieralli, F., Synthesis, Structures, and Magnetic Properties of Fe_2, Fe_{17}, and Fe_{19} Oxo-Bridged Iron Clusters: The Stabilization of High Ground State Spins by Cluster Aggregates, *J. Am. Chem. Soc.* 1995, *117*, 2491.

18 See, for example, Cotton, F.A., Wilkinson, G., Murillo, C.A., Bochmann, M. (1999) *Advanced Inorganic Chemistry*, 6th Edn. John Wiley & Sons, New York.

19 Kurata, T., Uehara, A., Hayashi, Y., Isobe, K. Cyclic Polyvanadates Incorporating Template Transition Metal Cationic Species: Synthesis and Structures of Hexavanadate $[PdV_6O_{18}]^{4-}$, Octavanadate $[Cu_2V_8O_{24}]^{4-}$, and Decavanadate $[Ni_4V_{10}O_{30} (OH)_2(H_2O)_6]^{4-}$, *Inorg. Chem.* 2005, *44*, 2524.

20 Khan, M.I., Ayesh, S., Doedens, R.J., Yu, M., O'Connor, C.J. Synthesis and characterization of a polyoxovanadate cluster representing a new topology, *Chem. Comm.* 2005, 4658.

21 Shi, Y., Yang, W., Xue, G., Hu, H., Wang, J. A novel crystal coexisting with two kinds of polyoxomolybdates: [n-$Bu_4N]_8[Mo_6O_{19}]_2[\alpha\text{-}(Mo_8O_{26})]$, *J. Mol. Struct.* 2006, *784*, 244.

22 (a) Muller, A., Kogerler, P., Dress, A.W.M. Giant metal-oxide-based spheres and their topology: from pentagonal building blocks to keplerates and unusual spin systems, *Coord. Chem.Rev*, 2001, *222*, 193; (b) Müller, A., Roy, S. En route from the mystery of molybdenum blue via related manipulatable building blocks to aspects of materials science, *Coord. Chem. Rev.* 2003, *245*, 153; (c) Muller, A., Roy, S. Multifunctional metal oxide based nanoobjects: spherical porous capsules/artificial cells and wheel-shaped species with unprecedented materials properties, *J. Mater. Chem.* 2005, *15*, 4673; (d) Muller, A., Roy, S. Linking Giant Molybdenum Oxide Based Nano-Objects Based on Well-Defined Surfaces in Different Phases, *Eur. J. Inorg. Chem.* 2005, 3561.

23 Müller, A., Zhou, Y., Bögge, H., Schmidtmann, M., Mitra, T., Haupt, E. T.K., Berkle, A. "Gating" the Pores of a Metal Oxide Based Capsule: After Initial Cation Uptake Subsequent Cations Are Found Hydrated and Supramolecularly Fixed above the Pores, *Angew. Chem. Int. Ed.* 2006, *45*, 460.

24 Müller, A., Todea, A.M., Bögge, H., van Slageren, J., Dressel, M., Stammler, A., Rusu, M. Formation of a "less stable" polyanion directed and protected by electrophilic internal surface functionalities of a capsule in growth: $[\{Mo_6O_{19}\}^{2-} \subset \{Mo^{VI}_{72}Fe^{III}_{30}O_{252}(ac)_{20}(H_2O)_{92}\}]^{4-}$, *Chem. Comm.* 2006, 3066.

25 Müller, A., Todea, A.M., van Slageren, J., Dressel, M., Bögge, H., Schmidtmann, M., Luban, M., Engelhardt, L., Rusu, M. Triangular Geometrical and Magnetic Motifs Uniquely Linked on a Spherical Capsule Surface, *Angew. Chem. Int. Ed.* 2005, 44, 3857.

26 Müller, A., Krickemeyer, E., Bögge, H., Schmidtmann, M., Peters, F., Menke, C., Meyer, J. An Unusual Polyoxomolybdate: Giant Wheels Linked to Chains, *Angew. Chem. Int. Ed.* 1997, 36, 484.

27 Müller, A., Beckmann, E., Bögge, H., Schmidtmann, M., Dress, A. Inorganic Chemistry Goes Protein Size: A Mo_{368} Nano-Hedgehog Initiating Nanochemistry by Symmetry Breaking, *Angew. Chem. Int. Ed.* 2002, 41, 1162.

28 Long, D-L., Abbas, H., Kögerler, P., Cronin, L. A High-Nuclearity "Celtic-Ring" Isopolyoxotungstate, $[H_{12}W_{36}O_{120}]^{12-}$, That Captures Trace Potassium Ions, *J. Am. Chem. Soc.* 2004, 126, 13880.

29 Sakai, Y., Yoza, K., Kato, C.N., Nomiya, K. A first example of polyoxotungstate-based giant molecule. Synthesis and molecular structure of a tetrapod-shaped Ti–O–Ti bridged anhydride form of Dawson tri-titanium(IV)-substituted polyoxotungstate, *J. Chem. Soc., Dalton Trans.* 2003, 3581.

30 Shivaiah, V., Das, S.K. Polyoxometalate-Supported Transition Metal Complexes and Their Charge Complementarity: Synthesis and Characterization of $[M(OH)_6Mo_6O_{18}\{Cu(Phen)(H_2O)_2\}_2][M(OH)_6Mo_6O_{18}\{Cu(Phen)(H_2O) Cl\}_2] \cdot 5H_2O$ (M = Al^{3+}, Cr^{3+}) *Inorg. Chem.* 2005, 44, 8846.

31 Reinoso, S., Vitoria, P., Gutierrez-Zorrilla, J.M., Lezama, L., Madariaga, J. M., San Felices, L., Iturrospe, A. Coexistence of Five Different Copper(II)-Phenanthroline Species in the Crystal Packing of Inorganic-Metalorganic Hybrids Based on Keggin Polyoxometalates and Copper(II)-Phenanthroline-Oxalate Complexes, *Inorg. Chem.* 2007, 46, 4010.

32 Anderson, T.M., Cao, R., Hill, C. L., et al, A Palladium-Oxo Complex. Stabilization of This Proposed Catalytic Intermediate by an Encapsulating Polytungstate Ligand, *J. Am. Chem. Soc.* 2005, 127, 11948.

33 Wang, J., Ma, P., Shen, Y., Niu, J. A Novel Polyoxotungstate $[Ni_4(H_2O)_2(\alpha-NiW_9O_{34})_2]^{16-}$ Based on an Old Structure with a New Component, *Cryst. Growth Des.*, 2007, 7, 603.

34 (a) Zimmermann, M., Belai, N., Butcher, R.J., Pope, M.T., Chubarova, E.V., Dickman, M.H., Kortz, U. New Lanthanide-Containing Polytungstates Derived from the Cyclic P_8W_{48} Anion: $\{Ln_4(H_2O)_{28}[K(P_8W_{48}O_{184}(H_4W_4O_{12})_2Ln_2(H_2O)_{10}]^{13-}\}_x$, Ln = La, Ce, Pr, Nd, *Inorg. Chem.* 2007, 46, 1737; (b) Bassil, B.S., Dickman, M.H., von der Kammer, B., Kortz, U. The Monolanthanide-Containing Silicotungstates $[Ln(\beta_2-SiW_{11}O_{39})_2]^{13-}$ (Ln = La, Ce, Sm, Eu, Gd, Tb, Yb, Lu): A Synthetic and Structural Investigation, *Inorg. Chem.* 2007, 46, 2452.

35 Long, D-L., Burkholder, E., Cronin, L. Polyoxometalate clusters, nanostructures and materials: From self assembly to designer materials and devices, *Chem. Soc. Rev.* 2007, 36, 105.

36 Tolis, E.I., Timco, G.A., Wernsdorfer, W., Winpenny, R.E.P., et al. Studies of an Fe_9 Tridiminished Icosahedron, *Chem. Eur. J.* 2006, 12, 8961.

37 Christou, G. Manganese Carboxylate Chemistry and Its Biological Relevance, *Acc. Chem. Res.* 1989, 22, 328.

38 Tasiopoulos, A.J., Vinslava, A., Wernsdorfer, W., Abboud, K. A., Christou. G. Giant Single-Molecule Magnets: A $\{Mn_{84}\}$ Torus and Its Supramolecular Nanotubes, *Angew. Chem. Int. Ed.* 2004, 43, 2117.

39 Saalfrank, R.W., Deutscher, C., Sperner, S., Nakajima, T., Ako, A.M., Uller, E., Hampel, F., Hienemann, F.W. Six-Membered Metalla-coronands. Synthesis and Crystal Packing: Columns,

Compartments, and 3D-Networks, *Inorg. Chem.* 2004, *43*, 4372.

40 Cañada-Vilalta, C., O'Brien, T.A., Pink, M., Davidson, E.R., Christou, G. Methanolysis and Phenolysis Routes to Fe_6, Fe_8, and Fe_{10} Complexes and Their Magnetic Properties: A New Type of Fe_8 Ferric Wheel, *Inorg. Chem.* 2003, *42*, 7819.

41 Taft, K.L., Delfs, C.D., Papaefthymiou, G.C., Foner, S., Gatteschi, D., Lippard, S.J. $[Fe(OMe)_2(O_2CCH_2Cl)]_{10}$, a Molecular Ferric Wheel, *J. Am. Chem. Soc.* 1994, *116*, 823.

42 (a) Raptopoulou, C.P., Tangoulis, V., Devlin, E. $[\{Fe(OMe)_2[O_2CC(OH)Ph_2]\}_{12}]$: Synthesis and Characterization of a New Member in the Family of Molecular Ferric Wheels with the Carboxylatobis(alkoxo) Bridging Unit, *Angew. Chem. Int. Ed. 2002*, 41, 2386; (b) King, P., Stamatos, T.C., Abboud, K.A., Christou, G. Reversible Size Modification of Iron and Gallium Molecular Wheels: A Ga_{10} "Gallic Wheel" and Large Ga_{18} and Fe_{18} Wheels, *Angew. Chem. Int. Ed.* 2006, *45*, 7379.

43 McInnes, E.J.L., Anson, C., Powell, A.K., Thomson, A.J., Poussereau, S., Sessoli, R. Solvothermal synthesis of $[Cr_{10}(\mu\text{-}O_2CMe)_{10}(\mu\text{-}OR)_{20}]$ "chromic wheels" with antiferromagnetic (R = Et) and ferromagnetic (R = Me) $Cr(III)\cdots Cr(III)$ interactions, *Chem. Commun.* 2001, 89.

44 Papaefstathiou, G.S., Manessi, A., Raptopoulou, C.P., Terzi, A., Zafiropoulos, T.F., Methanolysis as a Route to Gallium(III) Clusters: Synthesis and Structural Characterization of a Decanuclear Molecular Wheel, *Inorg. Chem.* 2006, *45*, 8823.

45 Armentano, D., Mastropietro, T.F., Julve, M., Rossi, R., Rossi, P., De Munno, G. A New Octanuclear Copper(II)-Nucleoside Wheel, *J. Am. Chem. Soc.* 2007, *129*, 2740.

46 Dearden, A.L., Parsons, S., Winpenny, R.E.P. Synthesis, Structure, and Preliminary Magnetic Studies of a Ni_{24} Wheel, *Angew. Chem. Int. Ed. Eng.* 2001, *40*, 151.

47 Affronte, M., Carretta, S., Timco, G.A., Winpenny, R.E.P. A ring cycle: studies of heterometallic wheels, *Chem. Comm.* 2007, 1789.

48 Ni, Z-H., Zhang, L-F., Tangoulis, V., Wernsdorfer, W., Cui, A-L., Sato, O., Kou, H-Z. Substituent Effect on Formation of Heterometallic Molecular Wheels: Synthesis, Crystal Structure, and Magnetic Properties, *Inorg. Chem.* 2007, *46*, 6029.

49 Papaefstathiou, G.S., Escuer, A., Mautner, F. A., Raptopoulou, C., Terzis, A., Perlepes, S.P., Vicente, R. Use of the Di-2-pyridyl Ketone/Acetate/Dicyanamide "Blend" in Manganese(II), Cobalt(II) and Nickel(II) Chemistry: Neutral Cubane Complexes, *Eur. J. Inorg. Chem.* 2005, 879.

50 Papaefstathiou, G.S., Perlepes, S.P. Families of polynuclear manganese, cobalt, nickel and copper complexes stabilized by various forms of di-2-pyridyl ketone, *Commun. Inorg. Chem.* 2002, *23*, 249.

51 Winpenny, R.E.P. Serendipitous assembly of polynuclear cage compounds, *J. Chem. Soc., Dalton Trans.* 2002, 1.

52 Parsons, S., Winpenny, R.E.P. Structural Chemistry of Pyridonate Complexes of Late 3d-Metals, *Acc. Chem. Res.* 1997, *30*, 89.

53 (a) Hwang, S-H., Shreiner, C.D., Moorefield, C.N., Newkome, G.R. Recent progress and applications for metallodendrimers, *New J. Chem.* 2007, *31*, 1192; (b) Stoddart, F.J., Welton, T. Metal-containing dendritic polymers, *Polyhedron*, 1999, *18*, 3575.

54 Newkome, G.R., Moorefield C.N., Vögtle, F. (1996) *Dendritic Molecules: Concepts, Syntheses, Perpectives*, VCH, Weinheim.

55 Leclaire, J., Dagiral, R., Pla-Quintana, A., Caminade, A-M., Majoral, J-P. Metallated Phthalocyanines as the Core of Dendrimers – Synthesis and Spectroscopic Studies, *Eur. J. Inorg. Chem.* 2007, 2890.

56 Balzani, V., Campagna, S., Denti, G., Juris, A., Serroni, S., Venturi, M. Designing Dendrimers Based on Transition-Metal Complexes. Light-Harvesting Properties and Predetermined Redox Patterns, *Acc. Chem. Res.* 1998, *31*, 26.

57 Angurell, I., Rossell, O., Seco, M., Ruiz, E. Dendrimers containing two metallic layers. Chloride migration from peripheral gold, palladium, or rhodium metals to internal ruthenium atoms, *Organometallics*, 2005, *24*, 6365.

58 Yang, H-B., Das, N., Huang, F., Hawkridge, A.M., Muddiman, D.C., Stang, P.J. Molecular Architecture via Coordination: Self-Assembly of Nanoscale Hexagonal Metallodendrimers with Designed Building Blocks, *J. Am. Chem. Soc.* 2006, *128*, 10014.

59 Langereis, S., Dirksen, A., Hackeng, T.M., van Genderen, M.H.P., Meijer, E.W. Dendrimers and magnetic resonance imaging, *New J. Chem.* 2007, *31*, 1152.

60 Nishiyama, N., Jang, W-D., Kataoka, K. Supramolecular nanocarriers integrated with dendrimers encapsulating photosensitizers for effective photodynamic therapy and photochemical gene delivery, *New J. Chem.* 2007, *31*, 1074.

61 Andronov, A., Fréchet, J.M.J. Light-harvesting dendrimers, *Chem. Comm.* 2000, 1701.

62 Lu, S-M., Alper, H. Synthesis of Large Ring Macrocycles (12–18) by Recyclable Palladium-Complexed Dendrimers on Silica Gel Catalyzed Intramolecular Cyclocarbonylation Reactions, *Chem. Eur. J.* 2007, *13*, 5908.

63 Helms, B., Meijer, E.W. Dendrimers at work, *Science*, 2006, *313*, 929.

64 Muller, A., Sarkar, S., Shah, S.Q.N., Bogge, H., Schmidtmann, M., Sarkar, S., Kogerler, P., Hauptfleisch, B., Trautwein, A.X., Schunemann, V. Archimedean Synthesis and Magic Numbers: "Sizing" Giant Molybdenum-Oxide-Based Molecular Spheres of the Keplerate Type, *Angew. Chem. Int. Ed.* 1999, *38*, 3238.

65 Kong, X-J., Ren, Y-P., Long, L-S., Zheng, Z., Huang, R-B., Zheng, L-S. A Keplerate Magnetic Cluster Featuring an Icosidodecahedron of Ni(II) Ions Encapsulating a Dodecahedron of La(III) Ions, *J. Am. Chem. Soc.* 2007, *129*, 7016.

66 Bai, J., Virovets, A.V., Scheer, M. Synthesis of inorganic fullerene-like molecules, *Science*, 2003, *300*, 781.

6
Metal–Metal Bond and Metal Clusters

6.1
Introduction: Definition and Main Characteristics

According to the definition of Cotton [1], a cluster includes all molecules which contain a definite number of metal atoms which, in an essential proportion, are linked by metal–metal bonds. In addition, they can be bonded to other ligands. The diagnostic criteria for the presence of M—M bonds are numerous:

- Thermochemical data. The chemical relevance of M—M bonds should be reflected in the thermochemical bond enthalpy.

- Bond distances d(M—M) from X-ray diffraction. In the evaluation of a bond distance the oxidation states of the metals involved as well as the nature of the ligands have to be taken into account. Some typical examples in molybdenum clusters are: 2.72 Å (single bond), 2.42 Å (double), 2.22 Å (triple), 2.10 Å (quadruple), 2.78 Å (metal).

- Spectroscopic data. A measure of the bond strength is also provided by the position of the metal–metal stretching mode in the Raman spectrum. A quadruple Cr—Cr bond shows a $\bar{v}_{M-M}$ at 556 cm^{-1}, an Mo—Mo bond at 420 cm^{-1} and a Re—Re bond at 285 cm^{-1}. Considering the large masses, these are surprisingly high values.

Apart from their fascinating structures, metal atom clusters possess properties which are highly interesting for practical applications: homogeneous catalysis, photochemical properties, mimicking important redox enzymes (Fe_4S_4 in ferredoxine), etc.

6.2
Transition Metal Clusters

The bond energies of transition metals present a maximum at the center of each series, corresponding to half of the d electrons. In order to emulate this trend the first transition elements form clusters with, above all, π-donor ligands that provide

Coordination Chemistry. Joan Ribas Gispert
Copyright © 2008 WILEY-VCH Verlag GmbH & Co. KGaA, Weinheim
ISBN: 978-3-527-31802-5

extra electrons, while the final metals form clusters with π-acid ligands which "eliminate" part of the electron surplus. Therefore, there are two types of clusters with transition metals: (i) with halido, oxido, alkoxido and other π-base ligands and (ii) with carbonyl, and other π-acid ligands.

The first type is formed with metals with relatively high oxidation states (2+, 3+). These metals are usually located on the left-hand side of the Periodic Table, mainly in the second and third transition series. The second type is formed with metals with low oxidation states ($n-$ to 1+), from the right-hand side of the transition series.

6.3
Transition Metal Clusters with π-Donor Ligands

Let us give a brief summary of these clusters following their nuclearity:

6.3.1
Dinuclear Clusters

Cotton's discovery of the Re–Re quadruple bond in the $[Re_2Cl_8]^{2-}$ anion signified the recognition of the δ bond, resulting in a bond of order four [1].

6.3.1.1 Dinuclear Clusters Based on ML$_4$ or ML$_5$ Fragments

Each transition metal center has locally five or four metal–ligand σ bonds. The metal–metal interaction between the two fragments gives rise to four ($1\sigma + 2\pi + 1\delta$) bonding orbitals and the corresponding four antibonding orbitals, as illustrated in Figure 6.1. The occupation of the four M–M bonding orbitals gives the optimal metal total d-electrons count of eight, forming the '*famous*' quadruple bonds. Deviation from this optimal occupation number results in a smaller bond order, and hence longer M–M bond lengths and weaker bond strengths. Ignoring

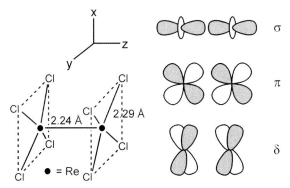

Figure 6.1 σ, π and δ molecular orbitals in a dinuclear complex with multiple metal–metal bond.

the molecular orbitals in which the ligands are involved, the MO diagram for the core orbitals is as shown in Figure 6.2, together with complexes with variable occupation, which leads to a different bond order.

The occupation of δ bonding orbitals in the complex $[Re_2X_8]^{2-}$ results in an eclipsed conformation. Occupation of the δ* orbitals cancels the effect of δ bonding, and such complexes are expected to have a staggered conformation due to the steric effect of the ligands. However, the energy splitting of the δ and δ* orbitals is small such that, for 10-electron complexes, in which the δ* orbital is occupied, both the eclipsed $[Re_2Cl_4(PMe_3)_4]$ and staggered $[Os_2Cl_8]^{2-}$ conformations are observed.

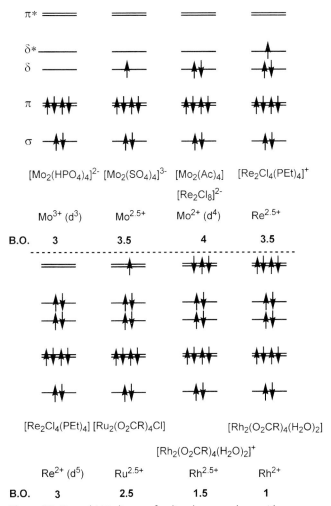

Figure 6.2 General MO diagram for dinuclear complexes with multiple M—M bond (B.O. = bond order).

A series of compounds having the core $[Re_2^{III}(hpp)_4]^{2+}$ (hpp = anion of a pyrimidine derivative) with two axial ligands has shown than the σ and π donating ability of the axial ligands is of great importance in determining the metal–metal distance. As the interaction between the axial ligand and the rhenium atom increases, there is an increase in the metal–metal bond distance thus reducing the Re—Re interaction. When the axial ligands are the weakly coordinating triflate anions, the Re—Re distance is only 2.156 Å, the shortest known for any quadruply bonded species [2].

For late transition metals, the metal d electrons become contracted. Hence the overlap of d orbitals becomes poor, and the corresponding bonding and antibonding orbitals do not split as much as in the earlier transition metals, thus allowing the δ* and π* orbitals to be occupied. Group 10 clusters (M^{2+}) with 16 metal d electrons do not have net metal–metal bonding, and their stabilities are the result of chelating effects from the bridging bidentate ligands. However, metal–metal bonding is present in oxidized $[Ni_2]^{5+}$ complexes (Ni—Ni bond order = 0.5) as well as in the electrochemically obtained $[Ni_2]^{6+}$ core, in solution, which has a single Ni—Ni σ bond [3].

6.3.1.2 Dinuclear Clusters Based on Face-sharing Bioctahedra

Symmetry requirements demonstrate that the orbital diagram is that of Figure 6.3 (symmetry labels of a C_{3v} point group). The optimal electronic structure has, thus, the three bonding orbitals fully occupied. A typical example is $[Cr_2^{III}F_6(\mu\text{-}F)_3]^{3-}$.

6.3.1.3 L₃MML₃ Unbridged Dinuclear Clusters

Clusters belonging to this class are formed mainly by W^{3+} or Mo^{3+} with a metal–metal triple bond. Electronic considerations demand that the ligands for these clusters are normally strong π-donor ligands, such as —OR and —NR₂. The current structures are ethane-like, such as in $[\{(CH_3)_2N\}_3W\text{-}W\{N(CH_3)_2\}_3]$.

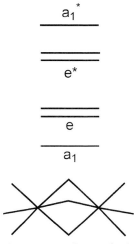

Figure 6.3 MO diagram for dinuclear clusters based on face-sharing bioctahedra.

6.3.1.4 A New and Unique Case: a Fivefold Cr—Cr Bonding

Power et al. [4] have reported the first stable compound with fivefold bonding between two chromium(I) centers. It is the [Ar'—Cr—Cr—Ar'] (Ar'=C_6H_3-2,6(C_6H_3-2,6-Pr^i_2)$_2$ and Pr^i = isopropyl). The ligands used to stabilize multiply bonded metal centers reduce the number of valence orbitals available to form metal–metal bonds. Thus, the number of ligands must be minimized, in order to achieve the highest bond order possible in an isolated compound. Moreover, the ligands must be sufficiently bulky to inhibit intermolecular reactions that yield polymers with lower bond orders. The Ar' ligand has these characteristics. In a simplified molecular-overlap diagram with the assumption of local C_{2h} symmetry, five metal–metal bonding molecular orbitals can be visualized. This fivefold Cr—Cr interaction is supported by structural, magnetic and theoretical data. The Cr—Cr distance is extremely short, 1.8351 Å.

Relativistic density functional computations have recently shown that the $U_2 \subset C_{60}$ (dimetalloendofullerene) has an unprecedented U—U multiple bond consisting of sixfold ferromagnetically coupled one-electron-two-center bonds with the electronic configuration $(5f\pi_u)^2(5f\sigma_g)^1(5f\delta_g)^2(5f\phi_u)^1$, which are dominated by the uranium 5f atomic orbitals [5]. This bonding scheme is completely distinct from the metal–metal bonds discovered thus far in the d- and f-block polynuclear metal complexes.

6.3.2
Trinuclear Clusters

The most typical example is the classic cluster $Cs_3[Re_3^{III}Cl_{12}]$ with $[Re_3Cl_9(\mu\text{-}Cl)_3]^{3-}$ core (Figure 6.4) which leads to the recognition of metal–metal multiple bonds [1]. The cluster can be viewed as consisting of three ML_5 fragments, each having four frontier fragment orbitals ($\sigma + t_{2g}$). The result of orbital interactions among the 12 orbitals gives 6 bonding and 6 antibonding molecular orbitals. The occupation of the six bonding molecular orbitals gives the optimal electronic structure with each M—M bond having a bond order of two. The Re(III), (d^4) complex provides an example of this optimal electron count of 12.

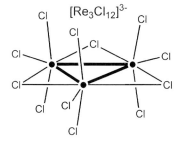

Figure 6.4 Structure of the $[Re_3^{III}Cl_{12}]^{2-}$ anion (see text).

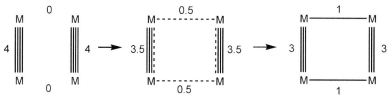

Figure 6.5 Some M_4 clusters formally derived from M_2 complexes with a bond order of 4.

6.3.3
Tetranuclear Clusters

For tetranuclear clusters three types of regular structures are reported: square (or rectangular) planar, edge-sharing triangular ("butterfly" or rhombus) and tetrahedral.

Some M_4 clusters can be formally derived from M_2 starting materials with a bond order of 4. Chisholm et al. [6a], have studied these clusters as formed when pairs of metal–metal quadruply bonded complexes undergo [2 + 2] cycloaddition with loss of the δ components to leave two triply bonded dinuclear units joined by single M—M bonds along the long edges of the rectangle (Figure 6.5). The clusters $[M_4^{II}Cl_4(PMe_3)_4(\mu\text{-}Cl)_4]$ and $[M_4^{II}Cl_4(MeOH)_4(\mu\text{-}Cl)_4]$ have this rectangular structure (M=Mo, W) [6a]. Very recently Cotton et al. have reported a similar complex but with an average bond order of only 0.25, assigned to the bond between the two Mo atoms along the long edges of the rectangle [6b].

Other tetranuclear clusters of Mo and W are derived from M_2 starting materials with a bond order of 3 or 2. Clusters belonging to tetrahedral geometry have four μ_3-L ligands. These clusters are also known as cubane-type clusters since the arrangement of metal and L atoms form approximately a cube. The relevant molecular orbital interaction diagram has been provided in the literature and will not be discussed in more detail here [7]. Examples are the $[Fe_4S_4]$ clusters, such as in $[Fe_4S_4(SPh)_4]^{2-}$. It has been considered that the Fe—Fe interactions are very weak in these clusters.

6.3.4
Hexanuclear Clusters

The octahedral metal clusters of the lower halide chemistry of niobium, tantalum, molybdenum and tungsten represent paradigmatic clusters. There are two important structural types, containing M_6X_8 and M_6X_{12} units, respectively. The former, typified by $[Mo_6Cl_8]^{4+}$, have a triply bridging halogen atom X, located over the center of each of the eight octahedral faces. The latter, typified by $[Nb_6Cl_{12}]^{2+}$, have 12 doubly bridging halogen atoms located over the centers of the twelve octahedral edges. They may have terminal ligands on each metal ion, giving the general formula of $[M_6L_6(\mu_3\text{-}X)_8]$ and $[M_6L_6(\mu\text{-}X)_{12}]$ (M = transition metal; X = bridging ligand and L = terminal ligand) (Figure 6.6). All of them can present a variable oxidation state of the metallic ions.

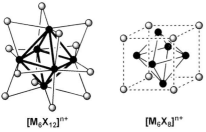

Figure 6.6 [M_6X_8] and [M_6X_{12}] units in hexanuclear clusters.

In a molecular orbital picture, the metal–metal orbital interactions are derived from the combination of the frontier orbitals of six square-pyramidal ML_5 fragments. Since each transition metal in these clusters has five localized metal–ligand σ-bonds (including also the M–L_{apical}) four frontier orbitals remain for the formation of the metal-metal bonds of the cluster: a σ bond and the t_{2g} set.

The bonding of the above mentioned octahedral clusters via the external halogens or through condensation processes involving the sharing of vertices, edges or faces gives rise to a large family of 1D, 2D and 3D compounds. The 2D MoX_2 nets, for example, are formed by octahedral clusters linked by four Cl^- bridges. The octahedral molybdenum clusters became most popular when the superconducting properties of the related chalcogenides $M_xMo_6Y_8$ (Y = chalcogenide), known as Chevrel phases, were discovered.

6.3.5
Centered Clusters

Scandium and zirconium halide cluster phases [$Y \subset Zr_6X_{12}$] (X = Cl, Br) are centered by Y = Be, B, Al, C and Si. [$C \subset W_6Cl_{18}$]$^{n-}$ (n = 0–3) and [$N \subset W_6Cl_{18}$]$^-$ are carbon- or nitrogen-centered W_6 trigonal prisms surrounded by twelve edge-bridging chloride anions and six radially extended terminal chloride ligands [8]. An extraordinary multiple centered unit has been recently reported: the 'nanomolecule' [$Sc_{24}C_{10}I_{30}$] has an outer envelope of 30 iodine atoms which surrounds 20 scandium atoms, which is filled by 10 carbon atoms, which, in turn, is filled by a tetrahedron of four Sc atoms. The cluster is, thus [$Sc_4 \subset C_{10} \subset Sc_{20} \subset I_{30}$]. This structure is reminiscent of a Russian doll, or, in other words, the molecule has an onion-like structure [9].

6.3.6
Structures Containing More Than One Metal–Metal Bonded Unit [1]

The introduction of metal–metal bonded units into assembled structures has expanded the field. Cotton et al. have reported several clusters containing more than one M–M bond [10], as well as heterometallic clusters containing [Mo_2]$^{4+}$ species coordinated to other metal units [11]. Zhou et al. have reported the synthesis and structure of cuboctahedral and anticuboctahedral cages containing 12 quadruply bonded dimolybdenum units with 1,3-benzenedicarboxilate derivatives as bridging ligands (see more examples in Chapter 16) [12].

6.3.7
Extended Metal Atom Chains (EMACs)

The field of extended metal atom chains (EMACs) began when, in 1991, the structure of $[Ni_3Cl_2(dpa)_4]$ (dpa = di(2-pyridyl)amide) was determined [13]. In the past decade, increasing attention has been given to EMACs due to their importance in the fundamental understanding of metal–metal interactions and in potential applications such as molecular metal wires and switches. With different metal ions it was found that in some cases the M—M distances were equal, but in others rather unsymmetrical. The $[Cr_3Cl_2(dpa)_4]$, for example, is unsymmetrical in the solid state [14]. The tricobalt compound $[Co_3Cl_2(dpa)_4]$ unequivocally exists in two different and well-characterized forms, symmetrical and unsymmetrical [15]. Thus, it is important to provide some insight into the factors which are ultimately responsible for the appearance of a symmetrical vs. an unsymmetrical chain and to see if the molecular geometry could be tuned by changing the electronic effects of the axial and equatorial ligands (effect of substituents on the dpa-derivative ligands). For example, in complexes of the type $[Cr_3XY(dpa)_4]$ when X ≠ Y the Cr—Cr distances become so different that the $[Cr_3]^{6+}$ chain could be described as consisting of a diamagnetic $[Cr_2]^{4+}$ quadruply bonded unit tethered to a high spin Cr^{2+} ion at a nonbonding distance [14]. When X = Y, recent studies demonstrate that more basic ligands appear to favor symmetrical $[Cr_3]^{6+}$ chains, whereas less basic ligands favor unsymmetrical compounds. The symmetrical complexes show reversible $[Cr_3]^{7+}/[Cr_3]^{6+}$ redox couples. Thus, symmetrical $[Cr_3]^{6+}$ complexes could be used as molecular switches which can be turned on or off via an applied potential [14].

The longest structurally characterized EMAC molecules obtained up to now contain nine chromium [16] and seven cobalt atoms [17], showing significant metal–metal interactions. A new type of EMAC, in which the ligand exchange at the axial position cannot occur because the axial ligands are part of the entire ligand, has been recently reported by Cotton et al. [18].

6.4
Clusters with π-Acid Ligands: Structure and Bonding

The study of clusters with π-acid ligands is one of the areas of coordination chemistry to have attracted the most interest over the last twenty years. The large majority of these clusters are of the metal-carbonyl kind, although there are also a considerable number with NO, CNR, PR_3 and H ligands. A generalized bond treatment which embraces all types of clusters and metal–metal bonded species is not possible at present. From the earliest days of cluster chemistry, theoretical descriptions of the bonding in polyhedral clusters of transition metal elements have been divided into three main types:

1. Localized bonding treatments where edge (2-center-2-electron) or face (3-center-2-electron) localized bonds are invoked to describe the bonding in the cluster skeleton. The cluster size is small, M_n ($n = 2–4$).

2. Delocalized bonding treatments where the cluster valence electrons occupy molecular orbitals, delocalized over the entire cluster framework. The cluster size is medium, M_n ($n = 5–9$).

3. Analogy between high nuclearity clusters and bulk metals. The cluster size is large, M_n ($n > 10$).

6.4.1
Localized Bonding Treatments: 18 Valence Electron (VE) Rule

In an octahedral complex the π-acid ligands stabilise the t_{2g} orbitals (see Chapter 1). This means that there are 12 electrons from the full bonding orbitals (resulting almost entirely from the six ligands) + 6 electrons in the t_{2g} orbitals. The total number is therefore 18 electrons. Non-π-acid ligands neither stabilise the t_{2g} orbitals nor cause the anti-bonding e_g^* to have high energy. Consequently, these ligands overcome the 18-electron rule.

Many perfectly stable neutral species, such as $[Cr(CO)_6]$, $[Fe(CO)_5]$ and $[Ni(CO)_4]$, abide by the 18-electron rule. Due to this tendency, if a species or fragment does not follow the 18-electron rule it tends to yield polynuclear clusters with metal–metal bonds, and thus almost always ends up following the rule. Consider the example of the triangular clusters $[M_3(CO)_{12}]$ (M=Fe, Ru, Os): each metal provides 8 electrons, to which must be added the 24 of the 12 carbonyls, thus giving 48 electrons. If there were no other additional bond each M would have 16 electrons, that is, two short of the 18 required. However, these two electrons are "gained" by forming three M–M bonds according to the vertices of a triangle (Figure 6.7A). *In these clusters it is not possible to predict whether the carbonyls will be terminal or bridge.* Figure 6.7B shows another cluster in which, in order to follow the 18-electron rule, it is necessary to assume there is a double Os–Os bond in the cluster (assuming that each H is a hydride ion that provides two electrons).

A common geometry for M_4 clusters is the *metallatetrahedrane* structure. Figure 6.8 shows two tetrahedral clusters in which the 18-electron rule is also followed for each metal center. Given the formula $[M_4(CO)_{12}]$ (M=Co, Rd, Ir, d^9), $\Sigma VE = 60$ electrons. Each metal therefore has 15 electrons. In order to abide by the 18-

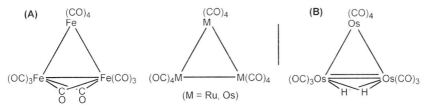

Figure 6.7 Some trinuclear clusters with localized bonding treatments: 18 VE rule (see text).

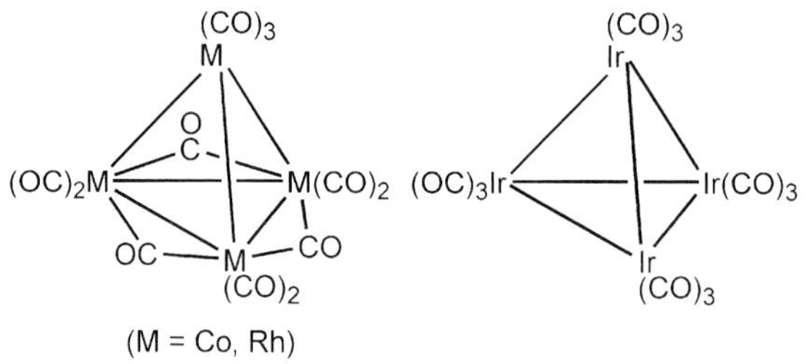

Figure 6.8 $[M_4(CO)_{12}]$ with localized bonding treatments: 18 VE rule (see text).

electron rule it must form three M—M bonds, and this is indeed what occurs. Although their empirical formula is identical, clusters of Co and Rh have bridge carbonyls whereas that of Ir only has terminal carbonyls.

6.4.2
Isolobal Analogies

This important approach was devised by R. Hoffmann [19] who calculated the bonding characteristics of cluster fragments ML_n in the frontier orbital region and compared them with the units CH_l and BH_m. In his 1982 Nobel lecture [19], Hoffmann described molecular fragments as isolobal,

> "if the number, symmetry properties, approximate energy and shape of their frontier orbitals as well as the number of electrons occupying them are similar – not identical but similar"

To illustrate this definition, it is useful to compare *fragments* of methane, CH_4, with *fragments* of an octahedrally coordinated transition metal complex, ML_6, considering only σ interactions between the metal and the ligands in this complex. In MO theory the orbitals for ML_5 complexes (C_{4v}), ML_4 (butterfly, C_{2v}) and ML_3 (pyramidal, C_{3v}) can be obtained from those of an octahedral ML_6 complex, by removing one, two or three ligands, respectively. In certain aspects, each of these complexes can therefore be considered as a *fragment* of an octahedron. The same procedure can be applied to molecules of the type AH_n. In particular, starting from tetrahedral methane, one can obtain the organic fragments pyramidal CH_3 (C_{3v}), bent CH_2 (C_{2v}) and CH ($C_{\infty v}$), by removing one, two, or three atoms of hydrogen.

This approach uses the theory of valence bonds, through the formation of the corresponding hybrids (Figures 6.9 and 6.10). Methane may be considered to use sp^3 hybrid orbitals in bonding, with eight electrons occupying bonding pairs. By similar reasoning, the metal in ML_6 uses d^2sp^3 (d_{z^2} and $d_{x^2-y^2}$) hybrids to bond to the ligands, with 12 electrons occupying bonding orbitals and six non-bonding electrons occupying d_{xy}, d_{xz} and d_{yz} orbitals.

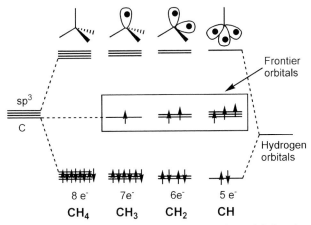

Figure 6.9 Tetrahedral fragments (from CH_4) used in isolobal analogy.

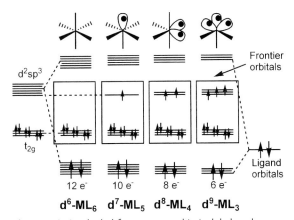

Figure 6.10 Octahedral fragments used in isolobal analogy.

Molecular fragments containing fewer ligands than the parent polyhedron can now be described. For the purpose of the analogy, these fragments will be assumed to preserve the original geometry of the remaining ligands. For example, in the 7-electron fragment CH_3, three of the sp^3 orbitals of carbon are involved in σ bonding with the hydrogens. The fourth hybrid is singly occupied and at higher energy than the σ pairs of CH_3 (Figure 6.9). The frontier orbitals of the 17-electron fragment $[Mn(CO)_5]$ are similar to those of CH_3 (Figure 6.10). The σ interactions between the ligands and Mn in this fragment may be considered to involve five of the metal d^2sp^3 hybrid orbitals. The sixth hybrid is singly occupied and at higher energy than the five σ orbitals. These orbitals are sufficiently similar to meet Hoffmann's isolobal definition.

Similarly, 6-electron CH_2 and 16-electron ML_4 are isolobal. Each of these fragments represents the parent polyhedron with a single electron occupying hybrid

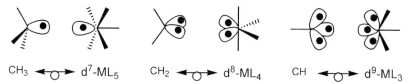

CH$_3$ ◀━○━▶ d^7-ML$_5$ CH$_2$ ◀━○━▶ d^8-ML$_4$ CH ◀━○━▶ d^9-ML$_3$

Figure 6.11 Isolobal relationships between tetrahedral and octahedral fragments.

Table 6.1 Isolobal fragments.

	Organic	Inorganic	Organometallic example	Vertices missing from parent polyhedron
Parent:	**CH$_4$**	**ML$_6$**	**Cr(CO)$_6$**	**0**
Fragments	CH$_3$	ML$_5$	Mn(CO)$_5$	1
	CH$_2$	ML$_4$	Fe(CO)$_4$	2
	CH	ML$_3$	Co(CO)$_3$	3

orbitals at otherwise vacant sites; each fragment also has two electrons less than the filled shell octet or 18-electron configuration. The isolobal character of CH$_3$ and d^7-ML$_5$ fragments as well as CH$_2$ and d^8-ML$_4$ or CH and d^9-ML$_3$ are depicted in Figure 6.11 and summarized in Table 6.1.

Isolobal fragments can be formally combined into molecules. 'Carbon' fragments tend to react to acquire 8 valence electrons; metallic fragments tend to react to acquire 18 valence electrons. For example, two CH$_3$ fragments, when linked, form ethane and two [Mn(CO)$_5$] fragments form the dimeric [Mn$_2$(CO)$_{10}$] in which there is a single Mn–Mn bond. Furthermore, organic and organometallic fragments can be intermixed; an example is H$_3$C–Mn(CO)$_5$, also a known compound.

6.4.2.1 Extensions of the Analogy
The concept of isolobal fragments can be extended to include charged species, a variety of ligands other than CO, and organometallic fragments based on structures other than octahedral:

- In general, the isolobal definition may be extended to isoelectronic fragments having the same coordination number, such as [Fe(CO)$_5$]$^+$, [Mn(CO)$_5$], [Cr(CO)$_5$]$^-$, all of them isolobal with CH$_3$.

- Gain or loss of electrons from two isolobal fragments yields isolobal fragments: [Mn(CO)$_5$] (17-electrons fragment) isolobal with CH$_3$ (7-electrons fragment); thus, [Mn(CO)$_5$]$^+$ (16-electrons fragment) isolobal with CH$_3^+$ (6-electrons fragment) and [Mn(CO)$_5$]$^-$ (18-electrons fragment) isolobal with CH$_3^-$ (8 electrons fragment).

- Other two-electron donors are treated similarly to CO: for example, PR$_3$, NCR, X$^-$. η^5-Cp is considered to occupy three coordination sites and to be a six-electron donor (as C$_5$H$_5^-$).

6.4.2.2 Examples of Isolobal Analogies Derived from Octahedral Fragments

CH_3 and d^7-ML_5 The fragments [Mn(CO)$_5$], CH_3, Cl and H are isolobal. Indeed, they dimerize spontaneously; the first three capture protons to form the corresponding acid; the possible combinations between them have been described: RMn(CO)$_5$, RCl, RCH$_3$.

Other d^7-ML_5 examples are: [M(CO)$_5$]$^-$ (M = Cr, Mo, W); [M(CO)$_5$] (M = Mn, Tc, Re); [M(CO)$_5$]$^+$ (M = Fe, Ru, Os); [Co(CO)$_5$]$^{2+}$.

The CH_3^+ fragment is also isolobal with BH$_3$ (6 electrons) and CH_3^- is isolobal with NH$_3$ (8 electrons).

CH_2 and d^8-ML_4 The following species are isolobal with respect to one another and with the fragment CH_2 (SnR$_2$): [M(CO)$_4$], (M=Fe, Ru Os); [Co(CO)(Cp)], [Re(CO)$_4$]$^-$. The CH_2 fragment is isolobal with BH$_2^-$ (BH$_3$). Many of the reactions of the CH_2 group are also given by the corresponding metal fragments. Some of the species formed are shown schematically in Figure 6.12.

CH and d^9-ML_3 The CH fragment is isolobal with [M(CO)$_3$] (M=Co, Rh, Ir); [M(CO)$_3$]$^+$ (M=Ni, Pd, Pt); [NiCp]. Furthermore, the CH fragment is isolobal with the BH$^-$ fragment.

6.4.2.3 Isolobal Analogies from Non-octahedral Geometry
The reasoning derived above from octahedral geometry is valid for any other geometry except square-planar, as in this case the molecules tend to abide by the 16-electron rule. Table 6.2 shows the metal fragments for coordination numbers 9 to 4, comparing them with the same organic fragments studied for octahedral geometry.

Figure 6.13 shows a number of rhenium organometallic molecules that can be understood in terms of the isolobal analogy of non-octahedral fragments. The

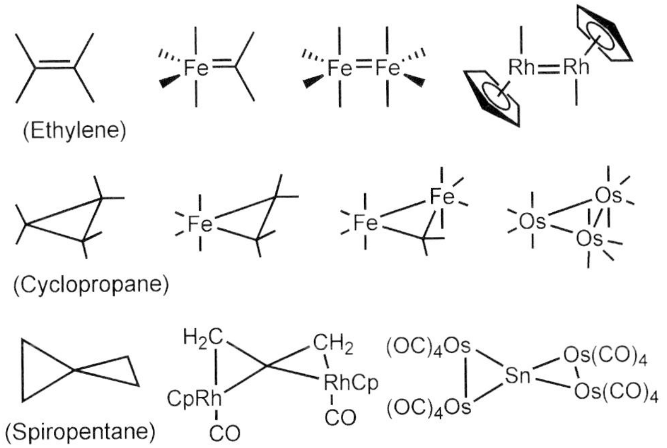

Figure 6.12 Some examples related to the isolobal analogy between CH_2 and d^8-ML_4.

Table 6.2 Isolobal metal fragments with organic fragments, for coordination number 9–4.

Organic fragment	Electrons metal fragment	c.n. = 9	c.n. = 8	c.n. = 7	c.n. = 6	c.n. = 5	c.n. = 4
CH_4	18	$d^0\text{-}ML_9$	$d^2\text{-}ML_8$	$d^4\text{-}ML_7$	$d^6\text{-}ML_6$	$d^8\text{-}ML_5$	$d^{10}\text{-}ML_4$
CH_3	17	$d^1\text{-}ML_8$	$d^3\text{-}ML_7$	$d^5\text{-}ML_6$	$d^7\text{-}ML_5$	$d^9\text{-}ML_4$	$d^{11}\text{-}ML_3$
CH_2	16	$d^2\text{-}ML_7$	$d^4\text{-}ML_6$	$d^6\text{-}ML_5$	$d^8\text{-}ML_4$	$d^{10}\text{-}ML_3$	$d^{12}\text{-}ML_2$
CH	15	$d^3\text{-}ML_6$	$d^5\text{-}ML_5$	$d^7\text{-}ML_4$	$d^9\text{-}ML_3$		

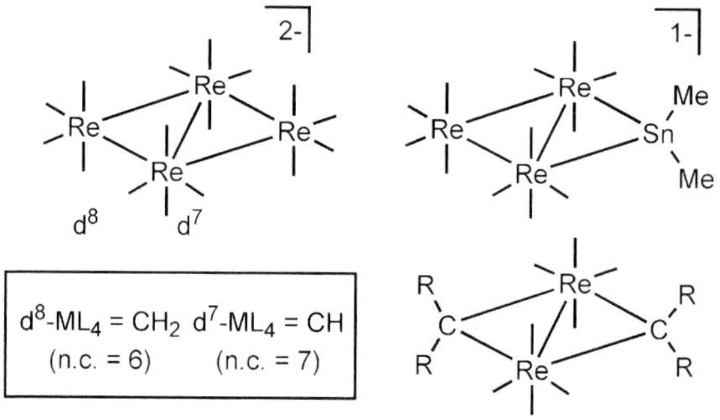

Figure 6.13 Some examples related to the isolobal analogy for six- and more than six-coordinated geometries.

octahedral Re must be Re^{-1} ($d^8\text{-}ML_4$) isolobal with CH_2, while the heptacoordinated rhenium atoms are Re^0 ($d^7\text{-}ML_4$) isolobal with CH. Thus, the central Re—Re bond is formed.

The situation resulting from square-planar coordination (16 electrons) is shown in Figure 6.14. The $[Co^0L_3]$ fragment is isolobal to CH_3; the $[Ni^0L_2]$ or $[Pt^0L_2]$ fragments (L = PR_3, CO) ($d^{10}\text{-}ML_2$) are isolobal with CH_2; $[Cu^0L_3]$ is isolobal with CH.

6.4.3
Delocalized Bonding Treatments: Structure and Bonding in Clusters of 'Medium' Nuclearity

The VE rules no longer apply in the case of clusters with octahedral frameworks. A new theory is thus necessary.

6.4.3.1 Delocalization and Deltahedra: Polyhedral Skeletal Electron Pair Theory (PSEPT or SEP)

If the vertices of the polyhedron have fewer electrons than they should, a completely delocalized electron diagram must be used. The extent of the delocalization reaches a maximum with the polyhedra known as deltahedra (solely triangular faces, which have the maximum number of edges for a given number of vertices)

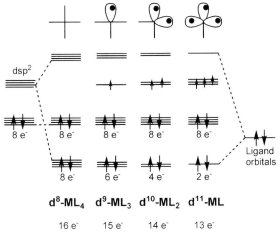

Figure 6.14 Isolobal analogies in square-planar complexes.

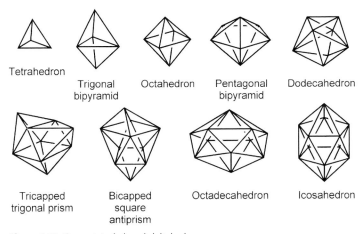

Figure 6.15 Geometrical closed deltahedra.

(Figure 6.15). These deltahedra give rise to what are termed clusters with delocalized electrons. Boron hydrides and the so-called 'neat' clusters of heavy elements, (see below), are the most representative examples of this type of cluster. The deltahedra may be *closo* (complete), *nido* (one vertex eliminated) and *arachno* (two vertices eliminated) (Figure 6.16).

In 1971, Kenneth Wade [20] recognized that the stability of the deltahedral *closo*-boranes $[B_nH_n]^{2-}$ and isoelectronic species was related to the presence in them of $2n + 2$ *skeletal electrons* (SEs), i.e., $n + 1$ *skeletal electrons pairs* (SEPs) (Table 6.3). Likewise, *nido*- and *arachno*-boranes usually possess $n + 2$ and $n + 3$ SEPs, respectively. In order to determine the number of SEs in boranes the following contributions need to be considered: (i) each B–H unit of the cluster is considered to contribute with 2 SEs; (ii) each additional H atom contributes with one SE; (iii) ion charges are included as SEs in the count.

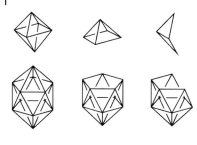

Closo   *Nido*   *Arachno*

Figure 6.16 *Closo, nido* and *arachno* deltahedra.

Table 6.3 Summary of Wade's rules.

Borane	Cluster valence electrons (CVE)	Number of skeleton electrons (SE)[a]	Number of skeleton electron pairs (SEP)[a]	Structure type	Example
$B_nH_{n+2} = (BH)_nH_2$	$4n + 2$	$2n + 2$	$n + 1$	*closo*	$[B_6H_6]^{2-}$
$B_nH_{n+4} = (BH)_nH_4$	$4n + 4$	$2n + 4$	$n + 2$	*nido*	B_5H_9
$B_nH_{n+6} = (BH)_nH_6$	$4n + 6$	$2n + 6$	$n + 3$	*arachno*	B_4H_{10}

[a] $SE = CVE - 2n = 2n + n'$(hydrogens); $SEP = SE/2$ (n = number of polyhedron corners).

It needs to be recognized that in boron hydride clusters, each B–H unit provides three atomic orbitals from the boron atom to the formation of the skeletal bonding. The two SEs furnished by the B–H unit are located on these orbitals, and correspond in fact to the (cluster)valence shell electrons (VEs or CVEs) of this unit that are not engaged in peripheral bonds. Here the number of VEs is four (three from B and one from H) of which, two are employed for the B–H bond. This allows for a more general method to count the SEs of any polyhedral borane, which consists of summing up the VEs of the whole cluster and subtracting the number of electrons engaged in peripheral bonds to hydrogen, or present as lone pairs.

Subsequent work beyond the chemistry of boranes has shown that these rules are also applicable to clusters of other main group elements and of transition metals that follow the 18-electrons rule [20, 21]. In all cases the SEs may be obtained using the general method described above, where, for transition metal clusters, the peripheral bonds will be to other species other than H (for example, to carbonyl or, in general, to π-acid ligands). Again, for a cluster with n core atoms, the number of SEPs is obtained as SEs/2, and the species will be categorized as *closo, nido* or *arachno*, for SEPs of $n + 1$, $n + 2$ and n + 3, respectively.

The number and type of atomic orbitals involved in structures with transition metals differ from the case of main group clusters. Thus, each transition element exhibits nine valence shell orbitals (one s, three p and five d), of which, three will participate in the skeletal bonding. The remaining six atomic orbitals host a total of twelve electrons (either from ligands or provided by the metal) which are con-

Table 6.4 Carbonyl cluster and SEP theory.

Framework Structure	Example	CVE	SEP = [CVE − 12n]/2	Structural type	Borane equivalent
Tetrahedron	$[Ir_4(CO)_{12}]$	60	12 e = 6 pairs	*nido*	
Trigonal bipyramid	$[Os_5(CO)_{15}]^{2-}$	72	12 e = 6 pairs	*closo*	$[B_5H_5]^{2-}$
Octahedron	$[Os_6(CO)_{18}]^{2-}$	86	14 e = 7 pairs	*closo*	$[B_6H_6]^{2-}$
Octahedron	$[Rh_6(CO)_{16}]$	86	14 e = 7 pairs	*closo*	$[B_6H_6]^{2-}$
Square pyramid	$[C \subset Ru_5(CO)_{15}]$	74	14 e = 7 pairs	*nido*	
Icosahedron	$[Sb \subset Rh_{12}(CO)_{27}]^{3-}$	170	26 e = 13 pairs	*closo*	$[B_{12}H_{12}]^{2-}$

n = number of polyhedron corners.

sidered as non-skeletal, since they are either involved in metal-ligand bonds or forming lone pairs of electrons. Therefore, the number of SEPs in a transition metal cluster can be calculated as follows; (i) the total cluster VEs are counted by adding the corresponding electrons per ligand, all the VEs from the metals in the oxidation state zero, and the charges of the cluster; (ii) to the total cluster VEs are subtracted the non skeletal electrons, i.e. twelve times the number of metals of the polyhedron.

The extension of Wade's rules to other main group, or transition metal clusters is called *polyhedral skeleton electron pair theory* (PSEPT) or else *Wade-Mingos* rules. As indicated by Mingos [21] *"such a scheme is not a replacement for accurate molecular orbital calculations on specific compounds, but it provides a simple way to understand the intriguing structural diversity shown in polynuclear molecules ..."*. In Table 6.4 some examples of carbonyl clusters are given where the rules of PSEPT are applied.

Important remarks: Interstitial atoms from the p-block provide all their valence electrons for cluster counting.

6.4.3.2 **Summary**

The general method to obtain the number of SEPs in a cluster (and therefore, determine whether it is *closo, nido* or *arachno*) consists of adding up the total number of VEs of the cluster (from skeletal atom, from their peripheral bonds to other species and from additional charges), subtracting the non skeletal electrons (2n if the cluster if formed by n main group elements, or 12n if it is formed by n transition metals) and dividing the resulting number by two. If the number of SEPs is $n + 1$, the cluster is *closo*, if it is $n + 2$, the cluster is *nido* and for $n + 3$, the species is *arachno*. This general method is summarized in Table 6.5.

A complementary method of calculation the SEPs is to calculate the SEs of each cluster-unit (vertex of the deltahedron) and multiply by the number of vertices, as was done originally by Wade for boron hydrides. From the considerations given above it is evident that the SEs of each cluster unit are $(v + x - 2)$ and $(v + x - 12)$ for main-group elements and transition elements, respectively, being v = number of the valence shell electrons of the atom; x = number the electrons from ligand(s)

Table 6.5 Summary of Wade's rules. Comparison between clusters of main-group elements vs. transition metal clusters (with π-acid ligands; 18 electron rule).

	Main group element clusters	Transition-metal clusters (with π-acid ligands)
Vertices	n	n
Valence atomic orbitals (each vertex)	4	9
Skeleton atomic orbitals (each vertex)	3	3
SEPs for *closo*	$n + 1$	$n + 1$
SEPs for *nido*	$n + 2$	$n + 2$
SEPs for *arachno*	$n + 3$	$n + 3$

Table 6.6 Selected clusters and its geometry based on the electron counting of each unit.

Formula	Division by units	SEs for each unit $(v + x - 2)$ or $(v + x - 12)$	SEPs	$(n + i)$	Type
$[Sb \subset Rh_{12}(CO)_{27}]^{3-}$	Sb 12$[Rh(CO)_2]$ 3CO	$5 + 12(9 + 4 - 12) + 6 +$ $3(\text{charge}) = 26$	13	$n + 1$	*closo*
$[C_2B_9H_{11}]^{2-}$	2(CH) 9(BH)	$(2 \times 3) + (9 \times 2) + 2$ $(\text{charge}) = 26$	13	$n + 1$	*closo*
$[Ge_9]^{4-}$	9[Ge]	$(9 \times 2) + 4 \,(\text{charge}) = 22$	11	$n + 2$	*nido*
$[Ir_4(CO)_{12}]$	4$[Ir(CO)_3]$	$4(9 + 6 - 12) = 12$	6	$n + 2$	*nido*
$[C \subset Fe_4(CO)_{12}]^{2-}$	C 4$[Fe(CO)_3]$	$4 + 4(8 + 6 - 12) + 2$ $(\text{charge}) = 14$	7	$n + 3$	*arachno*

and 2 (or 12) the number of electrons not involved in the cluster skeleton. This simple rules serve to describe easily the geometry of the vast group of compounds constituting the boron hydrides, Zintl anions (and its analogs, see Section 6.5.2) and many transition metal clusters (especially carbonyl clusters). Selected examples are shown in Table 6.6.

6.4.3.3 The Capping Principle

In the complex chemistry of transition metal cluster structures the metal atom polyhedra may be capped by further complex fragments. Mingos [21] introduced the Capping Principle to deal with clusters corresponding to capped polyhedra. He demonstrated that capping a face of a polyhedron *leads to no change* in the number of skeletal bonding MOs. This is because the frontier orbitals of the capping fragments are matched in symmetry by orbitals of the parent cluster which are already bonding. This means that the three cluster valence orbitals (deltahedron) participate in the bonding, and the anti-bonding MOs have too much energy to be able to accept electrons.

Table 6.7 Examples of the capping principle.

Cluster	CVE	14n + 2 rule for closo M_7, M_8 and M_{10}	(14n + 2) − CVE	$SEP^{[a]}$ = (CVE − 12n)/2	Number of capped fragments
$[Os_7(CO)_{21}]$	98	100	2 (1 pair)	7	1
$[Os_8(CO)_{22}]^{2-}$	110	114	4 (2 pairs)	7	2
$[Os_{10}C(CO)_{24}]^{2-}$	134	142	8 (4 pairs)	7	4

[a] SEP = 7 pairs (14 electrons) is the characteristic number for an octahedron (n + 1).

Thus, a cluster complex whose metal framework is a capped polyhedron has the same number of bonding skeleton orbitals as the non-capped polyhedron. The capping fragment only contributes to the cluster valence electron number but not to the framework electrons. Therefore the number of SEP remains unchanged in comparison with the basic polyhedron.

From the formula of a cluster it can be decided whether a capped cluster can be present or not. If the cluster valence electron (CVE) number is lower than $14n + 2$, it must be a mono- or poly-capped cluster. For each electron pair less than $14n + 2$ a capping fragment must be taken into account. This feature can be seen in the following examples (Table 6.7).

The capping rule does not permit any prediction of the relative positions of the capping fragments with respect to the basic polyhedron.

6.4.3.4 The Principle of Polyhedral Fusion (Condensed Polyhedra)

In recent years a large number of high-nuclearity metal carbonyl clusters have been synthesized, most of which have structures corresponding to the condensation of smaller tetrahedral, octahedral and trigonal prismatic cluster fragments. Mingos derived the rule that "*The total electron count in a condensed cluster is equal to the sum of the electron counts for the parent polyhedra A and B minus the electron count characteristic of the atom, pair of atoms, or face of atoms common to both polyhedra*" [21]. The number of electrons (CVE) is given for each polyhedron in Table 6.4. The electron count for various shared units are: 18 for a vertex, 34 for an edge and 48 for a triangular face (50 when one, or both, of the fused polyhedra are not 3-connected). Many examples of these condensed polyhedra and the corresponding electron counting can be found in Ref. [21].

6.4.4
Structure and Bonding in High Nuclearity Clusters

All previous concepts fail in the case of very large clusters where the metal–metal bonds require a totally delocalized description. The MO theory by Lauher is a good approach to this problem [22].

Higher nuclearity metal clusters (arbitrarily designated to contain at least 10 atoms) are of considerable current interest because of their special intermediate

Table 6.8 High-nuclearity clusters with close packing structure.

Cluster	Structure	Cluster	Structure
$[Os_{10}H_4(CO)_{24}]^{2-}$	ccp	$[Rh_{15}(CO)_{30}]^{3-}$	hcp
$[Ni_{12}H(CO)_{21}]^{3-}$	hcp	$[Pt_{26}(CO)_{32}]^{3-}$	hcp
$[Rh_{13}H_5(CO)_{24}]$	hcp	$[Rh_{28}H_xN_4(CO)_{41}]^{4-}$	ccp

positions at the diffuse boundary between monometallic species and colloidal/bulk metals. The ligand coordination to large metal cores of clusters may model ligand coordination on metal surfaces: they thereby function in an analogous fashion to homogeneous and heterogeneous metal catalysts. From a structural point of view, they tend to form closest packed structures mimicking the metallic state, giving structures similar to those in the close packing found in metals.

If a central metal atom is surrounded by 12 others of the same size in a cubic (ccp) or hexagonal close packed (hcp) structure, the first 'magic' number, 13, is realized. Forty-two atoms form a second shell around the first 12 atoms, yielding the second 'magic' number, 55. The nth shell consists of $10n^2 + 2$ atoms.

We could consider, thus, $n = 13$ as the starting point for these metal clusters. However, because little is known about naked clusters it is necessary to stabilize them with certain ligands. These ligands allow the synthesis not only of metal clusters with less than 13 atoms but also of many clusters with nuclearity different from these theoretical 'magic' numbers. Table 6.8 provides a list of some of these clusters.

In $[Rh_{13}H_5(CO)_{24}]$ or $[Rh_{13}H_3(CO)_{24}]^{2-}$ the 13 Rh atoms lie in three parallel planes displaying hexagonal closest packing. This cluster may be regarded as the smallest conceivable metal crystal with carbon monoxide chemisorbed at its surface (*cluster surface analogy*). The connection to heterogeneous catalysis is obvious. The rhodium cluster $[Rh_{28}H_xN_4(CO)_{41}]^{4-}$ shows packing similar to the compact cubic structure: three compact clusters of 6, 12 and 10 rhodium atoms in an *abc* sequence. The four nitrogen atoms occupy octahedral gaps between the three layers. Some of them have mixed structures. An example is the $[Ni_{16}Pd_{16}(CO)_{40}]^{4-}$ (Figure 6.17) cluster, which has an *abcbca* sequence of compact layers [23].

The 'magic' number 55 is achieved in the series $[M_{55}Cl_xL_{12}]$ (M = Au, Rh, Ru, Pt, Co and L = different phosphanes). They have a cubic close packed structure (ccp) with the outer geometry of a cuboctahedron.

Although the termed *encapsulated clusters* will be discussed later, it is worth mentioning here that even some large encapsulated clusters show the typical metal packed structures. Some polynitrido clusters are good examples, such as $[N_3 \subset Co_{14}(CO)_{24}]^{3-}$, $[N_4 \subset Rh_{23}(CO)_{38}]^{3-}$ and $[N_4 \subset Rh_{28}H_x(CO)_{41}]^{4-}$. The latter has a crystal structure consisting of a slightly distorted fragment of *ccp* lattice (Figure 6.18). The four interstitial nitrogen atoms are all located in octahedral cavities [24].

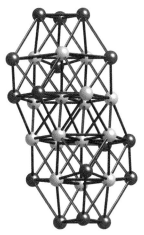

Figure 6.17 Structure of the $[Ni_{16}Pd_{16}(CO)_{40}]^{4-}$ cluster with an *abcbca* sequence of compact layers (see text).

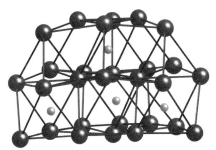

Figure 6.18 Structure of the $[N_4 \subset Rh_{28}H_x(CO)_{41}]^{4-}$ cluster consisting of a slightly distorted fragment of *ccp* lattice (see text). (small grey spheres = N atoms).

6.5
Main Group Metal Clusters

6.5.1
Alkali and Alkaline earth Metal Clusters [25]

Li atoms form various polynuclear complexes in which the Li–Li distance is less than the sum of the Van der Waals radii. A noteworthy example is that of Li_4Me_4 compounds, in which the four Li^+ ions form a tetrahedron. The bond with the four methyl groups yields a cubane-type structure. Li_{26} clusters exist in the compound the stoichiometry of which is $[Li_{13}Na_{29}Ba_{19}]$ [26]. The Li_{26} cluster is formed from a central Li_4 tetrahedron capped by 4, 6 and 4 × 3 Li atoms above its faces, edges and vertices, respectively. It can be considered as formed from four interpenetrating icosahedra.

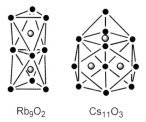

Rb_9O_2 $Cs_{11}O_3$

Figure 6.19 Scheme of $[Rb_9O_2]$ and $[Cs_{11}O_3]$ suboxides.

There is early evidence for the formation of metal-rich oxides, so-called *suboxides*, of rubidium and caesium. With alkaline earth metals several stable *subnitrides* are also known. All structures contain clusters formed from an octahedral M_6 unit with oxygen or nitrogen atoms at its center. Such M_6O or M_6N octahedra are linked via common faces and edges, respectively, resulting in discrete clusters or one- and two-dimensional structural units. These discrete and condensed clusters are bare, a fact which leads to extended inter-cluster bonding and results in electronic delocalization and metallic properties for all known compounds (color, for example). The distances between metals are similar to those found in the metals themselves.

The most well known Group 1 clusters are the $[Rb_9O_2]$ and $[Cs_{11}O_3]$ suboxides and their derivatives (Figure 6.19). The description of bonding for these clusters as $[(Rb_9O_2)^{5+} \cdot 5e^-]$ and $[(Cs_{11}O_3)^{5+} \cdot 5e^-]$ has been demonstrated. The five additional electrons of the cluster end up in the conduction band, both intra- and inter-cluster, and this gives them metallic characteristics. These suboxides form many structures with additional alkali metals in the network.

6.5.2
Metal Clusters of Groups 13, 14 and 15 [27, 28]

Some hints of these clusters are found in the older literature. The first hint was the Zintl ions (or phases), first obtained by Zintl and coworkers in 1930. The Zintl phases and anions are formed with alkali metals and the moderately electronegative (P, As, Sb, Se), or metalloidal (Ge, Ga), or even metallic (Tl, Sn, Pb) main group elements. They exhibit an extraordinary structural variety with zero-, one-, two-, or three dimensional linkages in which there is an essentially complete transfer of valence electrons from the alkali metals to the cluster. In general, discrete Zintl anions can often be crystallized from the initial Zintl phase in solution by substituting the alkali metal cations with bulky tetraalkylammonium ions or by encapsulating the latter in a cryptand such as cryptate-2.2.2 [27].

The second hint was the discovery of the long-known "bismuth monochloride" in the early 1960s. It actually has a Bi_6Cl_7 composition and contains $[(Bi_9^{5+})_2(BiCl_5^{2-})_4(Bi_2Cl_8^{2-})]$ [25]. The (Bi_9^{5+}) cations have distorted tricapped trigonal prism shapes. Various cationic bismuth clusters, $[Bi_n]^{x+}$ have been prepared since

1962, pointing out the close relationship between these species and borane clusters.

Wade's rules provide valuable guidelines regarding geometric–electronic interrelations for many types of polyhedra of main group metal clusters.

6.5.2.1 Examples

'Neat' or Bared Clusters   The term metalloid cluster was coined by Schnöckel in the context of Al (Group 13) chemistry to designate those metal clusters in which the *"number of metal–metal contacts exceeds the number of metal–ligand contacts"* and which contain *"metal atoms which participate exclusively in metal–metal interactions"* [29]. These clusters cannot be understood via a simple electron-counting rule. A qualitative description of the topology can often be obtained by comparison with modifications of the elements themselves. Gallium, with its seven known modifications, is a nearly ideal case in order to make plausible the concept that metalloid clusters are intermediate during the formation of the element.

With few exceptions isolated clusters of the heavier Group 14 elements (Si–Pb) are of the Zintl type, $[E_9]^{n-}$ (E = Si, Ge, Sn, Pb; n = 2, 3, 4). The first 'neat' phase discovered (in 1997) was the tetragonal Cs_4Ge_9 with the shape of a monocapped square antiprism. In 1998 the discovery of K_4Pb_9 was reported. The isolated $[Pb_9]^{4-}$ clusters are of two different geometries, a monocapped square antiprism (C_{4v} symmetry) and an elongated tricapped trigonal prism (D_{3h}). The dichotomy involving these two configurations is rather spread in these phases. The $[Ge_9]^{4-}$ (and Sn, Pb) are known to be fluxional on the NMR time scale with a pathway D_{3h}–C_{2v}–C_{4v} process that converts a tricapped trigonal prism into a monocapped antiprism, and vice versa.

$[Ge_9]^{4-}$ Zintl ions react with Au^+ compounds producing different anions such as $[Au_3Ge_{18}]^{5-}$, composed of two $[Ge_9]$ clusters linked by a gold triangle and $[Au_3Ge_{45}]^{9-}$ made up of three Au atoms and four polyhedral $[Ge_9]$ units, which are interconnected to nine further Ge atoms [30].

Sevov et al. [31] showed that deltahedral Zintl ions can also connect to each other, such as in the trimer $[Ge_9=Ge_9=Ge_9]^{6-}$ and in the tetramer $[Ge_9=Ge_9=Ge_9=Ge_9]^{8-}$ (Figure 6.20). The four Ge_9 clusters are quite similar and can be viewed as tricapped trigonal prisms. The alkylation of $[Ge_9]^{4-}$ yields the dialkylated (functionalized, see next section) dimers of Ge_9 clusters $[R-Ge_9-Ge_9-R]$ [32].

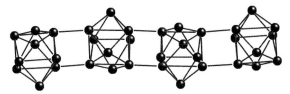

Figure 6.20 Structure of the tetramer $[Ge_9=Ge_9=Ge_9=Ge_9]^{8-}$ (see text).

In 2006, two novel examples of larger deltahedral Zintl anions were reported: the *closo*-$[Pb_{10}]^{2-}$ (bicapped square antiprism) and the icosahedron *closo*-$[Pb_{12}]^{2-}$ [33]. Other known species are non-*closo* derivatives, such as $[Sn_5]^{6-}$ and $[Pb_5]^{6-}$ (*arachno* structures). They are analogous to the cyclopentadienyl anion $C_5H_5^-$ [34].

Homoatomic Group 15 elemental clusters have a long standing history, and tetrahedral white phosphorus P_4 was the first representative discovered as early as 1669. P_4 and As_4 are typical non-metallic cages. The yellow form of the Sb is thought to be built from isolated Sb_4 cages. No structural data are, however, available for Sb_4 or Bi_4.

The elements of the Group 15 all react with alkali or alkaline earth elements to form $[E_n]^{m-}$ anions, such as many of the Zintl-type and other types. With the countercations indicated in Section 6.5.2. it is easy to crystallize new species such as $[K(2,2,2\text{-crypt})]_2Bi_4$ and $[K(2,2,2\text{-crypt})]_2Bi_2$ (with a Bi—Bi double bond). Many cluster species of Bi are cationic: $[Bi_5]^{3+}$ (D_{3h}), $[Bi_5]^+$ (C_{4v}), $[Bi_6]^{2+}$ (C_{2v}), $[Bi_8]^{2+}$ (D_{4d}), $[Bi_9]^{5+}$ (D_{3h}). They follow the Wade and Mingos rules, suggesting delocalized electron deficient clusters.

Functionalized Clusters In $[E_nR_n]^{x-}$ ($x = 0, 1, 2$) the clusters of Al, Ga, In and Tl have an equal number of cluster atoms and substituents. With few exceptions, their structures resemble those of the deltahedra boron compounds. The oxidation numbers of the elements in these clusters are +1 or deviate only slightly from this value. The first characterized examples were $[Al(C_5Me_5)]_4$, $[In(C_5Me_5)]_6$ and the icosahedral $K_2[Al_{12}{}^iBu]_{12}$. Very recently, a theoretical study has been reported on $[E_{12}Cl_{12}]^{0,2-}$ (E = B, Al, Ga) clusters, to determine which of the two structural patterns (icosahedral E_{12}-units or a cluster with an $[E_6(EX_2)_6]$), is energetically favored for the particular element. B is icosahedral; the difference between icosahedral and a metalloid cluster is insignificant in Al, and the metalloid isomer is preferred for Ga. The title of the article is very striking: Quo vadis, Wade's rules? [35].

However, the large majority of metalloid clusters reported for the two elements Al and Ga have the formula $[E_nR_{m<n}]$. The prototype is $[Al(Al_3R_3)_2]^-$ (or $[Al_7R_6]^-$) with the topology of the seven atoms in the form of vertex-sharing tetrahedral $[Al_4]$ moieties. This structure is unique for metal atom clusters and has raised the still unanswered question: Can the central 'naked' Al atom with its six directly bound neighbours be regarded as a section of the face-centered cubic (fcc) structure of solid Al . . . or is it better understood as an Al^{3+} ion stabilized by two aromatic $[Al_3R_3]^{2-}$ moieties in a sandwich type arrangement. A detailed discussion of this question can be found in Ref. [36].

More conventional examples are: (i) $[Al_{22}Br_{20} \cdot 12THF]$ which formula is better described as $[Al_{12}(AlBr_2 \cdot THF)_{10}] \cdot 2THF$ [37]. The 20 bromine atoms are arranged above the triangular faces of the icosahedron in such a way that they adopt the geometry of a distorted pentagonal dodecahedron (Figure 6.21A). (ii) $[Al_{20}Cp_8X_{10}]$ (X = Cl, Br) with icosahedral Al_{12} centers (Figure 6.21B) [38]. Structurally the central Al_{12} icosahedron consists of four Al^+ atoms, which are coordinated terminally by a halogen atom, and eight Al^0 atoms, which are coordinated exclusively

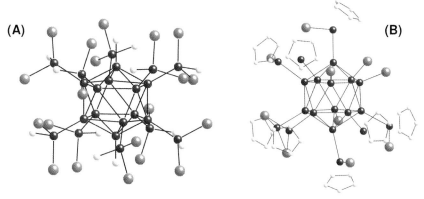

Figure 6.21 (A) [Al$_{12}$(AlBr$_2$·THF)$_{10}$]·2THF and (B) [Al$_{20}$Cp$_8$X$_{10}$] (X = Cl, Br) structures (see text). (Al = black spheres; Cl, Br = gray spheres).

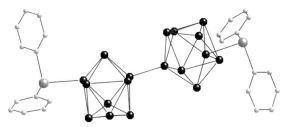

Figure 6.22 Structure of the [Ph$_2$Sb—Ge$_9$—Ge$_9$—SbPh$_2$]$^{4-}$ cluster (see text). (Sb = large gray spheres).

to an exohedral Al atom. Each of the eight exohedral Al atoms bears a Cp ligand. (iii) The [Al$_{77}$R$_{20}$]$^{2-}$ cluster is the largest aluminum cluster so far. It consists of a central Al atom, which is surrounded by a series of shells, first by 12 Al atoms arranged in the form of a distorted icosahedron, then by 44 Al atoms and, finally, by a further 20 Al^{I}R units [39]. With gallium, the largest structurally characterized cluster contains 84 Ga atoms and it is protected by the same kind and number of ligands N(SiMe$_3$)$_2$ as the Al$_{77}$ cluster [40]. This cluster exhibits many singularities, such as the Ga$_2$ unit in the center of the cluster with a very short Ga—Ga distance. This central Ga$_2$ unit is remarkable and unique in this entire field of chemistry.

A review on metalloid group 14 cluster compounds of the general formula [E$_n$R$_{m<n}$] has been published recently [41]. Peculiar and unique examples are [Ph$_2$Sb—Ge$_9$—Ge$_9$—SbPh$_2$]$^{4-}$ (Figure 6.22) synthesized from K$_4$Ge$_9$ with SbPh$_3$ [42], and the cluster [AuGe$_{18}$R$_6$]$^-$ (R=Si(SiMe$_3$)$_3$), which consist of two [Ge$_9$R$_3$]units connected by a central gold atom [43].

6.5.2.2 Metallated Zintl Clusters

The first metallated Zintl ion, a [Sn$_9^{4-}$—Pt(PPh$_3$)$_x$] complex, was detected in a reaction between *nido*-[Sn$_9$]$^{4-}$ and [Pt(PPh$_3$)$_4$] [44]. The *closo*-[Sn$_9$Cr(CO)$_3$]$^{4-}$ is another interesting example: the formation of this complex can be viewed as the insertion of a neutral $[Cr(CO)_3]$ fragment (zero electrons) into the open square face of a *nido-* $[Sn_9]^{4-}$ ion (22 electrons) producing a *closo* 22-electron deltahedral complex [45].

6.6
Clusters with Interstitial Atoms

We can distinguish three cases:

6.6.1
'Non-metallic' Atoms in d-Block Clusters

H, C, N, P and As atoms may be located in the cluster's internal cavity, giving rise to new clusters with interstitial atoms. The radius of the cavity depends essentially on the nuclearity and geometry of the cluster. In low nuclearity clusters Wade rules can be applied in several cases. Some examples are shown in Figure 6.23. An interesting review on these clusters has been reported by Housecroft [46]. Let us summarize the main aspects, considering that interstitial atoms provide *all* their valence electrons for cluster bonding.

- A good number of carbido, borido and nitrido species exist in octahedral cavities. Currently they followWade's electron counting rules. A cluster such as $[C \subset Ru_6(CO)_{17}]$ is the counterpart of $[B_6H_6]^{2-}$. *Nido* and *arachno* clusters, such as $[C \subset Fe_5(CO)_{14}]^{2-}$ (*nido*) and $[C \subset Fe_4(CO)_{12}]^{2-}$ (*arachno*) also follow Wade's rules, as well as many capped octahedral clusters. Some of the most beautiful examples are anions containing an M_{10}-cage in which four faces of an octahedron are capped. $[C \subset Os_{10}H(CO)_{24}]^-$ and $[N \subset Ru_{10}(CO)_{24}]^-$ can be highlighted [46].

- The trigonal prism: a problematic geometry? There is no trigonal prismatic borane prototype on which to base an isolobal cluster analog. However, $[C \subset Co_6(CO)_{15}]^{2-}$ and $[N \subset Co_6(CO)_{15}]^-$ possess a trigonal prismatic metal framework although they might be expected to be *arachno* clusters. $[C \subset W_6Cl_{18}]^{n-}$ ($n = 0$–3) and $[N \subset W_6Cl_{18}]^{n-}$ ($n = 1$–3) also present this trigonal prismatic structure [47]. The sodium salt when $n = 1$ represents an expansion of the NiAs structure.

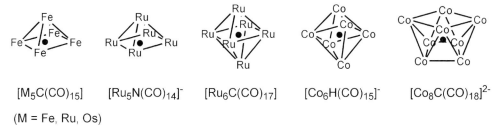

$[M_5C(CO)_{15}]$ $[Ru_5N(CO)_{14}]^-$ $[Ru_6C(CO)_{17}]$ $[Co_6H(CO)_{15}]^-$ $[Co_8C(CO)_{18}]^{2-}$

(M = Fe, Ru, Os)

Figure 6.23 Some clusters with non-metallic interstitial atoms.

High-nuclearity clusters are able to incorporate either more voluminous atoms such as Si, P, As and Sb or more than one interstitial N atom. Representative examples of the first class are $[Si \subset Co_9(CO)_{21}]^{2-}$ (monocapped square antiprism); $[P \subset Rh_9(CO)_{21}]^{2-}$ (monocapped square antiprism); $[P(As) \subset Rh_{10}(CO)_{22}]^{3-}$ (bicapped square antiprism); $[Sb \subset Rh_{12}(CO)_{27}]^{3-}$ (icosahedron). $[(N)_2 \subset Co_{10}(CO)_{19}]^{4-}$, $[(N)_3 \subset Co_{14}(CO)_{26}]^{3-}$, $[(N)_4 \subset Rh_{23}(CO)_{38}]^{3-}$ and some mixed Co–Rh clusters, such as $[(N)_2 \subset Co_{10}Rh(CO)_{21}]^{3-}$ and $[(N)_2 \subset Co_{10}Rh_2(CO)_{24}]^{2-}$ are representative examples of the second class [48].

6.6.2
d-Metal Atoms in p-Block Clusters

According to the calculations with d^n ions, only the s and p orbitals on the interstitial metal mix with appropriate cluster orbitals.

The synthesis, characterization and solution dynamics of platinum derivatives of $[Sn_9]^{4-}$, such as $[Pt \subset Sn_9H]^{3-}$, $[Pt_2 \subset Sn_{17}]^{4-}$ have been reported recently [49]. They do not obey Wade's rules; they adopt structures more akin to the subunits in alloys such as $PtSn_4$.

The highly stable empty cage $[Sn_{12}]^{2-}$, similar to $[B_{12}H_{12}]^-$, forms endohedral stannaspheres $[M \subset Sn_{12}]^-$ (M = Ti, V, Cr, Fe, Co, Ni, Cu, Y, Nb, Gd, Hf, Ta, Pt, Au). All anions can be described as $[M^+ \subset Sn_{12}^{2-}]$ [50].

The anion $[Pt \subset Pb_{12}]^{2-}$ [51], is a 26-electron polyhedron with a highly regular *closo* icosahedral structure and the anion $[Ni \subset Pb_{10}]^{2-}$ is a 22-electron polyhedron, i.e. a *closo* bicapped square antiprism centered by a Ni atom [52]. These clusters follow Wade's rules (Pt and Ni atoms are d^{10} and, thus, donate zero electrons to the cluster skeleton). The $[Pd_2 \subset Ge_{18}]^{4-}$ cluster possesses the 18-vertex deltahedron (Figure 6.24A), with a rather oblong, prolate, shape and with an inversion center. The two Pd atoms occupy the foci of the cluster and do not bring additional cluster bonding electrons. This new cluster stays intact in solution in DMF [53].

Another interesting case is the *'linear triatomic nickel filament enclosed in a dimer of nine-atom germanium clusters'*: $[K(2,2,2\text{-crypt})]_4[(Ni\text{–}Ni\text{–}Ni) \subset (Ge_9)_2]$ (Figure 6.24B) [54]. The Ge_9 unit can be described as distorted tricapped trigonal prisms. The anions remain intact in solution in DMF. The relation between the structure and electron counting is reported by the authors.

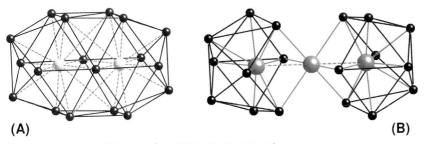

(A)   **(B)**

Figure 6.24 Structure of $[Pd_2 \subset Ge_{18}]^{4-}$ and $[(Ni\text{–}Ni\text{–}Ni) \subset (Ge_9)_2]^{4-}$ clusters (see text). (white spheres = Pd; gray spheres = Ni).

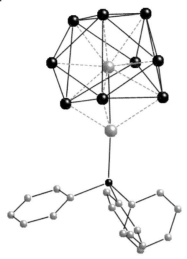

Figure 6.25 Structure of $[Pt \subset Sn_9(PtPPh_3)]^{2-}$ cluster (see text) (large gray spheres = Pt).

'Not-naked' encapsulated clusters are also known, such as $[(Ni \subset Ge_9Ni\text{-}CO]^{2-}$, $[(Ni \subset Ge_9Ni\text{-}(en)]^{3-}$, $[(Ni \subset Ge_9Ni\text{-}CCPh]^{3-}$, $[Ni \subset Sn_9Ni(CO)]^{2-}$ and $[Pt \subset Sn_9(PtPPh_3)]^{2-}$ [55, 56]. The latter (Figure 6.25) comprises an elongated tricapped Sn_9 trigonal prism with a capping $PtPPh_3$ and an interstitial Pt atom.

6.6.3
Special Cases

The $[Co_{11}Te_7(CO)_{10}]^{1-,2-}$, $[Co_{11}Te_5(CO)_{15}]^-$ clusters anions are special. They belong to the rare class of pentagonal prismatic (PP) metal clusters, with an additional metal atom lying in the interstitial cavity $[Co \subset Co_{10}]$. The metal skeleton is surrounded by a shell of a carbonyl and main group ligands. They exhibit several electrochemically reversible or quasi-reversible redox systems without structural rearrangement of the cluster skeleton. The behavior of these clusters has been termed by the authors 'electron-sponge behavior' [57].

Eichhorn reported a very peculiar interpenetrating As_{20} fullerene and Ni_{12} icosahedra in the 'onion-skin' $[As \subset Ni_{12} \subset As_{20}]^{3-}$ ion [58]. The outer layer consists of a regular As_{20} dodecahedron. Removing this dodecahedron gives a Ni_{12} icosahedron, which leaves behind only a single central arsenic atom. The $[Zn \subset (Zn_8Bi_4) \subset Bi_7]^{5-}$ ion is another special case [59]. It consists of a central zinc atom within a Zn_8Bi_4 icosahedron, capped by seven bismuth atoms.

6.7
Gold Clusters and Nanoparticles

The chemistry of gold nanoparticles exemplifies the fundamental difference between material in its bulk and in its cluster state. Whereas bulk gold is chemically inert, deposited gold nanoparticles and clusters can have a highly size-specific activity. Indeed, gold clusters and nanoparticles have received significant attention in cluster science because of their potential applications in nanotechnology. The strong relativistic effect of gold results in Au clusters exhibiting many unique properties that are different from the other coinage metals.

The discovery of unexpected catalytic properties and applications in the nano-technology of nanosized gold particles supported on substrates has rekindled extensive interest in the chemical and physical properties of gold clusters. One prerequisite for such applications is a detailed knowledge of the structures of the clusters (size and shape). Size-selected ligand-free gold clusters with diameters less than 2 nm can be routinely generated: in the last few years, a variety of synthetic strategies for metal nanostructures has been developed in aqueous or non-hydrolitic media. Using these strategies, a myriad of shape-controlled Au (and Ag) nanostructures, such as wires, rods, sheets, prisms, belts, polyhedral nanocrystals (cubes, tetrahedra, octahedra, decahedra, icosahedra etc.) have been prepared in large quantities in recent years [60]. One of the most remarkable results has been the discovery of planar gold cluster anions of up to 12 gold atoms and the 2D to 3D transition for clusters with n larger than 12 [61].

For most applications and fundamental studies, it is desirable to be able to tailor the nanoparticles by controlling not only the size and composition of the nanoparticle core but also the chemical nature of the stabilizing ligand shell. The ligand shell composition allows one to tailor chemical properties such as solubility, chemical reactivity, surface chemistry and binding affinity.

6.7.1
Naked Clusters

Most of the medium-sized gold clusters are space-filled compact structures. However, a recent study of the structures of $[Au_n]^-$ cluster anions in the medium-size range ($n = 15$–18) has shown that clusters with $n = 16$–18 possess unprecedented empty cage structures. In particular $[Au_{16}]^-$ and $[Au_{17}]^-$ cluster anions have a tetrahedral structure, confirmed by electron diffraction. $[Au_{32}]^-$, on the contrary, is a low-symmetry compact 3D structure [62].

6.7.2
Functionalized Clusters

The number of structurally characterized gold cluster compounds has risen steadily since the first investigation of the $[Au_{11}(SCN)_3(PPh_3)_7]$ cluster (1969). Other similar clusters $[Au \subset Au_{10}(PMe_2Ph)_7X_3]$ (X = halogen, pseudohalogen), showing an approximate centered bicapped square antiprism, were later reported [63a]. All these types have a *central gold* surrounded by AuL units at a distance 2.6–2.9 Å, compatible with metal–metal bonding not only between the central gold atoms and the surrounding gold atoms but also between the surrounding atoms themselves (Figure 6.26A). The cations $[Au_{13}Cl_2(PR_3)_{10}]^{3+}$ and $[Au_9M_4Cl_4(PR_3)_8]^+$ (M = Cu, Ag, Au) contain centered icosahedral metal frameworks. A gold atom sits at

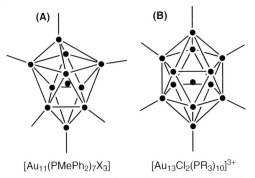

(A) $[Au_{11}(PMePh_2)_7X_3]$

(B) $[Au_{13}Cl_2(PR_3)_{10}]^{3+}$

Figure 6.26 Structures of $[Au \subset Au_{10}(PMe_2Ph)_7X_3]$ and $[Au \subset Au_{12}Cl_2(PR_3)_{10}]^{3+}$ clusters (see text).

the interstitial site. The other twelve metallic atoms have a terminal halide or phosphane ligands (Figure 6.26B) [63b].

Due to their quantum size, Au_{55} clusters are the most promising particles for application in future nanoelectronics. Furthermore, $[Au_{55}Cl_6(PPh_3)_{12}]$ can be incorporated within mesoporous silica. Several derivatives can be synthesized by ligand exchange reactions, which allow one to vary the solubility in different solvents and the cluster diameter. The reaction of $[Au_{55}Cl_6(PPh_3)_{12}]$ with several thiols yields interesting thiolate monolayer-protected Au_{75} clusters [64]. Many gold nanoparticles stabilized by thiolates have been synthesized and their electrical and optical properties have been investigated owing to their potential application to nanotechnology.

Ligand-exchange of phosphane-stabilized precursor particles with functionalized thiols, provides a convenient and general approach for the rapid preparation of large families of thiol-stabilized, subnanometer (d_{core} ~0.8 nm) particles [65]. Subnanometer-sized gold clusters protected by thiolates, referred to as '*thiolate gold clusters*', and '*thiolate monolayer-protected Au clusters*' (Au-MPC), exhibit unique physicochemical properties arising from the small core size and heavy ligation by the thiolate.

6.7.3
Doping Golden Clusters

The large empty space inside the cages of naked clusters suggested that they could be doped with a foreign atom to produce a new class of endohedral gold cages, analogous to endohedral fullerenes. $M \subset Au_{12}$ (M = V, Nb, Ta, Mo, W, Mo, Re) (icosahedral clusters) were the first to be confirmed experimentally. It is interesting that whereas the hollow cage structure of $[Au_{12}]^-$ as well as its neutral counterpart is unstable, the icosahedral core/shell clusters are highly stable. $[Cu \subset Au_{16}]^-$ and $[Cu \subset Au_{17}]^-$ have also been reported recently [66]. The Au_{20} cage is larger and can accommodate a small cluster. Similarly to the Au_{12} cage, the Au_{20} cage is unstable, but can be stabilized with a highly stable cluster such as $[Co_{13}]^-$ or $[Mn_{13}]^-$ with the same I_h symmetry, $[M_{13} \subset Au_{20}]^-$ [67].

6.7.4
Non-gold Clusters

Recently, nanosized clusters of one or more metals other than gold have attracted much attention because of their properties, as catalysts, chemical sensors, and devices. In a recent article Dyson and McIndoe postulated that *"part of the renaissance in molecular cluster chemistry has been driven by the development of nanotechnology . . . that is, catalysis, material science, but also in diagnostics and medicine"* [68].

Some of the known catalytic applications are: CO oxidation by molecular oxygen and the treatment of groundwater polluted with nitrate. The pollution of groundwater with nitrate is a widespread problem, because excess intake of nitrate is harmful for humans. Some Cu–Pd clusters with active carbon are effective as catalysts in the hydrogenation of nitrate in water, giving nitrite, nitrogen or ammonia [69].

6.7.4.1 **Clusters and Hydrogen Storage**
One of the most extensively studied methods for hydrogen storage is to keep the gas in a 'hydride' form. In this approach an alloy absorbs and holds a large amount of hydrogen by chemically bonding with the gas to form metal hydride compounds. But although a hydrogen-storage alloy can absorb and release hydrogen without compromising its own structure, heating is required to promote the release, and this process requires energy. The alloys studied so far usually require a temperature of about 300 °C to provide hydrogen at 1 atmosphere pressure.

Palladium is well-known for its remarkable capacity for hydrogen absorption, hydrogenation and dehydrogenation reactions, etc. Most of these applications are related to the adsorption of hydrogen on Pd *nanoparticles*. Pd is a face-centered cubic (fcc) noble metal, and Pd nanoparticles may take three different shapes when fast nucleation and growth are involved: octahedra, decahedra and icosahedra. A recent simulation revealed that Pd icosahedra can absorb a larger quantity of H_2 than their cubic analogs. Pd icosahedral nanoparticles can be easily obtained in high yields in the presence of citric acid [70]. The size of the nanoparticles can also be controlled by adjusting the reaction time and temperature [70].

Weller et al. [71] reported that hydrogen may be stored and released using molecular clusters in a process that is easily controlled by a simple chemical reaction or by an electric current, without a large input of energy. Their system is based on an organometallic compound that contains a core of six rhodium atoms, as part of a complex that also includes 12 hydrogen atoms. This cluster absorbs two H_2 molecules to produce a compound holding 16 hydrogen atoms. The absorption process takes 10 min at room temperature under 1 atm of hydrogen. The absorbed hydrogen molecules are retained at room temperature for weeks without any external hydrogen pressure (under an inert atmosphere of argon). The authors have found that hydrogen release can be accelerated simply by changing the cluster's oxidation state, by adding a reducing agent or electrochemically. Unfortunately, the hydrogen-storage capacity, expressed as the ratio of the mass of releasable hydrogen to that of the storage system is only 0.1%. For practical applications, the US Department of Energy wants a capacity of 6% by 2010. Nevertheless, this unique cluster provides a well-defined molecular model and a worthwhile strategy for the development of hydrogen storage materials with high efficiency and convenience [72].

References

1 Cotton, F.A., Murillo, C.A., Walton R.A. (Eds.) (2005), *Multiple Bonds Between Metal Atoms*, 3rd Edn., Springer New York.

2 Cotton, F.A., Dalal, N.S., Huang, P., Ibragimov, S.A., Murillo, C.A., Piccoli, P.M.B., Ramsey, C.M., Schultz, A.J., Wang, X., Zhao, Q. Better Understanding of the Species with the Shortest Re_2^{6+} Bonds and Related Re_2^{7+} Species with Tetraguanidinate Paddlewheel Structures, *Inorg. Chem.* 2007, *46*, 1718.

3 Berry, J.F., Bothe, E., Cotton, F.A., Ibragimov, S.A., Murillo, C.A., Villagran, D., Wang, X., Metal-Metal Bonding in Mixed Valence Ni_2^{5+} Complexes and Spectroscopic Evidence for a Ni_2^{6+} Species, *Inorg. Chem.* 2006, *45*, 4396.

4 (a) Nguyen, T., Sutton, A.D., Brynda, M., Fettinger, J.C., Long, G.J., Power, P.P. Synthesis of a stable compound with fivefold bonding between two chromium(I) centers, *Science*, 2005, *310*, 844. (b) Wolf, R., Ni, C., Nguyen, T., Brynda, M., Long, G.J., Sutton, A.D., Fischer, R.C., Fettinger, J.C., Hellman, M., Pu, L., Power, P.P. Substituent Effects in Formally ArCrCrAr compounds (Ar = Terphenyl) and Related Species, *Inorg. Chem.* 2007, *46*, 11277.

5 Wu, X., Lu, X. Dimetalloendofullerene $U_2@C_{60}$ Has a U-U Multiple Bond Consisting of Sixfold One-Electron-Two-Center Bonds, *J. Am. Chem. Soc.* 2007, *129*, 2171.

6 (a) Chisholm, M.H., Macintosh, A.M. Linking multiple bonds between metal atoms: Clusters, dimers of "dimers", and higher ordered assemblies, *Chem. Rev.* 2005, *105*, 2949; (b) Cotton, F.A., Li, Z., Liu, C.Y., Murillo, C.A., Zhao, Q. Transition from a Nonbonding to a Bonding Interaction in a Tetranuclear $[Mo_2]_2(\mu\text{-}OR)_4$ Cluster, *Inorg. Chem.* 2006, *45*, 6387.

7 Toan, T., Teo, B.K., Ferguson, J.A., Meyer, T.J., Dahl, L.F. Electrochemical Synthesis and Structure of the Tetrameric Cyclopentadienyliron Sulfide Dication, $[Fe_4(\eta^5\text{-}C_5H_5)_4(\mu_3\text{-}S)_4]^{2+}$: a Metal Cluster Bonding Description of the Electrochemically Reversible $[Fe_4(\eta^5\text{-}C_5H_5)_4(\mu_3\text{-}S)_4]^n$ System ($n = -1$ to $+3$), *J. Am. Chem. Soc.* 1977, *99*, 408.

8 Welch, E.J., Crawford, N.R.M., Bergman, R.G., Long, J.R. New Routes to Transition Metal-Carbido Species: Synthesis and Characterization of the Carbon-Centered Trigonal Prismatic Clusters $[W_6CCl_{18}]^{n-}$ ($n = 1, 2, 3$), *J. Am. Chem. Soc.* 2003, *125*, 11464.

9 Jongen, L., Mudring, A-V., Meyer, G. The Molecular Solid $Sc_{24}C_{10}I_{30}$: A Truncated, Hollow T4 Supertetrahedron of Iodine Filled with a T3 Supertetrahedron of Scandium That Encapsulates the Adamantoid Cluster Sc_4C_{10}, *Angew. Chem. Int. Ed.* 2006, *45*, 1886.

10 Cotton, F.A., Murillo, C.A., Yu, R., Zhao, Q. A Rare and Highly Oxidized $Mo_2^{5.5+}$ Unit Stabilized by Oxo Anions and Supported by Formamidinate Bridges, *Inorg. Chem.* 2006, *45*, 9046.

11 Cotton, F.A., Jin, J-Y., Li, Z., Liu, C.Y., Murillo, C.A. A deliberate approach for the syntheses of heterometallic supramolecules containing dimolybdenum Mo_2^{4+} species coordinated to other metal units, *Dalton Trans.* 2007, 2328.

12 Ke, Y., Collins, D.J., Zhou, H-C. Synthesis and Structure of Cuboctahedral and Anticuboctahedral Cages Containing 12 Quadruply Bonded Dimolybdenum Units, *Inorg. Chem.* 2005, *44*, 4154.

13 Aduldecha, S., Hathaway, B. Crystal structure and electronic properties of tetrakis[μ_3-bis(2-pyridyl)amido]dichlorotri nickel(II)–water–acetone (1/0.23/0.5), *Dalton Trans.* 1991, 4, 993. In fact the first report on [Ni$_3$Cl$_2$(dpa)4] was published in 1968 by Hurley, T.J. Robinson, M.A. Nickel(II)-2,2′-dipyridylamine system. I. Synthesis and stereochemistry of the complexes, *Inorg. Chem.* 1968, 7, 33.

14 Berry, J.F., Cotton, F.A., Lu, T., Murillo, C.A., Roberts, B.K., Wang, X. Molecular and Electronic Structures by Design: Tuning Symmetrical and Unsymmetrical Linear Trichromium Chains, *J. Am. Chem. Soc.* 2004, 126, 7082.

15 Clérac, R., Cotton, F.A., Daniels, L.M., Dunbar, K.R., Kirschbaum, K., Murillo, C.A., Pinkerton, A.A., Schultz, A.J., Wang, X. Linear Tricobalt Compounds with Di(2-pyridyl)amide (dpa) Ligands: Temperature Dependence of the Structural and Magnetic Properties of Symmetrical and Unsymmetrical Forms of Co$_3$(dpa)$_4$Cl$_2$ in the Solid State, *J. Am. Chem. Soc.* 2000, 122, 6226.

16 Ismayilov, R.H., Wang, W.-Z., Wang, R-R., Yeh, C-Y., Lee, G-H., Peng, S-M. Four quadruple metal–metal bonds lined up: linear nonachromium(II) metal string complexes, *Chem. Comm.* 2007, 1121.

17 Wang, W-Z., Ismayilov, R.H., Lee, G-H., Liu, I.P-C., Yeh, C-Y., Peng. S-M. The nano-scale molecule with the longest delocalized metal–metal bonds: linear heptacobalt(II) metal string complexes [Co$_7$(μ_7-L)$_4$X$_2$], *Dalton Trans.* 2007, 830.

18 Cotton, F.A., Chao, H., Murillo, C.A., Wang, Q. A trimetal chain cocooned by two heptadentate polypyridylamide ligands, *Dalton Trans.* 2006, 5416.

19 Hoffmann, R. Building Bridges Between Inorganic and Organic Chemistry (Nobel Lecture), *Angew. Chem. Int. Ed. Eng.* 1982, 21, 711.

20 Wade, K. Sturctural and Bonding Pattern in Cluster Chemistry *Adv. Inorg. Chem. Radiochem.* 1976, 18, 1.

21 (a) Mingos, D.M.P. Polyhedral Skeletal Electron Pair Approach, *Acc. Chem. Res.* 1984, 17, 311; (b) Manson, R., Thomas, K.M., Mingos, D.M.P. Stereochemistry of

Octadecacarbonylhexaosmium(0). A Novel Hexanuclear Complex Based on a Bicapped Tetrahedron of Metal Atoms, *J. Am. Chem. Soc.* 1973, 95, 3802; (c) Mingos, D.M.P. General theory for cluster and ring compounds of the main group and transition elements *Nature, Phys. Sci.* 1972, 236 99, (d) Rudolph, R. W. Boranes And Heteroboranes – Paradigm For Electron Requirements Of Clusters, *Acc. Chem. Res.* 1976, 9, 446; (e) Mingos, D.M.P., Johnston, R.L. Theoretical models of cluster bonding *Struct. Bond.*, 1987, 68, 31–82.

22 Lauher, J.W. The bonding capabilites of transition metal clusters, *J. Am. Chem. Soc.* 1978, 100, 5305 and Bonding capabilites of transition metal clusters. 2. Relationship to bulk metals, *J. Am. Chem. Soc.* 1979, 101, 2604.

23 Femoni, C., Iapalucci, M.C., Longoni, G., Svensson, P.H., Wolowska, J. Homoleptic Carbonyl Ni-Pd Clusters: Synthesis of [Ni$_{16}$Pd$_{16}$(CO)$_{40}$]$^{4-}$ and [Ni$_{26}$Pd$_{20}$(CO)$_{54}$]$^{6-}$ and Structural Characterization of [NnBu$_4$]$_4$[Ni$_{16}$Pd16(CO)$_{40}$], *Angew. Chem. Int. Ed.* 2000, 39, 1635.

24 Fumagalli, A., Martinengo, S., Bernasconi, G., Ciani, G., Proserpio, D.M., Sironi, A. [Rh$_{28}$N$_4$(CO)$_{41}$H$_x$]$^{4-}$, a Massive Carbonyl Cluster with Four Interstitial Nitrogen Atoms, *J. Am. Chem. Soc.* 1997, 119, 1450.

25 Driess, M., Nöth. H. (Eds.) (2004), *Molecular Clusters of the Main Group Elements*, Wiley-VCH, Weinheim.

26 Smetana, V., Babizhetskyy, V., Vajenine, G.V., Simon, A. Li$_{26}$ Clusters in the Compound Li$_{13}$Na$_{29}$Ba$_{19}$, *Angew. Chem. Int. Ed.* 2006, 45, 6051.

27 Corbett, J.D. Polyanionic Clusters and Networks of the Early p-Element Metals in the Solid State: Beyond the Zintl Boundary, *Angew. Chem. Int. Ed.* 2000, 39, 670.

28 Corbett, J.D. Diverse Naked Clusters of the Heavy Main-Group Elements. Electronic Regularities and Analogies. *Struct. Bond.*, 1997, 87, 157

29 Schnöckel, H. Metalloid Al- and Ga-clusters: a novel dimension in organometallic chemistry linking the molecular and the solid-state areas? *Dalton Trans.* 2005, 3131.

30 Spiekermann, A., Hoffmann, S.D., Fassler, T.F., Krossing, I., Preiss, U. $[Au_3Ge_{45}]^{9-}$-A Binary Anion Containing a $\{Ge_{45}\}$ Cluster, *Angew. Chem. Int. Ed,* 2007, *46,* 5310.

31 Ugrinov, A., Sevov, S.C. $[Ge_9=Ge_9=Ge_9=Ge_9]^{8-}$: A Linear Tetramer of Nine-Atom Germanium Clusters, a Nanorod, *Inorg. Chem.* 2003, *42,* 5789.

32 Hull, M.W., Ugrinov, A., Petrov, I., Sevov, S.C. Alkylation of Deltahedral Zintl Clusters: Synthesis of $[R\text{-}Ge_9\text{-}Ge_9\text{-}R]^{4-}$ ($R = {}^tBu, {}^sBu, {}^nBu, {}^tAm$) and Structure of $[{}^tBu\text{-}Ge_9\text{-}Ge_9\text{-}{}^tBu]^{4-}$, *Inorg. Chem.* 2007, *46,* 2704.

33 (a) Spiekermann, A., Hoffmann, S.D., Fässler, T.F. The Zintl Ion $[Pb_{10}]^{2-}$: A Rare Example of a Homoatomic closo Cluster, *Angew. Chem. Int. Ed.* 2006, *45,* 3459; (b) Esenturk, E.N., Fettinger, J., Eichhorn, B. The Pb_{12}^{2-} and Pb_{10}^{2-} Zintl Ions and the $M@Pb_{12}^{2-}$ and $M@Pb_{10}^{2-}$ Cluster Series Where M=Ni, Pd, Pt, *J. Am. Chem. Soc.* 2006, *128,* 9178.

34 (a) Todorov, Y., Sevov, S.C. Heavy-Metal Aromatic Rings: Cyclopentadienyl Anion Analogues Sn_5^{6-} and Pb_5^{6-} in the Zintl Phases Na_8BaPb_6, Na_8BaSn_6, and Na_8EuSn_6, *Inorg. Chem.* 2004, *43,* 6490; (b) Todorov, Y., Sevov, S.C. Heavy-Metal Aromatic and Conjugated Species: Rings, Oligomers, and Chains of Tin in $Li_{9-x}EuSn_{6+x}$, $Li_{9-x}CaSn_{6+x}$, $Li_5Ca_7Sn_{11}$, $Li_6Eu_5Sn_9$, $LiMgEu_2Sn_3$, and $LiMgSr_2Sn_3$, *Inorg. Chem.* 2005, *44,* 5361.

35 Koch, K., Burgert, R., Schnockel, H. From Icosahedral Boron Subhalides to Octahedral Metalloid Aluminum and Gallium Analogues: Quo vadis, Wade's Rules?, *Angew. Chem. Int. Ed.* 2007, *46,* 5795.

36 Yang, P., Koppe, R., Duan, T., Hartig, J., Hadiprono, G., Pilawa, B., Keilhauer, I., Schnockel, H. $[Al(Al_3R_3)_2]$: Prototype of a Metalloid Al Cluster or a Sandwich-Stabilized Al Atom?, *Angew. Chem. Int. Ed,* 2007, *46,* 3579.

37 Klemp, C., Köppe, R., Weckert, E., Schnöckel, H. $Al_{22}Br_{20}\cdot12$ THF: The First Polyhedral Aluminum Subhalide – A Step on the Path to a New Modification of Aluminum?, *Angew. Chem. Int. Ed.* 1999, *38,* 1740.

38 Vollet, J., Burgert, R., Schöckel, H. $Al_{20}X_{10}$ (X=Cl, Br): Snapshots of the Formation of Metalloid Clusters from Polyhedral Al_nX_m Molecules?, *Angew. Chem. Int. Ed.* 2005, *44,* 6956.

39 Ecker, A., Weckert, E., Schnöckel, H. Synthesis and structural characterization of an Al_{77} cluster, *Nature,* 1997, *387,* 379.

40 Schnepf, A., Schnöckel, H. Synthesis and Structure of a $Ga_{84}R_{20}^{4-}$ Cluster – A Link between Metalloid Clusters and Fullerenes?, *Angew. Chem. Int. Ed.* 2001, *40,* 711.

41 Schnepf, A. Metalloid group 14 cluster compounds: An introduction and perspectives to this novel group of cluster compounds, *Chem. Soc. Rev.* 2007, *36,* 745.

42 Ugrinov, A., Sevov, S.C. Derivatization of Deltahedral Zintl Ions by Nucleophilic Addition: $[Ph\text{-}Ge_9\text{-}SbPh_2]^{2-}$ and $[Ph_2Sb\text{-}Ge_9\text{-}Ge_9\text{-}SbPh_2]^{4-}$, *J. Am. Chem. Soc.* 2003, *125,* 14059.

43 Schenk, C., Schnepf, A. $[AuGe_{18}\{Si(SiMe_3)_3\}_6]^-$: A Soluble Au-Ge Cluster on the Way to a Molecular Cable? *Angew. Chem. Int. Ed,* 2007, *46,* 5314.

44 Teixidor, F., Luetkens, M.L., Rudolph, R.W. Transition-Metal Insertion into Naked Metal Cluster Polyanions, *J. Am. Chem. Soc.* 1983, *105,* 149.

45 Eichhorn, B.W., Haushalter, R.C., Pennington, W.T. Synthesis and Structure of closo -$Sn_9Cr(Co)_3^{4-}$: The First Member in a New Class of Polyhedral Clusters, *J. Am. Chem. Soc.* 1988, *110,* 8704.

46 Housecroft. C.E. Clusters with Interstitial Atoms from the p-Block: How do Wade's Rules Handle Them? *Struct. Bond.,* 1997, *87,* 137.

47 Welch, E.J., Yu, C.L., Crawford, N.R.M., Long, J.R. Synthesis and Characterization of the Nitrogen-Centered Trigonal Prismatic Clusters $[W_6NCl_{18}]^{n-}$ ($n = 1\text{–}3$), *Angew. Chem. Int. Ed.* 2005, *44,* 2549.

48 Costa, M., Pergola, R.D., Fumagalli, A., Laschi, F., Losi, S., Macchi, P., Sironi, A., Zanello, P. Mixed Co-Rh Nitrido-Encapsulated Carbonyl Clusters. Synthesis, Solid-State Structure, and Electrochemical/ EPR Characterization of the Anions $[Co_{10}Rh(N)_2(CO)_{21}]^{3-}$, $[Co_{10}Rh_2(N)_2(CO)_{24}]^{2-}$, and $[Co_{11}Rh(N)_2(CO)_{24}]^{2-}$, *Inorg. Chem.* 2007, *46*, 552.

49 Kesanli, B., Halsig, J.E., Zavalij, P., Fettinger, J.C., Lam, Y-F., Eichhorn, B. W. Cluster Growth and Fragmentation in the Highly Fluxional Platinum Derivatives of Sn_9^{4-}: Synthesis, Characterization, and Solution Dynamics of $Pt_2@Sn_{17}^{4-}$ and $Pt@Sn_9H^{3-}$, *J. Am. Chem. Soc.* 2007, *129*, 4567

50 Cui, L-F., Huang, X., Wang, L-M., Li, J., Wang, L-S. Endohedral Stannaspherenes $M@Sn_{12}^-$: A Rich Class of Stable Molecular Cage Clusters, *Angew. Chem. Int. Ed*, 2007, *46*, 742.

51 Esenturk, E.N., Fettinger, J., Lam, Y-F., Eichhorn, B. $[Pt@Pb_{12}]^{2-}$, *Angew. Chem. Int. Ed.* 2004, *43*, 2132.

52 Esenturk, E.N., Fettinger, J., Eichhorn, B. The *closo*-Pb_{10}^{2-} Zintl ion in the $[Ni@Pb_{10}]^{2-}$ cluster, *Chem. Comm.* 2005, 247.

53 Goicoechea, J.M., Sevov, S.C. $[(Pd-Pd)@Ge_{18}]^{4-}$: A Palladium Dimer Inside the Largest Single-Cage Deltahedron, *J. Am. Chem. Soc.* 2005, *127*, 7676.

54 Goicoechea, J.M., Sevov, S.C. $[(Ni-Ni-Ni)@(Ge_9)_2]^{4-}$: A Linear Triatomic Nickel Filament Enclosed in a Dimer of Nine-Atom Germanium Clusters, *Angew. Chem. Int. Ed. Eng.* 2005, *44*, 4026.

55 Goicoechea, J.M., Sevov, S.C. Deltahedral Germanium Clusters: Insertion of Transition-Metal Atoms and Addition of Organometallic Fragments, *J. Am. Chem. Soc.* 2006, *128*, 4155.

56 Kesanli, B., Fettinger, J., Gardner, D.R., Eichhorn, B. The $[Sn_9Pt_2(PPh_3)]^{2-}$ and $[Sn_9Ni_2(CO)]^{3-}$ Complexes: Two Markedly Different Sn_9M_2L Transition Metal Zintl Ion Clusters and Their Dynamic Behavior, *J. Am. Chem. Soc.* 2002, *124*, 4779.

57 Cador, O., Cattey, H., Halet, J-F., Meier, W., Mugnier, Y., Wachter, J., Saillard, J-Y., Zouchoune, B., Zabel, M. Electron-Sponge Behavior and Electronic Structures in Cobalt-Centered Pentagonal Prismatic $Co_{11}Te_7(CO)_{10}$ and $Co_{11}Te_5(CO)_{15}$ Cluster Anions, *Inorg. Chem.* 2007, *46*, 501.

58 Moses, M.J., Fettinger, J.C., Eichhorn, B.W. Interpenetrating As_{20} fullerene and Ni_{12} icosahedra in the onion-skin $[As@Ni_{12}@As_{20}]^{3-}$ ion, *Science*, 2003, *300*, 778.

59 Goicoechea, J.M., Sevov, S.C. $[Zn_9Bi_{11}]^{5-}$: A Ligand-Free Intermetalloid Cluster, *Angew. Chem. Int. Ed.* 2006, *45*, 5147.

60 Li, C., Shuford, K.L., Park, Q-H, Cai, W., Li, Y., Lee, E.J., Cho, S.O. High-Yield Synthesis of Single-Crystalline Gold Nano-octahedra, *Angew. Chem. Int. Ed,* 2007, *46*, 3264.

61 Häkkinen, H., Yoon, B., Landman, U., Li, X., Zhai, H-J., Wang, L-S. On the electronic and atomic structures of small Au_N^- (N = 4–14) clusters: A photoelectron spectroscopy and density-functional study, *J. Phys. Chem. A*, 2003, *107*, 6168.

62 (a) Xing, X., Yoon, B., Landman, U., Parks, J.E. Structural evolution of Au nanoclusters: From planar to cage to tubular motifs, *Phys. Rev. B*, 2006, *74*, 165423; (b) Ji, M., Gu, X., Li, X., Gong, X., Li, J., Wang, L-S. Experimental and Theoretical Investigation of the Electronic and Geometrical Structures of the Au_{32} Cluster, *Angew. Chem. Int. Ed.* 2005, *44*, 7119.

63 (a) Cariati, F., Naldini, L., Trianioneoptakis (triarylphosphine)undecagold cluster compounds, *Inorg. Chim. Acta*, 1971, *5*, 172; (b) Briant, C.E., Theobald, B.R.C., White, J.W., Bell, L.K., Mingos, D.M.P., Welch, A.J. Synthesis and *X*-ray structural characterization of the centered icosahedral gold cluster compound $[Au_{13}(PMe_2Ph)_{10}Cl_2](PF_6)_3$; the realization of a theoretical prediction, *J. Chem. Soc., Chem. Commun.* 1981, 201.

64 Balasubramanian, R., Guo, R., Mills, A.J., Murray, R.W. Reaction of $Au_{55}(PPh_3)_{12}Cl_6$ with Thiols Yields Thiolate Monolayer Protected Au_{75} Clusters, *J. Am. Chem. Soc.* 2005, *127*, 8126.

65 Shichibu, Y., Negishi, Y., Tsukuda, T., Teranishi, T. Large-Scale Synthesis of Thiolated Au_{25} Clusters via Ligand Exchange Reactions of Phosphine-Stabilized Au_{11} Clusters, *J. Am. Chem. Soc.* 2005, *127*, 13464.

66 Wang, L-M., Bulusu, S., Zhai, H-J., Zeng, X-C., Wang, L-S. Doping Golden Buckyballs: $Cu@Au_{16}^-$ and $Cu@Au_{17}^-$ Cluster Anions, *Angew. Chem. Int. Ed.* 2007, *46*, 2915.

67 Wang, J., Bai, J., Jellinek, J., Zeng, X.C. Gold-Coated Transition-Metal Anion [$Mn_{13}@Au_{20}$]$^-$ with Ultrahigh Magnetic Moment, *J. Am. Chem. Soc.* 2007, *129*, 4110.

68 Dyson, P.J., McIndoe, J.S. Hydrogen Sponge? A Heteronuclear Cluster That Absorbs Large Quantities of Hydrogen, *Angew. Chem. Int. Ed.* 2005, *44*, 5772.

69 (a) Sakamoto, Y., Nakata, K., Kamiya, Y., Okuhara, T. Cu–Pd bimetallic cluster/ AC as a novel catalyst for the reduction of nitrate to nitrite, *Chem. Lett.* 2004, *33*, 908; (b) Sakamoto, Y., Nakamura, K., Kushibiki, R., Kamiya, Y., Okuhara, T. A two-stage catalytic process with Cu-Pd cluster/active carbon and Pd/β-zeolite for removal of nitrate in water, *Chem. Lett.* 2005, *34*, 1510.

70 Xiong, Y., McLellan, J., Yin, Y., Xia, Y. Synthesis of Palladium Icosahedra with Twinned Structure by Blocking Oxidative Etching with Citric Acid or Citrate Ions, *Angew. Chem. Int. Ed.* 2007, *46*, 790.

71 Brayshaw, S.K., Green, J.C., Hazari, N., McIndoe, J.S., Marken, F., Raithby, P.R., Weller, A.S. Storing and Releasing Hydrogen with a Redox Switch, *Angew. Chem. Int. Ed.* 2006, *45*, 6005.

72 Takimoto, M., Hou, Z. Hydrogen at the flick of a switch, *Nature*, 2006, *443*, 400.

7
Thermodynamic and Non-redox Kinetic Factors in Coordination Compounds

Thermodynamic parameters indicate whether a complex is *stable* or not; kinetic parameters indicate the rate of a reaction, from which it is possible to determine the mechanism of the reaction. If the equilibrium constant of a complex is very high, the complex is termed *stable*; if the equilibrium constant is low it is termed *unstable*. On the other hand, a stable complex in solution may reach its equilibrium with its components at different rates: the complex will be named *labile* or *inert* depending on this rate. The stability term is, thus, strictly thermodynamic, that is, the *thermodynamic stability* is expressed by *positive free dissociation enthalpies (= negative free complex formation enthalpies) of the complexes* (which define the *equilibrium constants*). When interest centers on the reactivity of coordination compounds, the *kinetic stability* of the compounds is of importance. This is expressed by strong *positive activation energy* for the reaction. In order to avoid mixing up the concepts, the term pairs *stable/unstable* are used to express thermodynamic stability, whereas kinetic stability is termed *inert/labile*.

In order to distinguish between labile and inert complexes, the following arbitrary limit has been established: those coordination compounds whose ligand exchange at 25 °C and 0.1 M concentration has finished within 1 min are termed as labile, whereas those with a longer reaction duration are inert. For the octahedral complexes of the first transition metal series, it can be generalized that all species are substitution labile, with the exception of the electron configurations d^3 and d^6 (for example Cr^{3+} or Co^{3+}) and ligands with a strong ligand field (for example, CN^-). In the case of coordination compounds of metals from the second or third period, the proportion of inert systems is far larger, which is a consequence of the stronger metal–ligand bonds (and thus the higher activation energies for ligand dissociation).

An example: both $[Ni(CN)_4]^{2-}$ and $[Cr(CN)_6]^{3-}$ are thermodynamically extremely stable. However, when the rate of ligand exchange with radioisotopically labelled $^{14}CN^-$ is measured, a fast exchange for the nickel complex ($t_{1/2} = 30$ s) is found, whereas the half life period of the exchange for the Cr^{3+} complex is about 24 days. So the nickel complex is an example of a thermodynamically stable, but kinetically labile complex.

7.1
Thermodynamic Stability

The thermodynamic stability of coordination compounds is expressed by the equilibrium constants of the *formation* or *dissociation* equilibrium:

$$M^{n+} + L^- \leftrightarrow [ML]^{(n-1)+}; K = ([ML]^{(n-1)+})/[M^{n+}][L^-] \equiv [ML]/[M][L]$$

In principle, such a reaction equilibrium can be determined for all complexes. In the case of certain compound types, however, there can be practical difficulties when the ligands in their free form are unstable.

7.1.1
Steps Equilibria

The description of complex stability is easiest when only one complex is formed or when different complex types are formed whose existence areas are drastically different. This is usually not the case with the Werner type complexes, which is why, instead, a series of coupled complex formation equilibria has to be considered. Therefore, in general for the coordination compound $[ML_n]$ the following system of coupled equilibria is present

$$M + L \leftrightarrow ML \qquad\qquad K_1 = [ML]/[M][L]$$

$$ML + L \leftrightarrow ML_2 \qquad\qquad K_2 = [ML_2]/[ML][L]$$

$$ML_2 + L \leftrightarrow ML_3 \qquad\qquad K_3 = [ML_3]/[ML_2][L]$$

$$\cdots \qquad\qquad\qquad \cdots$$

$$ML_{n-1} + L \leftrightarrow ML_n \qquad\qquad K_{n(f)} = [ML_n]/[ML_{n-1}][L]$$

The equilibrium constants K_n are the *individual stability constants (individual formation constants)* of the $[ML_n]$ complexes. There will be n equilibria, where n is the maximum coordination number for this metal ion. For example, for Co^{2+} and Cl^-, n is 4, $[CoCl_4]^{2-}$, whereas for Co^{2+} and NH_3, n is 6, $[Co(NH_3)_6]^{2+}$.

When the equilibrium equation is defined as a whole, $M + xL \leftrightarrow [ML_x]^{n+}$. To calculate the equilibrium constant of the final product, (ML_x), $\beta_n = [ML_x]/[M][L]^x$ is used, where $\beta_n = K_1 \cdot K_2 \cdot K_3 \cdot \ldots K_n$.

The K_i values are called *step formation constants*, and β is the *global formation constant* (or *global stability constant*). $1/K_f$ is called the *dissociation or instability constant*, K_d. Like other equilibrium constants, the individual stability constants are in most cases tabulated logarithmically as pK_n values: $pK_n = -\log K_n$.

The stability constants K_n and β_n always refer to the ligands *as they are present in the complex*. If the reaction has been made in water, the concentration of water, $[H_2O]$, is not included, because it is assumed a great dilution, being thus a constant. When for a ligand, K_f is great, indicates a great stability. When for a ligand, K_f is small, it does indicate that it is a ligand weaker that the water, not the absolute magnitude of the strength.

7.1.1.1 Variation of K_i

When a ligand L is added to a solution of a metal ion the species ML is easily formed; by adding more ligand, the concentration $[ML_2]$ increases and $[ML]$ decreases. Later it is the turn of ML_3 and so on. Finally, the final ML_n species is formed, if the concentration $[L]$ is high enough. Each individual reaction possesses an equilibrium constant. The size of the individual stability constants usually decreases in the following order: $K_1 > K_2 > K_3 > \ldots > K_n$. This follows simply from a statistical consideration of the ligand dissociation and reassociation step. Steric hindrance also increases with an increase in the number of ligands, provided that the ligands are more voluminous than the H_2O. Electrostatic factors are also important if the ligands are charged.

If the stability constants of the single complex species in solution are known, then the *percentage distribution curves* of the individual complexes in dependence on the concentration of the ligand can be calculated. It is relatively easy to develop a mathematical model to calculate the different concentrations of each complex in solution. Let us assume the step formation of four amino complexes of Cu^{2+}: $[Cu(NH_3)]^{2+}$, $[Cu(NH_3)_2]^{2+}$, $[Cu(NH_3)_3]^{2+}$ and $[Cu(NH_3)_4]^{2+}$. Defining the fraction (α_i) of each complex in solution as:

$$\alpha_1 = \frac{[[Cu(NH_3)]^{2+}]}{[Cu^{2+}]_{total}}; \quad \alpha_2 = \frac{[[Cu(NH_3)_2]^{2+}]}{[Cu^{2+}]_{total}}; \quad \alpha_3 = \frac{[[Cu(NH_3)_3]^{2+}]}{[Cu^{2+}]_{total}}; \quad \alpha_4 = \frac{[[Cu(NH_3)_4]^{2+}]}{[Cu^{2+}]_{total}}$$

It can be demonstrated that:

$$\alpha_i = \frac{\beta_i[NH_3]^i}{1 + \beta_1[NH_3] + \beta_2[NH_3]^2 + \beta_3[NH_3]^3 + \beta_4[NH_3]^4}$$

For $i = 0$, $\beta_0[NH_3]^0 = 1$. The distribution curves obtained by plotting the α_i parameters in front of the total NH_3 concentration offer a very clear representation of the coupled equilibria and the stability areas of the respective coordination compounds. In this model we always consider there to be an excess of the ligand, so we could consider the amount of coordinated ligand to be negligible. Given the K_i and β_i values (Table 7.1), the distribution curves for the different $Cu^{2+}/H_2O/NH_3$, and $Cd^{2+}/H_2O/NH_3$ complexes at 18 °C are plotted in Figures 7.1A and B, respectively.

In some particular cases, the gradation of K_i does not exist: some intermediate K_i is anomalously great or small. In general, when the relation $K_n > K_{n+1}$ is not given, this can be considered a sign of a significant change in the structure type. This can be due to (i) an abrupt change in the coordination number, (ii) steric effects only important from a certain number of steps onward, (iii) an abrupt change in the structure of the metallic ion in a given step.

Table 7.1 K_i and β_i values for four different equilibrium reactions.

	K_i	β_i		K_i	β_i
$[Cu(H_2O)_4]^{2+} + NH_3$	$K_1 = 10^{4.3}$	$\beta_1 = 10^{4.3}$	$[Cd(H_2O)_4]^{2+} + NH_3$	$K_1 = 10^{2.65}$	$\beta_1 = 10^{2.65}$
	$K_2 = 10^{3.3}$	$\beta_2 = 10^{7.6}$		$K_2 = 10^{2.10}$	$\beta_2 = 10^{4.75}$
	$K_3 = 10^{2.9}$	$\beta_3 = 10^{10.5}$		$K_3 = 10^{1.44}$	$\beta_3 = 10^{6.19}$
	$K_4 = 10^{2.3}$	$\beta_4 = 10^{12.8}$		$K_4 = 10^{0.93}$	$\beta_4 = 10^{7.12}$
$Hg^{2+} + Cl^-$	$K_1 = 10^{6.7}$	$\beta_1 = 10^{6.7}$	$[Cd(H_2O)_6]^{2+} + Br^-$	$K_1 = 10^{1.56}$	$10^{1.56}$
	$K_2 = 10^{6.5}$	$\beta_2 = 10^{13.2}$		$K_2 = 10^{0.54}$	$10^{2.1}$
	$K_3 = 10^{0.9}$	$\beta_3 = 10^{14.1}$		$K_3 = 10^{0.06}$	$10^{2.16}$
	$K_4 = 10^{1.0}$	$\beta_4 = 10^{15.1}$		$K_4 = 10^{0.37}$	$10^{2.53}$

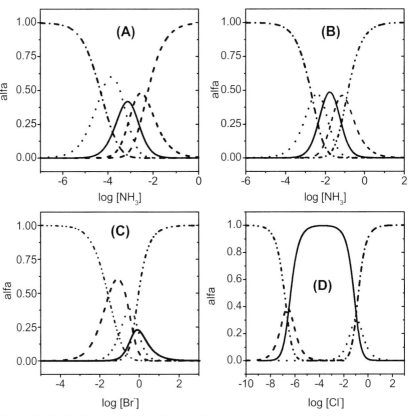

Figure 7.1 Distribution curves for the four multi-step complexation reactions given in Table 7.1. (A): Cu^{2+}/NH_3; (B): Cd^{2+}/NH_3; (C): Cd^{2+}/Br^-; (D): Hg^{2+}/Cl^-.

For example, the stability constants, in aqueous solution, for the reaction $[Cd(H_2O)_6]^{2+} + Br^-$ are given in Table 7.1. $K_4 > K_3$ because the last reaction step, $[CdBr_3(H_2O)_3]^- + Br^- \rightarrow [CdBr_4]^{2-} + 3H_2O$ (Figure 7.1C) involves the formation of three water molecules and is, therefore, entropically favorable. The higher stability of $HgCl_2$ is manifested in Figure 7.1D in form of a large plateau.

7.1.2
Factors that Affect the Stability of Complexes

Given the thermodynamic relationship $\Delta G^0 = -RT \ln K = \Delta H^0 - T\Delta S^0$ the factors that affect the stability of complexes are *enthalpic* and *entropic*. Some of the most important factors that affect the stability of complexes are given in Table 7.2.

7.1.3
Stability Trends for Metal–Ligand Interactions

7.1.3.1   The Irving–Williams Series
In the earliest days of coordination chemistry, some basic trends concerning the stability of complexes in certain ligand–central atom combinations were discovered. A very instructive example was given by Irving and Williams in 1953 (*Irving–Williams series*) [1]. These results have been formulated in the form of stability series, or by classification of central atoms and ligands. Indeed, when studying the stability constants of the $[ML_6]^{2+}$ cations in octahedral complexes, an increase in complex stability is found: $Ba^{2+} < Sr^{2+} < Ca^{2+} < Mg^{2+} < Mn^{2+} < Fe^{2+} < Co^{2+} < Ni^{2+} < Cu^{2+} > Zn^{2+}$.

Table 7.2 Main factors that affect the stability of complexes in solution.

Enthalpic effects	*Entropic effects*
variation of the bond strength: (σ) for $M^{2+} \sim$ $-200\,kJ\,mol^{-1}$; for $M^{3+} \sim -350\,kJ\,mol^{-1}$	number and size of chelate rings
effects of the ligand field	entropy of the metal ions and ligands in solution: movement freedom, essentially translation
steric and electrostatic repulsions between ligands in the complex	changes in solvation in the formation of the complex
desolvation enthalpy of the metallic ion and ligands when the complex is formed	entropy variations in non-coordinated ligands
enthalpy effects related to the conformation of the non-coordinated and coordinated ligands (mutual repulsion of ligands)	effects that arise from the configuration entropy differences of the ligands in the complex
"charge neutralization", mainly for combinations of positive and negative ions	

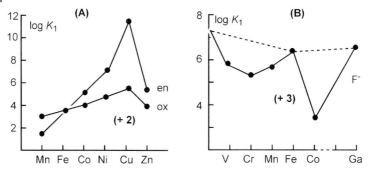

Figure 7.2 log K_1 for several $M^{2+} + L \leftrightarrow [ML]^{n+}$ reactions (M and L are indicated in the figure) (see text).

This series reflects the decrease in ionic radius from left to right, i.e., the stability constants' trend is basically an electrostatic effect. This order is relatively little affected by L. In other words, one of the most important factors is the ratio charge/radius (or the polarizing effect) of the cation. Similarly, for ions of different charge and similar radius, the polarizing power is essential.

However, when studying the Irving–Williams series for different ligand types, it is seen that the stability of complexes cannot be explained *solely* as a charge density effect. For example, beyond Mn^{2+} there is an abrupt increase in K_1 for the rest of the divalent cations and a sharp decrease for Zn^{2+}. These ions possess an additional ligand field stabilization energy (LFSE) (see below) when moving from d^5 to d^9. Considering that most complexes are obtained from aqueous solutions, to have a positive LFSE the ligand L needs to have a greater Δ_o (see Figure 1.4) than that created by H_2O, as occurs with ethylenediamine and oxalate (Figure 7.2A). If the situation was reversed (the case of F^-), the LFSE would be negative with regard to water, because Δ_o is smaller (Figure 7.2B).

Cu^{2+} complexes with chelate ligands present an additional stabilization energy compared to those of Ni^{2+} in spite of the supplementary electron placed in an antibonding e_g orbital. This anomaly is due to the stabilizing influence of the Jahn–Teller effect, which increases the K_1 value. In a tetragonally distorted complex, there is a strong bond with the four ligands in the plane. For this reason, the values of K_1 and K_2 follow this order (Cu^{2+} complexes being more stable). In contrast, K_3 for Cu^{2+} with three chelate ligands is much smaller, because it is formally impossible to place the three ligands in cis-positions.

7.1.3.2 Classification of Central Atoms and Ligands

The former, purely empirical type classification of central atoms (ions) was based on the observation that certain ligands form their most stable compounds with metal ions like Pb^{2+}, Cu^+, Ag^+, Hg^{2+} and Pt^{2+}, whereas other ligands prefer coordination with Al^{3+}, Co^{3+}, Ti^{4+}, Sc^{3+} or Fe^{3+}. Ahrland, Chatt and Davies [2] classified metals belonging to the latter type as class *a*, which includes alkali, alkaline earth metals and the light transition metals in high oxidation levels. H^+ may also be included.

Table 7.3 Stability series.

Stability series for metal–ligand atom interaction of class *a* ions	Stability series for metal–ligand atom interaction of class *b* ions
N >> P > As > Sb > Bi	N << P > As > Sb > Bi
O >> S > Se > Te	O << S ≤ Se ≈ Te
F >> Cl > Br > I	F < Cl < Br << I

Table 7.4 Significant cations and anions according to their hard/soft character.

Hard acids	Intermediates	Soft acids	Hard bases	Intermediates	Soft bases
H^+, alkali, alkaline earths, lanthanides(3+) Ce^{4+}, Th^{4+}, U^{4+}, UO_2^{2+}, Ti^{4+}, Zr^{4+}, Hf^{4+}, VO^{2+}, Cr^{3+} Mn^{2+}, Fe^{3+}, Co^{3+}	1st transition(2+) M^{3+} (Rh,Ir,Ru) M^{2+} (Zn, Sn, Pb, Fe, Co, Ni, Cu, Ru, Os)	M^{2+} (M = Pd Pt, Cd, Hg) M^+ (M = Cu, Ag, Au, Tl, Hg)	NH_3, N_2H_4, amines, OH^- H_2O, O^{2-}, ROH, R_2O, CH_3COO^-, CO_3^{2-}, NO_3^-, PO_4^{3-}, SO_4^{2-} F^-, ClO_4^-	N_2, N_3^-, Br^- Cl^-, pyridine NO_2^-, SO_3^{2-}	H^-, R^-, I^-, C_2H_4, C_6H_6, CN^-, CO, RCN, SCN^-, PR_3, AsR_3, R_2S, RSH, RS^-

On the other hand, the ions of the above first group belong to class ***b***, especially the heavy transition metals in low oxidation levels (Cu^+, Pb^{2+}, Hg^{2+}, etc.).

Depending on whether ligands prefer to coordinate with ions of class ***a*** or ***b***, they are also classified as class ***a*** or ***b*** ligands. Taking into account only the ligand atoms, the trends shown in Table 7.3 are found.

Summarizing, this classification means that metal ions and ligands *of the same class* form the most stable complexes.

7.1.3.3 The Concept of Hard and Soft Acids and Bases

The first division of metal ions and ligands was refined and amplified by Pearson in 1963 [3]. The concept comprises areas of acid–base interactions which go far beyond the limits of coordination chemistry. However, it is especially suitable for metal complexes, since the complex formation can also be understood as an interaction between Lewis acids (central atom/ion) and bases (ligands). Class ***a*** ions and ligand atoms are qualitatively small (in the case of cations, often with higher charges) and only slightly polarizable species, which Pearson named *hard acids and bases*. On the other hand, those of class ***b*** are larger polarizable metal ions and ligands, which Pearson named *soft acids and bases*. For the bonding of hard and soft acids and bases the following Pearson's rule applies: (a) *Hard bases B prefer bonding to hard acids A*; (b) *Soft bases B prefer bonding to soft acids A*.

In Table 7.4, the most important metal cations and ligands are divided into the three most frequently used categories (hard acid/base, soft acid/base, borderline).

In general it is not possible to define exact limits between these categories.

Remark   Acids: The class of a given element is not constant, but varies with oxidation state. A rule is that hard (class *a*) character increases with increasing positive oxidation state and vice versa for soft (class *b*) behavior. Bases. In some cases, it is impossible to classify them by using the ligand atom alone. Examples are nitrogen donor ligands. The pyridine ligand, for example, is softer than NH_3 and it has to be classified as borderline. This is due to the capability of pyridine to maintain π interactions with the central atom (see symbiosis effect).

Solvent Effect   The effect of water as solvent is to decrease the basicity of small (hard) anions with respect to related large (soft) anions. This is due to the negative hydration heats for anions. For neutral bases, the influence of the water is small. Solvents other than water would give effects in the same direction, but smaller in magnitude. The same effect is found in acids: solvents tend to increase class *b* (soft) character for acids compared to the gas phase.

Symbiosis Effect   "Symbiosis" means that the coordination of hard ligands at a central atom increases the latter's hardness. In the opposite case, the coordination of soft ligands increases the capability of a metal center to bond further soft ligands [4]. Whereas the cobalt centre in $[Co(NH_3)_5]^{3+}$ is a hard ligand acceptor, the complex fragment $[Co(CN)_5]^{2-}$ is a soft acid. $[CoX(NH_3)_5]^{2+}$ is more stable for X = F^- than I^-, whereas $[CoX(CN)_5]^{3-}$ is most stable with X = I^- or H^-. This difference is made more evident in the bonding-isomer coordination of the thiocyanato ligand: $[Co(NH_3)_5(NCS)]^{2+}$ but $[Co(CN)_5(SCN)]^{3-}$.

The concept of symbiosis can be generalized: the presence of a polarizable substituent may convert a hard acid into a softer one; the presence of an electron-withdrawing substituent reduces the soft character of the species.

The Theoretical Foundation of the Concept   The two most classical theories are: (i) the ionic-covalent theory which is the oldest and usually the most obvious explanation; (ii) the π-bonding theory, soft acids can form π-bonds by donation to suitable ligands (CO, ligands with P, As, S, etc., atoms).

Absolute Hardness   A property called absolute hardness, η, was defined by Parr and Pearson (1983) [5]. Let $E(N)$ be a ground-state electronic energy as a function of the number of electrons N. The *chemical potential* μ (or the negative value of the *absolute electrogenativity* χ) is defined as $\mu = (\partial E / \partial N)_Z = -\chi$.

Let us remember that Mulliken's definition of absolute electronegativity is, $\chi = (I + A)/2$ (where I and A are the ionization potential and electron affinity).

Parr and Pearson define the corresponding second derivative as the *hardness* [5]: $2\eta = (\partial^2 E / \partial N^2)_Z$. Chemically, they express η by $\eta \approx (I - A)/2$.

A low value of η means high softness and vice versa. When we consider the reaction $S + S \rightarrow S^+ + S^-$, low hardness (i.e. low η values) means that the electron transfer from S to S is easy, which certainly is the case when S is polarizable, that is, "soft". Maximum softness means no energy change associated with the previous disproportionation reaction.

Nevertheless, there are important limitations: strictly speaking, for example, the quantitative application of the relations giving μ and η is only possible in the case of neutral molecules as, in other cases, the approximation loses its validity. Most central atoms and ligands in coordination compounds are ionic, so this is a considerable limitation of the quantitative applicability of the theory which has been described. To date, there is no general theory of absolute hardness which includes all classes.

Frontier Orbitals Theory: Relationship to Polarizability From the beginning of the hard–soft acid–base (HSAB) theory, the importance of the frontier orbitals (HOMO and LUMO) has been emphasized. Let us remember that according to Koopmans theorem [6], the frontier orbital energies are given by: $-\varepsilon_{HOMO} = I$ and $-\varepsilon_{LUMO} = A$.

The definitions of absolute hardness can, thus, be understood from the point of view of molecular orbital theory. In Figure 7.3, the frontier orbital diagram for two systems of different hardness but equal electronegativity is given. The figure shows that hard molecules are characterised by a great HOMO–LUMO gap, whereas it is small for soft molecules. Thus, the gap between the frontier orbitals is a measure of the hardness (or softness).

Furthermore, from the expression for polarizability, $\alpha^{1/3} \approx 1/(I - A)$, it follows that molecules with a small HOMO–LUMO gap have high polarizability. Thus, the concept of hardness is much related to the polarizability concept, in agreement with the original empirical ideas of hardness: soft acids and bases have a high polarizability, α, which means the facility of an ion or molecule to be distorted by an electric field. Hard acids and bases are non-polarizable and form essentially ionic compounds, while soft acids and bases are polarizable and form essentially covalent compounds.

The relation between the softness of a ligand and the HOMO–LUMO gap also explains why ligands with low-energy acceptor orbitals (π acids) are examples of soft ligands. This also explains why, in comparative studies of the softness (hardness) of ligands, not necessarily those with the heavier (i.e, the more polarizable!)

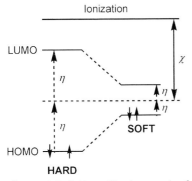

Figure 7.3 Definition of hardness and softness, from the gap between the frontier orbitals HOMO and LUMO.

ligand atoms form the more stable metal–ligand bonds with softer metal centers. Phosphorus and sulfur, for example, show the strongest soft behavior among the elements of these groups, as they have better capabilities for π interactions with soft metal centers than their heavier counterparts.

As is well known, the π bonding capacity decreases as follows: CS > CO $\approx$ PF$_3$ > N$_2$ > C$_2$H$_4$ > PR$_3$ $\geq$ AsR$_3$ $\approx$ R$_2$S > CH$_3$CN > py. This series can be compared with that of decreasing absolute softness.

Empirical Relationships Another way of looking at hard/soft properties is to consider empirical equations such as Drago's equation [7], $-\Delta H_{AB} = E_A E_B + C_A C_B$ where E are electrostatic (ionic) parameters for species A and B, and C are covalent parameters for A and B. A stable adduct requires a large value for $-\Delta H_{AB}$. With a hard species, E will be large and C will be small, while with a soft species C will be large and E will be small. Therefore, hard–hard and soft–soft interactions are favored. The reader interested in this series of parameters can consult many Inorganic Chemistry textbooks [1].

7.1.3.4 Stability Due to the Multidentate Character of Ligands

The possibility of "pre-programming" the structure, *stability* and reactivity of a coordination compound motivated the development of new multifunctional multidentate ligands (*"ligand design"*; see Chapter 2). Summarizing, the role of multidentate ligands can be described in two basic principles: (i) the determination of the coordination geometry of a complex (Chapter 2) and therefore of its reactivity; (ii) the thermodynamic stabilization of a coordination compound (*chelate effect; macrocyclic effect*). The general tendency is illustrated, as a typical example, in Figure 7.4.

The Chelate Effect, an Enthalpy and Entropy Effect The chelate effect refers to the increase in the stability of a complex containing chelate rings in comparison with similar complexes without chelate ligands. Let us consider the following equilibrium constants:

$$[Ni(H_2O)_6]^{2+} + 6\,NH_3 \leftrightarrow [Ni(NH_3)_6]^{2+} + 6\,H_2O; \qquad \log K = 8.61$$

$$[Ni(H_2O)_6]^{2+} + 3\,en \leftrightarrow [Ni(en)_3]^{2+} + 6\,H_2O; \qquad \log K = 18.28$$

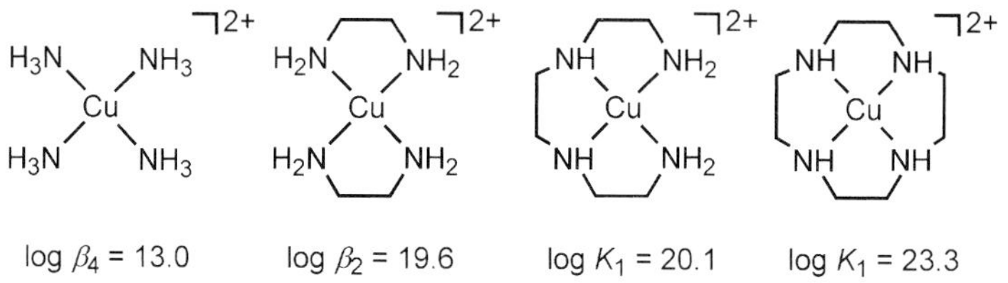

$\log \beta_4 = 13.0 \qquad \log \beta_2 = 19.6 \qquad \log K_1 = 20.1 \qquad \log K_1 = 23.3$

Figure 7.4 Variation of K_i according to the chelate and macrocyclic effect.

This difference of ca. 10 units cannot be attributed to the differences in energy between metal and ligands because they are very similar.

In the thermodynamic analysis of the chelate effect, we differentiate between the *enthalpy* and the *entropy* parts of the free energy of the chelate complexes. Considering, for example, ammine and chelate–amine complexes, the better donor capacity of the N ligands has to be taken into account with increasing alkylation level of the nitrogen atoms. Going from left to right in the complex series in Figure 7.4, the Lewis basicity and thus the donor capacity of the N ligands increases. Both effects are pure enthalpy effects. Nevertheless, the most important aspects come from the entropic effects, which can be understood in two different ways.

Let us first consider the number of reactants and products in the two reactions mentioned above, Ni^{2+} with NH_3 or en. The greater number of particles randomly distributed in the solution represents a state of greater probability or entropy. As a consequence, the second reaction is favored with respect to the first one: the reaction from left to right is performed by *increasing the particle number* in solution and thus *increasing the entropy* of the system.

The entropy role in the greater stability of chelate complexes can also be explained qualitatively by a microscopic (local) observation. The dissociation of, for example, an NH_3 ligand and a donor atom of ethylenediamine is studied. After the dissociation of NH_3 from the complex, it diffuses rapidly from the solvent cage of the complex. It is very improbable that this same molecule will be involved again as a ligand in the same complex. On the other hand, when a donor atom of the en ligand is dissociated, the NH_2 function can only move a few Angstroms from the coordination position because it is fixed to the hydrocarbon chain. The probability of being incorporated again in the same complex is therefore high. This situation is often termed "*effective (local) concentration*" of the second ligand. In other words, the concentration of the second Lewis base in the neighborhood of the metal ion is greater when the second group is linked to the first by a *relatively* short carbon chain. Thus, in this case the reaction is more likely and the equilibrium constant is greater.

A simple equation relating the formation constant of an *n*-dentate chelating ligand to that of the analogous complex containing monodentate ligands has been proposed [8]:

$$\log K_1(\text{polydentate}) = \log \beta_n(\text{monodentate}) + (n-1)\log 55.5$$

For polyamine complexes a factor of 1.152 must be added to $\log \beta_n$, which is related to the inductive *effect* of an alkylamine with regard to NH_3; the $(n-1)$ log 55.5 value concerns the entropic effect. Some calculated and experimental values are given in Table 7.5.

This theory is only approximate. It does not work when the steric factors are important.

The Size of the Chelate Ring   The chelate effect is decisively influenced by the *relation of the chelate ring size to the radius* of the central metal. Calculations with

Table 7.5 Calculated and experimental values for several complexes with polydentate amines.

Polyamine	en	dien	trien	tetren	penten
dentate character (n)	2	3	4	5	6
log K_1 (Ni^{2+}) calc. (exp)	7.58(7.47)	11.37(10.96)	14.67(14.4)	17.25(17.4)	19.16(19.1)
log K_1 (Cu^{2+}) calc. (exp)	10.76(10.54)	15.92(15.09)	20.20(20.1)	21.28(22.8)	—
log K_1 (Fe^{2+}) calc. (exp)	4.38(4.34)	6.82(6.23)	8.67(7.76)	10.02(9.85)	10.87(11.1)
log K_1 (Pb^{2+}) calc. (exp)	4.92(5.04)	7.51(7.56)	9.95(10.35)	11.18(10.5)	12.26(—)

molecular mechanical methods show, for example, that the ring strain of the coordinated en (1,2-ethanediamine) ligand is minimized when M–N bonding lengths of 2.50 Å and a N–M–N angle of 69° are given. On the other hand, the ideal metric parameters for the homologous ligand tn (1,3-diaminepropane) are: M–N = 1.6 Å and (N–M–N) = 109.5 °. If, therefore, the ring strain effects are taken into account, it can be seen that the en ligand coordinates preferentially on metals with a large ion radius (La^{3+}, Pb^{2+}); in contrast, for the homologous tn, the coordination on metal ions with a small ion radius is favorable.

In general, small cations will prefer, thus, six-membered ring ligands while larger cations will be stabilized by five-membered ring ligands. This knowledge is used for the development of *ion-selective bonding positions* in multidentate ligands. For example, comparing the stabilities of $[Ni(2,2,2-tet)(H_2O)_2]^{2+}$ and $[Ni(2,3,2,-tet)(H_2O)_2]^{2+}$ complexes, shows that the complex with the 2,3,2-tet ligand that forms a six-membered ring is more stable than the 2,2,2-tet complex. This feature can be generalized to other cations. For cations with large ionic radius, the complex with the ligand 2,2,2-tet is more stable. However, no general extrapolation is possible because of other steric effects.

Aromaticity of the Chelate bpy and phen ligands form five-membered rings when complexing metals, giving very stable complexes, likely due to their ability to act as σ donor and π acceptor ligands (through the ring π^* orbitals). Certain non-aromatic chelate ligands provide a strong stabilization due to resonance or delocalization effects, acquiring a marked aromatic character: this is the case for the acac (acetylacetone) ligand, which usually is coordinated as enolate (acetylacetonate). With trivalent cations it forms very stable complexes such as $[M(acac)_3]$ (M = Al, Ti, Cr, Co, etc). In all of them, all M–O and C–O distances are identical.

Some Practical Examples: Complexones The stability of chelate complexes is used in the quantitative analysis of metal ions (*complexometry*). This technique resulted first from studies on the stability of iminodiacetic acid complexes and rapidly led to the use of a whole series of analogous chelate ligands, named *complexones* (Figure 7.5). The most important complexone is, in practice, ethylenediaminetetraacetic acid (edta) in the form of the commercially available salt $Na_2[H_2edta] \cdot 2H_2O$.

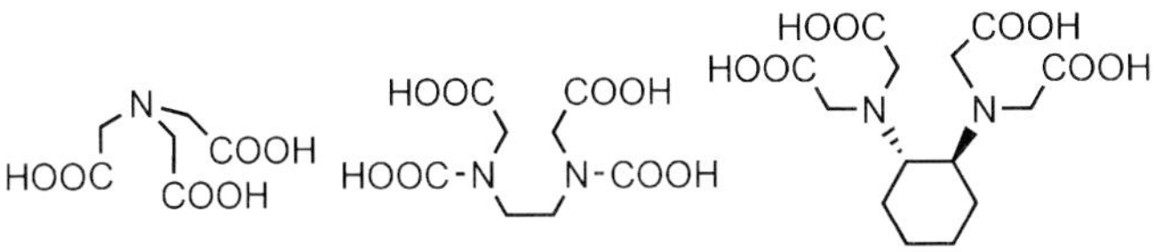

Figure 7.5 Some complexones (nta = nitrilotriacetic acid, cdta = 1,2-diaminocyclohexane-*N*,*N*,*N*′,*N*′-tetraacetic acid, edta = ethylenediaminetetraacetic acid).

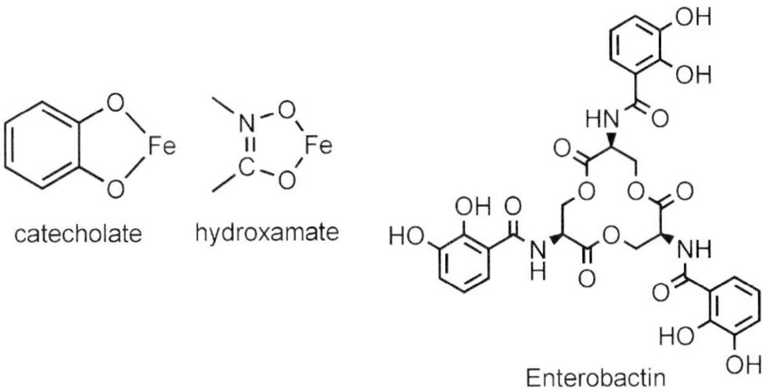

Figure 7.6 Some siderophores.

Some Biological Examples: Siderophores   The important role played by metal ions in living organisms will be reviewed in Chapter 17. Considering the transport of metal ions in organisms, a great number of macro and low-molecular weight carrier molecules are responsible. Relatively little is known about carrier molecules which serve as ligands for transition metal ions, with the exception of the iron-bonding systems. The low-molecular weight iron carriers, the *siderophores*, can be divided into two classes, the hydroxamates and the catecholates (Figure 7.6).

The enterobactin (a catecholate), forms the most stable complexes with Fe^{3+}. The high stability constant of 10^{52} facilitates the easy dissolution of Fe^{3+} under physiological conditions (at pH 7, the theoretical concentration of Fe^{3+} in aqueous solution is only 10^{-16} M). The siderophore ligands are important in "*chelate therapy*"(see Chapter 17). Enterobactin has not yet been surpassed by any other open-chain chelate ligand. A further stabilization of an iron complex can only be achieved by still more *preorganisation* of the metal bonding positions. A ligand which has been improved in this way is a *macrocyclic ligand* system, called *siderand*. The iron(III) complex formed in this way not only has a stability constant of 10^{59}, but the ligands themselves are far more robust than the chemically labile enterobactin. The reason for this additional complex stabilization is called the *macrocyclic effect* and is discussed in the next section.

The Macrocyclic Effect   The main types of macrocycles have been studied in Chapter 2. The first theory of the additional stability, termed the macrocyclic effect, was based on the idea of the perfect fit between the cavity of the macrocycle and the size of the cation. Nevertheless, it was rapidly realized that this theory is not perfect in many cases because the influence of the solvent is also very important.

The macrocyclic effect can be manifested by comparing the enthalpic effect with similar open ligands with the same donor atoms. The enthalpies of two different examples with open or closed ligands are given in Table 7.6.

The enthalpic effect allows us to deduce that complexes with macrocyclic ligands are more stable than those with similar but open ligands. In this case the entropic influence is almost nil: the number of particles in solution before and after the formation of the complex is the same. The current theory on the origin of the macrocyclic effect is based on the following factors:

1. *Ligand preorganization*. There are a given number of conformers, some of them with the required conformation for complexation with the metal ion. Two aspects are important: (i) the number of conformers is greater for the open ligands, but (ii) the preferred conformers in the solid state or in solution are currently different. For example, the preferred conformers for the 18-crown-6 ether in the solid state or in solution are depicted in Figure 7.7A. In the solid state, some of the oxygen atoms are in an *exo*-position to avoid steric tensions. Molecular mechanics calculations demonstrate that to convert all the oxygen atoms to an *endo*-position is energetically unfavorable. However, in aqueous solution the solvation weakens the repulsion interactions and allows a preorganization of the ligand to matching it perfectly to a metal-ion.

A paradigmatic example of the influence of the ligand preorganization and the stability of the complex formed can be observed with the two ligands shown in Figure 7.7B. The complex with the closed ring is 10^{21} (!) times more stable than the complex with the open ligand. The open ligand cannot exist as depicted in Figure 7.7B, because the piperazine needs to adopt the chair shape, keeping in this manner the two NH_2 groups as far apart as possible.

In flexible macrocycles the ligand preorganization is, thus, one of the most important elements in the macrocyclic effect, since the two factors (entropic and enthalpic) in the conformation of any non-macrocyclic ligand are not favorable (Figure 7.8). The other factors are approximately the same.

Non-flexible macrocycles, derived from benzenic rings, such as porphyrins, present greater stability than the more flexible analogous (14-aneN4), owing to

Table 7.6 Comparison of ΔH for open and closed (macrocyclic) ligands.

	Ligand	Na^+	K^+	Ba^{2+}
ΔH (kcal mol^{-1})	18-crown-6	−8.4	−13.4	−10.4
	pentaethylene glycol	−4.0	−8.7	−5.6
		Cu^{2+}	Ni^{2+}	Zn^{2+}
ΔH (kcal mol^{-1})	cyclam	−32.4	−24.1	−14.8
	2,3,2-tet	−26.5	−17.9	−11.6

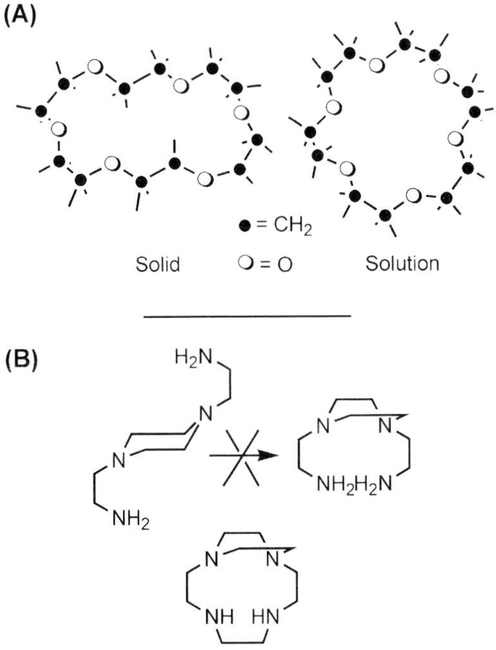

(A)
● = CH_2
○ = O
Solid
Solution
(B)
H_2N
N
N
NH_2
N
N
NH_2H_2N
N
N
NH
HN

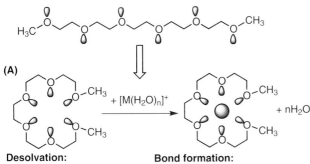

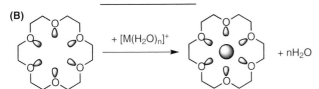

Figure 7.7 Ligand preorganization in macrocyclic effect (see text).

H_3C
CH_3
(A)
CH_3
CH_3
+ $[M(H_2O)_n]^+$
CH_3
CH_3
+ nH_2O
Desolvation:
enthalpy not favorable
entropy favorable
Conformation:
enthalpy and
entropy not favorable
Bond formation:
enthalpy and entropy favorable
(B)
+ $[M(H_2O)_n]^+$
+ nH_2O
Desolvation:
enthalpy not favorable
entropy favorable
Bond formation:
enthalpy and entropy favorable

Figure 7.8 Preorganization is the most important factor of the macrocyclic effect. In this Figure the macrocycle (B) is already 'preorganized' for coordination.

their greater preorganization. In general, any factor that reduces the flexibility of the macrocycle, increases the stability and selectivity of it complex.

2. *Inductive basicity* created by the additional ethylene groups in the macrocycle. As well as the ethylenediamine ligand being more basic than two NH_3 due to the alkylation of the basic nitrogens, the greater alkylation is the cause of the greater stability of macrocyclic ligands compared to their open analogs. This effect can be seen when comparing the ΔH^0 values for the reaction of Cu^{2+} with four NH_3, two ethylenediamine, one open 2,3,2-tet ligand and the cyclam macrocycle (Figure 7.9).

For this reason many macrocyclic ligands stabilize "unusual" oxidation states, such as Cu^{3+}, Cu^+, Ni^{3+}, Ni^+, Co^+, Ag^{2+}, Pd^{3+}, Pt^{3+}. However, this inductive effect cannot be extrapolated. For example the methylation of the cyclam to give the tetramethylcyclam, causes a destabilization of the complexes formed with divalent cations. This is due to the steric repulsions of the four new methyl groups that tend to fold the macrocycle, coordinating the cations in the cis-position instead of in trans- (normal) geometry.

Selectivity of Macrocyclic Ligands   Selectivity due to the size of the macrocycle cavity is important. A very good example is the study of the stability of complexes derived from crown-ethers, which will be discussed in Chapter 14.

By means of molecular mechanics calculations it has been possible to establish the adaptation of certain ions into the cavity of the tetrazamacrocycles, as indicated in Table 7.7.

In general it is possible to state that the selectivity for different cation sizes in these macrocycles is not perfectly explained *only* by the idea of size/cavity match. There are also conformation changes when the size varies. This effect is visualized in Figure 7.10 in which the meaning of $+ + + +$, $+ + - -$ and $+ - + -$ is indicated

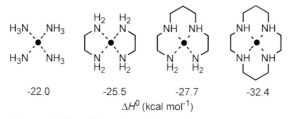

-22.0   -25.5   -27.7   -32.4
$$\Delta H^0 \text{ (kcal mol}^{-1}\text{)}$$

Figure 7.9 Effect of the inductive basicity (see text).

Table 7.7 The most favorable geometry in tetraazamacrocycles.

	12-aneN4	14-aneN4	16-aneN4
Geometry (the most favorable)	C_{4v}	D_{4h}	T_d
Conformation	$+ + + +$	$+ + - -$	$+ - + -$

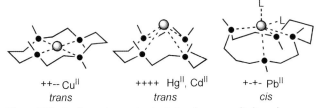

++-- Cu^{II} ++++ Hg^{II}, Cd^{II} +-+- Pb^{II}
trans   trans   cis

Figure 7.10 Conformation and selectivity of macrocyclic ligands.

according to the position of the hydrogen atoms of the amine group. From this figure it is deduced that one macrocycle (in this case the 14-ane4) can stabilize cations of different volume, adapting its conformation to the size of the cation. Thus, it will give trans $+ + - -$ conformation for Cu^{2+} and small cations, whereas it will adopt the trans $+ + + +$ and/or cis $+ - + -$ conformation for larger cations.

The Cryptate Effect   In spite of all the peculiarities mentioned above, the general tendency is to match the cavity and the cation. This tendency is very dominant in the case of the macropolycycles termed *cryptands* (see Chapter 2). In this case the cation needs to match exactly the cavity because the mobility of the macrocycle is reduced.

The oxygen cryptands with a small cavity (such as [1,1,1]-cryptand) are very selective for the proton; the medium cavity cryptands (such as [2,2,1]-cryptand) are selective for Na$^+$ and Ca^{2+} (ionic radius close to 1 Å) whereas the large cavity [2,2,2]-cryptand is selective for K$^+$ and Ba^{2+} (ionic radius close to 1.35 Å). These features will be studied in Chapter 14.

7.1.4
Steric Factors: Cone Angle

There are many steric effects that affect the stability of the complexes. Let us comment on the most important one, the *cone angle*. Tolman studied in a special manner many phosphane ligands, developing the concept of *cone angle*, defined in terms of an idealized M-PX$_3$ system [9]. The cone angle has its vertex in the metal and is a tangent to the van der Waals surface of the atoms of the ligand (Figure 7.11), Table 7.8.

An example of the importance of the cone angle may be seen in the "*umbrella*" effect on the [Ru(PMe$_2$Ph)$_4$(S$_2$CH)]$^+$ isomer. The reaction of [RuH(PMe$_2$Ph)$_5$]PF$_6$ with CS$_2$ produces an orange complex, [Ru(PMe$_2$Ph)$_4$(S$_2$CH)]$^+$. However, when the reaction is carried out in boiling methanol a different, purple, isomer is obtained. In this second isomer, the phosphane ligand is bound to the dithioformate leaving a vacant position on the ruthenium. This is due to the great value of the cone angle of the phosphane. This rearrangement is reversible if the product is heated in the

Cone angle

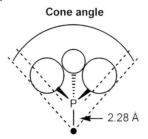

Figure 7.11 Scheme of the cone angle.

Table 7.8 Cone angle (in degrees) for several ligands.

Ligand	Cone angle	Ligand	Cone angle	Ligand (phosphane)	Cone angle	Ligand (phosphane)	Cone angle
H	75	Br, C$_6$H$_5$	105	PH$_3$	87	PCl$_3$	124
Me	90	I	107	PF$_3$	104	PEt$_3$	132
F	92	i-Pr	114	P(OMe)$_3$	107	PPh$_3$	145
CO,CN	95	t-Bu	126	P(OEt)$_3$	108	P(i-Pr)$_3$	160
Cl, Et	102	C$_5$H$_5$	136	PMe$_3$	118	P(t-Bu)$_3$	182

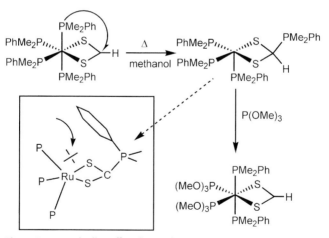

Figure 7.12 "Umbrella" effect due to the cone angle (see text).

presence of P(OR)$_3$ (very small cone angle). Reducing, thus, the steric hindrances, the Ru–P bond is favored. This reaction is illustrated schematically in Figure 7.12, which shows the structure of the pentacoordinated complex in which the benzenic ring acts as an "*umbrella*" avoiding the six-coordination of the ruthenium.

7.2
Non-redox Reaction Mechanisms

7.2.1
Introduction

Many aspects of chemical reactivity need to be interpreted in terms of reaction mechanisms. The detailed mechanism of a reaction is frequently not known: all possibilities must be considered. Fortunately, in any reaction there is, currently, a step that implies the lowest energy interchange and hence is the decisive step, at least at the macroscopic average level. We can think of a reaction mechanism at two different levels:

1. The reaction may occur through a series of distinct steps each of which can be written as a chemical equation. This series of steps is the "*stoichiometric mechanism*".

2. We can also consider what is happening during each of these individual steps. These details constitute the "*intimate mechanism*" of the reaction. Usually it is only the rate determining step that is examined.

Thus, when talking about the *mechanism* of a chemical reaction, this means the *subdivision of the total reaction into a series of elementary steps* which have a *sequential* or *parallel* relation to each other. In most cases, it is not possible to identify the single elementary steps directly. Therefore one tries to derive a model for the individual steps from the way in which the total reaction can be influenced through the variation of certain system parameters. The most important starting point is the *reaction rate* and its *dependence* on different characteristic variables. This is a subject of chemical kinetics, which will be explained in the next section. The most important basic reaction types in complexes are: (i) ligand substitution, (ii) ligand transformation, (iii) redox reaction and (iv) photoreaction. We will dedicate separate chapters for reactions (iii) and (iv); in this chapter we will deal mainly with reaction (i).

7.2.2
Summary of Formal Kinetics

The main goal of kinetics is the study of reaction rates and of the influence that several factors, such as temperature, pressure, concentration, etc., exert on the rate. The kinetic study leads to a rate law. This is simply an *empirical* equation that shows how the reaction rate varies as a function of the above-mentioned factors. Each step in the stoichiometric mechanism has a rate constant associated with it. In the following we will use the units: k (first order) = s^{-1}; k (second order) = $M^{-1} s^{-1}$; $\Delta H^{\neq}$ = kJ mol^{-1}; $\Delta S^{\neq}$ = J mol^{-1} K^{-1}; $\Delta V^{\neq}$ = cm^3 mol^{-1}.

7.2.2.1 The Rate Law
The rate law of a chemical reaction is the rate of change of the concentration of a reactant or a product.

$$\text{rate} = -d[\text{reactants}]/dt = d[\text{products}]/dt$$

It depends, in general, on the temperature, pressure and composition of the system: reactants, products, solvent, catalysts, etc. Assuming constant temperature, the most important measure which influences the reaction rate is the *concentration* of the reactants.

In many cases, the rate law has the form, rate $= k[A]^a[B]^b \dots [L]^l$ (A, B = reactants, products of reactions or other species that may affect the rate, such as catalysts). In these cases, k is the *rate constant*, whereas the exponents a, b, etc, determine the *reaction order*. When $a = 1$, the reaction is of *first order* with regard to [A]. The sum of the exponents defines the *total reaction order*. These reactions are called *first-order reactions, second-order reactions* etc. As the experimental studies in most cases comprise the collection of time-dependent concentration data for reactants and products, integration of the rate law is necessary. In principle, the observed dependence of [A], [B], etc on time is compared with the integrated rate laws, which leads to the determination of the reaction order and the rate constant.

7.2.2.2 Integrated Rate Laws of Simple Reactions
First-order reactions
Let us assume A → B

$$\text{rate} = k[A] = -d[A]/dt, \text{ thus } -d[A]/[A] = k dt$$

If $[A]_o$ is the initial concentration then, on integrating

$$-\ln[A_t] + \ln[A_0] = kt$$

which is the equation of a straight line with $y = \ln[A_t]$, $x = t$ and $b = \ln[A_0]$. Thus, a plot of $\ln[A_t]$ vs. time for a first-order process produces a line with slope equal to the rate constant, k.

Mathematically, $[A_t] = [A_0] e^{-kt}$. A plot of concentration of A will vary exponentially with time. When $[A_t] = 1/2[A_0]$, t is called the half life, $t_{1/2}$. Mathematically, $t_{1/2} = 0.693/k$.

Second-order reactions
Let us assume A + B → C, being a reaction of first order in the two reactants.

$$\text{rate} = k[A][B]$$

This rate law can be integrated (as can be found in specialized books on Kinetics and Mechanisms, see Bibliography), but the reaction is commonly run with either A or B in large excess, $([B_0] \gg [A_0])$. In this case [B] will not change significantly during the reaction.

$$\text{rate} = k[A][B] = k_{obs}[A] \; (k_{obs} = k[B]).$$

This is called a study of reaction kinetics under conditions of *pseudo-first-order*. A plot of ln [A] vs. t will yield a straight line of slope k_{obs}. Repeating the reaction for different concentrations of [B] (always in excess) and plotting k_{obs} vs. [B], the value of k may be obtained from the slope.

7.2.2.3 Activation Parameters and Reaction Mechanism

Currently, a reaction mechanism cannot be deduced unambiguously from a rate law. The *activation parameters* are usually essential and are obtained from empirical relationships. The known Arrhenius equation $k = Ae^{-Ea/kT}$ was established many years ago. However, the dependence of the reaction rate (rate constant) on temperature is better described by the *activated complex theory* (Eyring theory).

$$k = k_B T/h \exp[-\Delta H^{\neq}/RT] \cdot \exp[\Delta S^{\neq}/R] \quad \text{or}$$

$$\ln(k/T) = -\Delta H^{\neq}/RT + \Delta S^{\neq}/R + \text{constant}$$

where k_B is the Boltzmann constant = 1.38×10^{-23} J K^{-1}; h is the Planck constant = 6.626×10^{-34} J s).

When the reaction rate is measured at different temperatures, the plot of ln (k/T) versus $1/T$ gives $\Delta H^{\neq}$ from the slope and $\Delta S^{\neq}$ from the intercept. This type of plot is called an Eyring plot. The activation enthalpy $\Delta H^{\neq}$ is not generally very meaningful with regard to the reaction mechanism, however, the activation entropy $\Delta S^{\neq}$ is quite useful in determining the mechanism. A positive $\Delta S^{\neq}$ value would be expected, for example, in a dissociative substitution mechanism. In an associative case, bonds are formed between attacking ligands and complex in the activated complex, and therefore a negative $\Delta S^{\neq}$ is the expected result.

Even better than the activation entropy with regard to determining the reaction mechanism is the *activation volume*. This volume is determined by measuring the pressure dependence of the rate constant: $[d\ln k/dP]_T = -\Delta V^{\neq}/RT$.

The experimentally determined value of $\Delta V^{\neq}$ consists of two parts, an *intrinsic activation volume* $\Delta V_{intr}^{\neq}$ which arises from modifications of the bonding lengths and angles and is a measure of the compressibility difference between the ground state and transition state; and a part due to modification of the solvation in the activated complex, $\Delta V_{solv}^{\neq}$. The intrinsic activation volume is the decisive parameter for the mechanistic details of bond formation, allowing one to distinguish between dissociative and associative reaction steps.

7.2.2.4 Reaction Profiles: Potential Energy Curves

The energy changes for a given reaction are described graphically with a *reaction profile*. This shows the relative energies of the reactants, products, intermediates and transition states as a function of the reaction coordinate. Two typical cases are shown in Figure 7.13. The energy difference between the reactants and the highest energy point (the *transition state*) is the *activation energy* (or *activation barrier*). A minimum in the reaction profile indicates the existence of an *intermediate* whose stability depends on the depth of the relative minimum.

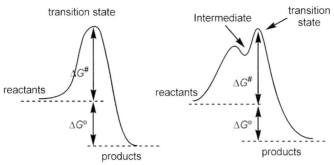

Figure 7.13 Reaction profiles: potential energy curves.

If the free energy is negative – spontaneous reaction – the products are more stable than the reactants. However, to get the final products, the reactants need to overcome the transition state, with higher energy. The reaction rate is due, in part, to the height of the energy barrier.

7.2.3
Generalities on Ligand Substitution Reactions

Ligand substitutions are important in almost any area of coordination chemistry, and knowledge of their kinetics is decisive for knowing the ideal conditions for the synthesis of a complex or for homogeneous catalysis. Substitutions on octahedral and square planar complexes have been the most intensively studied and, therefore, we will restrict ourselves to these systems.

7.2.3.1 Classification of Substitution Mechanisms and their Experimental Differentiation

For some reactions the simple division into *dissociative* (D) or *associative* (A) is sufficient. In other words, for a D mechanism an intermediate of reduced coordination number is formed, it has already largely lost X (the leaving group); whereas for an A mechanism an intermediate of increased coordination number is formed, that includes both X (the leaving group) and Y (the entering group) bonded to metal. These two types are the extremes of the possible reaction forms. When the substitution proceeds more or less concertedly, with partial association and dissociation, this is called a mechanism of *interchange* (I) type. For an I mechanism no intermediate is formed. In practice, however, this symmetrical case is seldom found. Usually, the entering or leaving groups in the activated complex are either more strongly embedded into the coordination sphere (I_a = associative interchange), or more weakly embedded (I_d = dissociative interchange). The cases I_a and I_d represent, therefore, continuous mechanistic possibilities between the extreme cases described at the beginning. These different types can be shown schematically as in Figure 7.14.

In the D case, the energy necessary to break the bond is the energy that determines the activation energy and hence the rate. Bond breaking occurs during the

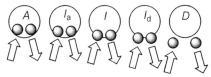

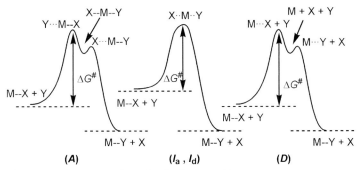

Figure 7.14 Scheme of the classification of substitution mechanisms.

Figure 7.15 The energy scheme for the four possible processes (activation profiles) in substitution mechanisms.

rate limiting step. In the A case, this energy comes from the new bond formed. Bond making occurs during the rate limiting step. In these two extreme cases, the reaction intermediate can be detected owing to its "long" half-life. In I_a and I_d mechanisms this possibility does not exist: the potential well of the intermediate becomes increasingly narrow until it disappears. In I_a the reactants accumulate sufficient energy to break M–X bonds, but before they break, new M–Y bonds are already formed. In I_d the bond with the entering group begins to be formed before the breaking of the bond of the leaving group. The energy schemes for the four possible processes (activation profiles) are given in Figure 7.15.

The identification of an A or I_a mechanism is by the fact that the rate constant is sensitive to the entering group (nature and concentration). The identification of a D or I_d mechanism is by the fact that the rate constant is sensitive to the leaving group (nature and concentration). D vs. I_d and A vs. I_a may be distinguished if it is possible to detect an intermediate with lower or greater coordination number, respectively.

Remark No mechanism can be taken as absolute. Currently, it is only possible to propose the most probable mechanism.

7.2.4
Substitutions on Square Planar Complexes

Among the square planar complexes with d^8 configuration, those of platinum have been most extensively studied, because ligand substitutions on platinum(II) take

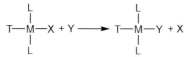

Figure 7.16 Scheme of a substitution reaction in a square planar complex.

place very slowly. These studies were expanded later to the square planar complexes of other d^8 metals (Rh^+, Ir^+, Ni^{2+}, Pd^{2+}, Au^{3+}) when techniques for the study of faster reactions became available,

7.2.4.1 Rate Law

The first step in elucidation of the mechanism is the experimental determination of the rate law. A substitution reaction in a square planar complex is shown schematically in Figure 7.16. Y is the entering group and X the leaving ligand. T is the ligand trans to the leaving group X. In the experimental procedure the conditions are, usually, of pseudo-first order, [Y] is in very great (excess) and practically constant during the reaction.

Assuming that the reverse process is not important, the rate law is:

$$rate = -d[ML_2TX]/dt = k_1[ML_2TX] + k_2[ML_2TX][Y] \quad or$$

$$rate = (k_1 + k_2[Y])[ML_2TX] = k_{obs}[ML_2TX], \quad where \quad k_{obs} = k_1 + k_2[Y]$$

While for different nucleophiles Y the value of k_1 remains unchanged, k_2 depends on the attacking ligand. Thus, the rate law of the substitution on square planar complexes usually consists of two competing pathways.

1. The reaction path expressed by the term $k_1[ML_2TX]$ corresponds to a dissociative reaction (I_d or D), where the intermediate product after the dissociation step is a low-coordinated species. This three-coordinated species reacts finally with Y. However, the role of the solvent must be emphasized in this mechanism. Indeed, there are many evidences indicating that this step is *associative*, going via solvent substitution of the leaving group. Indeed, the solvent is, in general, nucleophilic and will compete with Y to form ML_2TS (S = solvent). $k_1[ML_2TX]$ can be rewritten as $k'[ML_2TX][S]$, where $k'[S] = k_1$ ([S] is constant due its great excess as a solvent).

2. The $k_2[ML_2TX][Y]$ term (or direct path) is of first-order with regard to ML_2TX and Y (thus second order). This term implies a nucleophilic attack of Y on ML_2TX. If this attack is important (formation of the new M–Y bond) the rate of the reaction will depend markedly on the concentration and nature of Y. This will indicate an I_a mechanism.

We can rewrite the rate law as: rate = $(k_S + k_Y[Y])[ML_2TX]$

7.2.4.2 Mechanisms of Nucleophilic Substitution

The substitution reactions of square planar complexes are usually associative. The associative mechanism for ligand attack (Y) and for solvent attack (S) on a square-planar complex is shown in Figure 7.17. All experimental evidence supports this mechanism. Assuming a nucleophilic attack of Y (or S) on a square-planar complex, Y (or S) can be coordinated to M through an empty p_z orbital to form a square-pyramidal intermediate. There is then a rearrangement of the square-pyramid to give a trigonal bipyramid, with three ligands in the equatorial plane (Y(S), T, X) and two axial (L_1 and L_2). When X is released from the coordination, the T–M–Y angle is opened and the geometry comes again to square-pyramid. After the attack of the ligand Y(S), a pentacoordinated species is formed which is either a transition state or an intermediate product in this reaction (Figure 7.18).

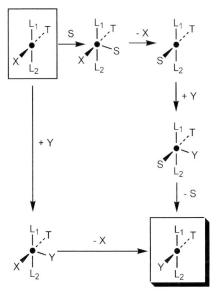

Figure 7.17 The associative mechanism for ligand attack (Y) and for solvent attack (S) on a square-planar complex.

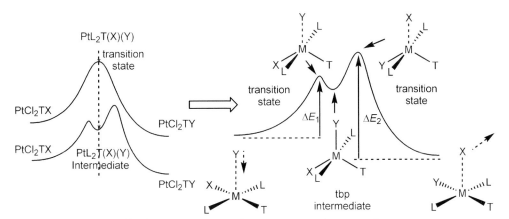

Figure 7.18 Reaction profiles for an associative mechanism in square planar complexes.

The two possible mechanisms, A or I_a are, thus, very difficult to distinguish. A long-lived intermediate would indicate an A mechanism. In reactions of Pd^{2+} and Pt^{2+} complexes no intermediate step has yet been found, which is why it is assumed that the pentacoordinated intermediate state is a transition state.

7.2.4.3 Proofs of the Associative Mechanism

Activation Magnitudes $\Delta S^{\#}$ or $\Delta V^{\#}$ are negative in substitution reactions of square-planar complexes indicating an associative mechanism.

Effect of the Entering Group Rates are strongly influenced by the nucleophile. The rate is faster according to the series: $PR_3 > I^-$, $SCN^- > N_3^- > NO_2^- > Br^- > py > NH_3 > Cl^- > H_2O > OH^-$.

We should certainly expect a dependence on the *nucleophilicity* of the entering ligand. Nucleophilicity is a *kinetic* measure which, therefore, can only be quantified kinetically. For platinum complexes, a frequently used scale of nucleophilicity of different ligands, the *nucleophilic reactivity constant*, is $n_{Pt} = \log [k_Y/k_S]$, where k_Y is the rate constant for the reaction of an entering nucleophile and k_S is the rate constant for the attack by solvent (methanol). The n_{Pt} values of several ligands are summarized in Table 7.9.

Increase in the value of n_{Pt} implies greater reactivity. For the reactivity series $I^- > Br^- > Cl^-$ the most important factor seems to be the polarizability or "softness". In general, nucleophiles with second row donors are much more effective. It is seen that soft nucleophiles such as PR_3 and CN^-, which can act as donors as well as acceptors, are better entering groups that hard nucleophiles, such as F^-, which are only donors.

Role of the Leaving Group In all mechanisms, even if predominantly associative, the stronger the bond with the leaving group, the slower the substitution. For associative reactions the effect of the leaving group depends on the extent of bond breaking in the *transition state*. For example it has been empirically demonstrated that the basicity of the leaving ligand makes substitution difficult. In general, the

Table 7.9 Rate constants (k, $M^{-1}s^{-1}$) of the reactions of *trans* $[PtCl_2(py)_2]$ with different nucleophiles in methanol, and the corresponding n_{Pt} values [10].

Nucleophile	10^3k	n_{Pt}	Nucleophile	10^3k	n_{Pt}
CH_3OH	0.00027	0	$(CH_3)_2S$	21.9	4.87
Cl^-	0.45	3.04	I^-	107	5.46
NH_3	0.47	3.07	SCN^-	180	6.65
N_3^-	1.55	3.58	CN^-	4000	7.14
Br^-	3.7	4.18	$P(C_6H_5)_3$	250000	8.80

good entering groups (good nucleophiles) are poor leaving groups with a spread of over 10^6 in rate across the series.

Effect of Steric Hindrance The presence of centers with strong steric hindrance inhibits the A or I_a reactions and facilitates the D or I_d reactions because the reduction of the coordination number may relieve the steric tensions in the activated complex, leading to so-called *steric acceleration*. From an experimental point of view, the square-planar complexes experience difficulties for approaching voluminous nucleophiles (A or I_a).

An example that can help to demonstrate the steric influence is found when comparing the rate of substitution of Cl^- by pyridine in $[PtCl(dien)]^+$ or $[PtCl(Et_4dien)]^+$ (dien = diethylenetriamine) complexes. The dien complex allows the substitution of Cl^- at room temperature, whereas the presence of the voluminous Et_4dien ligand requires a temperature close to 80 °C.

7.2.4.4 The *trans*-Effect [11]

The *trans-effect* is defined as the *"effect of a ligand upon the rate of ligand replacement of the group trans to itself"*. For discussion of the trans-effect, we will restrict ourselves to square planar complexes.

Empirically, the ligands trans to the leaving group (T ligands) can influence the reaction rate by several orders of magnitude (10^5–10^6). This phenomenon can be used for the directed synthesis of certain complexes. The synthesis of the two isomers of $[PtCl_2(NH_3)_2]$, which was already developed in the 19th century, constitutes the paradigm of the trans-effect. In the first substitution step all four leaving ligands are equal. In the second step, in principle, the two isomers can be formed, however, in each case only the secondary substitution trans to a chlorido ligand is observed (Figure 7.19). This is due to the greater trans-effect of the chloride relative to the ammonia.

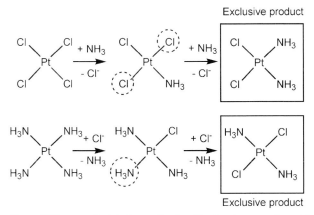

Figure 7.19 trans-effect in the synthesis of the two isomers *cis*- and *trans*-$[PtCl_2(NH_3)_2]$.

This trans-directing effect of a ligand seems to be *kinetically controlled,* as not in every case is the thermodynamically more stable isomer formed. By many similar substitutions an experimental relative series of the trans-directing properties of ligands in square planar complexes was obtained:

$$F^-, H_2O < OH^- < NH_3 < \text{amines} < \text{py} < Cl^- < Br^- < SCN^- < I^- \sim NO_2^- \sim C_6H_5^- <$$
$$CH_3^- < \text{phosphanes} \sim \text{arsines} < H^- < \text{olefins} < CN^- < CO, NO$$

In general, the rate of substitution of X by Y is faster if T is a strong σ donor (H^-, for example) or a good π-acceptor (CN^-, CO, NO, C_2H_4, etc.). However, relatively hard pure σ donors like NH_3, H_2O and OH^- only have a weak trans-directing effect. This apparent contradiction seems to indicate that several factors are responsible.

Two different effects can be differentiated, distinguishing between the trans-influence and the trans-effect:

1. trans-influence. The weakening of a trans-positioned ligand in its *ground state* has been called a *trans-influence,* which *is purely a thermodynamic phenomenon.* For example, for a series of *trans*-[ML$_2$**T**X] complexes in which we change the ligand T, the M–X length is very sensitive to the electronegativity of T: if the electronegativity decreases, the distance M–Cl increases. This increase in length weakens the bond in the ground state and leads to an increase in the reactivity of the dissociative process (trans-influence). The overall process can be visualized in terms of the bonds shared by T and X (Figure 7.20A). If T forms a strong σ bond, the M–X bond is weakened. As a consequence, the molecule is destabilized. Examples of strong σ-donor ligands are: $H_2O < OH^- < NH_3 < R-NH_2 < \text{py} < Cl^- < Br^- < CN^- < I^- < SCN^- < PR_3 < H^-$ (Figure 7.20A).

2. trans-effect. This is a kinetic effect. The trans-effect concerns the effect of the ligand on the *rate* of substitution of the ligand that is trans. *It arises from the relative stabilization of the transition state.* In an associative process it is necessary to take into consideration the transition state (or intermediate) whose geometry is trigonal bipyramid (in the cases studied here). A number of explanations have been offered for the trans-effect. The best consider both the σ-donor and π-acceptor capabilities of the ligand:

   – The σ-donor competition mentioned in relation to the trans-influence between T and X now decreases considerably because the ligands are placed at 120° not at 180° (Figure 7.20B), assuming a tbp transition state (Figure 7.20B). The trans-effect arising from a σ-donation follows, logically, the order of the trans-influence. A relative σ-donor scale for the rate of substitution based on the trans ligand T is: $Cl^- = 1$, $Ph^- = 2$; $CH_3^- = 3$; $H^- = 4$.

   – Chatt and Duncanson, were the first to consider that if T is a π-acceptor ligand, it increases the rate of the substitution process of the ligand positioned

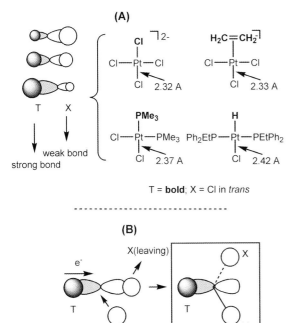

(A)

(B)

X(leaving)

e⁻

T

Y(entering)

Figure 7.20 *trans*-influence, as a thermodynamic effect.

trans to T. In this case, the state modified is not the ground state but the transition state (tbp) (Figure 7.21). π-Acid ligands remove the excess electron density acquired by the metal ion when the fifth ligand is bound. *This feature stabilizes the transition state.* The π-trans-directing series of a few selected ligands are as follows: $C_2H_4 > CO > CN^- > PR_3 > AsR_3$.

Summary (i) *a good σ donor* T (alkyl and hydrides) will bond more strongly to the metal and thus stabilize (weakly) the transition state (trans-effect of good σ donors); (ii) *a good π acceptor* ligand can absorb the electron density on the metal, which has increased due to an increase in the coordination number. The resulting strengthening of the metal T bond again stabilizes the transition state (trans-effect of good π acceptors, for example CN^-, CO, NO); (iii) finally, the empty d_{x2-y2} orbital in the transition state on the metal can receive π electron density and can form a weak π bond with a π donor ligand, thus stabilizing the transition state (trans-effect of π donor ligands, for example halides).

The trans-influence and the trans-effect achieve the same goal: they reduce the $\Delta G^{\#}$ value. However, the way in which they do this is completely different. The trans-influence assumes a thermodynamic mechanism because it destabilizes the ground state whereas the trans-effect is absolutely kinetic, because it stabilizes the transition state (Figure 7.22).

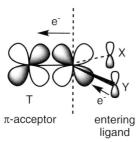

π-acceptor   entering
                    ligand

Figure 7.21 *trans*-effect due to a π-acceptor ligand.

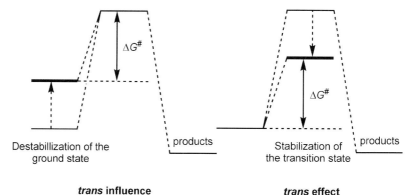

Figure 7.22 The energetic difference between *trans*-influence and *trans*-effect.

7.2.5
Substitution Mechanisms in O_h Complexes

7.2.5.1 Introduction

The replacement of one or more solvent molecules in an O_h solvated species by a ligand or ligands is a very common and important reaction. Such reactions were not, however, much studied in the early days of coordination chemistry since they were generally too fast to monitor by the long-established conventional techniques. Only ligand substitution at $[Cr(H_2O)_6]^{3+}$ species was slow enough for kinetic studies. The situation changed with the advent of stopped-flow techniques. In general, it has been demonstrated that the substitution mechanism in octahedral complexes is D–I_d, for the most studied complexes, mainly those of Co^{3+}.

7.2.5.2 Kinetics: Rate Laws

Under pseudo first-order conditions the rate depends on the concentration of L (entering ligand) but only for low ligand concentration. If [L] is very high the rate law is compatible with the Eigen–Wilkins mechanism, that comprises pre-association of the reactants in a rapidly established equilibrium followed by rate-limiting replacement (or interchange) of solvent ligand by the entering ligand.

The first step is, thus, the encounter between the complex and the incoming group, L, giving a kind of "outer-sphere" association. In most current solvents the half-life of this encounter complex is approximately 1 ns. This 'encounter complex formation' can thus be treated as a pre-equilibrium in any reaction occurring in some ns. This behavior is consistent with the following two reactions and the rate law derived from them:

$$M\text{-}H_2O + L \overset{K_e}{\leftrightarrow} [M\text{--}H_2O, L]$$

$$[M\text{--}H_2O, L] \overset{k}{\rightarrow} [M\text{--}L, H_2O]$$

$$\text{rate} = \frac{kK_e[M-H_2O][L]}{1+K_e[L]}$$

It is possible to carry out many experiments over a wide range of concentrations to find the rate law. For small concentrations of the entering group, such as for $K_e[L] \ll 1$, the rate law is reduced to the following second-order law:

$$\text{rate} = k_{obs}[M-H_2O][L] \text{ being } k_{obs} = kK_e$$

It is possible to find the k value because k_{obs} can be measured, and K_e estimated by means of the Fuoss and Eigen equation:

$$K_e = 4\pi a^3/3 \cdot N_A e^{-V/RT}$$

Where a is the distance of closest approach between the reactants and V is the electrostatic interaction energy between the complex and the nucleophile.

7.2.5.3 Mechanisms

Substitution reactions in O_h complexes with an A or D mechanism are scarce. Most of them are I_d (and some I_a). For example, for Co^{3+} complexes the mechanism is I_d, since the incoming group L has very little influence and does not show the *trans*-effect, typical of an associative process. In contrast, the Cr^{3+}, Rh^{3+} and Ir^{3+} substitutions are I_a.

As previously indicated, in ligand exchange reactions on solvate complexes, the rate law is not the most useful criterion for the formulation of a reaction mechanism as the concentration of one of the species, the solvent, is constant throughout the reaction. Therefore one depends basically on the interpretation of the *activation parameters*, especially the activation volumes. As the reactants and products of the reaction are equally solvated, the measured activation volumes can be interpreted directly as the theoretical activation volumes of the activated complexes. In Table

Table 7.10 Kinetic data for the H_2O exchange of some hexaaqua complexes of divalent metals, $[M(H_2O)_6]^{2+}$.

	V^{2+}	Mn^{2+}	Fe^{2+}	Co^{2+}	Ni^{2+}	Ru^{2+}
k, s^{-1}	89	2.1×10^7	4.4×10^6	3.2×10^6	3.2×10^4	1.8×10^{-2}
$\Delta H^{\#}$, kJ mol^{-1}	+62	+33	+41	+47	+57	+88
$\Delta S^{\#}$, J mol^{-1}k^{-1}	−0.4	+6	+21	+37	+32	+16
$\Delta V^{\#}$, cm^3 mol^{-1}	−4.1	−5.4	+3.8	+6.1	+7.2	−0.4
Electron configuration	t_{2g}^3	$t_{2g}^3 e_g^2$	$t_{2g}^4 e_g^2$	$t_{2g}^5 e_g^2$	$t_{2g}^6 e_g^2$	t_{2g}^6
Ionic radius, Å	0.79	0.83	0.78	0.74	0.69	0.73

Table 7.11 Rate constants k_{obs} for (a) the reaction of the complex $[Co(H_2O)(NH_3)_5]^{3+}$ with X^{n-} (at 45 °C) and (b) the hydrolysis of $[Co(NH_3)_5X]^{m+}$.

(a)		(b)	
X^{n-}	k, M^{-1}s^{-1}	X^{n-}	k, M^{-1} s^{-1}
NCS$^-$	1.3×10^{-6}	NCS$^-$	5.0×10^{-10}
$H_2PO_4^-$	2.0×10^{-6}	$H_2PO_4^-$	2.6×10^{-7}
Cl$^-$	2.1×10^{-6}	Cl$^-$	1.7×10^{-6}
NO$_3^-$	2.3×10^{-6}	NO$_3^-$	2.7×10^{-5}
SO$_4^{2-}$	1.5×10^{-6}	SO$_4^{2-}$	1.2×10^{-6}

7.10, the kinetic data for the water exchange of some aqua complexes of divalent metals are given.

For a pure A or D mechanism, activation volumes of about $-10\,cm^3\,mol^{-1}$ and $+10\,cm^3\,mol^{-1}$, respectively, are to be expected. Intermediate values allow deduction of an I_a or I_d mechanism. The increasingly positive $\Delta S^{\#}$ and $\Delta V^{\#}$ values when going from V^{2+} to Ni^{2+} (except $\Delta V^{\#}$ for Mn^{2+}) are an indication of an increasingly dissociative exchange mechanism, which is why the correspondences I_a (V^{2+}, Mn^{2+}), I (Fe^{2+}) and I_d (Co^{2+}, Ni^{2+}) can be made.

Dissociation of a ligand from an octahedron leads to a five-coordinate species. There is very little difference between the two possibilities for five-coordinated complexes, trigonal bipyramidal and square pyramidal. Both geometries are suggested in different octahedral substitution reactions.

7.2.5.4 Factors That Affect the Rate of O_h Complexes

Influence of the Entering and Leaving Ligands   As expected, the influence of the entering group is small. For example, the substitution of the aqua ligand in $[Co(H_2O)(NH_3)_5]^{3+}$ by anionic ligands following an I_d mechanism becomes evident from the small dependence of the reaction rate on the type of the incoming group (Table 7.11(a)).

Given the character I_d (or D) in most of the substitution reactions in O_h complexes, the nature of the leaving group will be more important than that of the entering group. As an example, the reaction rate of the hydrolysis of $[CoX(NH_3)_5]^{m+}$ is strongly influenced by the properties of the leaving ligand (Table 7.11(b)).

These results are supported by the observed influence of the space required by the "non-participating" ligands on the substitution rate: the solvolysis of $[CoCl(NH_2Me)_5]^{2+}$ to $[Co(H_2O)(NH_2Me)_5]^{3+}$ is faster by more than one order of magnitude than that of the pentaammine complex. The greater space requirements of the methylamine ligand favor the dissociation of the chlorido ligand.

Effect of the Metal Ion: Influence of the Crystal Field Activation Energy (CFAE) Some metal complexes are labile regardless of the ligand environment, whereas complexes of other metals are inert. One would anticipate that electrons in the e_g orbitals that point at the ligands would cause an extra destabilization of the metal–ligand bond. This concept has been semi-quantitatively applied by considering the CFAE. We can define: CFAE = CFSE (activated complex) – CFSE (O_h complex). The activation data, assuming the activated complex to be square-pyramidal (I_d mechanism, c.n. = 5) or monocapped trigonal prism (I_a mechanism; c.n. = 7) are given in Table 7.12.

More positive (or less negative) CFAE will give a greater inertia for the corresponding configurations. As a consequence, the d^3, d^6 and d^8 are the most inert complexes. Furthermore, the most likely structure of the intermediate, for a given configuration, will be that with the lower CFAE. This feature will indicate the more favorable reaction pathway.

Table 7.12 CFAE for several d^n and different activated complexes (in Dq units).

d^n	Activated complex structure			
	Strong field		Weak field	
	Square pyramidal	Monocapped trigonal prism	Square pyramidal	Monocapped trigonal prism
d^1	−0.57	−2.08	−0.57	−2.08
d^2	−1.14	−0.68	−1.14	−0.68
d^3	2.00	1.80	2.00	1.80
d^4	1.43	−0.26	−3.14	−2.79
d^5	0.86	1.14	0	0
d^6	4.00	3.63	−0.57	−2.08
d^7	−1.14	−0.98	−1.14	−0.68
d^8	2.00	1.80	2.00	1.80
d^9	−3.14	−2.79	−3.14	−2.79

Acid and Base Catalysis   Additional components in the reaction medium, as for example H^+, OH^- or metal ions, can accelerate significantly the substitution reactions on complexes. These species either modify one of the reactants or participate in the transition state of the substitution. When studying their influence it is important to distinguish between a pure "medium effect" of the added component (for example, ion strength variation of the aqueous solution) and an alternative reaction path.

Hydrolysis in an Acid Medium   Studies on the substitution of monodentate ligands in acid solutions have been carried out, especially on inert complexes, for example $[CrX(H_2O)_5]^{2+}$. In these cases, in a first step, there is partial protonation of the leaving anionic ligand X^-. The protonation weakens the M–X bond and facilitates X^- removal as HX. In some cases, the protonized intermediate product is formed in higher concentration (for example, with $X^- = CH_3COO^-$), so that k_{obs} presents a non-linear dependence on $[H^+]$ and a complex relationship is found.

Hydrolysis in a Basic Medium   The substitution reactions in octahedral complexes are generally not very sensitive to the entering group, with one exception: the OH^- group. When *at least one ligand in the complex contains acid hydrogen atoms*, it is possible that it may be deprotonated by a base, and thus another, more reactive intermediate of the complex for the substitution, its *conjugated base*, may be formed. The classical example of a base-induced acceleration is the hydrolysis of the cobalt(III) complex, $[CoX(NH_3)_5]^{2+}$, in alkaline solution. The rate law is: rate = $k[CoCl(NH_3)_5][OH]$.

The studies carried out with OH^- demonstrate that the role of this group is to act not only as an entering group but also as a Brönsted base.

$$[CoCl(NH_3)_5]^{2+} + OH^- \rightarrow [CoCl(\mathbf{NH_2})(NH_3)_4]^+ + H_2O$$

$$[CoCl(NH_2)(NH_3)_4]^+ \rightarrow [Co(NH_2)(NH_3)_4]^{2+} + Cl^- \quad \text{(slow)}$$

$$[Co(NH_2)(NH_3)_4]^{2+} + H_2O \rightarrow [Co(\mathbf{OH})(NH_3)_5]^{2+} + NH_3 \quad \text{(rapid)}$$

This mechanism was controversial for a long time, but now it has been established unequivocally as a *mechanism of the internal conjugated base*. In a previous equilibrium an NH_3 ligand, polarized by the coordination on the trivalent cobalt center and therefore more acid than free NH_3, is converted by OH^- into the NH_2^- amide complex (NH_2^- is a strong π-donor). The release of Cl^- is accelerated. Via a reaction path over a pentacoordinated intermediate step, the anionic ligand is substituted by H_2O; through the subsequent internal proton transfer, the isolated $[Co(OH)(NH_3)_5]^{2+}$ product is obtained.

According to this mechanism, ligands such as pyridine or CN^- cannot withstand the attack of the OH^- group and, hence, the $[CoCl_2(py)_4]^+$ or $[CoCl(CN)_5]^{3-}$ species do not present the basic hydrolysis typical of the amminated and similar complexes.

Proof that the basic hydrolysis proceeds via the elimination of a proton is given by "trapping" the pentacoordinated intermediate by adding a great quantity of an anion instead of OH^-. This occurs, for example, in the complex $[CoX(NH_2R)_5]^{2+}$ with N_3^- as sequestering agent. If the hydroxo complex was the first to be formed, it would be impossible to detect the presence of the azido-complex, as happens during the course of the reaction.

Other Cases   Added metal cations can play a similar role to H^+ in ligand substitutions, in which case the efficiency of the metal ions depends on a series of factors. The acceleration of the solvolysis through M^{n+} is closely dependent on the bonding capacity of the metal ion to the leaving ligands (and naturally on the capacity of the ligand to fulfil a bridge function). For example, hard metal ions like Be^{2+} and Al^{3+} accelerate, like H^+, the abstraction of hard coordinated ligands, like F^-, whereas soft metal ions (Ag^+, Hg^{2+}) are most effective when the leaving ligand is also soft (Cl^-, Br^-).

References

1 The Irving–Williams series can be found in almost all Inorganic Chemistry textbooks. A very interesting and complete explanation is given in Huheey, J.E., Keiter, E.A., Keiter, R.L. (1993), *Inorganic Chemistry. Principles of Structure and Reactivity*. Harper Collins College Publishers, New York.

2 Ahrland, S., Chatt, J., Davis, N.R. The Relative Affinities of Ligand Atoms for Acceptor Molecules and Ions, *Quart. Rev. (London)*, 1958, *12*, 265. (The interested reader can find good explanations in most Inorganic Chemistry books).

3 Pearson, R. J. Hard and Soft Acids and Bases, *J. Am. Chem. Soc.* 1963, *85*, 3533.

4 Jorgensen, C.K. "Symbiotic" Ligands, Hard and Soft Central Atoms, *Inorg. Chem.* 1964, *3*, 1201.

5 Parr, R.G., Pearson, R.G. Absolute hardness: companion parameter to absolute electronegativity, *J. Am. Chem. Soc.* 1983, *105*, 7512.

6 Atkins P., de Paula, J. (2002), *Physical Chemistry*, 7th edn. Oxford University Press. Oxford.

7 Drago, R.S., Wayland, B.B. A Double-Scale Equation for Correlating Enthalpies of Lewis Acid–Base Interactions, *J. Am. Chem. Soc.* 1965, *87*, 3571.

8 Martell, A.E., Handcock, R.D. (1996), *Metal Complexes in Aqueous Solution*, Plenum Press, New York.

9 Tolman, C.A. Steric Effects of Phosphorus Ligands In Organometallic Chemistry And Homogeneous Catalysis, *Chem. Rev.* 1977, *77*, 313.

10 Pearson, R.G., Sobel, H., Songstad, J. Nucleophilic Reactivity Constants toward Methyl Iodide and trans-[Pt(py)$_2$Cl$_2$], *J. Am. Chem. Soc.* 1968, *90*, 319.

11 Kauffmann, G.B. Il'ya Il'ich Chernyaev (1893–1966) and the trans-effect, *J. Chem. Educ.* 1977, *54*, 86.

Bibliography

Constable, E.C. (1996) *Metals and Ligand Reactivity*, VCH, Weinheim.

Tobe, M.L., Burgess, J. (1999) *Inorganic Reaction Mechanisms*, Longman, Essex.

Atwood, J.D. (1997) *Inorganic and Organometallic Reaction Mechanisms*, 2nd edn., VCH, New York.

Part Two   Electronic Properties

8
Crystal Field Theory and Spin–Orbit Coupling: Energy Terms and Multiplets

8.1
Introduction: The Atomic/Electronic Hamiltonian

The energy of a d^n atom or ion containing unpaired electrons or non-zero spin nuclei is calculated by means of the following Hamiltonian:

$$\hat{H} = \hat{H}_{EL} + \hat{H}_{CF} + \hat{H}_{LS} + \hat{H}_{SS} + \hat{H}_{ZE} + \hat{H}_{HF} + \hat{H}_{ZN} + \hat{H}_{II} + \hat{H}_Q$$

$\hat{H}_{EL}$ is the electronic Hamiltonian, comprising the sum of the kinetic energy of each electron, their potential energy and the electron repulsion energies, e^2/r_{ij}. Its order of magnitude is 10^4 to $10^5\,cm^{-1}$.

$\hat{H}_{CF}$ is the crystal field Hamiltonian which perturbs and splits the above energy. Its effect on the d orbitals has been studied in Chapter 1. Although it is a complex Hamiltonian, in the case of highly symmetrical local configurations (O_h, T_d, D_{4h}) it takes on particularly simple forms. Its order of magnitude is similar to that of $\hat{H}_{EL}$.

$\hat{H}_{LS}$ is the spin–orbit Hamiltonian. It expresses the interaction between the spin of an electron and its own orbital angular momentum. Within an atomic term, it takes the form of a function of the operators $\hat{\boldsymbol{L}}$ and $\hat{\boldsymbol{S}}$ (orbital and spin angular momenta), $\hat{H}_{LS} = \lambda \hat{\boldsymbol{L}} \cdot \hat{\boldsymbol{S}}$, where λ is the spin–orbit splitting parameter. Its energy is in the order of $10^2\,cm^{-1}$ for elements of the first transition row, higher for elements of the second and third transition rows, and very high for lanthanide and actinide elements.

$\hat{H}_{SS}$ is the spin-spin interaction Hamiltonian. Within the "spin Hamiltonian formalism" (wavefunctions will *only* depend on $|S,M_s\rangle$) it takes the form $\hat{H}_{SS} = D[\hat{S}_z^2 - S(S+1)/3] + E(\hat{S}_x^2 - \hat{S}_y^2)$ where D and E are the *zero field splitting* parameters. The order of magnitude of this energy is 0 to $20\,cm^{-1}$.

Working with these four Hamiltonians, we will get to the final splitting of the multiplets. Thus, even the D parameter (see below) appears naturally with these Hamiltonians.

$\hat{H}_{ZE}$ is the Zeeman electronic Hamiltonian, which arises when an external magnetic field is applied. It is expressed as $\hat{H}_{ZE} = \mu_B g \boldsymbol{H} \cdot \hat{\boldsymbol{S}}$. Its magnitude is of the order of 0 to $1\,cm^{-1}$ and it depends on $\boldsymbol{H}$ (the magnetic field).

Coordination Chemistry. Joan Ribas Gispert
Copyright © 2008 WILEY-VCH Verlag GmbH & Co. KGaA, Weinheim
ISBN: 978-3-527-31802-5

Table 8.1 Atomic/electronic Hamiltonians and their corresponding spectroscopic techniques.

Hamiltonian	Spectroscopic technique
$\hat{H}_{EL} + \hat{H}_{CF}$	electron spectroscopy
$\hat{H}_{LS}$	electron spectroscopy and magnetism
$\hat{H}_{SS}$	magnetism and electron paramagnetic resonance
$\hat{H}_{ZE}$	magnetism and electron paramagnetic resonance
$\hat{H}_{HF}$	electron paramagnetic resonance
$\hat{H}_{ZN} + \hat{H}_{II} + \hat{H}_{Q}$	nuclear magnetic resonance
$\hat{H}_{Q}$	nuclear quadrupole resonance and Mössbauer spectroscopy

$\hat{H}_{HF}$ is the hyperfine interaction Hamiltonian. It is expressed as $\hat{H}_{HF} = \sum_i A_i \hat{S}_i \cdot \hat{I}_i$ and contains the interactions between the unpaired electrons and the adjacent nuclei. If there are no unpaired electrons then $\hat{H}_{LS}$, $\hat{H}_{SS}$, $\hat{H}_{ZE}$ and $\hat{H}_{HF}$ are zero as $S = 0$.

The three remaining Hamiltonians ($\hat{H}_{ZN}$ = Zeeman nuclear; $\hat{H}_{II}$ = internuclear interaction; and $\hat{H}_{Q}$ = nuclear quadrupole effect) are less important in coordination chemistry and correspond to nuclear magnetic resonance, nuclear quadrupole resonance, and the Mössbauer effect. Their energies are the smallest.

As will be seen in subsequent chapters each of the above energy operators has a specific effect on the electronic structure of coordination compounds. It is helpful to know from the outset which spectroscopic techniques will provide the information required about the coordination compound under study (Table 8.1).

8.2
Application of Atomic and Spin Hamiltonians to Many-electron Wavefunctions: Terms, Multiplets and Magnetic States

8.2.1
Basic Concepts

In Chapter 1 we studied the effect of the crystal field Hamiltonian on monoelectronic wavefunctions (orbitals). Here, the d and f orbitals were split into different energy groups depending upon the symmetry of the crystal field applied. This procedure would be sufficient if all the coordination complexes were from monoelectronic ions. However, transition ions are normally many-electron systems. Therefore, in energy calculations it is necessary to introduce the interelectron repulsions that give rise to the concept of *energy term*. The application of successive Hamiltonians gives rise to further energy splitting that leads to what are known as *energy multiplets* and *magnetic states* (Figure 8.1).

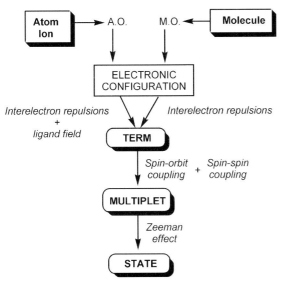

Figure 8.1 Scheme showing the concept of term, multiplet and state from the electron configuration.

Applying the first two Hamiltonians ($\hat{H}_{EL} + \hat{H}_{CF}$) gives rise to what in spectroscopy is known as the *spectroscopic term*. Application of the Hamiltonian $\hat{H}_{LS}$ gives rise to what are termed *multiplets* and application of the Hamiltonian $\hat{H}_{ZE}$ leads to *magnetic states*.

Readers can find some confusion in the literature about the names term, multiplet and magnetic state. In general, nowadays there is a tendency to use the word "term" to describe what arises from an appropriate treatment of electron repulsion, whereas the word "state" is used to describe something which is observable experimentally.

Coordination compounds may be treated at two levels, the atomic (with special emphasis on the central ion d^n) and the molecular (studying the real molecule by calculating molecular orbitals). At the atomic level the energy terms are calculated by means of two treatments: the *weak field method* and the *strong field method*. Both treatments are perturbational. The weak field model assumes that interelectron repulsion is more significant than the crystal field, and therefore they are applied in this order. The strong field method proposes the opposite: first, the d orbitals are split through the effect of the crystal field, before moving on to include the electron repulsions. Both methods can lead to what are known as *crystal field terms*. A molecular level treatment assumes knowledge of the energy diagram of the molecular orbitals for the complex studied. This diagram can be used to calculate the energy terms using the *direct product* method for the symmetry species of the corresponding molecular orbitals. Its mathematical treatment is very similar to

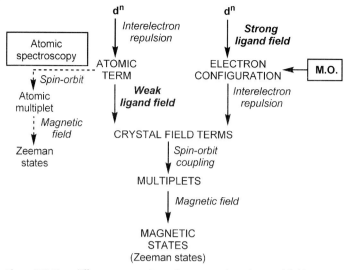

Figure 8.2 Two different perspectives of strong and weak crystal fields.

the strong field method. Figure 8.2 shows these approaches. In the next section the most important aspects of the direct product in calculating the energy terms, are discussed.

8.2.2
Direct Product Method for Calculating Terms

This method is based on the quantum notion that the energy term to which a given electron configuration corresponds can be directly derived from the direct product of the symmetry species of the electron configuration in question. The spin function is also calculated by means of the direct product of the spin components. It is worth remembering a principle of quantum chemistry that is very important for an overall understanding of the above argument:

> *Pauli Exclusion Principle*: The total electron wavefunction is antisymmetric with respect to the interchange of any two electrons.

A wavefunction has two components: $\psi = \psi_{orbital} \cdot \psi_{spin}$. If the spin wavefunction is symmetric (as in the case of the triplet state, $S = 1$) then the orbital wavefunction must be antisymmetric; if the spin function is antisymmetric (as in the singlet state, $S = 0$) the orbital function must be symmetric.

8.2.3
Different Cases for the Direct Product Calculation

The representations of any direct product are given in the Character Tables for the point symmetry groups, and these can be found in many books on group theory [1].

Case 0 *A set of fully occupied orbitals belongs to the totally symmetric species of the point group considered* ($^1A_{1g}$ in O_h). Thus, only the partially occupied orbitals are important in calculating the electronic terms. The direct product must be carried out *only* on these.

Case 1 *One electron in an orbital.* The energy term belongs to the same representation as the corresponding orbital. The spin multiplicity will be $2S + 1 = (2 \times 1/2) + 1 = 2$ (a doublet). This situation is equivalent to a set of degenerate orbitals which lack *a single electron* to be fully occupied (positive hole formalism). In the term symbol the orbital part is written in capitals, while the orbital component is rewritten in the lower case. Example: $(t_{2g})^1$ or $(t_{2g})^5 \rightarrow {}^2T_{2g}$.

Case 2 *Two electrons in a degenerate orbital.* Example $(t_{2g})^2$. The corresponding term will have two possible spin parts: $1/2 + 1/2 = 1$ or $1/2 - 1/2 = 0$. The multiplicity $2S + 1$ will be either 3 (triplet) or 1 (singlet). The orbital part is calculated by means of the direct product $t_{2g} \times t_{2g} = A_{1g} + E_g + [T_{1g}] + T_{2g}$. The square brackets around the T_{1g} term appear in the tables of direct products and indicate that, orbitally, it is the only antisymmetric term. Given the Pauli exclusion principle, the possible terms will be: ${}^3T_{1g} + {}^1A_{1g} + {}^1E_g + {}^1T_{2g}$.

Case 3 *Unpaired electrons in more than one group of orbitals.* Example $(t_{2g})^1(e_g)^1$. The terms of each group of orbitals are calculated separately, and then the direct product is carried out again. In this case the direct product of the spin components must be taken into consideration, as they are independent of the point symmetry group to which the molecule belongs (Table 8.2). These direct products come from the theory of double groups [1].

In the proposed example, $(t_{2g})^1(e_g)^1 \rightarrow {}^2T_{2g} \times {}^2E_g = {}^{1,3}T_{1g} + {}^{1,3}T_{2g}$

As they are electrons in different orbitals, there is no restriction on the combinations of the spin and orbital parts. All the possible terms are valid: ${}^3T_{1g} + {}^3T_{2g} + {}^1T_{1g} + {}^1T_{2g}$.

Case 4 *m electrons in m degenerate orbitals (m > 2).* In the coordination chemistry of d elements the maximum value of m is 3, in the configuration $(t_{2g})^3$.

Spin part. This may be a doublet or a quadruplet, as shown in Figure 8.3.

For the orbital part we cannot rely on the symmetric nature of the triplet or the antisymmetric nature of the singlet, as there are other spin multiplicities (doublet and quadruplet). Group theory shows that the χ nature of the two spin multiplicities, for an operation of symmetry R, is: $\chi_{\text{DOUBLET}} = 1/3\{[\chi(R)]^3 - \chi(R^3)\}$; $\chi_{\text{QUADRUPLET}} = 1/6\{[\chi(R)]^3 - 3\chi(R)\chi(R^2) + 2\chi(R^3)\}$.

Table 8.2 Direct product of the spin components (independent of the specific point group).

Separate Terms	Product
singlet × singlet	singlet
singlet × doublet	doublet
singlet × triplet	triplet
singlet × quadruplet	quadruplet
doublet × doublet	singlet + triplet
doublet × triplet	doublet + quadruplet
doublet × quadruplet	triplet + quintuplet
triplet × triplet	singlet + triplet + quintuplet
triplet × quadruplet	doublet + quadruplet + sextuplet
quadruplet × quadruplet	singlet + triplet + quintuplet + septuplet

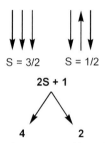

Figure 8.3 Doublet and quadruplet arising from the $(t_{2g})^3$ configuration.

As applying these formulae is usually mathematically laborious, other operative methods have been developed which enable the correspondence between the orbital and spin components to be identified. The simplest of these methods is known as the *symmetry descent method*. The reader can find this method in Ref. [1].

8.3
Weak Field Method

The weak field model assumes that the interelectron repulsions prevail over the crystal field. Therefore, the perturbational treatment must be applied in this order. This implies that we must first know the atomic terms derived from the interelectron repulsions of the free ion.

8.3.1
Atomic (Spectroscopic) Terms

For light atoms, the spin–orbit interaction is much weaker than the interelectron interactions. In this case the Russell–Saunders (LS) coupling takes place for which the orbital and spin momenta of electrons l_i and s_i are summed separately into the total orbital momentum $L = \Sigma l_i$ and $S = \Sigma s_i$. In the LS coupling scheme the wavefunctions of the atom are characterized by the following quantum numbers, L, M_L, S, M_S and are written as: ^{2S+1}L.

When $L = 0$ the term is called S, for $L = 1$ the term is referred to as P, for $L = 2$ the term is known as D, and so on consecutively using the terms F, G, H, I, J, K, etc. as L increases by one. In other words, *we use the same terminology as that for atomic orbitals, which are written in lower case, as the orbital component of a wavefunction l behaves the same as an L term with respect to any symmetry operation of a given point group.*

The terms which are derived from any d^n configuration are known in atomic spectroscopy. Table 8.3 shows the energy terms for any d^n atomic configuration (obtained by means of the L–S, or Russell–Saunders, coupling formalism). The above treatment is described in more detail in many Quantum Chemistry books and in Ref. [2].

The energy ordering of the free-ions terms results after a detailed numerical computation. Only the ground term may be deduced from some simple rules due to Hund:

- the most stable (ground) term is the one with the highest spin multiplicity;
- in the case of equal spin multiplicity, the one with the highest L multiplicity.

Table 8.3 places the ground term first, followed by terms with the same multiplicity as the ground term. The other terms are not shown in order of energy.

Important Note   In coordination chemistry the terms of interest are, principally, the ground term and those which have the same spin multiplicity as the ground term.

Table 8.3 Energy terms of the d^n configurations.

Configuration	Terms	Total degeneracy
d^1, d^9	^2D	10
d^2, d^8	^3F, ^3P, ^1G, ^1D, ^1S	45
d^3, d^7	^4F, ^4P, ^2H, ^2G, ^2F, 2^2D, ^2P	120
d^4, d^6	^5D, ^3H, ^3G, 2^3F, ^3D, 2^3P, ^1I, 2^1G, ^1F, 2^1D, 2^1S	210
d^5	^6S, ^4G, ^4F, ^4D, ^4P, ^2I, ^2H, 2^2G, 2^2F, 3^2D, ^2P, ^2S	252
d^{10}	^1S	1

8.3.2
Energy of the Atomic Terms

The energy of the atomic terms is parametrized. The most widely used parameters in atomic spectroscopy are the Slater–Condon and/or Racah parameters. The Slater–Condon parameters are F_0 (which expresses the spherical repulsion, derived from the radial part of the wavefunction), F_2, F_4, etc. (which express the non-spherical repulsion). F_0 uniformly shifts all free-ion terms. Thus, only the two parameters F_2 and F_4 are required to define d terms splitting due to interelectron repulsion in a central field. Racah uses other parameters that are related to the Slater–Condon ones: $B = F_2 - 5F_4$ and $C = 35F_4$. As an example, Table 8.4 shows the term energies for the d^2 configuration.

The advantage of the Racah parameters is that the energy difference between the ground term and the excited terms of the same spin multiplicity only depends on B. Thus, by way of example, the energy differences between the terms of maximum spin multiplicity for any d^n configuration are given in Table 8.5.

The other d^n configurations have no excited term with the same spin multiplicity as the ground term.

Note The values of the parameters B and C are determined experimentally from the spectra of atoms or gaseous ions; $C \approx 4B$ [2]. Only the B and C values for ions from the first transition row are known with exactitude. In general, for a given oxidation state the values decrease by 50% when moving from the first to the second transition series, and by a lesser amount between the second and the third. There is a positive increase when moving from one element to another of the same period from left to right, and when the element's oxidation state is increased. The

Table 8.4 Atomic term energies for the d^2 configuration.

Term	Slater–Condon	Racah
^1S	$F_0 + 14F_2 + 126F_4$	$A + 14B + 7C$
^1G	$F_0 + 4F_2 + F_4$	$A + 4B + 2C$
^3P	$F_0 + 7F_2 - 84F_4$	$A + 7B$
^1D	$F_0 - 3F_2 + 36F_4$	$A - 3B + 2C$
^3F	$F_0 - 8F_2 - 9F_4$	$A - 8C$

Table 8.5 Energy differences between the terms of maximum spin multiplicity.

Configuration	Difference between:	Slater–Condon parameters	Racah parameters
d^2, d^8	$^3P - ^3F$	$15F_2 - 75F_4$	$15B$
d^3, d^7	$^4P - ^4F$	$15F_2 - 75F_4$	$15B$

Table 8.6 Splitting of the terms (multiplicity in brackets) due to the effect of a crystal field.

Term	O_h	T_d	D_{4h}
S (1)	A_{1g}	A_1	A_{1g}
P (3)	T_{1g}	T_1	$A_{2g} + E_g$
D (5)	$E_g + T_{2g}$	$E + T_2$	$A_{1g} + B_{1g} + B_{2g} + E_g$
F (7)	$A_{2g} + T_{1g} + T_{2g}$	$A_2 + T_1 + T_2$	$A_{2g} + B_{1g} + B_{2g} + 2E_g$
G (9)	$A_{1g} + E_g + T_{1g} + T_{2g}$	$A_1 + E + T_1 + T_2$	$2A_{1g} + A_{2g} + B_{1g} + B_{2g} + 2E_g$
H (11)	$E_g + 2T_{1g} + T_{2g}$	$E + T_1 + 2T_2$	$A_{1g} + 2A_{2g} + B_{1g} + B_{2g} + 3E_g$
I (13)	$A_{1g} + A_{2g} + E_g + T_{1g} + 2T_{2g}$	$A_1 + A_2 + E + T_1 + 2T_2$	$2A_{1g} + A_{2g} + 2B_{1g} + 2B_{2g} + 3E_g$

books by Sutton [2] and Figgis [3] include a list of these parameters for some ions, in different oxidation states.

8.3.3
Crystal Field Effect

Mathematically, L (terms) is equivalent to l (atomic orbital). *Therefore, for a crystal field of a given symmetry the atomic terms are split, qualitatively, like the atomic orbitals.* The splitting for any value of L is shown in Table 8.6 for the three geometries O_h, T_d, and D_{4h}, (compare with Tables 1.2 and 1.3 for the splitting of atomic orbitals).

Two points of clarification should be made: (i) all the energy terms of a d^n metal ion come from its d orbitals. These d orbitals are g (gerade). Therefore, in centrosymmetric environments all the crystal field terms coming from d^n must also be g. If the environment is not centrosymmetric (for example, T_d) no g (gerade) or u (ungerade) subscript is used ; (ii) the spin multiplicity of the energy term before applying the crystal field is not modified by the perturbation of the ligands. The crystal field does not affect the spin properties.

8.3.4
Observation Regarding the Ground Term

Once the crystal field terms are known (Table 8.6) it is necessary to determine the order of the respective energies from both the qualitative and quantitative point of view.

Qualitatively it is very useful to know which the ground term is. The symmetry label of the ground term in octahedral complexes is shown in Figure 8.4. The spin multiplicity is the same as that of the spectroscopic term of the free ion (term of a maximum spin multiplicity in Table 8.3). As a practical exercise it can also be deduced directly from Figure 8.4, simply by applying the expression $2S + 1$, where $S = \Sigma s$ ($s = 1/2$) counting only the unpaired electrons. Thus, the multiplicity varies from 1 to 6 and then falls to 1 for d^{10} (where all the electrons are paired). The calculation of the orbital part can be done intuitively.

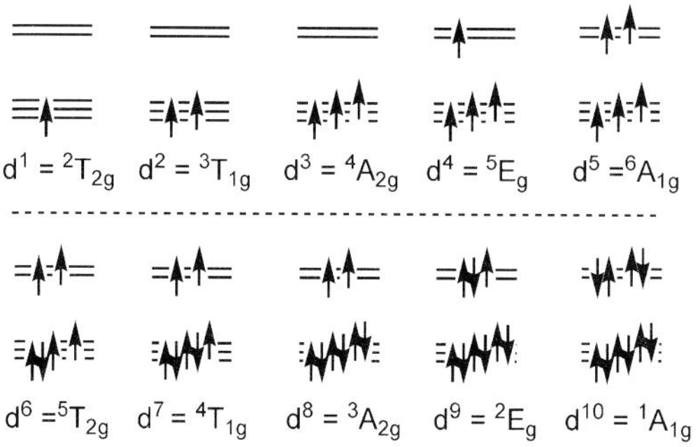

$d^1 = {}^2T_{2g}$ $d^2 = {}^3T_{1g}$ $d^3 = {}^4A_{2g}$ $d^4 = {}^5E_g$ $d^5 = {}^6A_{1g}$

$d^6 = {}^5T_{2g}$ $d^7 = {}^4T_{1g}$ $d^8 = {}^3A_{2g}$ $d^9 = {}^2E_g$ $d^{10} = {}^1A_{1g}$

Figure 8.4 d^n electron configurations in a weak field with their corresponding ground terms.

- When there is a single unpaired electron in the t_{2g} orbitals, $(t_{2g})^1$, the symmetry of the corresponding term will be the same as that of the orbital and is written in capitals (T_{2g}).

- When there are two unpaired electrons, $(t_{2g})^2$, there are three possible orbital occupations, and therefore it is a T (triply degenerate) term. It is not possible to know *a priori* if it is a T_{1g} or T_{2g} term and, therefore, it is necessary to carry out the direct product $(t_{2g}) \times (t_{2g})$. This direct product, as pointed out above, is $A_{1g} + E_g + [T_{1g}] + T_{2g}$. Given that the spin part is a *triplet,* and therefore symmetric, the orbital function must be antisymmetric. The only possible term is ${}^3T_{1g}$.

- When there are three unpaired electrons in the t_{2g} orbitals, $(t_{2g})^3$, there is only one possible occupation (there is no orbital degeneracy). It must be an A_g term, but is it A_{1g} or A_{2g}? The A_{1g} term is totally symmetric and can only correspond to five unpaired electrons in the five d orbitals. Therefore, by exclusion, it must be the A_{2g} term. A rigorous demonstration of this term can be carried out by applying the symmetry descent method.

- Provided there are three electrons in the t_{2g} layer its symmetry will be A_{2g}. When there is a fourth electron in the e_g orbitals there will be two orbital possibilities $(x^2 - y^2)^1$ and $(z^2)^1$. This double orbital possibility is symbolised by the term E_g (E, doubly degenerate according to group theory). The final term will be E_g, as $A \times E = E$, regardless of whether we are dealing with A_1 or A_2.

- The presence of a fifth electron in the e_g orbitals produces a configuration that is totally symmetric from the orbital point of view $(t_{2g})^3(e_g)^2$. Therefore, the term will be totally symmetric from the orbital point of view, ${}^6A_{1g}$. From here on, and taking into account that five unpaired electrons produce a totally symmetric term, the orbital part of the terms corresponding to the configurations d^6, d^7, d^8, d^9 and d^{10} will be the same as those for the configurations d^1, d^2, d^3, d^4 and d^5, respectively (Figure 8.4). The only thing that needs to be changed is, obviously, the spin multiplicity.

8.3.5
Energy Values of the Crystal Field Terms

8.3.5.1 **Effect of the Crystal Field on the Spectroscopic Terms**
An S term is orbitally non-degenerate. As the crystal field only affects the orbital part, it will not produce any splitting. This is the case of a d^5 ion: $^6S \rightarrow {}^6A_{1g}$.

The orbital part of the wavefunctions of a P term is the same as the orbital part of the wavefunctions of the p orbitals. Just as the components of a V_{oct} potential do not split the p orbitals, neither do they split the P terms.

The electron configurations which have a D term as their ground term are d^1, d^4, d^6 and d^9 (Table 8.3). The orbital part of the wavefunctions of a D term is the same as the orbital part of the wavefunctions of the d orbitals. Therefore, they are split orbitally into $T_{2g} + E_g$.

In the four cases the energy is measured with respect to the unsplit term. In these simple cases there is no need for a rigorous demonstration, as the separation between the d orbitals upon applying an O_h symmetry field is precisely $10Dq = \Delta_o$. The energy difference between the two terms, T_{2g} and E_g, is, thus, $10Dq = \Delta_o$. Intuitively it can be seen that the value Δ_o will be distributed according to the *respective multiplicities* of these terms. The energy calculation for a d^1 ion thus gives the following results:

d^1 (^2D): $\quad T_{2g}$ (ground term) at energy: $\quad -4Dq\,(3 \times 4 = 12)$

E_g (excited term) at energy: $\quad 6Dq\,(2 \times 6 = 12)$

d^9 (^2D): applying the positive hole formalism naturally produces the same energy values but with the signs changed: $Dq(d^9) = -Dq(d^1)$

E_g (ground term) at energy: $\quad -6Dq$

T_{2g} (excited term) at energy: $\quad 4Dq$

The other two configurations which have a D ground term are d^4 and d^6. The energy diagram of a d^4 ion is the same as the energy diagram of a d^9 (d^{5+4}) ion, while the diagram for a d^6 (d^{5+1}) ion is the same as the diagram for a d^1 ion. These splitting relationships for D ground terms are summarized in Figure 8.5.

The electron configurations with an F ground term are d^2, d^3, d^7 and d^8 (Table 8.3). Comparing f orbitals with F terms, as for d orbitals/D terms, the splitting of f orbitals in an octahedral field is $a_{2u} + t_{2u} + t_{1u}$. All f orbitals are odd (ungerade) and so are labelled with the u subscript. F terms split analogously but, because we are here considering F terms arising from d^n configuration, the many-electron wavefunctions are built from products of d orbitals of g symmetry. Hence, the octahedral-field terms arising are necessarily of g symmetry and so we get the results $F \rightarrow T_{1g} + T_{2g} + A_{2g}$. The ground term for d^2, d^3, d^7 and d^8 configurations is known from Figure 8.4. Thus, the *qualitative* splitting can be deduced from the

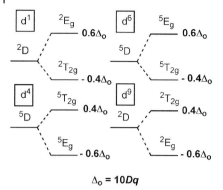

$$\Delta_o = 10Dq$$

Figure 8.5 The effect of an octahedral crystal field on the D terms of d^1, d^6, d^4 and d^9 configurations. The energy of each term is given on the right-hand side.

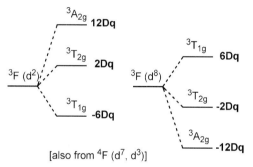

[also from ^4F (d^7, d^3)]

Figure 8.6 The effect of an octahedral crystal field on the ^3F term of d^2 and d^8configuration. The energy of each term is given on the right-hand side.

ground term and the splitting of f orbitals, without considering the ordering of the two excited terms. This ordering needs a quantitative calculation. The energy calculation (applying the corresponding Hamiltonian to the respective wavefunctions) is not immediate and its demonstration is beyond the scope of this book. The interested reader can find some approach to this demonstration in the books of Figgis [3] and Ballhausen [4]. The calculation establishes the splitting for these d^2, d^7 and d^8, d^3 cases as that shown in Figure 8.6. The energy is measured, in each case, with respect to the unsplit term. When the orbital degeneracy of the resulting terms is included the total energy is, thus, zero.

Conclusion (for D and F terms)
1. The splitting for $d^n = d^{n+5}$, as d^5 is orbitally a spherical shell.
2. The splitting for d^{5-n} or d^{10-n} is inverted with respect to the splitting for a d^n (positive hole quantum formalism for *complementary* configurations).
3. The splitting of the terms does not depend on their spin multiplicity

Table 8.7 Excited terms with the same spin multiplicity as the ground term.

Configuration	Free ion term	Crystal field terms (by energy order)	Excited term with same multiplicity as the ground term	Crystal field terms
d^1	^2D	$^2T_{2g} + {}^2E_g$		
d^2	^3F	$^3T_{1g} + {}^3T_{2g} + {}^3A_{2g}$	^3P	$^3T_{1g}$
d^3	^4F	$^4A_{2g} + {}^4T_{2g} + {}^4T_{1g}$	^4P	$^4T_{1g}$
d^4	^5D	$^5E_g + {}^5T_{2g}$		
d^5	^6S	$^6A_{1g}$		
d^6	^5D	$^5T_{2g} + {}^5E_g$		
d^7	^4F	$^4T_{1g} + {}^4T_{2g} + {}^4A_{2g}$	^4P	$^4T_{1g}$
d^8	^3F	$^3A_{2g} + {}^3T_{2g} + {}^3T_{1g}$	^3P	$^3T_{1g}$
d^9	^2D	$^2E_g + {}^2T_{2g}$		

Important Remark The splitting of the atomic terms in the T_d symmetry is the same as, but inverted with respect to, that of the O_h symmetry, without the g subscripts. The splitting diagrams are, thus, inverted and the splitting is changed by the appropriate factor, $-4/9$ [3].

8.3.5.2 Excited Terms with Same Spin Multiplicity as Ground Term: Configuration Interaction

The d^2, d^3, d^7 and d^8 configurations (O_h symmetry) have an excited term with the same spin multiplicity as the ground term (Table 8.3). Table 8.7 shows the spectroscopic terms of maximum spin multiplicity ("free ion term" column) and the excited term of the same spin multiplicity, as well as the splitting generated by the crystal field.

The four configurations d^2, d^3, d^7 and d^8 have, as their ground term, ^3F, ^4F, ^4F and ^3F, respectively. Above these terms there is a P term with the same spin multiplicity. When an octahedral field is applied, this P term, rather than being split, is converted into T_{1g}. The splitting of the F terms also gives rise to a T_{1g} term. Therefore, there may be configuration interaction between the two T_{1g} terms. The configuration interaction of the two terms of the same symmetry produces a variation in their energies. Figure 8.7 shows schematically the splitting of the F and P terms for the four configurations, d^2, d^3, d^7 and d^8, together with the influence of the configuration interaction on the energy of the terms.

For a d^2 (d^7) ion it is demonstrated that the perturbation energies of the two terms $T_{1g}(F)$ and $T_{1g}(P)$ are the solutions to the following equation [2, 3]:

$$(-6Dq - E)(15B - E) = (4Dq)^2$$

On expanding this equation, we obtain:

$$E^2 + (6Dq - 15B)E - 16(Dq)^2 - 90DqB = 0$$

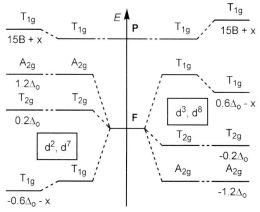

Figure 8.7 Energy diagram for the d^2, d^7, d^3 and d^8 configurations, assuming a weak field. Note the interaction between the T_{1g} terms derived from F and P free-ion terms.

At the limit of the weak field ($Dq = 0$) the solutions are $E = 0$ and $15B$, as would be expected (Table 8.5). At the limit of the strong field, Dq is very large compared to B, and so it can be ignored. The roots will be $-8Dq$ and $2Dq$ (which correspond to the energies at the limit of the strong field, as will be seen in a subsequent section). For any other intermediate field the quadratic equation must be solved, knowing Dq and B. These arguments can be extended to d^3 and d^8 configurations, in O_h symmetry, for which the following equation must be used instead:

$$(6Dq - E)(15B - E) = (4Dq)^2$$

The energy of the terms in octahedral and tetrahedral complexes for d^2, d^3, d^7, and d^8 configurations, derived for the previous equations, are given in Tables 8.8 and 8.9.

Table 8.8 Energy of the terms arising from a d^2 (d^7) configuration [5, 6].

Term	Energy	Energy (ground state = 0)
$T_{1g}(F)$	$7.5B - 3Dq - 0.5F$	0
T_{2g}	$2Dq$	$-7.5B + 5Dq + 0.5F$
$T_{1g}(P)$	$7.5B - 3Dq + 0.5F$	F
A_{2g}	$12Dq$	$-7.5B + 15Dq + 0.5F$

$$F = [225B^2 + 180BDq + 100Dq^2]^{1/2}$$

8.3.5.3 Orgel Diagrams

Using the results obtained for any electron configuration and knowing the energy values for each term, it is possible to draw what are known as Orgel diagrams.

Table 8.9 Energy of the terms arising from a d^3 (d^8) configuration [5, 6].

Term	Energy	Energy (Ground State = 0)
A_{2g}	$-12Dq$	0
T_{2g}	$-2Dq$	$10Dq$
$T_{1g}(F)$	$7.5B + 3Dq - 0.5G$	$7.5B + 15Dq - 0.5G$
$T_{1g}(P)$	$7.5B + 3Dq + 0.5G$	$7.5B + 15Dq + 0.5G$

$$G = [225B^2 - 180BDq + 100Dq^2]^{1/2}$$

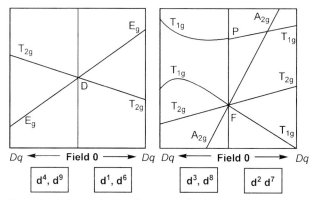

Figure 8.8 Scheme showing the relationship between the cubic field term energies of the lower lying terms of d^n configurations with D ground terms (left) and F ground terms (right).

These semi-quantitative diagrams express the energy variations of the ligand field terms as a function of Dq and are shown schematically in Figure 8.8. It can be seen how the energies of the T_{1g} terms are curved strongly for the d^3 and d^8 configurations. This is a result of the configuration interaction: the term energies tend to cross one another, but this is prohibited in quantum theory as the terms belong to the same symmetry species (non-crossing rule). The same does not occur with ions of d^2 and d^7 configurations (Figure 8.8), as the two T_{1g} terms are much further apart and their energies tend not to cross.

Although they are qualitatively very useful, these diagrams have a serious drawback when studying transition ions whose electron configuration possesses a ground term for the weak field that is different from that of the strong field (d^4, d^5, d^6 and d^7). As the value of the ligand field (Δ_o) increases, a point is eventually reached when the ground term alters; this is not taken into account in the diagrams. This fact reflects a more general issue: in those cases in which Δ_o is large the previous treatment is not suitable as the validity criterion for the weak field method is based on the hypothesis that the splitting of the spectroscopic terms is *much less* than the energy separation between them. When the Δ_0 values are large the approach is no longer valid, and a new treatment must be applied, namely, the *Strong Field Method*.

8.4
Strong Field Method

We now assume that the interelectron repulsion is small compared to the crystal field effects. A crystal field that is large enough to cause splitting of such magnitude is said to be *strong*. In a perturbational treatment of this kind, the crystal field operator must be applied first, followed by the repulsion energy operator. Therefore, the first step is to apply the crystal field and consider which are the possible electron configurations for a d^n ion upon applying this field (Table 8.10).

The number of configurations is the same for a d^n and a d^{10-n} ion. Moreover, the configurations for a d^n ion are equivalent to those for a d^{10-n} ion, due to the rule of *complementarity*. For example, the first configuration in a d^8 ion is $(t_{2g})^6(e_g)^2$, which is equivalent to $(e_g)^2$ in a d^2 ion as $(t_{2g})^6$ is totally symmetric from both an orbital and spin point of view. The next configuration of a d^8 ion, $(t_{2g})^5(e_g)^3$, is equivalent to the next one of a d^2 ion, $(t_{2g})^1(e_g)^1$, and the configuration $(t_{2g})^4(e_g)^4$ (d^8) is equivalent to $(t_{2g})^2$ (d^2).

The energy of each one of the configurations is a function of $10Dq$. The t_{2g} orbitals are stabilised by $-4Dq$ and the e_g orbitals are destabilised by $6Dq$ (see Chapter 1). Therefore, for a d^2 ion the energy of each configuration will be:

$$(t_{2g})^2 = 2\times(-4Dq) = -8Dq$$

$$(t_{2g})^1(e_g)^1 = -4Dq + 6Dq = 2Dq$$

$$(e_g)^2 = 2\times 6Dq = 12Dq$$

Table 8.10 Strong field electron configurations for d^n ions.

d^n	Strong field electron configurations				
	Ground	Excited			
d^1	$(t_{2g})^1$	$(e_g)^1$			
d^2	$(t_{2g})^2$	$(t_{2g})^1(e_g)^1$	$(e_g)^2$		
d^3	$(t_{2g})^3$	$(t_{2g})^2(e_g)^1$	$(t_{2g})^1(e_g)^2$	$(e_g)^3$	
d^4	$(t_{2g})^4$	$(t_{2g})^3(e_g)^1$	$(t_{2g})^2(e_g)^2$	$(t_{2g})^1(e_g)^3$	$(e_g)^4$
d^5	$(t_{2g})^5$	$(t_{2g})^4(e_g)^1$	$(t_{2g})^3(e_g)^2$	$(t_{2g})^2(e_g)^3$	$(t_{2g})^1(e_g)^4$
d^6	$(t_{2g})^6$	$(t_{2g})^5(eg)^1$	$(t_{2g})^4(e_g)^2$	$(t_{2g})^3(e_g)^3$	$(t_{2g})^2(e_g)^4$
d^7	$(t_{2g})^6(e_g)^1$	$(t_{2g})^5(e_g)^2$	$(t_{2g})^4(e_g)^3$	$(t_{2g})^3(e_g)^4$	
d^8	$(t_{2g})^6(e_g)^2$	$(t_{2g})^5(e_g)^3$	$(t_{2g})^4(e_g)^4$		
d^9	$(t_{2g})^6(e_g)^3$	$(t_{2g})^5(e_g)^4$			

Table 8.11 Energy terms derived from the strong crystal field.

Electron configuration		Energy terms
Free ion	O_h symmetry	
d^1, d^9	$(e_g)^1 (t_{2g})^6(e_g)^3$	^2E_g
	$(t_{2g})^1 (t_{2g})^5(e_g)^4$	$^2T_{2g}$
d^2, d^8	$(e_g)^2 (t_{2g})^6(e_g)^2$	$^3A_{2g}$, $^1A_{1g}$, ^1E_g
	$(t_{2g})^1(e_g)^1 (t_{2g})^5(e_g)^3$	$^3T_{1g}$, $^3T_{2g}$, $^1T_{1g}$, $^1T_{2g}$
	$(t_{2g})^2 (t_{2g})^4(e_g)^4$	$^3T_{1g}$, $^1A_{1g}$, ^1E_g, $^1T_{2g}$
d^3, d^7	$(e_g)^3 (t_{2g})^6(e_g)^1$	^2E_g
	$(t_{2g})^1(e_g)^2 (t_{2g})^5(e_g)^2$	$^4T_{1g}$, $2\,^2T_{1g}$, $2\,^2T_{2g}$
	$(t_{2g})^2(e_g)^1 (t_{2g})^4(e_g)^3$	$^4T_{1g}$, $^4T_{2g}$, $^2A_{2g}$, $2\,^2T_{1g}$, $2\,^2T_{2g}$, $2\,^2E_g$, $^2A_{1g}$
	$(t_{2g})^3 (t_{2g})^3(e_g)^4$	$^4A_{2g}$, ^2E_g, $^2T_{1g}$, $^2T_{2g}$
d^4, d^6	$(e_g)^4 (t_{2g})^6$	$^1A_{1g}$
	$(t_{2g})^1(e_g)^3 (t_{2g})^5(e_g)^1$	$^3T_{1g}$, $^3T_{2g}$, $^1T_{1g}$, $^1T_{2g}$
	$(t_{2g})^2(e_g)^2 (t_{2g})^4(e_g)^2$	$^5T_{2g}$, ^3E_g, $3\,^3T_{1g}$, $2\,^3T_{2g}$, $^3A_{2g}$, $2\,^1A_{1g}$, $^1A_{2g}$, $3\,^1E_g$, $^1T_{1g}$, $3\,^1T_{2g}$
	$(t_{2g})^3(e_g)^1 (t_{2g})^3(e_g)^3$	^5E_g, $^3A_{1g}$, $^3A_{2g}$, $2\,^3E_g$, $2\,^3T_{1g}$, $2\,^3T_{2g}$, $^1A_{1g}$, $^1A_{2g}$, ^1E_g, $2\,^1T_{1g}$, $2\,^1T_{2g}$
	$(t_{2g})^4 (t_{2g})^2(e_g)^4$	$^3T_{1g}$, $^1A_{1g}$, ^1E_g, $^1T_{2g}$
d^5	$(t_{2g})^3(e_g)^2$	$^6A_{1g}$, $^4T_{1g}$, $^4A_{2g}$, $2\,^4E_g$, $^4A_{1g}$, $^4T_{2g}$, $2\,^2A_{1g}$, $^2A_{2g}$, $3\,^2E_g$, $4\,^2T_{1g}$, $4\,^2T_{2g}$
	$(t_{2g})^4(e_g)^1 (t_{2g})^2(e_g)^3$	$^4T_{1g}$, $^4T_{2g}$, $^2A_{1g}$, $^2A_{2g}$, $2\,^2E_g$, $2\,^2T_{1g}$, $2\,^2T_{2g}$
	$(t_{2g})^5 (t_{2g})^1(e_g)^4$	$^2T_{2g}$

In the strong field method, once the crystal field perturbation (which gives us the electron configurations mentioned above) has been introduced, it is necessary to work with the perturbation of the interelectron repulsions in each one of the $(t_{2g})^n(e_g)^m$ configurations. Therefore, the energy terms must be calculated by means of the *direct product* rules for the symmetry point group to which the molecule belongs. In general, if we know the terms of a given electron configuration, then the terms of the new configuration formed by adding an electron belong to the irreducible representations contained in the direct product of the terms of the initial configuration multiplied by the terms of the added electron. These terms are summarized in Table 8.11.

It is not easy to predict the order of energies of the strong-field terms. As with the weak field method, the only term that is easy to predict is the ground term corresponding to that with the maximum total multiplicity (spin times orbital).

8.4.1
Considerations Regarding the Ground Term

The symmetry label of the ground term (from both the orbital and spin point of view) can be found directly, without any quantitative calculation, by placing the electrons in the t_{2g} and e_g orbitals and analyzing the term which is formed. This process is the same as that carried out for the weak field configurations described in Figure 8.4. It should be borne in mind that the ground terms for the

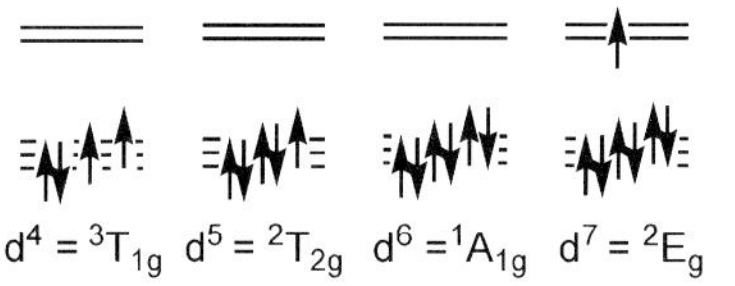

$d^4 = {}^3T_{1g}$ $d^5 = {}^2T_{2g}$ $d^6 = {}^1A_{1g}$ $d^7 = {}^2E_g$

Figure 8.9 d^n electron configurations in a strong field with their corresponding ground terms.

configurations d^1, d^2, d^3, d^8, d^9 and d^{10} are the same as in the weak field, since they present the same electron configuration. For d^4, d^5, d^6 and d^7 ions the method is as shown in Figure 8.9.

A d^4 ion has two unpaired and two paired electrons in the t_{2g} orbital. Therefore, the term will be identical to that of the d^2 ion ($^3T_{1g}$), as the two paired electrons do not influence the symmetry label. A d^5 ion has its five electrons in the t_{2g} orbital: two pairs of electrons and one unpaired electron. The four paired electrons will not affect the energy term, so that only the unpaired electron is important, here-fore, the term of a d^5 ion will be the same as that of a d^1 ion ($^2T_{2g}$). A d^6 ion has its six electrons paired in the t_{2g} orbitals and, therefore, its term will be the totally symmetric: $^1A_{1g}$. Finally, a d^7 ion has six paired electrons in the t_{2g} orbitals (which will not affect the symmetry label of the corresponding term) and one unpaired electron in the e_g orbitals: its symmetry will be ^2E_g.

8.4.2
Term Energies

Calculation of the energies produced by the interelectron repulsion on the strong-field terms is complicated, and is beyond the scope of this book. These calculations (starting from the strong field approach and going to the limit of the weak field, i.e. $Dq = 0$) were first reported by Tanabe and Sugano in 1954 [7]. The results are usually presented in the form of the so-called Tanabe–Sugano diagrams (see section 8.6).

8.5
Correlations Between the Energy Terms Derived from the Weak and Strong Fields: Intermediate Fields

As most complexes possess an intermediate crystal field the perturbational treat-ment must be done simultaneously, as the energies of the crystal field and the electron repulsion may be of the same order. A rigorous treatment for intermedi-ate fields is very complex and they are usually studied in a more simple way by extrapolating, in one direction or the other, the results obtained from the weak and strong field limit. Moving from a strong to a weak field for d^1–d^3 and d^8–d^9 complexes is merely a question of the magnitude of Dq. For d^4–d^7 complexes it is a more complex task than the magnitude of Dq. The ground term alters after a

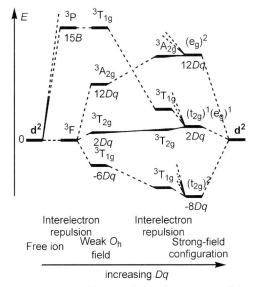

Figure 8.10 Simplified correlation diagram (weak field, strong field) for a d^2 configuration.

certain field value, as pointed out in previous sections, and if we wish to draw the correlation diagrams this fact must be taken into account.

Figure 8.10 shows a simplified correlation diagram for the d^2 configuration as an example of a d^n diagram in which the ground term is the same in both the weak and strong fields. This is a simplified diagram as only those terms whose spin multiplicity matches that of the ground term have been drawn. A number of features of this correlation diagram should be highlighted. For example, the energy separation between the three strong field configurations is $10Dq$, while the separation between the energy terms at the limit of the weak field is $8Dq$ ($^3T_{1g} - {}^3T_{2g}$) and $10Dq$ ($^3T_{2g} - {}^3A_{2g}$). Therefore, the separation between $^3T_{1g} \leftrightarrow {}^3T_{2g}$ varies from $8Dq$ (weak field limit) to $10Dq$ (strong field limit). This variation is due to the configuration interaction mentioned above.

The complete correlation diagrams for any d^n configuration can be found in more specific books, such as in Figgis–Hitchman [3]. Obviously, *the terms obtained through the weak or strong field methods must be the same.*

From these diagrams it can be seen that for d^4–d^7 electron configurations, the ground state for the weak field and the strong field are different, as commented previously. As an example, Figure 8.11 shows this difference in a simplified correlation diagram for d^5 ions.

Given the complexity of these diagrams, and bearing in mind that for most spectroscopic and magnetic properties only the ground term and excited terms of the same spin multiplicity are of interest, it is helpful to draw partial diagrams comprising only the terms of the maximum multiplicity in both fields. Figures

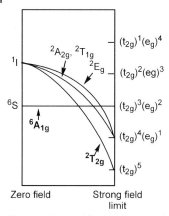

Figure 8.11 Simplified energy (Orgel) diagram for a d^5 ion in an O_h field.

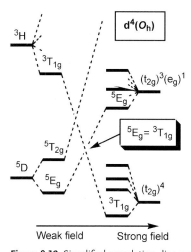

Figure 8.12 Simplified correlation diagram for a d^4 ion (strong and weak field).

8.12–8.15 show these partial, but much more useful, diagrams for the d^4, d^5, d^6 and d^7 configurations, respectively.

When the field value increases, the weak field ground term is destabilized while the strong field ground term is stabilized. Starting from an infinite field and reducing its magnitude produces the same phenomenon. This indicates that there is an intermediate field at which the two terms cross each other, and thus *they have the same energy*. This characteristic of d^4, d^5, d^6 and d^7 configurations is very important for magnetic phenomena, which will be studied in Chapter 10.

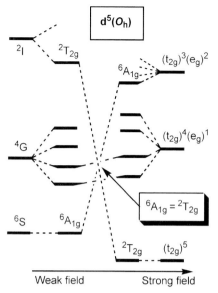

Figure 8.13 Simplified correlation diagram for a d^5 ion (strong and weak field).

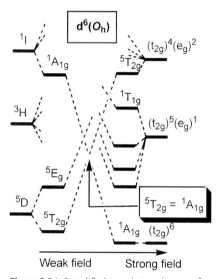

Figure 8.14 Simplified correlation diagram for a d^6 ion (strong and weak field).

8.6
Tanabe–Sugano Diagrams

The most widely used diagrams, *especially in electronic spectroscopy*, are the Tanabe–Sugano diagrams [7]. In 1954 Tanabe and Sugano solved the energy equations of

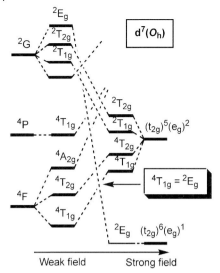

Figure 8.15 Simplified correlation diagram for a d^7 ion (strong and weak field).

the terms, starting from the strong field model. In their calculations they included all the interactions between terms with the same symmetry and spin multiplicity. The problem with these equations and, consequently, with the corresponding diagrams is that it is difficult to know *a priori* the crystal field parameters and impossible to generalize them. In a given complex, they differ with respect to the known free-ion parameters and, moreover, they are different for each free ion in its different oxidation states.

If the energy scale comprises units of B, then the energy values will only depend on $10Dq$ (Δ_o). Therefore, in Tanabe–Sugano diagrams the energies of each term are divided by B. Thus, instead of representing the energy with respect to the field, this approach represents E/B with respect to Δ_o/B for a given value of C/B which is assumed to be constant. In these diagrams the ground term is always taken as the x axis. With these considerations, Tanabe–Sugano diagrams are used for any ion of a given d^n configuration. Figures 8.16 and 8.17 show the Tanabe–Sugano diagrams for d^2 and d^3 (d^8) ions, respectively, in which only those terms with the spin multiplicity matching the ground term have been drawn. The *qualitative diagrams* with all the energy terms can be found in Appendix 1. The reader interested in more detailed and useful (for spectroscopic purposes) Tanabe–Sugano diagrams can consult some sites on the web. A very good web-site is http://wwwchem.uwimona.edu.jm:1104/courses/Tanabe-Sugano/ [8]. A calculation of Tanabe–Sugano diagrams by matrix diagonalization was reported in 2001 by Howald [9].

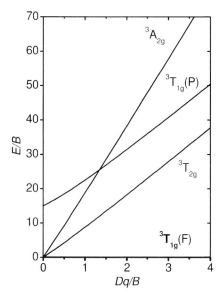

Figure 8.16 Simplified Tanabe–Sugano diagram for a d² (d⁷, weak field) configuration.

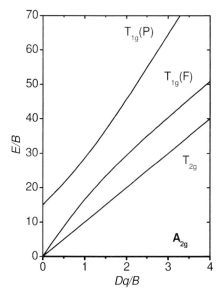

Figure 8.17 Simplified Tanabe–Sugano diagram for a d³ (d⁸) configuration.

8.7
Ligand Field

As indicated in Chapter 1, we have to make a distinction between *crystal-field* and *ligand-field* theory. *Crystal-field* theory is defined as an electronic theory of complexes which neglects the overlap of electrons associated with the metal with electrons associated with the ligands. The corresponding theory, with overlap not neglected, is termed *ligand-field* theory.

Many *"theoretical failures, together with the ever-increasing weight of experimental evidence of the importance of covalency has caused the electrostatic theory to be abandoned as a fundamental theory describing metal–ligand interactions"*. This statement was published in early 1973 by Gerloch and Slade [10].

Historically, the initial crystal field theory was substituted by the ligand field theory and, finally, by the MO theory. Ligand-field theory has been used for different studies in coordination compounds, such as electron spectroscopy, magnetic properties and EPR spectroscopy. In all these techniques the reduction of Racah's B parameter and of the spin–orbit coupling from their free-ion values are satisfactorily explained. The reduction of the B parameter is calleded the *nepheleuxetic* effect.

The nepheleuxetic effect describes the fact that the parameter values of inter-electron repulsion are smaller in complexes than in the corresponding free ions. If B and C are assumed to be reduced from their free-ion values on complexation by similar proportions, the value of B in a complex may be taken to represent the nepheleuxetic effect in any given case. Conventionally, we define β (nepheleuxetic *parameter*) as the ratio of B in the complex to B_0 in the free ion: $\beta = B/B_0$. Usually the termed nepheleuxetic effect is given as $(1 - \beta)$. Like other series (spectrochemical, for instance) the nepheleuxetic effect can be factorized into functions of ligand only and metal only: $(1 - \beta) = h$ (ligands) $\cdot$ k (central ion). Some h and k values can be found in some Inorganic Chemistry textbooks (see Ref. [11]).

The nepheleuxetic series for ligands, defined in order of increasing values of $(1 - \beta)$ is: (large B) $F^- < H_2O < NH_3 < en < ox^{2-} < NCS^- < Cl^- < CN^- < Br^- < I^-$ (small B).

The corresponding series for metals, less well defined, is: (large B) $Mn(II) < Ni(II) \approx Co(II) < Mn(III) < Fe(III) < Ir(III) < Co(III) < Mn(IV)$ (small B).

Let us summarize the main trends in the nepheleuxetic series:

1. In the ligand series, the ligands lie, to a reasonable approximation, in the order of *decreasing* electronegativity of the donor atoms: F, O, N, Cl, etc. That is, the less electronegative a donor atom, the more it reduces the repulsion between d electrons already on the metal. To reduce d electron repulsion (the B parameter) the d electrons must expand into a larger volume in the molecule than in the atom so they can stay farther apart.

2. Most frequently, while Dq values increase markedly along the series $3d^n < 4d^n < 5d^n$ for a given oxidation number, the nepheleuxetic effect $(1 - \beta)$ shows little change and in fact a slight decrease.

3. As far as the dependence of B (β) on oxidation state is concerned we can discern, qualitatively at least, that the nepheleuxetic effect for metals increases as their oxidizing power.

4. Nepheleuxetic effects in tetrahedral complexes have been reported as very similar, actually slightly larger, than in the corresponding octahedral complexes.

In general, as expected, the nepheleuxetic series resembles a chemical concept of increasing covalency. In other words, B is reduced when electrons delocalize from the original compact atomic orbital into diffuse molecular orbitals. Two main mechanisms have been described, referred to as "*central-field covalency*" and "*symmetry-restricted covalency*".

Central-field covalency refers to a transfer of negative charge from the ligands to the central metal ion, so tending to satisfy Pauling's electroneutrality principle and reducing the effective charge of the metal. The resulting expansion of the metal orbitals, is supposed to increase the average distance between metal electrons and hence decrease interelectron repulsion parameters. In summary, it describes changes in Z_{eff} on the metal.

Symmetry-restricted covalency refers to the formation of different molecular orbitals on complexation, describing electron delocalization effects. The nepheleuxetic character of a ligand is different for electrons in t_{2g} and e_g orbitals. Because the σ overlap of e_g is usually larger than the π overlap of t_{2g}, the cloud expansion is larger in the former case. Thus, the e_g orbitals suffer a greater nepheleuxetic effect than the t_{2g} ones. This effect might be expressed in terms of a "*differential expansion*" of the t_{2g} and e_g orbital functions.

8.8
Spin–Orbit Coupling

In this section (see also Section 8.1) we will give the more important qualitative features without a rigorous mathematical treatment, which is beyond the scope of this book. In any case, it is not possible to avoid dealing with quantum concepts.

8.8.1
Spin–Orbit Coupling in Atomic (Spectroscopic) Terms

The *single-electron* spin–orbit coupling parameter, ζ (or ξ), measures the strength of the interaction between the spin and orbital angular momentum of a single electron. Assuming spherical symmetry (free ion) the value of ζ is given by relativistic quantum theory and depends only on quantum numbers n and l and is always positive. The single-electron spin–orbit Hamiltonian can be written as:
$$\hat{H}_{SO} = +\zeta l \cdot s$$
For use in connection with terms it is more convenient to deal with a parameter that is a property of the term itself; and for this purpose the parameter λ is introduced. The corresponding operator is $\hat{H}_{LS} = \lambda \hat{L} \cdot \hat{S}$, (scalar product) where $\lambda = \pm\zeta/2S$.

λ, thus, can be either positive or negative. In atomic spectroscopy it is not difficult to demonstrate that the + sign in the formula holds for a shell that is less than half filled. The negative sign in the formula of λ holds for a more than half filled shell. Some λ parameters for d^n electronic configurations are given in Table 8.12.

Table 8.12 Spin–orbit splitting parameters, λ, for some free ions in the ground state.

Ion	d^n	Ground state	λ (cm^{-1})	Ion	d^n	Ground state	λ (cm^{-1})
Ti^{3+}	d^1	^{2}D	154	Mn^{3+}	d^4	^{5}D	85
V^{3+}	d^2	^{3}F	104	Fe^{2+}	d^6	^{5}D	−100
V^{2+}	d^3	^{4}F	55	Co^{2+}	d^7	^{4}F	−180
Cr^{3+}	d^3	^{4}F	87	Ni^{2+}	d^8	^{3}F	−335
Cr^{2+}	d^4	^{5}D	57	Cu^{2+}	d^9	^{2}D	−852

Free ion terms

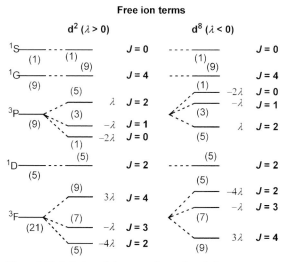

Figure 8.18 Splitting of the terms from d^2 and d^8 configurations by spin–orbit coupling (taken from Ref. [3] (multiplicities in parentheses)).

Spin–orbit coupling is considered as a small perturbation acting on the terms. In particular it applies quite well to the first transition series ions. The following treatment will, therefore, be valid only for d-elements (mainly 3d elements) but not for the lanthanides or actinides. This is the basis of the Russell–Saunders coupling-scheme in which each term is split into a number of multiplets, which are specified by the quantum number, J ($J = |L − S|$ to $|L + S|$). Successive multiplets are specified by values of J differing by units. To specify a particular multiplet that arises from a term, the value of J is added as a subscript to the term symbol. The symbol for a multiplet is, thus, $^{(2S+1)}L_J$

The value of the energy for a given set of J, L, S is:

$$E_J = \frac{1}{2}\lambda[J(J+1) - L(L+1) - S(S+1)]$$

The splitting of the terms from d^2 and d^8 by spin–orbit coupling is shown in Figure 8.18.

8.8.2
Spin–orbit Coupling in Coordination Compounds

In coordination compounds, the spin–orbit operator follows the crystal field, because it is assumed that λ is smaller than $\Delta = 10Dq$. This is totally valid for the d ions of the first row. Thus, it is now necessary to study how the spin–orbit Hamiltonian acts on the crystal field terms. Mathematic calculations use perturbational treatment. As a consequence, we have to take into account perturbations of first and second order.

First-order perturbation mixes directly the orbital and spin components of the crystal field terms. This phenomenon can be given only in those terms which are orbitally degenerate.

Second-order perturbation takes place even when the ground state is orbitally non-degenerate but there are some orbitally degenerate excited states that interact with these levels.

In this part we are going to study the energy terms once the crystal field (weak or strong) is applied. This feature will lead to some difficulties: the spin component of any crystal field term is exactly the same as the corresponding term for the free-ion ($2S + 1$) but the orbital component is not so simple to define. For atomic terms the orbital part is simply S, P, D, F, etc. that have 0, 1, 2, 3 etc L values, respectively. However, upon application of the crystal field, the meaning of the L value has been lost. What is, therefore, the actual meaning of L and in which cases is there an orbital contribution?

8.8.3
First-order Spin–Orbit Coupling

Let us remember an important feature: *The angular orbital momentum is associated with the free rotation of the electron from one orbital to another of the same symmetry and energy, with respect to the x, y, z axes.*

In a free ion all d orbitals are degenerate. Orbital angular momentum about an axis is associated with the ability of the electron to rotate about this axis to give an *identical* and *degenerate* orbital. In a free ion, for example, an electron in an orbital xy can freely rotate $45°$ around the z axis, to occupy the orbital $x^2 - y^2$. These two orbitals are degenerate. However, upon application of a crystal field, these two orbitals *are no longer degenerate*. In this case, any possible rotation is impeded and there is no angular momentum associated with it.

8.8.3.1 **Cubic Symmetry, O_h or T_d**
In the presence of a cubic ligand field the d_{xy} and d_{x2-y2} orbitals do not belong to the same degenerate manifold and no orbital angular momentum arises. However, the d_{xz} and d_{yz} (t_{2g}) orbitals remain degenerate after the application of the cubic field. As a consequence, orbital angular momentum remains to some extent within the $t_{2(g)}$ orbital set because, a rotation around the z axis transforms d_{yz} into d_{xz}. Furthermore, a rotation around the x or y axes transforms d_{xy} into d_{xz} or d_{yz},

respectively. No rotation can transform the d_{z2} orbital into the d_{x2-y2} orbital. There is, therefore, no orbital angular momentum associated with the $e_{(g)}$ set.

Group theory is very helpful for deciding which terms, for a given symmetry, present first-order spin–orbit coupling. As reported by Ballhausen [4], "*A term whose orbital symmetry is Γ, will exhibit properties of angular momentum and therefore propensities for spin–orbit coupling only if the direct-product representation $\Gamma \times \Gamma(L) \times \Gamma$ contains the identity representation*".

The irreducible representation(s) for L correspond to the representation(s) of R (given in all common tables), because the operator $\boldsymbol{L}$ ($= L_x + L_y + L_z$) possesses the symmetry properties of the rotation (R_x, R_y, R_z) about the three Cartesian axes. $\Gamma(L)$ is always an *antisymmetric* representation of the point group.

Instead of applying the triple product it is usually easier for the reader, to apply the direct product, which is divided in symmetric and antisymmetric parts. Thus, we can state: *Spin–orbit coupling is not-zero if the product $\{\Gamma \times \Gamma\}$ contains $\Gamma'(L)$*

In cubic groups,

$$E \times E = A_1 + A_2 + E.$$

$$T_2 \times T_2 = A_1 + A_2 + \{T_1\}(R_x, R_y, R_z) + T_2$$

Orbital momentum is thus quenched for E terms, but it is not quenched for T_2 terms. It is obvious that for A terms it is also quenched.

Notice: 1, 2, 4, 5 electrons in the t_{2g} (O_h) orbitals cause a Jahn–Teller distortion (Chapter 3), that removes the degeneracy, giving $B_{2g} + E_g$ terms. Thus, the Jahn–Teller effect and the spin–orbit coupling compete with each other. However, as we have already commented, the Jahn–Teller effect on t_{2g} orbitals is relatively small.

8.8.3.2 General Case: Non-cubic Symmetry

The reasoning given in the last section is not true when the symmetry is lowered, because there is then the possibility of E terms not necessarily coming from z^2 and $x^2 - y^2$ orbitals, but from other orbitals. Let us imagine a complex with D_{4h} symmetry. The corresponding crystal-field d-functions are e_g (xz, yz), a_{1g} (z^2), b_{2g} (xy), b_{1g} ($x^2 - y^2$). To check for the possibility of spin–orbit coupling for an unpaired electron in the e_g orbital (doubly degenerate) it is sufficient to find the antisymmetric product $e_g \times e_g = A_{1g} + \{A_{2g}\}(R_z) + B_{1g} + B_{2g}$. This product contains A_{2g}, which is the representation of R_z and thus there will be spin–orbit coupling.

8.8.3.3 Splitting of the Terms

From a quantitative point of view, the exact amount to which the orbital contribution to the magnetic moment occurs in the T terms is a matter of calculation in each individual instance. It is not trivial. Some of these calculations can be found in more specialized books [12].

In the book of Figgis and Hitchman a simpler approach is made [3]. They consider there to be a correspondence between the wavefunctions of a T term and those of a free-ion P term. In each case there is three-fold orbital degeneracy. The results for the P term are readily available ($L = 1$, as a true quantum number). Attention is restricted initially to strong field configurations where the wavefunctions of the T terms may be written in the form of $(t_{2g})^n (e_g)^m$ configurations. This method has been called T–P isomorphism, in which it can be demonstrated that $L(T) = -L(P)$ [3].

With this formalism, the L value for a T term is again equal to 1 and, thus, defining the total angular momentum $J = L + S$, the Hamiltonian for the spin–orbit coupling in a $^{2S+1}T_1$ or $^{2S+1}T_2$ term can be written as $\hat{H}_{SO} = -A\lambda L \cdot S$.

The minus sign appears to distinguish between the matrix elements of the orbital angular momentum operator calculated with the wavefunctions of the ground T term with those calculated with the use of the P-basis. A value arises for any T_{1g} in the weak field limit, in which there is mixing of the T_{1g} terms from the F and P free ion terms of d configurations, as previously discussed. It is not difficult to demonstrate that A lies between 1.0 and 1.5 for T_{1g} (weak field limit) terms. With this formalism, the multiplet energies can now be written as:

$$E_J = -\frac{1}{2} A\lambda [J(J+1) - L(L+1) - S(S+1)]$$

Within this formalism, for $(t_{2g})^n$ less than half occupied, λ is positive, and for $(t_{2g})^n$ more than half occupied, λ is negative. Now it must be considered $(t_{2g})^n$, not d^n.

8.8.3.4 Energy of the Terms in O_h Symmetry

Applying the previous formula it is very easy to calculate the energy for all multiplets for any configuration. The results for $^2T_{2g}$ (d^1, $(t_{2g})^1$, are given in Table 8.13.

The calculation of the energy multiplets in other O_h and T_d symmetries can be made exactly as we have done for d^1 complexes. It will be a good task for the reader. The corresponding energy diagrams for O_h configurations are shown in Figure 8.19.

8.8.4
Second-order Spin–Orbit Coupling

If the ground term is not T but E or A (assuming cubic symmetry), the spin–orbit coupling mixes the ground state with all excited states. As a result, the energies of the two states (ground and excited) undergo a slight variation. *The ground state*

Table 8.13 Energy for all multiplets for $^2T_{2g}$ configuration (in O_h symmetry).

	S	L	J	$J(J+1)$	$-L(L+1)$	$-S(S+1)$	Energy	Energy
								$\lambda > 0$
$^2T_{2g}$ (d^1, t_{2g}^1)								
	1/2	1	1/2	1/2 * 3/2	−1 * 2	−1/2 * 3/2	$-\lambda/2[-2] = \lambda$	λ
	1/2	1	3/2	3/2 * 5/2	−1 * 2	−1/2 * 3/2	$-\lambda/2[4/4] = -\lambda/2$	$-\lambda/2$

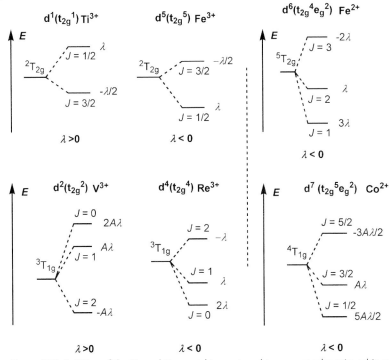

Figure 8.19 Splitting of the T_{1g} and T_{2g} crystal terms in cubic symmetry by spin–orbit coupling.

becomes more stable and the excited state more unstable. This effect, which has little importance in electron spectroscopy, is of great importance in magnetism and electron paramagnetic resonance.

Let us remember that, according to the perturbation theory, if E_g and E_e are the ground and excited terms, respectively, the "mixture" of both terms becomes affected by a second-order perturbation with magnitude:

$$\sum_f \frac{<\phi_e \mid \lambda_e \, L \cdot S \mid \phi_g><\phi_g \mid \lambda_g \, L \cdot S \mid \phi_e>}{E_g - E_e}$$

The summation is assumed over all excited states. If we assume, for example, that the ground term is $^4A_2$, there exists an excited term $^4T_2$ with greater energy. This would be the case for the Cr^{3+} (O_h) ion. The $^4A_2$ term has four wavefunctions while the $^4T_2$ term has twelve wavefunctions. It will be necessary then to calculate all $\langle \varphi_e \mid \lambda L \cdot S \mid \varphi_f \rangle$ integrals, knowing the wavefunctions of each term. Some of these tedious calculations can be found in Ref. [12].

It can be demonstrated that the first effect of the second-order spin–orbit coupling is the stabilization energy of the ground term. In a cubic ligand field, there are only four possibilities of ground terms without orbital angular momentum:

^{2}E, ^{3}A$_2$, ^{4}A$_2$ and ^{5}E. The stabilization energy, which depends only on the orbital part, is:

$$A \text{ terms} = -8\lambda^2/10Dq$$

$$E \text{ terms} = -4\lambda^2/10Dq$$

The two most important features that derive from the second-order spin–orbit coupling in coordination compounds are the **D** and **g** tensors (or D and g parameters) which will be developed more extensively in other chapters of this book. We will give here an introduction to this issue.

8.8.4.1 The Zero Field Splitting (ZFS)

This splitting takes place in systems with more than one unpaired electron and having some distortion with regard to the cubic symmetry (O_h or T_d). As indicated by Boca in his review on ZFS, "*The essence of the ZFS lies in a weak interaction of the spins mediated by the spin–orbit coupling . . . The ZFS case is met in transition metal complexes having S $\geq$ 1 and with no first-order contribution to the angular momentum*" [13].

A Qualitative Approach The ^{3}A$_{2g}$ ground term of the d^8 configuration in O_h symmetry cannot be split in *first order* by spin–orbit coupling or by a low-symmetry crystal field component. However, it is split in second order by interaction with the ^{3}T$_{2g}$ term lying higher, itself split by the low-symmetry component. The situation is illustrated qualitatively in Figure 8.20A.

From another perspective, let us apply, from the mathematical point of view, the spin–orbit effect in a term orbitally non-degenerate (A, for example), *without and with* distortion. The particular case of a Ni^{2+} (O_h) ion has been schematized in Figure 8.20B (O_h and D_3 point groups, respectively). The ground state is ^{3}A$_{2g}$. The original symmetry has been reduced from O_h to O because this is sufficient when working with rotational groups. The representation of the $S = 1$ (triply degenerate) for an O group is T$_1$. Assuming distorted geometry, D_3, T$_1$ becomes A$_2$ + E (Correlation Tables).

Applying the spin–orbit coupling before the distortion, the direct product of orbital and spin parts is A$_2 \times$ T$_1 = $ T$_2$. *Thus the T term remains unaltered because, for this effect, it is not important if it is T$_1$ or T$_2$*. Thus, as expected, being an A term, the spin coupling does not alter (split) the term. However, if we assume a certain distortion, such as D_3, the direct product of the orbital and spin part is: A$_2 \times$ (A$_2$ + E) = A$_1$ + E. A$_1$ is the totally symmetric species, for which it can be demonstrated that it corresponds to $M_S = 0$ (from $S = 1$), and E is a two-fold degenerate species which corresponds to ±1 components of $S = 1$. Thus, the effect of the spin–orbit coupling plus the loss of symmetry has created a new and important effect: without

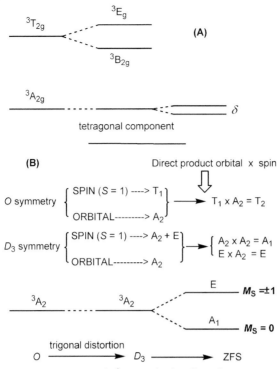

Figure 8.20 Splitting of a $^3A_2$ term by the effect of simultaneous distortion and second-order spin–orbit coupling.

magnetic field, the $S = 1$ spin state has been split into two multiplets. This phenomenon is called Zero Field Splitting.

A Quantitative Approach The rigorous general treatment of ZFS is one of the most complicated developments in quantum chemistry. The complete ZFS Hamiltonian is very complicated and includes the **D** tensor. However, as indicated by Boca, *"their experimental determination is an unrealistic task and thus some simplifications are desirable"* [13]. Usually, the quantitative approach to ZFS is made through a formal spin-Hamiltonian. Assuming a diagonal and traceless (by a correct choice of the axis) form of the **D**-tensor, the ZFS Hamiltonian can be rewritten in an equivalent form:

$$\hat{H}_{SS} = D[\hat{S}_z^2 - S(S+1)/3] + E(\hat{S}_x^2 - \hat{S}_y^2)$$

D and E, are now parameters. Assuming $E = 0$ (axial distortion, $x = y$) the Hamiltonian can be written as: $\hat{H}_{SS} = D[\hat{S}_z^2 - S(S+1)/3]$. D can be positive or negative. Assuming $D > 0$, the scheme of the energy values of the corresponding multiplets

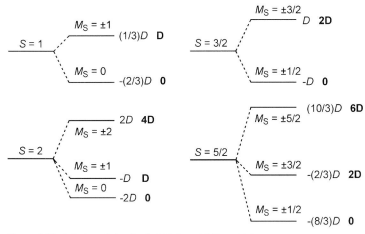

Figure 8.21 Splitting of the $S = 1$, 3/2, 2 and 5/2 multiplets due to the ZFS effect. Energy values (in bold assuming the energy of the lowest multiplet = 0).

for $S = 1$, 3/2, 2 and 5/2 is given in Figure 8.21. This effect of ZFS on the energy values will be extensively studied in the chapters devoted to magnetic properties and EPR spectroscopy.

8.8.4.2   The *g* Parameter Variations

The *g* parameter can be considered as a second-order property. Indeed the stabilization of the ground state described above is not simply stabilization, but also affects the effect produced when applying an external magnetic field (Zeeman effect).

The external magnetic field (H) splits each level in the function of $M_S g \mu_B H$, as will be discussed in later chapters. M_S is the S_z component of S, μ_B the Bohr magneton and H the applied magnetic field. *g* represents, then, the measurement of the major or minor splitting of the M_s states when applying the external magnetic field. *This splitting depends on the energy levels, modified by the second-order spin–orbit coupling.* The *g* value for a free electron, *without orbital contribution,* is 2.0023. However, considering not a free electron but one in a coordination compound, *the second-order spin–orbit coupling* modifies the *g* value, according to the following formula (assuming isotropic behavior):

$$g(\text{A terms}) = 2 - 8\lambda/10\,Dq)$$

$$g(\text{E terms}) = 2 - 4\lambda/10\,Dq)$$

In the two cases, the factor 2, is simply the g value for a free electron, without orbital contribution. T terms (in which the first-order spin–orbit coupling is important) also have second-order spin–orbit coupling, but this contribution is very small compared to the first-order one. As a consequence, it is not taken into consideration in quantum calculations.

8.9
Final Considerations on Spin–Orbit Coupling and Zero Field Splitting

Crystal field, spin–orbit coupling and zero field splitting have very different magnitude: close to $10^4 \, cm^{-1}$ for crystal field, $10^2 \, cm^{-1}$ for spin-orbit coupling in first row transition elements and close to $1 \, cm^{-1}$ for ZFS. The techniques, thus, that give information about each phenomenon are different.

Electronic Spectroscopy   This technique allows determination of the electronic transitions between terms. These transitions are from the ground term to the excited terms of the same spin multiplicity. In some cases, first-order spin–orbit coupling can be observed and deduced from electronic spectra (splitting of the bands, irregular shapes, etc.). However, the ZFS is always so small that it can neither be observed nor calculated from electronic spectra.

Magnetic Measurements   The effect of the external magnetic field is seen *only* on the thermally populated terms. In coordination chemistry, the greater attainable temperature is, usually, room temperature (ca. 300 K). kT is, thus, of the order of $0.695 \, cm^{-1} \, K^{-1} \times 300 \, K = 200 \, cm^{-1}$. *The excited terms of the same spin multiplicity are not populated at room or lower temperature. As a consequence they are not important in the first-order approach.* However, the spin–orbit coupling of a T term is of the same order as kT, being very important in magnetic phenomena. The ZFS will be only important at very low temperature when kT is of the order of $1–10 \, cm^{-1}$.

Electron Paramagnetic Resonance   As in the case of magnetic measurements, the EPR technique only reflects the thermally populated states. Considering that the energy of microwaves used in EPR is of the order of $0.3–1 \, cm^{-1}$ (conventional instruments), it will be important and necessary to take into account the ZFS, because it is of the same order of magnitude (see Chapter 11).

In conclusion, the scheme of three different cases is shown in Figures 8.22–8.24. The first corresponds to a d^3 ion, whose ground term is orbitally non-degenerate ($^4A_{2g}$) assuming a cubic symmetry (without any distortion) (Figure 8.22); the second to a d^1 ion with first order spin–orbit coupling in the ground state ($^2T_{2g}$) (Figure 8.23) and the third to a distorted d^3 ion in which the ground state ($^4A_{2g}$) possesses ZFS (Figure 8.24). The energetic aspects are shown in these figures, indicating their correspondence to electronic spectroscopy, magnetic measurements and electron spin resonance spectroscopy.

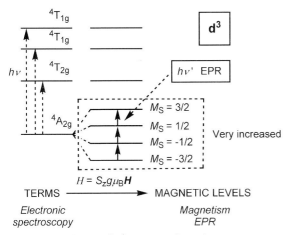

Figure 8.22 Splitting of a $^4A_{2g}$ term without distortion or spin–orbit coupling. The splitting of the right side corresponds to the Zeeman effect.

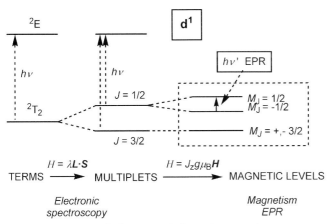

Figure 8.23 Splitting of a $^2T_2$ term with first-order spin–orbit coupling. The splittings of the far right side correspond to the Zeeman effect. (Note: In this case, the multiplet with $J = 3/2$ is not split by the magnetic field).

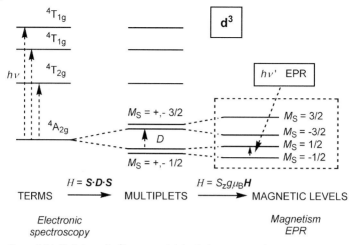

Figure 8.24 Splitting of a $^4A_{2g}$ term with both distortion and second-order spin–orbit coupling. The splittings of the far right side correspond to the Zeeman effect.

References

1 Cotton, F.A. (1990) *Chemical Applications of Group Theory*, 3rd edn., John Wiley, New York.

2 Sutton, D. (1975) *Electronic Spectra of Transition Metal Complexes*, Mc Graw-Hill, Berkshire.

3 Figgis, B.N., Hitchman, M.A. (2000) *Ligand Field Theory and its Applications*, Wiley-VCH, New York.

4 Ballhausen, C.J. (1979) *Molecular Electronic Structures of Transition Metal Complexes*, McGraw-Hill, New York.

5 Lever, A.B.P. (1997) *Inorganic Electronic Spectroscopy*, 2nd edn., Elsevier, Amsterdam, p. 126.

6 Lever, A.B.P. Electronic spectra of some transition metal complexes: Derivation of Dq and B, *J. Chem. Educ.* 1968, *45*, 711.

7 Tanabe, Y., Sugano, S. On the Absorption Spectra of Complex Ions, I and II, *J. Phys. Soc. Jpn.*, 1954, *9*, 753 and 766.

8 http://wwwchem.uwimona.edu.jm:1104/courses/Tanabe-Sugano/ created and maintained by R. J. Lancashire. The Department of Chemistry, University of the West Indies, Mona Campus, Kingston 7, Jamaica. Created Feb 1999.

9 Howald, R.A. Calculation of Tanabe-Sugano Diagrams by Matrix Diagonalization, *Chem Educ.*, 2001, 6, 78. See Supporting Material.

10 Gerloch, M., Slade, R.C. (1973) *Ligand-Field Parameters*, Cambridge Univesity Press, Cambridge.

11 Huheey, J.E., Keiter, E.A., Keiter, R.L. (1993) Inorganic Chemistry, Principles of Structure and Reactivity, 4th edn., Harper-Collins, New York.

12 Mabbs, F.E., Machin, D.J. (1973) *Magnetism and Transition Metal Complexes*. Chapman and Hall, London.

13 Boca, R. Zero-field splitting in metal complexes, *Coord. Chem. Rev.* 2004, *248*, 757.

9
Electronic Spectroscopy

9.1
Introduction

Most transition metal complexes have characteristic colors, that is, one of their properties is that they absorb certain wavelengths from the visible region of the electronic spectrum; the color of the complex is complementary to that absorbed. This absorption generally extends to the neighboring regions of the infrared and ultraviolet spectra. Thus, although we often refer to visible spectroscopy it is more accurate to use the term *electronic spectroscopy* as this covers the three above-mentioned regions.

The absorption of electromagnetic radiation from these regions is due to excitation of an electron from one molecular orbital to another of higher energy. If the radiation used has enough energy to remove an electron, and thus ionise the complex, the corresponding phenomenon is known as *photoelectronic spectroscopy*.

9.2
Electromagnetic Radiation

Ultraviolet, visible and infrared light, as well as X-rays, γ-rays, and microwaves, etc., are examples of electromagnetic radiation which differ solely in their energy. When studying the interaction between electromagnetic radiation and matter, we usually refer to radiation energy (E), frequency (ν), wavenumber ($\bar{\nu}$) and wavelength (λ).

The units for wavelength (λ) are ångströms (Å) ($=10^{-8}$ cm) or nanometers (nm), where $1\,\mathrm{nm} = 10\,\text{Å}$. The units for wavenumber (the inverse of wavelength, $1/\lambda$) are cm^{-1}. The unit of frequency is s^{-1}, cycles per second (Hertz). Finally, energy has the most varied units, as it can be measured in $\mathrm{kJ\,mol}^{-1}$, eV, etc.

The most widely used unit in electronic spectroscopy is neither wavelength nor energy, but wavenumber, and is expressed in cm^{-1}. One cm^{-1} is also known as a

Coordination Chemistry. Joan Ribas Gispert
Copyright © 2008 WILEY-VCH Verlag GmbH & Co. KGaA, Weinheim
ISBN: 978-3-527-31802-5

kayser (K), and kilokaysers (kK = 1000 kaysers = 1000 cm^{-1}) are often used. Table 9.1 shows the values of these units for the most common regions of electronic spectroscopy.

By focusing on the region of the visible spectrum we have the scheme shown in Figure 9.1, which illustrates the wavelength (nm) and wavenumber of the absorbed light (kK), along with the complementary color (the color seen by the observer). The complementary color can be identified by using the six-pointed star in Figure 9.1. In a first approach, each color has its complement in the opposite vertex of the star.

Table 9.1 Magnitudes of electromagnetic radiation in the regions of electronic spectroscopy.

	Near-UV	**Visible**			**Near-IR**
$\tilde{\nu}$ (cm^{-1})	50 000	26 300	to	12 820	3333
ν (s^{-1})	1.5×10^{15}	7.9×10^{14}	to	3.8×10^{14}	10^{14}
Energy (J mol^{-1})	6×10^5	3.1×10^5	to	1.5×10^5	4×10^4
(eV)	6.23	3.23	to	1.55	0.41
λ (Å)	2000	3800	to	7800	30 000

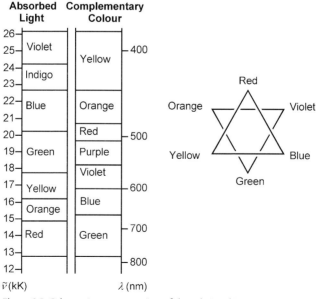

Figure 9.1 Schematic representation of the relationship between absorbed light and its complementary color.

9.3
Fundamentals of Spectroscopy: Selection Rules

One of the most important aspects of a given absorption band is its *allowed or forbidden* nature, as the intensity of the band depends on this feature. Intensity is measured with a spectrometer by means of what is known as *molar absorbance*. The value of molar absorbance (A) is $A = \varepsilon \cdot c \cdot l$ (A is the measured absorbance, c the molar concentration, l the path length of light in cm and ε the molar absorbance coefficient characteristic of each complex). This is the well-known Beer–Lambert's law.

Molar extinction coefficients (ε) generally vary considerably. In the case of intense bands ε_{max} values of 10^4–10^5 may be found, while for weaker bands the ε_{max} value may only be 10^{-2}. In order to understand these notable differences it is necessary to refer to the basic principles of electronic spectroscopy with respect to what are known as selection rules.

9.3.1
Basic Spectroscopic Principles

The *Born–Oppenheimer approach* assumes that the total energy of a system is the sum of three separate energies: electronic, vibrational and rotational. If rotational transitions are not taken into account, then we can draw the potential energy curves for the ground and excited electronic terms, in which the various vibrational states are considered (Figure 9.2).

The continuous line (arrow) is the electronic or vertical transition, while d_e is the equilibrium distance in the ground electronic state. As the excited electronic state has different bond lengths, its potential energy curve is not parallel to the potential energy curve of the ground state. Furthermore, the electronic transition

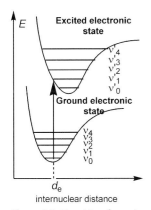

Figure 9.2 Transition from the ground electronic state to an excited electronic state, taking into account the vibrational energy of both the ground and the excited states.

is infinitely faster than the atomic movement that reorders the atoms and returns them to a new equilibrium position and, therefore, the electronic transition will be from the ground electronic and vibrational state to a vibrationally excited electronic state $(v_0 \rightarrow v_i')$. The study of this spectral characteristic has important consequences that will be considered later in Chapter 15.

9.3.2
Transition Moment

During the absorption of electromagnetic radiation there is an interaction between this radiation and the molecule which absorbs it: the molecule is "excited" and one of its electrons is shifted to an excited molecular orbital with a very short half-life (around 10^{-18} s). This is followed by relaxation to the ground state through the emission of heat, light, etc. (Chapter 15). Therefore, this constitutes an electronic transition between an initial (ground) state and a final (excited) state. *The probability that a given transition will occur depends on the initial and final states, as well as on the inductive properties of the light.*

The transition moment, M, is defined as

$$M = \langle \phi_{\text{ground}} | \text{light operator} | \phi_{\text{excit}} \rangle$$

where ϕ_{ground} and ϕ_{excit} represent the ground and excited state wavefunctions. The intensity of this transition – the probability that the transition takes place – is given by the square of the transition moment, $I \propto M^2$.

Light is an electromagnetic radiation that is comprised of two oscillating fields, one electric and the other magnetic, which propagate perpendicularly to one another in the direction of the propagation. Therefore, the corresponding operator will be a combination of an electric dipole and a magnetic dipole, although the latter has a very limited capacity (10^{-4} times less than that of the electric dipole) to produce a transition. If only the electric dipole part is involved the transition moment, M, can be written as $<\phi_{\text{ground}} | \hat{\mu} | \phi_{\text{excit}}>$ where $\hat{\mu}$ is the electric dipole moment operator. $\hat{\mu} = er$ (e = electron charge; r = radius vector).

Given that the wavefunctions have a space-only part and a spin-only part, and rewriting ϕ instead of ϕ_{ground} and ϕ^* instead of ϕ_{excit},

$$M = <\phi_{\text{space}}\phi_{\text{spin}} | \hat{\mu} | \phi_{\text{space}}^* \phi_{\text{spin}}^* >$$

As the operator $\hat{\mu}(er)$ is a function only of the space part, the above expression can be written as: $M = <\phi_{\text{space}} | \hat{\mu} | \phi_{\text{space}}^* ><\phi_{\text{spin}} | \phi_{\text{spin}}^* >$. The space part can be broken down into a vibrational component and an orbital component. In other words:

$$M = <\phi_v^* \phi_v><\phi_{orbital}|\hat{\mu}|\phi_{orbital}^*><\phi_{spin}|\phi_{spin}^*>$$

It is demonstrated that the vibrational integral (the Frank–Condon factor) is not zero. In contrast, the two other integrals may or may not be zero. In the event that one of the two is zero then the total integral, or transition moment, will be zero and there will be no absorption. The second integral is the basis of the orbital selection rule and the third integral is the basis of the spin selection rule. Let us consider these two selection rules in more detail.

9.3.3
Selection Rules

Spin Selection Rule   In the corresponding integral the $\hat{\mu}$ operator is not involved. Regardless of the nature of the space parts, M vanishes if $\phi_{spin} \neq \phi_{spin}^*$. It is demonstrated that for a transition to be allowed the spin multiplicity of the ground and excited terms must be the same ($\Delta S = 0$). Transitions between states of different spin are not allowed because light has no spin properties and cannot, thus, change the spin.

Orbital (Parity or Laporte) Selection Rule   This rule is derived from the integral $<\phi_{orbital}|\hat{\mu}|\phi_{orbital}^*>$ and may be expressed in many other complementary ways. The most spectroscopic definition is: *a transition is forbidden if it only implies the redistribution of electrons within the same type of orbitals.* This would mean that every d → d transition of a coordination compound is forbidden, which is true in theory but not experimentally.

The best way of describing this integral is through group theory. From this perspective, the integral $<\phi_{orbital}|\hat{\mu}|\phi_{orbital}^*>$ can be regarded as a triple product. For $M \neq 0$, this triple product must contain the totally symmetric species of the point group under consideration or, the direct product of the symmetry species of the ground and excited wavefunctions must be, or contain, the representation of the electric dipole term μ ($\mu = er = ex + ey + ez$).

$$<\phi_{orbital}|\hat{\mu}|\phi_{orbital}^*> = <\phi_{orbital}|x|\phi_{orbital}^*> + <\phi_{orbital}|y|\phi_{orbital}^*> + <\phi_{orbital}|z|\phi_{orbital}^*>$$

As this is a summed expression only one of the terms needs to be non-zero for the integral to have a value other than zero.

From the point of view of *atomic spectroscopy*, in which spherical symmetry is considered, the x, y, z coordinates belong to a u type symmetry species. Group theory demonstrates that the direct product of two species of the same symmetry, whether g or u, is always g (g × g = g; u × u = g). Therefore, assuming spherical symmetry, any transition between orbitals of the same kind will be forbidden as its direct product cannot contain symmetry species belonging to x, y or z (u). Thus, any excitation, s → s, p → p, d → d, f → f, is forbidden. This reasoning is known as the "parity" or "Laporte rule".

This rule can be complemented by saying that an electronic transition between orbitals is only allowed if $\Delta l = \pm 1$. Thus, s → p, p → d and d → f are allowed transitions, while s → s, p → p, d → d, f → f (in which $\Delta l = 0$) or s → d, p → f (in which $\Delta l = 2$) are forbidden transitions.

These spectroscopic rules are perfectly applicable to octahedral complexes: the d orbitals are centrosymmetric (g symmetry) while x, y, z present u symmetry (non-centrosymmetric). The direct product g × g can never yield u. The same occurs for any molecule belonging to a centrosymmetric point group. If the symmetry point group of the molecule is not centrosymmetric, some of the d orbitals belong to the same symmetry species as some of the p orbitals: therefore, there is a mixing of orbitals that enables electronic transitions which are no longer d → d but rather a mixture of d(p) → p(d). Thus, in these cases, it is necessary to calculate the direct product of the symmetry species of the ground and excited terms to see if it contains the representation of x, y, z.

9.4
Interpretation of the Selection Rules

It could be concluded from the above that d → d transitions should never be observed, and yet actually they are observed for all transition metal complexes. *In fact, these are the transitions responsible for the color in these complexes.* Although the parity or Laporte rule is a strict quantum rule that cannot be violated it must be interpreted properly. There are three main causes of an increase in the *allowed nature* of the bands, or which even *turn the forbidden bands into allowed ones.* Let us look at each of these in turn.

9.4.1
Mixture of d and p Orbitals of the Same Symmetry Species

Experimentally, the d → d transitions of a tetrahedral complex are more intense than those of an octahedral complex. Thus, for example, aqueous solutions of Co^{2+} have a pale pink color (small ε) due to the presence of the octahedral cation $[Co(H_2O)_6]^{2+}$, whereas in concentrated HCl solution a deep blue color is formed due to the presence of the tetrahedral anion $[CoCl_4]^{2-}$. The molar extinction coefficient of this complex is around 100 times greater than that of the $[Co(H_2O)_6]^{2+}$ complex in aqueous solution. This difference in tetrahedral vs. octahedral coordination is also manifested in simple Ni^{2+} complexes such as $[NiBr_4]^{2-}$ vs. $[Ni(H_2O)_6]^{2+}$.

If the transitions were purely d → d and were only centered on the transition ion then this would not make sense. Crystal field theory does not explain this apparent anomaly which is, however, easy to explain through molecular orbital theory (Chapter 1). Using the molecular orbital diagram of an O_h complex we saw how the symmetry representations of the d and p orbitals are different (t_{2g} and t_{1u}, respectively), whereas in a tetrahedral complex both types of orbital belong to the same symmetry species (t_2). In T_d geometry, any molecular orbital with t_2 symmetry, obtained from a d atomic orbital, contains a mixture of the p atomic orbital of the same symmetry. Thus, in the first case (O_h) a d $\rightarrow$ d transition is much "purer" than in the second case (T_d), in which there is a mixture of p and d orbitals. This p $\rightarrow$ d mixture makes the transition much more allowed, and therefore more intense.

From the point of view of group theory (direct product rule) the treatment is very straightforward. For example, in the case of a tetrahedral Co^{2+} ion, d^7, and referring solely to transitions between terms of the same spin multiplicity, there are three possible d $\rightarrow$ d transitions: $^4A_2 \rightarrow {}^4T_2$, $^4A_2 \rightarrow {}^4T_1(F)$ and $^4A_2 \rightarrow {}^4T_1(P)$ (see Tanabe–Sugano diagram for a d^3, O_h, Chapter 8).

$\mu(x, y, z)$ belongs to the T_2 symmetry species. The triple product $\langle A_2 \otimes T_2 \otimes T_1 \rangle$ is $A_1 + E + T_1 + T_2$. Therefore, the transition is allowed as it contains the totally symmetric species (A_1). In contrast, the triple product $\langle A_2 \otimes T_2 \otimes T_2 \rangle$ is $A_2 + E + T_1 + T_2$. The transition $A_2 \rightarrow T_2$ is forbidden as its triple product does not contain the totally symmetric representation. This "allowed" nature of a d $\rightarrow$ d transition will never arise in a "perfect" octahedral complex (or any symmetry with an inversion center).

9.4.2
Vibronic Coupling

Molecular vibrations can destroy the symmetry of a complex. In the case of O_h symmetry, for example, a T_{1u} vibration deforms the molecule and reduces its symmetry from O_h to C_{4v} (as *trans*-$ML_4L'L''$) (Figure 9.3). In this symmetry point group the d_{z^2}, p_z and s orbitals belong to the same symmetry species (a_1) and therefore they can be mixed. Similarly, the p_x, p_y and d_{xz}, d_{yz} orbitals also belong to the same symmetry species (e). Thus, the transition is no longer d $\rightarrow$ d but rather a mixture of d $\rightarrow$ d and p $\rightarrow$ d (d $\rightarrow$ p). Obviously, the intensity will depend on the extent of the mixing between the two types of orbitals.

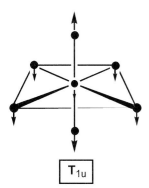

T_{1u}

Figure 9.3 T_{1u} vibration in O_h symmetry. Vibronic coupling (see text).

Once again, the best way of studying whether or not there is vibronic coupling in each case is to apply group theory. The integral that contains the vibrational part and the orbital part can be written as follows:

$$M = <\phi_e^* \phi_v^* |\hat{\mu}| \phi_e \phi_v >$$

where ϕ_e is the electronic (orbital) part and ϕ_v is the vibrational part. Therefore, it is only necessary to calculate the direct product of the symmetry species of the electronic part and the symmetry species of the normal vibrational modes of the molecule under consideration. If this new direct product contains the representations of $\mu(x, y, z)$ then the transition will be vibronically allowed.

Let us consider an octahedral Co^{3+} complex. The ground term is $^1A_{1g}$ and the first excited term is $^1T_{1g}$. The direct product $A_{1g} \otimes T_{1g}$ is T_{1g}. Given that x, y, z correspond to the triply degenerated species T_{1u}, absorption will be forbidden. Group theory demonstrates that the normal vibrational modes of an octahedral complex are $A_{1g} + E_g + T_{2g} + 2T_{1u} + T_{2u}$. The new direct product of T_{1g} must be calculated for any of the normal vibrational modes with u symmetry (as the final product has to be u). Both $T_{1g} \otimes T_{2u}$ and $T_{1g} \otimes T_{1u}$ contain the T_{1u} representation (that corresponds to x, y, z): *the transition will be vibronically allowed.*

9.4.3
Spin–Orbit Coupling

When referring to selection rules it was asssumed that the wavefunctions may be factorized into an orbital part and a spin part, concluding that each factor has its own selection rule. *This idea is false when used exclusively.* Spin–orbit coupling (Chapter 8) "mixes" the orbital part and the spin part such that, in many cases, the two cease to be meaningful concepts. Consider, for example, an orbitally degenerate term T, split into energy multiplets as a result of first-order spin–orbit coupling. In this case the transition moment integral must be written as $M = <\Gamma_{excit} |\hat{\mu}| \Gamma_{ground} >$ where Γ is the representation of these energy multiplets. A transition will be allowed if the direct product $\Gamma_{excit} \times \Gamma_{ground}$ contains the representations of x, y, z.

The spin–orbit coupling coefficients for the first row transition elements are small, the mixing coefficients are also small, and hence the intensities of these spin-forbidden transitions are very weak. However, this mixing is also roughly inversely proportional to the energy separation of the states being mixed in this way. Thus, in tetrahedral complexes this mixing will be stronger than in octahedral ones.

Spin–orbit coupling is the main cause behind the appearance of very weak bands that are forbidden by the spin rule. These bands are generally narrower than the bands allowed by the spin rule. Let us consider the $^3A_{2g} \rightarrow {}^1E_g$ transition of a d^8 ion which is observed in certain Ni^{2+} complexes as a weak, fine band. Due to spin–orbit *mixing* (being an A term we prefer the word *mixing* instead of *coupling*), $^3A_{2g}$ term becomes T_2 (direct product of $A_2 \otimes T_1$ ($S = 1$) and ^1E_g becomes E (direct product of $E \otimes A_1$ ($S = 0$)). The two terms, ground and excited, have been converted into T_2 and E, respectively. The direct product $T_2 \otimes E$ is equal to $T_1 + T_2$. The representation of x, y, z (in O symmetry) is T_1. The transition is allowed.

Final Note A final explanation for the existence of these forbidden bands lies in the magnetic dipole mechanism (that takes into account the magnetic field of the electromagnetic radiation). This contribution is small.

9.5
Types of Spectra for Transition Metal Complexes

The spectra of transition metal complexes may have different origins.

- Spectra mainly associated with the central ion, influenced by the presence of ligands. These are known as d → d spectra.
- Spectra associated with electronic transitions between the central ion and the ligands. These are known as charge-transfer (CT) spectra.
- Spectra mainly associated with the ligands (generally organic molecules; when they possess, for example, a π electron system they give rise to intense $\pi^*-\pi$ bands).

Given the nature of this book the discussion will be limited to the d → d and charge-transfer bands.

9.5.1
d → d Spectra

The bands observed in the visible region are usually the result of d → d transitions, allowed due to the mixing of d and p orbitals or to the existence of vibronic coupling. Naturally, both effects may be present simultaneously and thus reinforce the allowed nature of the transition. Various aspects of these bands may be considered.

9.5.1.1 Number of Bands that Appear and Their Assignation

The number of d → d transitions is easy to explain through Orgel or Tanabe–Sugano diagrams. For example, an octahedral Ni^{2+} complex will present three d → d transitions that are allowed due to the spin rule, $^3A_{2g} \rightarrow {}^3T_{2g}$, $^3A_{2g} \rightarrow {}^3T_{1g}(F)$ and $^3A_{2g} \rightarrow {}^3T_{1g}(P)$. In many Ni^{2+} complexes–as in most cations–the third *experimental* band is usually much more intense than the first two as, rather than being the third d → d transition, it is actually a charge-transfer band which has a more allowed nature than the d → d transition, which is masked.

The bands resulting from d → d transitions are different for each central ion, and depend on the surrounding ligands (different crystal field value, $10Dq$) and its geometry.

9.5.1.2 Intensity of the Bands

The intensity depends on the principle comments before on M and the possible "violation" of the selection rules. Sometimes, spin-allowed bands are much weaker than otherwise expected. One particular case is called "two-electron jumps". This is well illustrated by the spectra of octahedral cobalt(II) species. Three spin-allowed transitions are expected for these d^7 complexes, namely $^4T_{1g}(F) \rightarrow {}^4T_{2g}$, $^4T_{1g}(F) \rightarrow {}^4A_{2g}$ and $^4T_{1g}(F) \rightarrow {}^4T_{1g}(P)$. The second band, $^4T_{1g}(F) \rightarrow {}^4A_{2g}$ is usually very weak (shoulder) and can be missed. Looking at the d^7 TS diagram, at the strong-field limit the transition corresponds to a jump of two electrons from the t_{2g} subset $t_{2g}^5 e_g^2$ to the e_g subset $(t_{2g}^3 e_g^4)$. It is intrinsically less probable than a one-electron jump and so the $^4T_{1g}(F) \rightarrow {}^4A_{2g}$ is only weakly observed.

9.5.1.3 Study of the Bandwidth

Experimentally, $d \rightarrow d$ transitions are of great width and there are many reasons for this, the most important being:

Vibrational Structure As M–L bonds are constantly vibrating, light strikes the molecule in various vibrational positions. The molecule experiences a "breathing" during which Δ varies. *All this produces a degree of uncertainty regarding the value of Δ.* Given this uncertainty the bandwidth will depend on the different slope between the lines representing each energy term. This phenomenon is shown graphically in Figure 9.4.

As can be seen in the Tanabe–Sugano diagrams, the lines representing terms of the same spin multiplicity are usually far from parallel. The greater the divergence between the lines the greater is the bandwidth for the same value of vibrational uncertainty ($\delta\Delta$) (Figure 9.4).

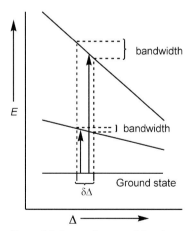

Figure 9.4 Energy diagram of the electronic terms as a function of the crystal field splitting (Δ), emphasizing the origin of the bandwidth in electronic spectra (the energy of the ground state has been arbitrarily set as constant).

In contrast, certain transitions that are forbidden by the spin rule (but which are observed very weakly as a result of spin–orbit coupling) are much narrower. This is the case in the electronic spectra of the $[Mn(H_2O)_6]^{2+}$ ion. The ground term is $^6A_{1g}$ and there is no other term with the same spin multiplicity. Therefore, there will be no transition allowed by this rule. Nevertheless, very weak bands attributable to $^6A_{1g} \rightarrow ^4A_{1g}$ transitions, and other similar ones, are observed. These two terms are quasi-parallel in the Tanabe–Sugano diagram as they both originate in the same electronic configuration $(t_{2g})^3(e_g)^2$, and thus the energy difference does not depend on Δ_o. In addition to being very weak, these bands are much more defined than normal d $\rightarrow$ d transitions.

These types of band may also be produced in transition ions presenting wide d $\rightarrow$ d transitions. In some Ni^{2+} (d^8) complexes the $^3A_{2g} \rightarrow ^1E_g$ transition (see Tanabe–Sugano diagram) can be observed. Both terms are derived from the $(t_{2g})^6(e_g)^2$ configuration. The only thing that has changed is the spin orientation of the two electrons of the e_g orbitals (a shift from parallel spin, $S = 1$, to anti-parallel spin, $S = 0$). This is named a "*spin–flip*" transition. These transitions do not involve any spatial rearrangement but only a spin change. We must remember that the crystal-field strength is a space-only property.

Spin–Orbit Coupling   When λ is small compared to $10Dq$ (first row transition elements) the splitting of terms is not enough to split the bands and simply adds a new factor to the bandwidth. For example, for Ti^{3+} ion the free ion value of the spin–orbit constant is $158 cm^{-1}$. So, given band half-widths of ca. $3000 cm^{-1}$, fine structure caused by spin–orbit coupling is most unlikely to be seen in the spectra of Ti^{3+} complexes.

Effect of Temperature   As the temperature rises, the population of excited vibrational states increases, causing widening of the band. At very low temperatures the bands not only become narrower but are also displaced to greater frequencies (displacement towards the blue region) as only the ground vibronic state, ν_0, of the ground electronic term is populated. Consequently, the transitions begin from the lowest energy possible and therefore ΔE is greater.

9.5.1.4   Band Splitting or Asymmetry

Anything that reduces the symmetry of the starting complex (assuming O_h) will split the ground and excited terms, making the bands wide, asymmetric *and even causing them to split*. The most important of these causes are: (i) The presence of different ligands which are far apart in the spectrochemical series. For example, complexes like $[Ni(NCS)_2(NH_3)_4]$ or $[Ni(NH_3)_4(NO_2)_2]$ are very different in terms of their real symmetry. The former deviates very little from octahedral symmetry as NCS^- and NH_3 produce similar Δ_o, whereas the latter is greatly distorted with respect to octahedral symmetry as NH_3 and NO_2^- (linked by the oxygen) exert a very different effect on the value of Δ_o. (ii) The Jahn–Teller effect which is illustrated in Figure 9.5 for a Cu^{2+} ($O_h \rightarrow D_{4h}$) complex. The tetragonal distortion splits the terms (ground and excited) into two new terms. Thus, from the ground term $^2B_{1g}$ there will be at least two transitions to the terms $^2B_{2g}$ and ^2E_g. If the splitting

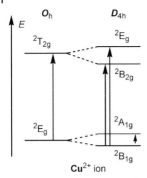

Figure 9.5 Energy diagram of the terms of Cu^{2+} ions and the effect of the Jahn–Teller distortion ($O_h \rightarrow D_{4h}$).

is not very pronounced, no band separation will be observed, simply an asymmetrization of the original band $^2E_g \rightarrow {}^2T_{2g}$. The asymmetry of the absorption band in the visible region of the $[Ti(H_2O)_6]^{3+}$ cation may be explained through Jahn–Teller distortion of the Ti^{3+} $(t_{2g})^1$ cation. As was pointed out in Chapter 3, the Jahn–Teller distortion in t_{2g} orbitals is not the same as that in e_g orbitals. In the first case (t_{2g}, Ti^{3+}) the distortion is small and therefore no band splitting will be observed, only asymmetrization; in the second case (e_g, Cu^{2+}) the distortion may be very large and therefore band splitting may be observed.

9.5.1.5 Band Polarisation: Dichroism

The intensity of absorption is proportional to the transition moment, which is equivalent to the dipolar moment ($\mu = \mu_x + \mu_y + \mu_z$). Therefore, in order for a band to be allowed it is sufficient for it to be so with respect to x, y, z. In high symmetries (O_h and T_d) x, y, z belong to the same symmetry species. As the symmetry is reduced the three coordinates cease to be degenerate, and thus the electronic spectra can be studied using polarized light in order to see if the transitions are allowed according to x, y or z. Let us consider two different cases: a complex *without an inversion centre* (no need for vibronic coupling) and a complex *with an inversion centre* (requires vibronic coupling).

1. A complex of D_3 symmetry, such as $[Cr^{III}(acac)_3]$ or $[Cr^{III}(ox)_3]^{3-}$, illustrates the first case. The splitting of the energy terms of a d^3 ion, upon changing from O_h to D_3 symmetry, is as shown in Figure 9.6. The allowed or forbidden nature of the transitions according to the three coordinates x, y, z, will be deduced from the direct product of the symmetry species of each term. These direct products are shown in Table 9.2.

2. A complex of D_{4h} symmetry, such as *trans*-$[CoCl_2(en)_2]^+$, illustrates the second case. The splitting of the energy terms for a d^6 ion, upon changing from O_h to D_{4h} symmetry, is as shown in Figure 9.7.

   The direct product of the g terms cannot give rise to the symmetry species of x, y, z (u). Therefore, it is necessary to check whether the transitions are

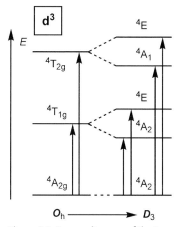

Figure 9.6 Energy diagram of the terms of a d^3 ion ([$Cr^{III}(acac)_3$] or [$Cr^{III}(ox)_3$]$^{3-}$) and their splitting upon changing from O_h to D_3 symmetry (see text).

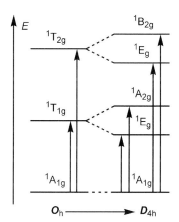

Figure 9.7 Energy diagram of the terms of a d^6 ion, such as *trans*-[$CoCl_2(en)_2$]$^+$ and their splitting upon changing from O_h to D_{4h} symmetry (see text).

Table 9.2 Allowed and polarized nature of d $\rightarrow$ d transitions in a d^3 ion (D_3 symmetry).

Transition	Direct Product	Polarisation	
		z (A₂)	(x, y) (E)
$A_2 \rightarrow A_2$	A_1	Forbidden	Forbidden
$A_2 \rightarrow E$	E	Forbidden	**Allowed**
$A_2 \rightarrow A_1$	A_2	**Allowed**	Forbidden

Table 9.3 Allowed and polarized nature of the d $\rightarrow$ d transitions of a d^6 (D_{4h} symmetry) ion.

Transition	Direct product	Direct product with A_{2u}	Direct product with E_u	Direct product with B_{2u}	Polarization
$^1A_{1g} \rightarrow {}^1E_g$	E_g	E_u	$A_{1u} + A_{2u} + B_{1u} + B_{2u}$	E_u	x, y, z
$^1A_{1g} \rightarrow {}^1A_{2g}$	A_{2g}	A_{1u}	E_u	B_{1u}	x, y
$^1A_{1g} \rightarrow {}^1B_{2g}$	B_{2g}	B_{1u}	E_u	A_{1u}	x, y

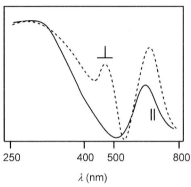

Figure 9.8 Electronic spectrum of a d^6 ion (D_{4h} symmetry) showing the polarized nature of the d $\rightarrow$ d transitions (see text).

allowed – and polarized – by means of vibronic coupling. A new direct product with the normal vibrational modes (u) of a D_{4h} symmetry complex has to be calculated. These normal vibrational modes are $A_{2u} + B_{2u} + 2E_u$. Table 9.3 summarises these calculations. In the D_{4h} point group, z belongs to the A_{2u} symmetry series, while x and y belong to the E_u symmetry series.

The spectrum corresponding to this complex, with polarised light parallel or perpendicular to z is shown in Figure 9.8. Assuming, as often occurs in this type of complex, that the $^1A_{1g} \rightarrow {}^1B_{2g}$ band is masked by charge-transfer bands, then we would only expect the two bands $^1A_{1g} \rightarrow {}^1E_g$ and $^1A_{1g} \rightarrow {}^1A_{2g}$. The first is allowed according to x, y, z (Table 9.3) and so it will be visible with polarized light perpendicular or parallel to the z (Cl–Co–Cl) axis, whereas the $^1A_{1g} \rightarrow {}^1A_{2g}$ transition is only allowed in the x, y direction (perpendicular to z), as can be seen in Figure 9.8.

9.5.2
Charge-Transfer Spectra

9.5.2.1 Definition and Types

Due to the covalence between the central ion and the ligands there is the possibility of new transitions involving the transfer of an electron from the central atom to

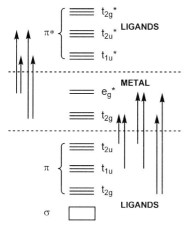

Figure 9.9 Energy diagram of the MOs of an octahedral complex showing the M → L and L → M charge-transfer bands.

the ligands, or vice versa. The bands so produced are much more intense than those resulting from d → d transitions as they are usually allowed by selection rules. In addition to their greater intensity these bands are usually located in the UV region, although in certain complexes they appear in the visible region. Examples of the latter case are the color of chromate (yellow) and dichromate ions (orange), as well as the strong violet color of the permanganate ion. These charge transitions must be considered as transitions between two molecular orbitals, one of which has a more metallic than ligand nature, and vice versa for the other one. The d → d transitions are also transitions between molecular orbitals but *entirely centered on the central ion*.

Figure 9.9 shows the possible transitions due to charge transfer (O_h symmetry). A transition between e_g^* and t_{1u}^* antibonding orbitals involves a parity change: g → u (or u → g). Provided other selection rules are satisfied, therefore, such transitions may be fully allowed.

It can be seen that there are two different types of charge-transfer electronic transitions: metal → ligand (M → L) and ligand → metal (L → M). They can also called metal to ligand charge-transfer transitions (MLCT) and ligand to metal charge-transfer transitions (LMCT). The name charge transfer arises from the fact that these transitions take place between t_{2g} or e_g orbitals, largely of metal character, and some of the bonding or antibonding sets that possess very much greater ligand character. Hence, these transitions involve a much greater displacement of charge between metal or ligand than do d–d transitions.

9.5.2.2 Intensity and Allowed Nature of CT Transitions

In general, the bands due to charge transfer come from transitions that are allowed by the spin and Laporte rules and, therefore, they are usually very intense. This is because, in the corresponding molecular orbital diagram, there are various orbitals with different symmetry labels, and thus it is easy for one of the direct products between the ground term and another excited one to contain the irreducible representation of x, y, z. By way of a practical example, let us study the allowed or forbidden nature of the transitions observed in a couple of cases, applying the direct product rule of the corresponding symmetry species. *It should be borne in mind that this study cannot rely on Tanabe–Sugano diagrams; here, the molecular orbital diagram must be drawn previously.*

1. A classic example for studying $L \rightarrow M$ transitions is the MnO_4^- (or CrO_4^{2-}) (d^0) ion. These are tetrahedral complexes whose simplified molecular orbital diagram is shown in Figure 9.10 (for both the ground configuration and the first excited configuration). The direct products between the ground term ($^1A_1$) and the two possible excited terms (T_1 and T_2) are $A_1 \otimes T_2 = T_2$ and $A_1 \otimes T_1 = T_1$. As the ground term is a spin singlet, only the $^1T_1$ and $^1T_2$ excited terms are of interest. The $^1A_1 \rightarrow {}^1T_2$ transition is allowed orbitally as T_2 is the representation of the x, y, z coordinates. In TiX_4 the phenomenon is the same. The promotion of electrons from halogen orbitals to the empty metal d orbitals gives rise to the charge-transfer spectra, and, for example, the orange color of $TiBr_4$.

2. An example that enables the $M \rightarrow L$ charge-transfer bands to be studied is the square-planar complex $[PtCl_4]^{2-}$. The simplified molecular orbital diagram is shown in Figure 9.11 (for both the ground configuration and the first excited configuration). It is assumed that the $M \rightarrow L$ transitions go from the HOMO b_{2g} molecular orbital (mainly centered on Pt^{2+}) to the three π-anti-bonding molecular orbitals of lower energy, which mainly belong to the halide ligands. Considering solely the excited singlets (the triplets do not permit transitions due to the spin rule), one of the direct products of Figure 9.11, E_u, belongs to

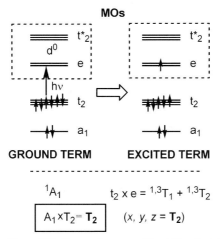

Figure 9.10 Energy diagrams of the MOs of a tetrahedral d^0 complex in the ground and excited terms showing the allowed nature of $L \rightarrow M$ charge-transfer transitions (metal centered MOs shown inside dashed boxes).

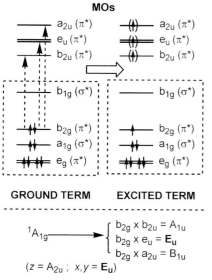

Figure 9.11 Energy diagrams of the MOs of a square-planar d^8 complex in the ground and excited terms showing the allowed nature of M → L charge-transfer transitions (metal centered MOs shown inside dashed boxes).

the same symmetry species as the *x, y* coordinates. Therefore, this band will be allowed by the orbital selection rule.

9.5.2.3 Band Energy: Relationship to Redox Potentials

The energy of a charge-transfer band depends on the magnitude of the energy difference between the d orbitals of the metal and the (σ or π) orbitals of the ligands. This energy difference is a function of the overlap between the starting orbitals. Thus, a greater overlap will, generally, correspond to charge-transfer bands of greater energy.

This effect is observed, for example, when relating the coordination number of the central ion with the energy of L → M charge-transfer bands. The greater the number of ligands surrounding the central ion, the more destabilised are the d-antibonding orbitals of the metal and, therefore, the greater is the transition energy (it is displaced toward the UV region). For example, the $CuCl_2$ (linear) complex absorbs at 546 nm, the $[CuCl_4]^{2-}$ (T_d) complex at 435 nm, and the $[CuCl_5]^{3-}$ (tbp) at 413 nm.

A noteworthy feature of these types of CT band is the relationship between their energy and the redox nature of the metallic ion and the ligands. In fact, because the charge-transfer process involves electron transfer the spectra are sometimes called *redox spectra* and, certainly, band positions can correlate with the ease of oxidation or reduction of the metal ion or ligand.

In L → M transitions the L ligand is, formally, oxidised and the metal M is reduced. Therefore, the more oxidizing M and the more reducing L, the less energy the corresponding CT band will have (it may appear in the visible region). As pointed out above, this is what occurs with chromate, dichromate and permanganate ions. In this way, various series of complexes can be compared. The greater the metal's oxidation state, the greater its oxidizing nature and thus the lower the CT energy will be. If we compare the complexes $[Os^{IV}X_6]^{2-}$ and $[Os^{III}X_6]^{3-}$ (X = halide), the former is more oxidizing and presents CT bands of lower energy. Table 9.4 shows some of these bands (for X = Cl^- and I^-).

In Table 9.4 it can be seen that for the same metallic ion and different halide ligands (Cl^-, I^-) the CT bands have less energy with the I^- than with the Cl^- ligand, as I^- has a greater reducing nature. The energies of the L → M transition in TiX_4 (d^0 ion) or $[Ni^{II}X_4]^{2-}$ (d^8 ion), for example, also increase in this order (Table 9.5).

If M is very oxidizing and L is very reducing then we are dealing with a true redox reaction rather than charge transfer: the complex becomes unstable. This occurs in moving from $[CuCl_4]^{2-}$ to $[CuBr_4]^{2-}$ and to $[CuI_4]^{2-}$: the charge transfer in the latter becomes irreversible, yielding Cu^+ and I_2.

As a consequence, if we study a transition ion that has no d → d transitions (d^5, Fe^{3+}), the colour variation can be related to the charge-transfer bands. Thus, $[FeF_6]^{3-}$ is virtually colourless (it absorbs in the UV region), $[FeCl_6]^{3-}$ is yellow (it absorbs in the violet region), $[FeBr_6]^{3-}$ is brownish, and $[FeI_6]^{3-}$ is unstable and yields I_2 and Fe^{2+}.

In M → L transitions, empty molecular orbitals centered on the ligands are required. These are the high-energy, antibonding orbitals situated above the d orbitals centered on the metal. Charge-transfer bands of this type are thus characteristics of complexes containing strong-field ligands such as CO, CN^- and PR_3.

Table 9.4 Effect of the oxidizing nature of the metal.

Complex	$\pi(L) \rightarrow t_{2g}(M)$		Complex	$\pi(L) \rightarrow t_{2g}(M)$	
	(cm^{-1})	(nm)		(cm^{-1})	(nm)
$[Os^{III}Cl_6]^{3-}$	35 450	282	$[Os^{III}I_6]^{3-}$	19 100	524
$[Os^{IV}Cl_6]^{2-}$	27 000	370	$[Os^{IV}I_6]^{2-}$	12 300	813

Table 9.5 Effect of the reducing nature of the ligand.

Complex	$\pi(L) \rightarrow t_2(M)$		Complex	$\pi(L) \rightarrow t_2(M)$	
	(cm^{-1})	(nm)		(cm^{-1})	(nm)
$TiCl_4$	35 400	280	$[NiCl_4]^{2-}$	35 500	282
$TiBr_4$	29 500	340	$[NiBr_4]^{2-}$	28 300	353
TiI_4	19 600	510	$[NiI_4]^{2-}$	12 300	509

Ligands with delocalised π orbitals, such as phen and bpy, will also provide suitable π^* orbitals.

In this type of transition the metal is, formally, oxidised (it loses electrons) while the ligand is reduced (it gains electrons). Therefore, the more reducing M and the more oxidizing L, the less energy the corresponding CT band will have (it may appear in the visible region). A Fe^{2+} complex will yield M $\rightarrow$ L charge-transfer bands of lower energy than those of a Fe^{3+} complex: Fe^{2+} can easily be oxidized to Fe^{3+} whereas Fe^{3+} is not readily oxidized to Fe^{4+}.

9.5.2.4 M $\rightarrow$ L or L $\rightarrow$ M Transitions?

In any given complex it can be difficult to know if a charge-transfer transition corresponds to M $\rightarrow$ L or L $\rightarrow$ M. A detailed study of the differences is beyond the scope of this book. In certain cases in which two or more highly similar complexes can be compared, the appearance or disappearance of a given band may indicate the type of transition for the band in question. For example, if we compare the ions $[Ir^{IV}Br_6]^{2-}$ (d^5) and $[Ir^{III}Br_6]^{3-}$ (d^6) a wide (split) band is observed for both ions in the 300–350 nm region; however, in the 500–800 nm region a set of bands appears *only* for the $[Ir^{IV}Br_6]^{2-}$ ion. Furthermore, the difference between the two bands corresponds approximately to the Δ_o of the complex. This phenomenon can be explained by assuming that the bands are of the L $\rightarrow$ M type, that is, they correspond to electronic transitions between molecular orbitals centered on ligands and molecular orbitals centered on the iridium ion. In the case of the $[Ir^{IV}Br_6]^{2-}$ (d^5) complex the occupied orbitals are $(t_{2g})^5$ and therefore two transitions, L $\rightarrow$ t_{2g} and L $\rightarrow$ e_g, may take place, whereas in the case of the $[Ir^{III}Br_6]^{3-}$ $(t_{2g})^6$ complex only a CT band corresponding to the L $\rightarrow$ e_g transition will be observed, the difference between the two being Δ_o.

When similar complexes are compared, a displacement of the bands towards lower energy as the oxidation number of the cation increases is indicative, as pointed out above, of L $\rightarrow$ M bands, whereas a displacement of the absorption toward higher energy corresponds to M $\rightarrow$ L bands.

Summary   Table 9.6 shows the intensities of bands in the electronic spectra of transition metal complexes, taking into account their main characteristics.

9.5.3
Intervalence Charge-Transfer Bands

In this chapter we have assumed transition metal ions to be individual, isolated along with their ligands. This hypothesis is not always valid and, in certain cases, we must treat polynuclear complexes which can give new and interesting electronic transitions. *Intermolecular* charge-transfer transitions are observable in certain polynuclear complexes containing more than one transition ion. Owing to their characteristics these bands are often called *intervalence charge-transfer bands*. Given the importance of this issue it will be dealt with later in Chapter 13.

Table 9.6 Relative intensities of bands in the electronic spectrum.

Types of transition	ε (approx.)
Forbidden by orbital and spin rules	0.1
Allowed by spin rule. Forbidden by orbital rule but allowed by vibronic coupling	10–100
Allowed by spin rule. Forbidden by orbital rule but allowed by mixing of p and d orbitals	>100
Allowed by Laporte and spin rules (charge transfer)	10^4

Table 9.7 Equations for calculating *Dq* and *B* for the A_2 ground state (d^3, d^8).

Observed transition	Dq	B
v_1, v_2, v_3	$v_1/10$	$(v_3 + v_2 - 3v_1)/15$
v_1, v_2	$v_1/10$	$(v_2 - 2v_1)(v_2 - v_1)/3(5v_2 - 9v_1)$
v_1, v_3	$v_1/10$	$(v_3 - 2v_1)(v_3 - v_1)/3(5v_3 - 9v_1)$
v_2, v_3	$\{9(v_2 + v_3) - [85(v_2 - v_3)^2 - 4(v_2 + v_3)^2]^{1/2}\}/340$	$(v_3 + v_2 - 30Dq)/15$

Table 9.8 Equations for calculating *Dq* and *B* for the T_2 ground state (d^2, d^7 weak field).

Observed transition	Dq	B
v_1, v_2, v_3	$(v_2 - v_1)/10$	$(v_3 + v_2 - 3v_1)/15$
v_1, v_2	$(v_2 - v_1)/10$	$-v_1(v_2 - 2v_1)/3(4v_2 - 9v_1)$
v_1, v_3	$\{[5v_3^2 - (v_3 - 2v_1)^2]^{1/2} - 2(v_3 - 2v_1)\}/40$	$(v_3 - 2v_1 + 10Dq)/15$
v_2, v_3	$\{[85v_3^2 - 4(v_3 - 2v_2)^2]^{1/2} - 9(v_3 - 2v_2)\}/340$	$(v_3 - 2v_2 + 10Dq)/15$

9.6
Calculating the Crystal Field Parameters from the Position of the d–d Bands

9.6.1
Numerical Method for d^2, d^3, d^7 (Weak Field) and d^8 Ions

In Chapter 8 we noted the equations of the ground and excited terms (same spin multiplicity) for some octahedral ions such as d^2 (d^7, weak field) and d^3 (d^8). From these equations Dou [1] published the solutions to the calculation of *Dq* and *B* for d^2, d^3, d^7 and d^8 octahedral configurations. For unknown reasons these solutions have not been reproduced in most books dealing with the electronic spectroscopy of coordination compounds. The calculated values of *Dq* and *B* are given in Tables 9.7 and 9.8.

Given that Dou's article [1] does not indicate the mathematical procedure for reaching the above-mentioned solutions, the values of Dq and B have been recalculated here using the Mathematica 2.2 program [2], the results being practically identical to those reported by Dou.

There is an important fact to be taken into account in d^2 and d^7 ions: in the Tanabe–Sugano diagram the transitions v_2 and v_3 cross each other for a value of $Dq/B = 1.35$. Therefore, the order of the energies is:

$$v_1 < v_2 < v_3 \text{ for } Dq/B > 1.350$$

$$v_1 < v_3 < v_2 \text{ for } Dq/B < 1.350$$

9.6.2
Graphical Methods for Any d^n Configuration

When studying the Tanabe–Sugano diagrams in Chapter 8 it was pointed out that these diagrams are useful to calculate the Δ and B parameters for any d^n electron configuration. These graphical methods are reported in some Inorganic Textbooks. It should be noted that the precision of this graphical method is limited by the student's ability to extract useful values from the Tanabe–Sugano diagrams as printed in textbooks. However, our advice to the reader interested in the spectroscopic utility of Tanabe–Sugano diagrams is to consult several sites on the web. To the best of my knowledge, the most useful and pedagogic web-site is Ref. [3], already mentioned in Chapter 8.

The reader will find – in a very easy manner – how to calculate the B and $10Dq$ parameters and all other forbidden transitions. Unfortunately, it is only possible for one C/B value (given by the author) [3]. Indeed, in general, this is the main important inherent limitation of the Tanabe–Sugano diagrams.

References

1 Dou, Y-S. Equations for Calculating Dq and B, *J. Chem. Educ.* **1990**, *67*, 134.
2 Jordi Casabó, UPC, Barcelona, 2006, personal communication.
3 http://wwwchem.uwimona.edu.jm:1104/courses/Tanabe-Sugano/ created and maintained by R. J. Lancashire, The Department of Chemistry, University of the West Indies, Mona Campus, Kingston 7, Jamaica. Created Feb 1999. Links checked and/or last modified 14th August 2006.

Bibliography

Lever, A.B.P. (1997) *Inorganic Electronic Spectroscopy*, 2nd edn., Elsevier, Amsterdam.

Sutton, D. (1968) *Electronic Spectra of Transition Metal Complexes*, McGraw-Hill, Berkshire.

Kettle, S.F.A. (1996) *Physical Inorganic Chemistry. A Coordination Chemistry Approach*, Spektrum, Oxford.

Gerloch, M., Constable, E.C. (1994) *Transition Metal Chemistry*, VCH, Weinheim.

Huheey, J.E., Keiter, E.A., Keiter, R.L. (1993) *Inorganic Chemistry. Principles of Structure and Reactivity*, Harper Collins College Publishers, New York.

10
Molecular Magnetism

10.1
Mononuclear Complexes

10.1.1
Introduction

In Chapter 8 we dealt with the characteristics of the terms and multiplets derived from crystal field and spin–orbit Hamiltonians. Now, we introduce the Zeeman Hamiltonian, $\hat{H}_{ZE}$, derived from the application of an external magnetic field. The main property of this Hamiltonian is to split the terms or multiplets from the crystal field to give the *magnetic states*.

For the treatment of this problem we have to take into consideration that the energy of the *excited terms* created by crystal fields is very high ($\approx 10^4 \, cm^{-1}$ above the ground state) which impedes its thermal population at room temperature or lower. In magnetic studies of coordination compounds it must be emphasized that, in first order, only the populated terms are important. The Zeeman effect for external magnetic fields within the conventional experimental limits, i.e. from 0 to 7 Tesla, is of the order of a few cm^{-1}. This allows us to assume that *only* the magnetic states derived from the ground term are populated and their population – following Boltzmann statistics – depends on their energy: the lower the energy the higher the population and vice versa.

From this hypothesis, molecular magnetism tries to answer the following questions:

1. What is the energy of each magnetic state originating from the splitting of the terms (multiplets) due to the effect of the external magnetic field?

2. In a macroscopic system, how many molecules have the same energy?

3. How can we measure the overall phenomenon?

Coordination Chemistry. Joan Ribas Gispert
Copyright © 2008 WILEY-VCH Verlag GmbH & Co. KGaA, Weinheim
ISBN: 978-3-527-31802-5

10.1.2
Quantum Magnetism: Magnetic Moment and Energy of the States

10.1.2.1 Generalities

We may state that all magnetic properties are fundamentally quantum mechanical in origin, although their development started with classical concepts. In coordination chemistry we are currently not interested in one electron but in polyelectronic terms, as studied in Chapter 8. The three principal 'magnetic' properties and their relationships in classical and quantum mechanics are given in Table 10.1.

It is convenient to drop the factor of $\hbar$ in expressions for angular momentum operators, which amounts to saying that these operators measure the angular momentum in units of $\hbar$. In textbooks of Physical Chemistry [1], the orbital and spin magnetic moments as well as the Hamiltonian operators are usually given as $\hat{\mu}_z = \gamma_e \hat{l}_z$, $\hat{\mu}_z = \gamma_e g_e \hat{s}_z$ and $\hat{H} = -\gamma_e g_e \hat{s}_z \mathbf{H}$. With this notation, the eigenvalues must contain the $\hbar$ parameter. However, in books and papers on Molecular Magnetism, the above operators are expressed in terms of μ_B instead of γ_e [2, 3]. Therefore, the resulting eigenvalues do not contain the $\hbar$ factor, because it is already contained in μ_B (see Table 10.1).

$\hbar = h/2\pi$; μ_B (Bohr magneton; the absolute value of the magnetic moment of the z-component of a *free* electron) $= e\hbar/2m_e = eh/(4\pi m_e)$. Units: S.I.: $9.273 \times 10^{-24} \, \mathrm{J\,T^{-1}}$ $(\mathrm{A\,m^2})$; emu-cgs: $9.273 \times 10^{-21} \, \mathrm{erg\,G^{-1}}$ (or $4.668 \times 10^{-5} \, \mathrm{cm^{-1}\,G^{-1}}$); (T = tesla, G = gauss; $1 \, \mathrm{T} = 10^4 \, \mathrm{G}$)

10.1.2.2 Orbital and Spin Angular Momentum

In quantum mechanics, the *orbital angular momentum* is an "observable" associated to the operator, $\hat{\mathbf{L}}$, formally derived from classical mechanics. There exists a complete set of eigenfunctions of $\hat{L}^2$ and one of the components, $\hat{L}_z$.

$\hat{L}_z Y_{LM}(\theta,\varphi) = M_L \hbar Y_{LM}(\theta,\varphi)$, with $M_L = L, (L-1) \ldots -L$.

Furthermore, when trying to compare the concepts of classical and quantum mechanics, a very important new concept arises: the *spin*. In addition to the angular momentum associated with the orbital motion (and, thus, with an "equiva-

Table 10.1 Angular momentum, magnetic moment and energy in classical and quantum mechanics; the magnetic field, *H*, is assumed to be applied in the z direction.

	Angular momentum	*Magnetic moment*	*Energy*
Classical mechanics	$m_e vr = l$	$\gamma_e l = -(e/2m_e)l = \mu^{[a]}$	$-\mu \cdot H$
Quantum operator	$\hat{L}_z + \hat{S}_z$	$\hat{\mu}_z = -\mu_B(\hat{L}_z + g_e \hat{S}_z)$	$\hat{H} = -\hat{\mu}_z H$
Eigenvalues	$(M_L + M_S)$	$-\mu_B(M_L + g_e M_S)$	$\mu_B(M_L + g_e M_S)H$

[a] γ = gyromagnetic (or magnetogyric) ratio. The γ_e value is $-(e/2m_e)$ (standard electromagnetic theory). The angular momentum of the electron $m_e vr$ is quantized in units of $\hbar$; so we can write, $m_e vr = n\hbar$. Furthermore, $\hbar \cdot \gamma_e = \mu_B$ where μ_B is the Bohr magneton.

lent" in classical mechanics) the electron possesses an *intrinsic angular momentum* called *spin* and denoted as $\hat{s}$ ($\hat{S}$ for the corresponding term). It cannot be visualized except as a vector analogous, in certain formal aspects, to the orbital angular momentum, $\hat{L}$. Indeed, it is well established that the component of the spin in a chosen direction (magnetic field, for example) is defined as Ms (see Table 10.1).

10.1.2.3 Electronic Magnetic Moment

Any magnetic moment is associated with an angular momentum. In quantum mechanics, the magnetic moment operator is, for many-electron systems,

$$\hat{\mu} = -\mu_B(\hat{L} + g_e\hat{S}) \tag{10.1}$$

The eigenvalues values in the z-direction (the most important for us) are, thus: $\mu_Z^l = -\mu_B M_L$ and $\mu_Z^S = -g\mu_B M_S$.

In the μ_Z^S formula the spin g-factor appears. This extremely important factor is well authenticated experimentally and may be explained by relativistic quantum mechanics. For a free electron $g_e = 2.002319\ldots$

10.1.2.4 Energy and Magnetization

Electrons possess magnetic moments as a result of their orbital and spin angular moments and these moments interact with an external magnetic field. Classically, the energy of a magnetic moment μ in a magnetic field H is equal to the scalar product $E = -\mu \cdot H$.

Quantum mechanically, we may write the corresponding Hamiltonian as

$$\hat{H} = -\hat{\mu}H \tag{10.2}$$

Knowing the operator of the magnetic moment (with its two components, orbital and spin), the corresponding energy operator (Hamiltonian), considering the interaction between the z-component of the magnetic moment with a magnetic field of magnitude H in the z-direction, will be:

$$\hat{H} = -(\mu_L + \mu_S)_z H = \mu_B(\hat{L}_z + g_e\hat{S}_z)H \tag{10.3}$$

The treatment of this Hamiltonian is very different if we are dealing with (A, E) or T terms, i.e. with or without first-order spin–orbit coupling.

10.1.3
The Hamiltonian

As indicated in Chapter 8, g will be 2.00 if the electron has only a spin component and not an orbital component in its wavefunction. However, this is not true in coordination compounds, in which the orbital component is important either in first-order (T terms) or in second-order (A or E terms).

10.1.3.1 Spin Hamiltonian

A or E terms are currently treated by applying the spin Hamiltonian formalism, which facilitates calculation with the spin quantum operators. In this formalism, only the spin component is considered. The orbital component is included in other terms that usually are parametrized.

Let us remember that the sum of the two Hamiltonians, spin–orbit and Zeeman can be written as:

$$\hat{H} = \lambda \hat{L} \cdot \hat{S} + \mu_B (\hat{L} + g_e \hat{S}) H \tag{10.4}$$

This Hamiltonian can be transformed, assuming polyelectronic systems with rhombic distortion, in

$$\hat{H} = g\mu_B H \cdot \hat{S} + D\hat{S}_z^2 + E(\hat{S}_x^2 - \hat{S}_Y^2) \tag{10.5}$$

where *g*, *D* and *E* are parameters that have been discussed in Chapter 8 (Section 8.4) in relation to the second-order spin–orbit coupling. This is the spin Hamiltonian. The demonstration of the derivation of Eq. (10.5) from Eq. (10.4) is beyond the scope of this book.

10.1.3.2 Hamiltonian for T Terms

The magnetic treatment for T terms is very complicated. The Hamiltonian is the general one (10.4). We will only give a short comment in Section 10.1.7.

10.1.4
Macroscopic Magnetism: Definitions and Units [2–4]

It is important to emphasize, from the beginning, that in molecular magnetism SI units are less frequently used than CGS-emu ones. Indeed, in this (and other chapters) we will deal with CGS-emu units because, so far, it is the preferred system in publications on molecular magnetism (Appendix 2).

The most important property in magnetochemistry is the magnetization (*M*), that may be defined as "*what happens inside the sample when a homogeneous magnetic field is applied on this sample*". Three kinds of magnetization have to be defined, related to volume, mass or molecular weight. These, with their units are:

- volume magnetization (*M*): G (gauss)
- mass magnetization (*M*$_\rho$): *M* /density = G cm^3 g^{-1}
- molar magnetization (*M*$_m$): (G cm^3 g^{-1}) (g mol^{-1}) = G cm^3 mol^{-1}

In molecular magnetism, it is very useful to deal with another quantity, the *susceptibility* (χ). Magnetic susceptibility is defined as $\chi = \dfrac{\partial M}{\partial H}$. Under certain conditions (*H* not too great and temperature not too low) χ = *M/H*. Given that there are three magnetization definitions, there will also be three susceptibility definitions:

- volume susceptibility (χ) : G / G (dimensionless)
- mass susceptibility (χ_ρ): $G\,cm^3\,g^{-1}$ / $G = cm^3\,g^{-1}$
- molar susceptibility (χ_m): $G\,cm^3\,mol^{-1}$ / $G = cm^3\,mol^{-1}$ (cm^3 = emu in cgs-units).

Remarks (i) In a paramagnetic sample the magnetization created by an external field is parallel to the field. From the physical viewpoint this implies that the lines of force are concentrated inside the sample, which creates an increase in the apparent weight of the sample when applying the magnetic field. This physical phenomenon can be studied in any elementary book of Physics. This was the origin of the Faraday balances that measure differences in weight before and after the application of the magnetic field. This Δm is then converted into magnetization and/or susceptibility. Today, the modern magnetometers (SQUIDs) measure other physical phenomena that occur within the sample when applying the magnetic field but always, in the end, these processes are expressed in terms of magnetization and susceptibility. (ii) All paramagnetic compounds (with external unpaired electrons) have their internal shells fully occupied. These closed shells are the origin of the universal phenomenon called *diamagnetism*. Diamagnetism arises from the motion of electrons, treated as charged particles, in the applied magnetic field. Diamagnetism is a very small effect, but, since all materials contain electrons, it is always present. It is of importance because the presence of a large number of diamagnetic atoms can make the diamagnetism an appreciable contribution to the susceptibility of a molecule which contains a paramagnetic ion. Diamagnetism is independent of the temperature and magnetic field. Table 10.2 summarizes the magnitudes, signs and origin of diamagnetism and paramagnetism.

The diamagnetism produces, thus, an effect opposite to the paramagnetism: it diverts the lines of force from the magnetic field toward the exterior of the sample. This phenomenon produces, from the physical point of view, a decrease in the weight of the sample when applying the magnetic field.

It is, thus, very important to take into consideration that in any magnetic experiment we are measuring the "total" susceptibility of the sample (paramagnetism + diamagnetism). However, the treatment of the data must be done only on the paramagnetic susceptibility. This paramagnetic susceptibility is calculated by subtracting the diamagnetic contribution from the measured bulk susceptibility. The diamagnetic contribution is negative, opposite to the paramagnetic contribution, which is positive.

Table 10.2 Characteristics of paramagnetism and diamagnetism.

Type of magnetic susceptibility	Sign	Approximate magnitude of χ	Origin
Diamagnetism	negative	1×10^{-6}	electronic charge
Paramagnetism	positive	$0 - 10^{-4}$	angular moment + electron spin

$$\chi_m(\text{measure}) = \chi_m(\text{param.}) - \chi_m(\text{diam.});$$

$$\chi_m(\text{param.}) = \chi_m(\text{measure}) + \chi_m(\text{diam.})$$

The reader can find in many books on magnetism the values of the diamagnetic corrections for ions and ligands. These tables are called Pascal Tables [2, 5].

10.1.5
Magnetic Measurements: Magnetization and Molar Susceptibility

Molar magnetization is the first measure of magnetochemistry. It gives very good information if working at low temperatures and high magnetic fields.

10.1.5.1 Boltzmann Statistics

According to Boltzmann statistics, the number of particles, N_i, in a sample of N particles that will be found in a non-degenerate state with energy E_i, when it is part of a system in thermal equilibrium at a temperature T is expressed by the *Boltzmann distribution*:

$$N_i = \frac{Ne^{-E_i/kT}}{q} \quad \text{where} \quad q = \sum_i e^{-E_i/kT}$$

where k (or k_B) is the Boltzmann constant ($0.695\,\text{cm}^{-1}\,\text{K}^{-1}$) and q is the partition function.

From this definition is also derived the Boltzmann *probability factor*:

$$P_i = \frac{N_i}{N} = \frac{e^{-E_i/kT}}{\sum_i e^{-E_i/kT}} \tag{10.6}$$

10.1.5.2 Overall Magnetization

In order to calculate the macroscopic magnetic moment (magnetization) of a macroscopic sample, it is necessary to sum all the individual moments of each state, weighted according to their Boltzmann distribution (population). The molar magnetization will thus be $M = N_A \langle\mu\rangle$, where N_A is the Avogadro number and $\langle\mu\rangle$ the average magnetic moment of one molecule. Applying the Boltzmann probability factor, $\langle\mu\rangle = \Sigma\mu_i P_i$

$$M = N_A \sum_i \mu_i P_i = N_A \frac{\sum_i \mu_i e^{-E_i/kT}}{\sum_i e^{-E_i/kT}} \tag{10.7}$$

10.1.5.3 Magnetization: Brillouin Function [6]

In the following formulae, the factor $g\mu_B H/kT = x$ appears. The numerator is the magnetic energy (*order*) whereas the denominator is the thermal energy (*disorder*).

Let us assume a term with an S value, without orbital component. From the magnetization formula given above and remembering that $x = g\mu_B H/kT$ together with the definition of the hyperbolic trigonometric functions, $\sinh x = \frac{1}{2}(e^x - e^{-x})$ and $\cosh x = \frac{1}{2}(e^x + e^{-x})$ it is easy to obtain, simply from mathematical calculation, the following expression:

$$<\mu> = g\mu_B \left[\left(S + \frac{1}{2} \right) \coth\left(S + \frac{1}{2} \right) x - \frac{1}{2} \coth\frac{x}{2} \right] \tag{10.8}$$

Thus, the general formula for the molar magnetization is

$$M = N_A <\mu> = N_A g\mu_B \left[\left(S + \frac{1}{2} \right) \coth\left(S + \frac{1}{2} \right) x - \frac{1}{2} \coth\frac{x}{2} \right] \tag{10.9}$$

N_A being the Avogadro number. This expression is usually written using the Brillouin function

$$B_S(x): \quad M = N_A g\mu_B S[B_s(x)] \tag{10.10}$$

where

$$B_S(x) = \frac{2S+1}{2S} \operatorname{ctgh}\left(\frac{2S+1}{2} x \right) - \frac{1}{2S} \operatorname{ctgh}\left(\frac{x}{2} \right) \tag{10.11}$$

From the chemical point of view it is very important to calculate the limits of the Brillouin function, according to the x values

$$x = \frac{g\mu_B S H}{kT} \begin{cases} >> 1 \Rightarrow B_S(x) = 1 & \text{(a)} \\ << 1 \Rightarrow B_S(x) = (S+1)\dfrac{x}{3} & \text{(b)} \end{cases}$$

(a) $\quad M = N_A g\mu_B S$

(b) $\quad M = H\dfrac{N_A g^2 \mu_B^2 S(S+1)}{3kT}$

Several Brillouin functions are shown in Figure 10.1. Two important concepts are deduced from these formulae:

1. In the limit $x >> 1$ ($g\mu_B H >> kT$, i.e. strong fields and/or low temperature) the molar magnetization approaches a constant value, $N_A g\mu_B S$, depending only on S (spin of the state). This value is the *saturation magnetization* and may be understood as due to the effect of the magnetic field aligning all the spins in a

$S = 1/2 ; 1 ; 3/2 ; 2 ; 5/2$

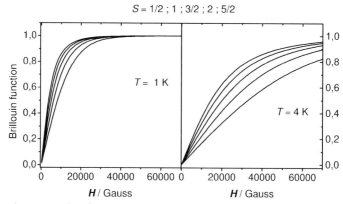

Figure 10.1 Plot of the Brillouin function vs. *H* for $S = 1/2, 1, 3/2, 2, 5/2$ at 1 K and 4 K.

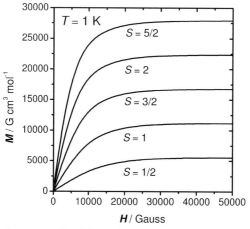

Figure 10.2 Plot of the molar magnetization (*M*) vs. *H* for $S = 1/2, 1, 3/2, 2$ and $5/2$ at 1 K.

parallel form. Under these conditions, the thermal energy cannot misalign the spins.

2. In the limit $x \ll 1$ ($g\mu_B H \ll kT$, i.e. small fields and/or high temperatures) the Curie law ($\chi = M/H = C/T$) appears, with the magnetization values depending not only on *S* but also inversely on *T*.

Graphical Representations For the best understanding of the previous concepts it is useful to represent the magnetization formula $M = N_A g \mu_B S[B_s(x)]$, varying the different parameters that participate:

1. Figure 10.2 shows the variation of the magnetization curves at very low temperature for transition metal ions with 1–5 unpaired electrons. There is

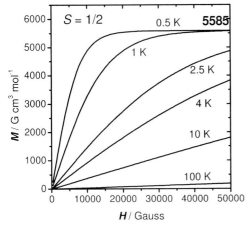

Figure 10.3 Plot of the molar magnetization vs. **H** for S = 1/2 at different temperatures.

a clear tendency to achieve a maximum value (saturation of the molar magnetization) when the magnetic field increases. Saturation magnetization indicates that all electrons are aligned parallel to the magnetic field. For small fields, the thermal energy does not allow total alignment. In the lowest limit (**H** = 0) the magnetization is zero because the electrons are randomly oriented at any temperature: their vector sum will be zero.

2. The shape of the curves depends on temperature. The greater the temperature the greater the difficulty for the electrons to be aligned parallel to the magnetic field, because the temperature creates thermal agitation opposed to the alignment. This phenomenon is shown in Figure 10.3, for the S = 1/2 case. The saturation magnetization value is, for g = 2.0, $5585 \, G \, cm^3 \, mol^{-1}$, which corresponds to the $N_A\mu_B$ value. For different S values the curves will be very similar, with different M_s values (= $N_A g \mu_B S$).

3. It is useful to represent $M/N_A\mu_B$ instead of **M**. We have already indicated that $N_A\mu_B$ is $5585 \, cm^3 \, mol^{-1} \, G$. Let us remember that the limit at low temperature and strong field is $M = N_A g \mu_B S$. Thus, $M/N_A\mu_B = gS$. If g = 2, the product 2S corresponds to $2n/2 = n$ (n being the number of unpaired electrons in the sample). As a consequence, $M/N_A\mu_B$ corresponds to the number of unpaired electrons in the ground state of the ion. Another form to demonstrate this fact is remembering that **M** is the molar magnetization and $N_A\mu_B$ is the magnetization for 1 mol of electrons. The values of $M/N_A\mu_B$ for S = 1/2 – 5/2 versus H/T (this abscissa allows a more general curve) are shown in Figure 10.4. The curves tend to 1, 2, 3, 4 and 5 electrons for high **H** values (large **H** or low T).

Finally, further important information deduced from the magnetization curves is knowledge of the region of the magnetic field (abscissa) in which **M** is linearly dependent on **H**, since in this region $\chi_m = M/H$. Usually this happens between 0 and < 1 T.

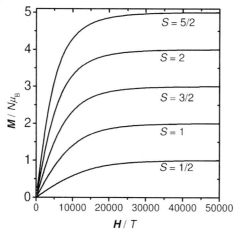

Figure 10.4 Plot of the reduced magnetization ($M/N\mu_B$) vs. H for $S = 1/2, 1, 3/2, 2$ and $5/2$ at 1 K.

10.1.5.4 Molar Susceptibility: van Vleck Equation

In 1932 Van Vleck used the perturbational method to calculate the equation of the molar susceptibility vs. T. He decomposed the energy levels of the system into a Taylor series

$$E_n = E_n^{(0)} + HE_n^{(1)} + H^2 E_n^{(2)} + \ldots$$

and assumed H not too high and T not too small. Applying subsequent approximations, the following expression for the susceptibility is obtained:

$$\chi_m = N_A \frac{\sum\limits_n \left[\frac{(E_n^{(1)})^2}{kT} - 2E_n^{(2)} \right] \exp\left(\frac{-E_n^{(0)}}{kT} \right)}{\sum\limits_n \exp\left(\frac{-E_n^{(0)}}{kT} \right)} \tag{10.12}$$

This equation can be simplified in some cases. For example, for all complexes that do not have an orbital contribution (A, E terms), $E_n^{(2)}$ contributions collapse to a constant term $N\alpha$.

Given the following mathematical relationships:

$$\sum (E_{|S,M_S>}^{(1)})^2 = g^2 \mu_B^2 [S^2 + (S-1)^2 + \ldots + (-S)^2] = g^2 \mu_B^2 [S(S+1)(2S+1)/3]$$

$$\sum_n \exp\left(-E_n^{(0)}/kT\right) = \sum (2S+1) \exp(-E_S^{(0)}/kT)$$

the Van Vleck formula transforms into a simplified form

$$\chi_m = \frac{N_A g^2 \mu_B^2}{3kT} \frac{\sum\limits_S S(S+1)(2S+1) \exp\left(\frac{-E_S^{(0)}}{kT} \right)}{\sum\limits_S (2S+1) \exp\left(\frac{-E_S^{(0)}}{kT} \right)} + N\alpha \tag{10.13}$$

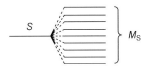

Figure 10.5 Regular splitting of an S state into its M_s owing to an external magnetic field (Zeeman effect).

We can apply the van Vleck formula to various cases:

1. Only the spin-degenerated term ^{2S+1}A, without interactions with other excited terms.

   This is the simplest case (Figure 10.5). Taking $E_S^{(0)} = 0$ as the level of energy reference, the formula of the molar susceptibility is simply:

$$\chi_m = \frac{N_A g^2 \mu_B^2}{3kT} \frac{S(S+1)(2S+1)}{(2S+1)} = \frac{N_A g^2 \mu_B^2}{3kT} S(S+1) = \frac{0.125 g^2}{T} S(S+1) \qquad (10.14)$$

   In molecular magnetism a frequent application concerns the effective *magnetic moment*, $\mu_{eff} = \mu/\mu_B = g[S(S+1)]^{1/2} = [n(n+2)]^{1/2}$, considering that $S = n/2$ (n being the number of unpaired electrons). Thus, $\chi_m T = 0.125 g^2[S(S+1)] = 0.125 \mu_{eff}^2 = \mu_{eff}^2/8$ (if $g = 2.00$). The values of $\chi_m T$ and μ_{eff} for 1–10 unpaired electrons are given in Table 10.3.

Table 10.3 $\chi_m T$ and μ_{eff} values for n unpaired electrons ($g = 2.00$).

n	$\mu_{eff} = [n(n+2)]^{1/2}$	$\chi_m T = \mu_{eff}^2/8$	n	$\mu_{eff} = [n(n+2)]^{1/2}$	$\chi_m T = \mu_{eff}^2/8$
1	$(3)^{1/2} = 1.73$	0.375	6	$(48)^{1/2} = 6.92$	6.00
2	$(8)^{1/2} = 2.82$	1.00	7	$(63)^{1/2} = 7.93$	7.875
3	$(15)^{1/2} = 3.87$	1.875	8	$(80)^{1/2} = 8.94$	10.00
4	$(24)^{1/2} = 4.90$	3.00	9	$(99)^{1/2} = 9.95$	12.375
5	$(35)^{1/2} = 5.92$	4.385	10	$(120)^{1/2} = 10.95$	15.00

2. Spin-singlet ground term, ^1A or ^1E, with non-thermally populated excited terms, with which it interacts.

   $E_n^{(0)}$ and $E_n^{(1)} = 0$ (there is no Zeeman effect, because the term is a singlet). In this case only $E_n^{(2)}$ is different from zero. Thus, $\chi = N_A \Sigma 2 E_n^{(2)} = \text{constant} = \text{TIP}$ (temperature independent paramagnetism). Generally expressed as $N\alpha$. This value is usually very small, because it depends only on the gap between E_i and E_j (energy of ground and excited terms).

3. Only-spin degenerated ground term, ^{2S+1}A, with non-thermally populated excited terms, with which it interacts.

   This is the sum of the two previous cases. It is the actual case of octahedral and tetrahedral complexes with A or E ground terms, having excited terms not ther-

mally populated (currently of the same spin multiplicity) with whom it interacts:
$\chi_m = C/T + N\alpha$

4. An *n*-fold degenerate ground term, with excited terms thermally populated.

In this case $E_i^{(0)} = 0$ but there are a series of levels with $E_j^{(0)} \neq 0$ that it will be necessary to take into account. This situation occurs in T terms of O_h or T_d symmetry, in which the first-order spin–orbit coupling splits the term into multiplets with such energies that can be populated at room temperature. In these cases $E_n^{(2)}$ can neither be omitted nor parametrized. The corresponding van Vleck equation cannot be simplified and it must be deduced for any particular case. Experimentally, these compounds follow (in certain cases) the empirical Curie–Weiss law:
$\chi_m = C/(T - \theta)$

10.1.5.5 Plots of χ_m, $\chi_m T$ and $1/\chi_m$ for Complexes that Follow Curie Law

Molar magnetic susceptibility is represented as χ_m, $\chi_m T$ or $1/\chi_m$ vs. *T*. Instead of $\chi_m T$, some authors prefer to use μ_{eff} because they are mathematically related ($\chi_m T = \mu_{eff}^2/8$). In this chapter, we prefer the $\chi_m T$ representations. The corresponding plots for $S = 1/2, 1, 3/2, 2$ and $5/2$ (assuming $g = 2.00$ in all cases) are shown in Figure 10.6.

The shape of the χ_m vs. *T* curves is very similar for any *S* value. The information given by these curves is not very accurate. The information given by $\chi_m T$ is much more intuitive. The $\chi_m T$ product is constant for each *S* value and corresponds to the values given in Table 10.3 (assuming $g = 2.00$). Finally, the information given by the $1/\chi_m$ vs. *T* curves for any *S* value is also intuitive. All $1/\chi_m$ vs. *T* curves are a straight line that crosses the coordinate origin, but with a different slope.

10.1.5.6 Plots for Complexes that Follow the Curie–Weiss Law

Plots of χ_m, $\chi_m T$ and $1/\chi_m$ vs. *T* for some cases that follow the Curie–Weiss law are shown in Figure 10.7. The three drawings correspond to $S = 3/2$ (taken as an

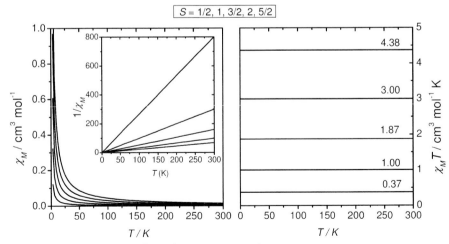

Figure 10.6 The three different forms (χ_m, $1/\chi_m$ and $\chi_m T$ vs. *T*) to represent the Curie law for $S = 1/2, 1, 3/2, 2$ and $5/2$.

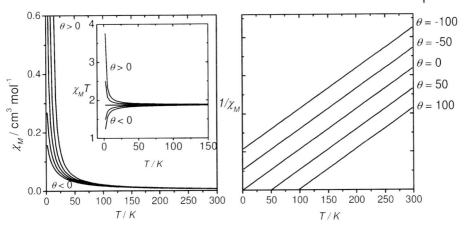

Figure 10.7 The three different forms (χ_m, $1/\chi_m$ and $\chi_m T$ vs. T) to represent the Curie–Weiss law ($\theta > 0$ and $\theta < 0$) for $S = 1/2, 1, 3/2, 2$ and $5/2$.

example) and different positive or negative θ values. For negative θ values the tendency of χ_m towards infinity is restrained; for positive θ values, χ_m tends asymptotically to infinity at temperatures above 0 K. It is much easier to visualize the $\chi_m T$ curves: for $\theta > 0$ $\chi_m T$ values increase instead of remaining constant while for $\theta < 0$ they decrease. Finally, the $1/\chi_m$ figures are also significant. For $\theta = 0$ (Curie law) the straight line crosses the origin of the coordinates. When $\theta < 0$, the corresponding lines cut the y axis at a positive T value. This means that $1/\chi_m$ will be 0 for negative T values (without physical meaning). When $\theta > 0$, the corresponding lines cut the abscissa axis at a positive temperature value.

10.1.6
Application of the Spin Hamiltonian to the A or E Terms

In the treatment of the spin Hamiltonian we have to distinguish between the *isotropic* (O_h, T_d) or *anisotropic* systems (symmetry lower than O_h or T_d). It is also necessary to distinguish between systems with one or more unpaired electrons ($S = 1/2$ or $S > 1/2$). Thus, from the magnetic point of view, four cases must be considered, these are shown in Figure 10.8.

10.1.6.1   Isotropic Systems (Mono- or Polyelectronic)
The M_s components of the **S** vector are split in a regular and identical manner when the external magnetic field is applied in any direction of the space (Figure 10.5). Thus, only a single g value will be observed. The corresponding spin Hamiltonian is simply $\hat{H} = g\mu_B H\hat{S}$.

However, it is important to note, once again, that $g \neq 2.00$. As indicated in Chapter 8 it is possible to demonstrate that for A terms: $g = 2 - (8\lambda/10Dq)$ and for E terms: $g = 2 - (4\lambda/10Dq)$. We achieve, once again, the Curie law, but with an effective g value.

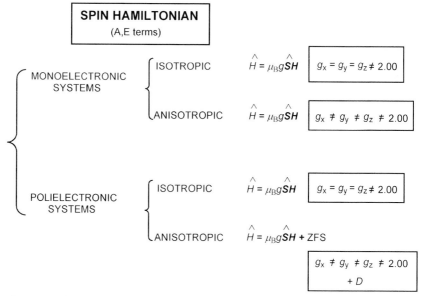

Figure 10.8 The spin Hamiltonians for A and E terms (in cubic field).

$$\chi_m T = \frac{N_A g^2 \mu_B^2}{3kT} S(S+1) \tag{10.15}$$

10.1.6.2 Anisotropic Systems

The M_s components of the vector S are split in a different manner according to the direction or the perturbation. It is necessary to distinguish between monoelectronic and polyelectronic systems.

Monoelectronic Systems   The splitting of the M_s manifolds will be different according to the direction of the external magnetic field. The corresponding three g values are g_x, g_y and g_z (in some cases, $g_x = g_y$). Following the same spin Hamiltonian formalism, the different g values are:

$$g = 2.0023 - \frac{m\lambda}{E(0) - E(n)} \tag{10.16}$$

m being different along the x, y or z axis. In all cases, $E(0)$ is the energy of the ground state and $E(n)$ that of the corresponding excited state. The m values will be treated in Chapter 11.

All these complexes follow the Curie law, but in an anisotropic manner. This means that although the g values are different for any direction, the corresponding magnetic measurements *in any direction follow the Curie law*. This feature could be demonstrated by a measurement on a single crystal. Currently in magnetism we are dealing with powder samples: the observed g value will be simply the average of the three g values, χ_m (average) = $(\chi_x + \chi_y + \chi_z) / 3$.

Polyelectronic Systems: Zero Field Splitting (ZFS) A polyelectronic anisotropic system must be treated with the complete spin-Hamiltonian, with g and D parameters. The problem is usually very complicated because the energies of the two perturbations (Zeeman effect and D, Zero Field Splitting) are very similar. Assuming that $|D| \gg$ Zeeman effect (*usually not true*), it is easier to calculate the corresponding energies and derive the corresponding van Vleck equation. The demonstration is beyond the scope of this book (the reader can consult Ref. [2]). The D parameter can be positive or negative. Working with powder samples, the average $\chi_m T$ values are almost the same for $D > 0$ or $D < 0$: the D value leads to a *decrease* in $\chi_m T$ values only at low temperatures.

To summarize, anisotropic polyelectronic ions follow the Curie law over a wide range of temperatures but at low temperatures the magnetic behavior deviates from the law. In Chapter 11 we will study g values in more depth since the effect of D is more important in EPR spectroscopy than in magnetism.

10.1.7
Magnetism in Orbitally Degenerate Terms (T in O_h or T_d)

As studied in Chapter 8 after applying the spin–orbit Hamiltonian, all terms are split into multiplets (with the corresponding J values). These systems must be studied with the general van Vleck equation. The quantum-mathematical treatment is very complicated and is beyond the scope of this book. The main consequence of this feature is that these complexes do not follow the Curie law. We will only deal, very schematically, with the most studied case: the octahedral Co^{2+} ion.

The splitting diagram for the weak field $^4T_1$ term under the action of spin–orbit coupling and a magnetic field is given in Figure 10.9 (see also Chapter 8).

Although not easy, it is possible to deduce the mathematical expression for χ_m vs. T, A and λ [5]. Let us remember that A is simply a crystal field parameter (already studied in Chapter 8) which is 1.5 for very weak ligand fields and 1.0 for intermediate ligand fields. The $\chi_m T$ formula is:

$$\chi_m T = 0.125 \frac{\frac{7(3-A)^2 x}{5} + \frac{12(2+A)^2}{25A} + \left\{ \frac{2(11-2A)^2 x}{45} + \frac{176(2+A)^2}{675A} \right\} \exp(-5Ax/2) + \left\{ \frac{(5+A)^2 x}{9} - \frac{20(2+A)^2}{27A} \right\} \exp(-4Ax)}{(x/3)[3 + 2\exp(-5Ax/2) + \exp(-4Ax)]}$$

$$\boxed{x = \lambda/kT}$$

(10.17)

Assuming $\lambda = -170\,cm^{-1}$ (free ion), the two limiting cases, for A = 1.5 and A = 1.0, are shown in Figure 10.10. The experimental curves will be intermediate.

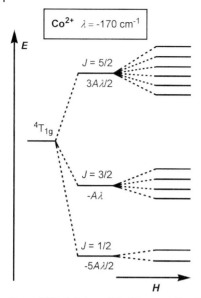

Figure 10.9 Splitting of the $^4T_{1g}$ term (O_h, Co^{2+} ion) into multiplets (J) due to the first-order spin–orbit coupling and, splitting of each multiplet due to the Zeeman effect (magnetic field).

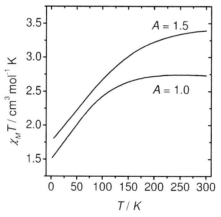

Figure 10.10 Plot of $\chi_m T$ vs. T for an isolated and strictly octahedral Co^{2+} ion. The two extreme A values (1.5 and 1.0) are the limits (see the text for the meaning of these values).

Here, neither the effect of the orbital reduction factor (very important when there is an orbital component) nor the distortions from the regular O_h geometry have been considered. Thus, a much more rigorous quantum treatment is necessary, which is not easy to develop. The reader can find a clear and pedagogical description in Ref. [7].

10.2
Polynuclear Complexes

10.2.1
Introduction

The study of the magnetic behavior of polynuclear complexes is the field related to molecular magnetism that has grown in the most spectacular manner during the past three decades. The studies in this field are important by themselves and, furthermore, have allowed modelling and understanding of the behavior of many bioinorganic processes and the synthesis of many new molecules whose packing in the crystal net has given rise to the so-called *molecule-based magnets*. Last but not least, a new series of discrete molecules was discovered at the end of the 20th century, with high spin and strong anisotropy and showing magnetic behavior intermediate between classical and quantum. These *single-molecule magnets* (SMM) have become the most popular subject in molecular magnetism.

To date, almost all studies in this field have been carried out with polynuclear compounds in which the corresponding ions do not show first-order spin–orbit coupling. Polynuclear complexes from ions with T ground terms need such a complicated quantum treatment that is very difficult to find a good solution.

We are mainly interested in the so-called *'exchange mechanism'* between two (or more) paramagnetic metal centers through diamagnetic bridges.

10.2.2
Magnetic Interpretation in a Cu^{2+} Dinuclear Complex

Let us start with the simplest case: a dinuclear copper(II) complex ($S = 1/2$) in which the bridging ligand between the two copper(II) ions is a monoatomic bridge participating with its p_z orbital (z being defined by the Cu—X—Cu vector) (Figure 10.11).

Group theory demonstrates that the linear combination of the three group orbitals gives three MOs: one of them strictly bonding, which is occupied by two electrons, another antibonding orbital, with high energy, and a third intermediate orbital, that we can call, in a first approach, non-bonding, in which the participation of the p_z bridge orbital is reduced. The four electrons, two from the ligand and one from each copper(II) ion, will be placed in the following way: (i) two in the bonding orbital, the most stable. These electrons allow the existence and stability of the molecule i.e. they are "*core*" electrons, but without important magnetic contribution. (ii) The other two electrons may go either both in the non-bonding orbital or one in the non-bonding and the other in the antibonding orbital. The corresponding gap between these '*magnetic*' orbitals (antibonding or non-bonding) is currently small. In d^n complexes this gap is of the order or 0–500 cm^{-1} approximately. As a consequence, the temperature will be very important in the thermal population of these orbitals and it will be possible to study this with Boltzmann statistics.

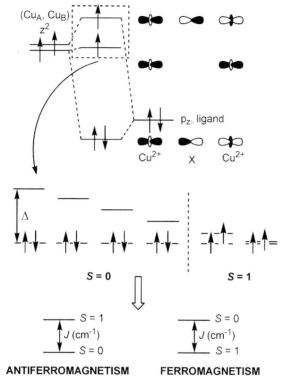

Figure 10.11 Relationship between the MOs of a $Cu^{2+}-X-Cu^{2+}$ system and the concepts of antiferro- and ferromagnetism (see text).

If the two electrons are placed in the same MO the final ground state is a spin singlet ($S_T = 0$); if the two electrons are placed in the two different orbitals the final ground state is a spin triplet ($S_T = 1$). When the ground state is the singlet ($S_T = 0$) the interaction of the "atomic" orbitals is called *antiferromagnetism*, AF, while if the ground state is the triplet ($S_T = 1$) the interaction is called *ferromagnetism*, F. J is a physical observable, the value of which depends on the theoretical model used to describe the phenomenon.

The greater the energy difference between the corresponding MOs, the greater is the energy gap between the singlet and the triplet. In the upper limit, the singlet–triplet gap will be so large that the triplet will not be thermally populated, even at room temperature: instead of antiferromagnetism we could speak of *diamagnetism*.

Let us assume the opposite situation: the two corresponding MOs with the unpaired electron are orthogonals or quasi-orthogonals (overlap integral $S = 0$). Then the two electrons will be placed in degenerate or quasi-degenerate molecular orbitals. In this case, following the Hund rule, each electron will be placed in one of the two orbitals: the triplet term will be stabilized with regard to the singlet term.

The bridge thus allows overlap between the atomic orbitals having the unpaired electrons of the metal centers and the frontier orbitals of the diamagnetic bridging ligand. As a consequence of this overlap the orbitals containing the unpaired spins are no longer localized metal d-orbitals, but are non-bonding or antibonding MOs that encompass both the metal atom and the intermediate bridging atom or atoms. These "new" orbitals are usually termed *'magnetic orbitals'* (see below).

10.2.3
Phenomenological Treatment

10.2.3.1   The Heisenberg–Dirac–Van Vleck Hamiltonian

The notion of "exchange energy" resulting from electrostatic interactions between the electrons shared by two atoms i and j, with spins S_i and S_j has been successfully represented by the Heisenberg–Dirac–van Vleck (HDVV) Hamiltonian. This Hamiltonian describes phenomenologically the interaction between several spins and allows the calculation of the interaction energy between two (or more) paramagnetic ions:

$$\hat{H} = -J_{ij}\hat{S}_i\hat{S}_j \tag{10.18}$$

J_{ij} is the exchange constant between spins S_i and S_j. J may be measured in cm^{-1} or K (1 K = 0.695 cm^{-1}; 1 cm^{-1} = 1.44 K).

The reader can find in the literature other definitions of this Hamiltonian, $\hat{H} = -2J_{ij}\hat{S}_i\hat{S}_j$ or $\hat{H} = J_{ij}\hat{S}_i\hat{S}_j$. It is very important, therefore, to indicate which is the Hamiltonian used. In this textbook we will use the Hamiltonian of Eq. (10.18). This Hamiltonian implies that $J < 0$ and $J > 0$ correspond to antiferro- and ferromagnetic coupling, respectively.

Let us assume two A and B ions, with spins S_A and S_B. If $S = S_A + S_B$, $S^2 = S_A^2 + S_B^2 + 2S_A \cdot S_B$, $S_A \cdot S_B = \frac{1}{2}(S^2 - S_A^2 - S_B^2)$, and, thus the HDVV Hamiltonian may be written as:

$$\hat{H} = -\frac{J}{2}(\hat{S}^2 - \hat{S}_A^2 - \hat{S}_B^2) \tag{10.19}$$

This Hamiltonian has an exact quantum solution with the following eigenvalues:

$$E(S, S_A, S_B) = -\frac{J}{2}[S(S+1) - S_A(S_A+1) - S_B(S_B+1)] \tag{10.20}$$

S_A and S_B are constant. Thus, the energy values will be rewritten for each S value:

$$E(S) = -\frac{J}{2}S(S+1) \tag{10.21}$$

10.2.3.2 Calculation of the Energies for Homodinuclear Complexes with d^n Configuration

Applying Eq. (10.21) it is easy to calculate the energies for dinuclear systems of Cu^{2+}, Ni^{2+}, Cr^{3+}, Mn^{3+}, Mn^{2+}. The calculated energies are shown in Figure 10.12. The first step will be, logically, the calculation of all S values. The arrows in Figure 10.12 indicate which are the S values (with their corresponding energies) that one has to introduce in the van Vleck equation to calculate the molar susceptibility. We have assumed, in an arbitrary form, that the coupling between the two centers is antiferromagnetic thus giving an $S = 0$ ground state. The diagram will be reversed for a ferromagnetic case.

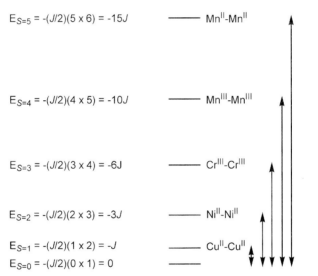

Figure 10.12 Energy values for $S = 0$ to $S = 5$ in homodinuclear complexes.

10.2.4
Molar Susceptibility Equations for Homodinuclear Complexes

If we consider a dinuclear copper(II) complex the S values are 1 and 0; the development of the simplified Van Vleck equation will be as follows:

$$
\begin{aligned}
\chi_m &= \frac{Ng^2\mu_B^2}{3kT} \times \frac{0(0+1)(2\times0+1)\exp(0/kT) + 1(1+1)(2\times1+1)\exp(J/kT)}{(2\times0+1)\exp(0/kT) + (2\times1+1)\exp(J/kT)} \\
&= \frac{Ng^2\mu_B^2}{3kT} \times \frac{6\exp(J/kT)}{1+3\exp(J/kT)} = \frac{Ng^2\mu_B^2}{kT} \times \frac{2\exp(J/kT)}{1+3\exp(J/kT)} \tag{10.22}
\end{aligned}
$$

The reader may easily derive the expressions for all the other homodinuclear d^n–d^n systems (Appendix 3).

10.2.5
χ_m and $\chi_m T$ vs. T Plots

In all the following plots it has been assumed artificially that $g = 2.00$.

10.2.5.1 $S_A = S_B = 1/2$ (Dinuclear Copper(II) Complex)

The χ_m plots for a Cu(II) dinuclear complex, assuming different J values (in cm^{-1}), are shown in Figure 10.13 left. When J is negative, the shape of the curve (mainly the temperature of the maximum) is clearly indicative of the magnitude of J. The more shifted this maximum to higher temperatures, the greater the J value. When J is positive, the divergence of the values of χ_m is very similar to that of the Curie law (no coupling). This is one of the reasons why the $\chi_m T$ vs. T plots are more convenient. Some of these $\chi_m T$ plots, both for $J > 0$ and $J < 0$ are shown in Figure 10.13 right.

If $J < 0$, $\chi_m T$ always tends to 0 because $S_{G.S.} = 0$ while if $J > 0$ $\chi_m T$ tends to 1 cm^3 mol^{-1} K (value for two coupled electrons, $S = 1$). At high temperature (300 K for example), the $\chi_m T$ value will depend on J. If J is small, the $\chi_m T$ value will be close to the expected value for two non-coupled electrons (0.75 cm^3 mol^{-1} K) (see Table 10.3).

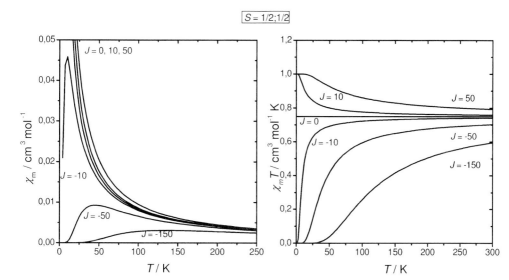

Figure 10.13 Plots of χ_m and $\chi_m T$ vs. T for dinuclear Cu—Cu complexes assuming different J values.

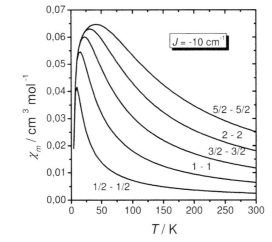

Figure 10.14 Plots of χ_m vs. T for dinuclear d^n–d^n complexes assuming $J = -10\,cm^{-1}$. The corresponding S values are indicated in each curve.

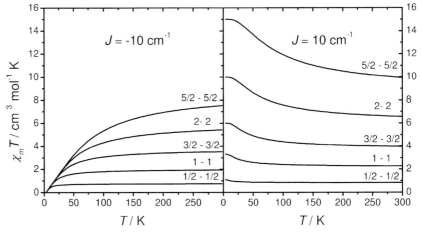

Figure 10.15 Plots of χ_m and $\chi_m T$ vs. T for dinuclear d^n–d^n complexes assuming $J = \pm 10\,cm^{-1}$. The corresponding S values are indicated in each curve.

10.2.5.2 Homodinuclear Complexes d^n–d^n ($S > 1/2$)

The corresponding χ_m vs. T plots for the d^n–d^n ($S > 1/2$) are shown in Figure 10.14, assuming $J < 0$. The most peculiar feature is the presence of a maximum of χ_m (as commented for the Cu^{2+}–Cu^{2+} case). In all cases, the susceptibility maximum is shifted to higher temperatures when the absolute value of J increases.

Usually, it is better to use $\chi_m T$ vs. T plots instead of χ_m vs T. The $\chi_m T$ plots for $J = 10$ and $-10\,cm^{-1}$ are shown in Figure 10.15. The shape of these plots can be generalized for different values of J. To summarize, all these curves: (i) start at room temperature with a $\chi_m T$ value close to twice the $\chi_m T$ value for two non-interacting ions; (ii) when $J < 0$ (AF coupling), $\chi_m T$ tends to zero at 0 K; (iii) when $J > 0$ (F coupling), $\chi_m T$ tends to a finite value that corresponds to $S_T = 2S$.

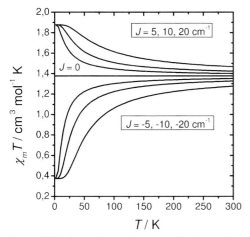

Figure 10.16 Plots of χ_m and $\chi_m T$ vs. T for heterodinuclear $Ni^{2+}–Cu^{2+}$ complexes assuming different J values.

10.2.6
Heterodinuclear Complexes

The treatment for heterodinuclear complexes is very similar. If $S_A \neq S_B$ the vector sum will span from $(S_A + S_B)$ to $(S_A - S_B)$. The coupling is ferromagnetic if $(S_A + S_B)$ is the ground state and antiferromagnetic if $(S_A - S_B)$ is the ground state. The most peculiar feature of these heterodinuclear complexes is that $(S_A - S_B)$ *cannot be zero*. Thus, even assuming a strong antiferromagnetic coupling, the $\chi_m T$ value at low temperature will tend to a non-zero value. The finite value will correspond to that anticipated for the n unpaired electrons of the $(S_A - S_B)$ term. This phenomenon of non-compensated antiferromagnetic coupling (different size of the interacting spins) is called *molecular ferrimagnetism*.

The reader will not have any difficulty in developing the Van Vleck equation for any possible heterodinuclear system, given that the procedure is exactly the same as that followed for the homodinuclear complexes (Appendix 3).

Once again, it is better to use $\chi_m T$ vs. T plots. Some of these plots for a $Cu^{2+}–Ni^{2+}$ heterodinuclear complex with different J values (positive or negative) are shown in Figure 10.16, and they are compared with those for $J = 0$.

10.2.7
Generalization to Other Polynuclear Compounds

The phenomenological treatment explained above for dinuclear complexes, can be generalized to other polynuclear systems. There are different methods reported in the literature for the simplest cases. The reader can consult Ref. [2].

On increasing the nuclearity (tri- tetra- penta- n-nuclear complexes) the system becomes more and more complex and the calculation of the energies becomes more difficult. It is necessary to change the methodology and to work with matrix-diagonalization calculations when the nuclearity increases. Fortunately, in many cases, to study some interesting magnetic properties, it is sufficient to know which the ground state is. Experimentally, the ground state is related to the $\chi_M T$ value at ~0 K. It is thus necessary to carry out magnetization measurements at the lowest temperature possible and at very high fields to measure the $M/N\mu_B$ saturation value.

Coming back to the central problem of this section, the difficulty in the calculation of J is even more complicated when the studied system is a low-dimensional one. In most of such cases there is no exact solution for the corresponding Hamiltonian. For example, an isotropic uniform chain of $S = 1/2$ ions has an exact quantum solution, but not if $S = 1$ or greater. In these cases – which are very frequent – it is necessary to work with rings of finite size and then to extrapolate the values to infinity. Currently, the results from a ring of 12 centers are very close to the exact value.

Exact solutions do not exist for 2D and 3D systems. It is possible to do some calculations with the Monte Carlo approach, for example, with good results.

10.2.8
Relations Between Molecular Orbitals and Magnetic Coupling

The phenomenological Hamiltonian studied in the previous section does not provide any information on the "microscopic" mechanism really involved so that it has no predictive character. Consequently, it becomes necessary to build up a new "microscopic" Hamiltonian, taking into account the kinetic energy of the unpaired electrons on A and B centers, the electron-core potential energies and the electrostatic interaction between the electrons of the system ($\propto e^2/r_{ij}$, r_{ij} being the interelectronic distance) [8].

10.2.8.1 The Nature of the Magnetic Exchange
The coupling mechanism was first introduced by Anderson (1959) for extended systems like oxides and fluorides. Anderson's model was crystallized in a simple set of rules by Goodenough and Kanamori (1959, 1963) [8].

From these first approaches, two different methodologies have evolved in the context of *molecular magnetism*: the Kahn [2] and the Hoffmann [9] models. In both cases, the J parameter is decomposed into two components: one antiferromagnetic and one ferromagnetic.

$$J = J_F + J_{AF}$$

In both theories, the delocalization of the magnetic orbitals is crucial. The main difference between the models lies in the method of calculating J_F and J_{AF}. The theoretical interpretation of these differences is beyond the scope of this book.

In Kahn's model, the definition of magnetic orbitals, is easy to visualize: "*The natural magnetic orbital 'a' in AXB is defined as the singly occupied molecular orbital*

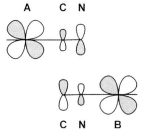

Figure 10.17 Magnetic orbitals for two ions (A and B) bridged by a CN⁻ group (see the text for the explanation).

for the AX fragment in its local ground state; 'b' is defined in the same way with respect to the XB fragment . . . The cutting of the AXB entity into two AX and XB fragments with a common bridging region is, in general, somewhat arbitrary, which is certainly the weak point of this approach" [2]. An example is shown in Figure 10.17. It correspond to two metal ions bridged by a cyanide, CN⁻, group.

As simplified results:

$J_{AF} \propto -\Sigma \Delta^2$. This term represents the antiferromagnetic factor favoring the singlet state. J_{AF} is proportional to Δ^2, the latter being the square of the energy gap between the corresponding symmetric and antisymmetric MOs constructed from the two magnetic orbitals (see Figure 10.11 and Section 10.2.9).

$J_F \propto k$, where k is the exchange integral between the two magnetic orbitals, k (or K_{ab}) = $<a(1)b(2)|e^2/r_{12}|a(2)b(1)>$.

In this context, Kahn's approach provides a more intuitive way of understanding the magnitude of J_F, based on the concept of the *orthogonality of the magnetic orbitals* and, at the same time, constraining the two magnetic orbitals to have a large *overlap density*.

Let us consider the bimetallic moiety $Cu^{II}–V^{IV}O$, in which the magnetic orbitals a and b are orthogonal (Figure 10.18A): copper(II) has a d_{x2-y2} magnetic orbital, while oxovanadium(IV) has a d_{xy} magnetic orbital ($x^2 - y^2$ and xy are orthogonal). The overlap density at the point i (x,y,z) is defined as the product of the two wavefunctions at this point.

$$\rho_i = a(i)b(i) \qquad [(\phi_{Cu}(i)\phi_{VO}(i)], i = \text{point } x,y,z; \text{Figure 10.18B]} \qquad (10.23)$$

The two magnetic orbitals have large electron delocalization on the bridging ligand, thus providing the required large overlap density for ferromagnetic coupling. Indeed, the Cu–VO complex presents two strongly positive lobes around one of the bridging atoms and two strongly negative nodes around the other bridge.

On the other hand, the two-electron exchange integral k can be written as

$$k = \int_{\text{space}} \frac{\rho_i(1)\rho_i(2)}{r(12)} dV_1 dV_2 \qquad (10.24)$$

(A)   **(B)**

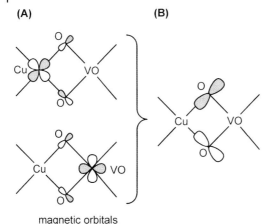

magnetic orbitals

Figure 10.18 (A) Magnetic orbitals in a Cu^{2+}–VO^{2+} system with O bridge; (B) Overlap density map in the bridging zone of the same system.

which indicates that the magnitude of k is governed by the overlap density in the zone of the space corresponding to the bridge. In such a bridging zone, the numerator of k is large and the denominator small. If the numerator is 0, there is no ferromagnetic contribution; if the denominator is large, the J_F values will be small.

Since the depth of the negative lobes exactly compensates the height of the positive holes (Figure 10.18B), the overlap integral, $S = \int_{space} \rho_i d\tau_i = 0$. The AF term is, thus, zero. Finally, it must be noted that the J_{AF} component (< 0) is currently much greater than the J_F (> 0) component. Thus, when the two components exist (the most frequent case) it is almost certain that the final coupling will be antiferromagnetic. Only when J_{AF} is nil or negligible will the global behavior have the possibility to be ferromagnetic.

Summary (i) The orthogonality of the magnetic orbitals explains why the triplet is the lowest but does not provide any information on the magnitude of the phenomenon. The magnitude is given by Eq. (10.24). (ii) if $J_{AF} = 0$ and $J_F = 0$ the global behavior is simply paramagnetic: there is no magnetic coupling, indicating that the two ions are isolated (from the magnetic point of view).

10.2.8.2 New Trends

The calculation of Δ is much easier than the other components. For this reason the MO theory has been widely employed for the calculation of J_{AF} using semiquantitative methods, actually not very accurate but very pedagogical, such as the extended-Hückel molecular orbitals calculation, the availability of which has been extraordinarily increased with the computer program CACAO [10].

The last ten years have witnessed the growth of several *ab initio* calculation methods. A technique that has gained considerable importance in recent years to become one of the most widely used techniques in molecular magnetism is *density functional theory* (DFT) [11]. The explanation of these methods is not easy and is beyond the scope of this book. All these theoretical calculations have been an extraordinary help to experimentalist chemists who work in the field of molecular magnetism. For example, these methods have allowed them to find many excellent magneto-structural correlations from which, more than the exact *J* value, it is interesting to know how to tune the coupling parameter when distances, angles or other structural parameters vary in a series of similar complexes. We will study some examples in Section 10.2.9.

10.2.8.3 Extrapolation to Different Polynuclear Systems

If we try to generalize the previous orbital feature to other kind of complexes with $S > 1/2$ or greater nuclearity, the qualitative treatment is similar. If we are dealing with Ni^{2+} (two unpaired electrons for each Ni^{2+}) there will be four non-bonding or antibonding MOs in a dinuclear complex; 6 MOs in a trinuclear complex; 8 MOs in a tetranuclear complex and so on. A similar procedure can be applied to polynuclear M^{n+} complexes.

Extrapolating these results to two Mn^{2+} ions (d^5–d^5), for example, it will be necessary to calculate the k and Δ^2 value for each $(z^2 - z^2)$, $(x^2 - y^2) - (x^2 - y^2)$, $(xy - xy)$, $(xz - xz)$, $(yz - yz)$ and crossed pairs.

As a consequence, trying to generalize for the best understanding of the phenomenon, homonuclear or heteronuclear complexes with magnetic orbitals of the same symmetry species (e.g. Cu^{2+}–Cu^{2+} or Cu^{2+}–Ni^{2+} in O_h geometry, with the unpaired electrons in the e_g orbitals) will show a tendency to have antiferromagnetic coupling. With these ions, the ferromagnetic coupling will be limited to those cases in which the disposition of the ligands gives the named *accidental orthogonality* (see later). If we are dealing with a pair of ions in which their magnetic orbitals are strictly orthogonal the magnetic coupling will necessarily be ferromagnetic (or simply paramagnetic if J_F is negligible). For example, dinuclear $Cr^{3+}(t_{2g})^3$–$Cu^{2+}(e_g)^3$ or $Cr^{3+}(t_{2g})^3$–$Ni^{2+}(e_g)^2$ complexes will be ferromagnetically coupled.

10.2.9
Qualitative Study of Some Antiferromagnetically Coupled Systems

10.2.9.1 Influence of the Geometry in Dinuclear Cu^{2+} Complexes with Halido, Oxido, Hydroxido or Alkoxido Bridges: a Paradigmatic Case

Qualitative analysis of the two MOs in the bridging zone (the most important part from magnetic point of view) can be made by simply using group theory. Assuming an ideal D_{2h} symmetry, the shape of the corresponding MOs can be schematized as shown in Figure 10.19. It is important to mention that: (i) the two MOs are antibonding from the metal–ligand point of view; (ii) the two MOs are essentially d orbitals of the metals, the contribution of the orbitals of the ligands being small.

It is easy to plot the energy of the two MOs, from Table 1.10 (Chapter 1) when the Cu—O—Cu angle varies. The Walsh diagram for a variation of this angle

(A) (B)

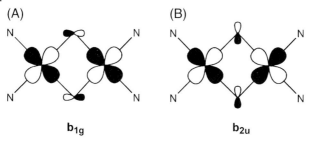

b_{1g} b_{2u}

Figure 10.19 MOs formed by the symmetric (A) and antisymmetric (B) combination of the corresponding magnetic orbitals for a dinuclear Cu–Cu complex with two monoatomic (X) bridging ligands.

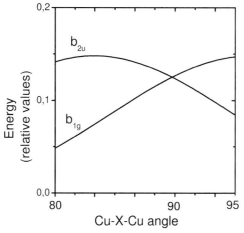

Figure 10.20 Walsh diagram for the two MOs of Figure 10.19, when the Cu–X–Cu angles vary from 80° to 95°.

between 80° and 95° is shown in Figure 10.20, The first important feature we see is that for values of the Cu–X–Cu angle of 90°, the two MOs have the same energy: they are *accidentally orthogonal*. The reader may easily deduce this result qualitatively by simply looking at the MOs in Figure 10.19: the overlap between the orbitals of the ions and the ligand orbitals is strictly the same for the b_{1g} and b_{2u} MOs, only when the geometry is strictly square.

In this qualitative approach we have deliberately omitted the, difficult to calculate, bielectronic integral k which gives the magnitude of J_F. More rigorous studies without omitting k, demonstrate that the ferromagnetic behavior predominates at ca. 95° [2].

Many of these complexes have been reported in the literature. The magnetic behavior is very different, depending on the Cu–X–Cu angle. In some cases the coupling is strongly antiferromagnetic (J close to -500 cm^{-1}) whereas in other cases the coupling is strongly ferromagnetic (J close to $+170$ cm^{-1}). Theoretical calculations show, indeed, that the major factor controlling the spin coupling between the $S = 1/2$ metal centers is the Cu–O(X)–Cu angle, although other less important factors also need to be considered [2, 12].

10.2.9.2 Tuning of J Value in Complexes with Antiferromagnetic Coupling

Given the direct relationship between Δ^2 and J_{AF}, the J value may be tuned by modifying the Δ value, i.e., the overlap between the magnetic orbitals. Let us study some illustrative examples:

Modification of the Bridging Ligand: Oxalato Bridge and Its Derivatives Oxalato and oxalato-like bridged dinuclear copper(II) complexes have been extensively studied. Some derivatives of the oxalato on which tuneable magnetic studies have been carried out are shown in Figure 10.21. The electronegativity of the atoms in these ligands and the more or less diffuse character of their orbitals are changed. This feature allows us to tune the magnetic behavior.

The two MOs derived from the linear combinations of the magnetic orbitals of the copper(II) ions are shown in Figure 10.22A. For a qualitative study of these

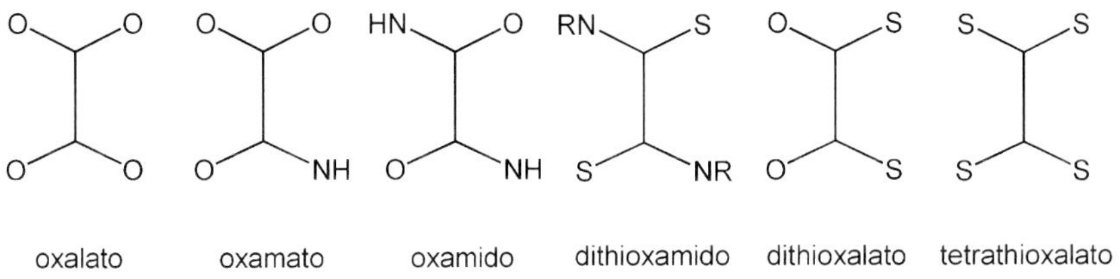

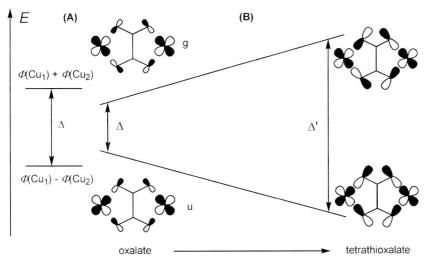

| oxalato | oxamato | oxamido | dithioxamido | dithioxalato | tetrathioxalato |

Figure 10.21 Oxalato and oxalato-like ligands than can act as bridge between two paramagnetic metal ion centers.

Figure 10.22 Variation of the MOs and their qualitative energy in a Cu–"oxalato-like"–Cu complex when passing from oxalato to tetrathioxalato bridging ligand.

orbitals it is sufficient, once again, to use group theory which allows calculation of the symmetric and antisymmetric combinations of the magnetic orbitals. The Δ value will depend on the overlap between the p orbitals of the oxygen atoms of the oxalato ligand. If we replace the oxygen atom by less electronegative atoms (N or S), the p orbitals will be more diffuse and, thus, the electronic density less attracted by the atom. As a consequence, the bonding overlap will be more bonding (more stable) and the antibonding one, more antibonding (more unstable). The main consequence of this will be that Δ will increase when going from the oxygen atom to the nitrogen or the sulfur. Having this in mind it is clear that the antiferromagnetic coupling will increase in the following order: oxalate < oxamato < oxamido < dithiooxalato < dithiooxamido < tetrathiooxalato. In the literature many complexes of copper(II) with this kind of ligand are described [13]. The *average* experimental J values agree perfectly with these predictions: $J \approx -385\,\text{cm}^{-1}$ (oxalato), $J \approx -425\,\text{cm}^{-1}$ (oxamato), $J \approx -580\,\text{cm}^{-1}$ (oxamido), $J \approx -730\,\text{cm}^{-1}$ (dithiooxamido), $J < -1000\,\text{cm}^{-1}$ (tetrathiooxalato).

The most interesting feature to emphasize in this kind of complex was predicted by Verdaguer [14]. He postulated, before its synthesis, the strong antiferromagnetic character of a hypothetical polynuclear copper(II) complex with tetrathiooxalato as a bridging ligand. A few years later, Vicente et al. synthesized the first tetrathiooxalato-bridged copper complex, showing, effectively, such a strong antiferromagnetic J value, that the compound was diamagnetic at room temperature [15]. The MO qualitative energy scheme, when passing from oxalato to tetrathiooxalato bridging ligand is shown in Figure 10.22B.

Modification of the Terminal Ligands   It is possible to modify the geometry of the environment of the metal ions, simply by changing the terminal ligands. This effect is very easy with the flexible Cu^{2+} ion, because it may show different geometries (see Chapter 3). When changing the geometry, we also change the nature of its magnetic orbital and, thus, the J value. When the copper(II) geometry is octahedral, square-planar or square-pyramidal, the magnetic orbital is the $d_{x^2-y^2}$; if the geometry is trigonal bipyramidal the magnetic orbital is the d_{z^2} (see Chapter 1).

Let us see what happens in oxalato-bridged dinuclear copper(II) complexes and different terminal amine ligands, according to the geometry of the copper(II) atom. The scheme for five different possibilities is shown in Figure 10.23. *Only* the antibonding orbitals with respect to the p orbitals of the oxalato bridging ligand have been arbitrarily chosen. Furthermore, the J values given in Figure 10.23 are mean values because they are averaged from those reported in the literature with the same or similar ligands.

When the terminal ligand is bpy or tetramethylethylenediamine, the magnetic orbitals $(x^2 - y^2)$ are in the same plane and, thus, the overlap with the p-orbitals of the oxygen atoms of the oxalato bridge is large. This leads to a strong antiferromagnetic coupling ($J = -400\,\text{cm}^{-1}$). By replacing one of the bpy ligands with one tridentate pentaethyldiethylenetriamine ligand (now the magnetic orbital is the z^2) there is still a good overlap through the major lobe of the z^2 orbital but the overlap

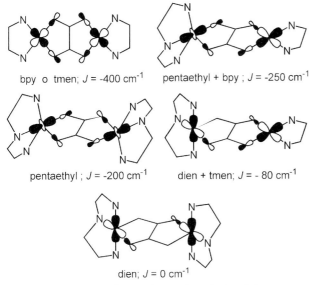

bpy o tmen; $J = -400$ cm^{-1} pentaethyl + bpy ; $J = -250$ cm^{-1}

pentaethyl ; $J = -200$ cm^{-1} dien + tmen; $J = -80$ cm^{-1}

dien; $J = 0$ cm^{-1}

Figure 10.23 Variation of J according to the variation of the geometric environment of Cu^{2+} ion by changing the terminal ligands.

through the small lobe of the same orbital is lower: the antiferromagnetic coupling diminishes considerably ($J = -250$ cm^{-1}). When the two magnetic orbitals are the z^2, the overlap is still lower, J decreasing to -200 cm^{-1}. If a terminal ligand is the diethylenetriamine, with a marked tendency to give square-pyramidal geometry "perpendicular" to the oxalato bridge, there will be good overlap only through a part of the bridge, giving a small coupling ($J = -80$ cm^{-1}). Finally, if the two terminal ligands are diethylenetriamine there is no overlap: J will thus be practically zero, such as indeed occurs experimentally.

10.2.10
Ferromagnetic Interactions

10.2.10.1 Theory and Experimental Data

The *best strategy* to get ferromagnetic coupling is by using orthogonal magnetic orbitals, as has been previously mentioned (Section 10.2.8.1). Let us again consider the Cu^{II}–V^{IV}O system analyzed in Section 10.2.8.1, but changing the bridging ligands. If the overlap density is diffuse, shared by many ligand atoms, the k value would be smaller and, at the limit, k could be zero and the coupling negligible: the two metals would be neither ferro- nor antiferro magnetically coupled. A very heuristic example is given by comparing two similar complexes of Cu^{2+} and VO^{2+}: the first with two small oxido bridges (J strongly ferro) and the second with a large oxalato bridge (J ferro but very small). The r_{12} distance (denominator in Eq. (10.24) is short in the first case but long in the second. The reader can consult Ref. [16] for a deeper analysis.

10.2.10.2 Tuning of *J* Value with Azido Bridging Ligands: Antiferromagnetism vs. Ferromagnetism

Azide anion is one of the most versatile ligands for linking divalent metal ions [17]. As a common factor, two typical coordination modes occur when N_3^- anion acts as a bridging ligand: end-to-end (EE or μ-1,3) and end-on (EO, μ-1,1) (Figure 10.24A). Normally, the EE coordination mode causes antiferromagnetic coupling while the EO leads to ferromagnetic coupling, although some exceptions are reported in the literature. It is important to note that the azido molecular orbitals, which are involved in the global molecular orbitals, are the π-antibonding orbitals of the anion (Figure 10.24B).

Focusing on the μ-1,1 coordination mode in copper(II) complexes, the two singly occupied molecular orbitals (SOMOs) are represented schematically in Figure 10.25A. The b_{1g} orbital is markedly Cu−N antibonding and is strongly destabilized upon increasing the Cu−N−Cu angle, as a result of an enhanced metal–nitrogen overlap. In contrast, the b_{2u} orbital has non-bonding Cu−N character and its energy is practically insensitive to changes in the Cu−N−Cu angle. Thus, on increasing this angle, the gap between b_{1g} and b_{2u} increases [18]. These results indicate the possibility of antiferromagnetic coupling for angles larger than 104°, in agreement with the experimental data.

With regard to the μ-1,3 coordination mode, it has been stated that it is always antiferromagnetic, but recent results indicate that this is not true in certain cases. Let us assume, for example, one of the most frequent experimental cases: a dinuclear or 1D system in which Cu^{2+} ions are linked by one μ-1,3 azido bridging ligand. The corresponding scheme of the SOMOs is shown in Figure 10.25B. Currently, all reported complexes are not linear, showing a noticeable gap between the non-bonding and the antibonding MOs, giving AF coupling. When the Cu−N−N angle diminishes, the antibonding overlap also decreases, giving a smaller

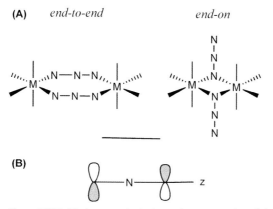

Figure 10.24 (A) the two principal coordination modes of the azido ligand acting as a bridge ; (B) the MOs of the azido ligand involved in the magnetic orbitals (p_x and p_y are equivalent, although only one has been drawn).

(A)

(B)

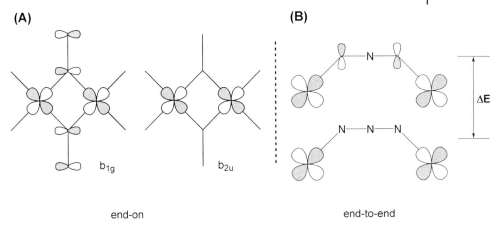

end-on

end-to-end

Figure 10.25 (A) The two SOMOs of a Cu—(μ-N₃)₂—Cu system, with azide in 1,1 coordination mode; (B) the two SOMOs of a Cu—(μ-N₃)₂—Cu system, with azide in 1,3 coordination mode.

gap. At the limit, 180° (or close) the two MOs will be degenerate and thus we could expect ferromagnetic coupling. This fact was demonstrated by Escuer et al. in a dinuclear μ-1, 3 azido complex in which the azide bridge is inside a cryptand, which constrains the azide to be almost linear with the two Ni^{2+} ions [19].

10.2.11
Towards Molecule-based Magnets

In this chapter, we have only explained molecular magnetism considering the molecules as isolated entities, without taking into account that at low temperature we cannot omit the intermolecular interactions among them. At this point, it is important to state that paramagnetic entities (mono or polynuclear) will present a *three-dimensional* ordering at a given finite temperature (T_c). This ordering is called *long-range order* and may be ferro- or antiferromagnetic.

This 3D order depends logically on the type of interaction that exists between the isolated entities: if it is only of van der Waals type, the ordering temperature will be very small (generally of the order of 10^{-2} K); hydrogen bonds somewhat increase this low long-range ordering temperature to some K; covalent or ionic bonds allow long-range order with T_c in the range of room temperature or even above.

The transition from a paramagnetic state to an ordered magnetic state is generally a phase transition: in the paramagnetic phase, above T_c, the orientation of each molecular spin is independent of the spin of its neighbor atoms; there is no correlation between the spins. At this stage, the length of the spin correlation is zero. When approaching T_c, this correlation length starts to be finite but not zero and at $T = T_c$ it becomes infinite. Below T_c total correlation among all the spins occurs.

If this correlation among the neighboring ions is ferromagnetic all spins will be oriented in a parallel manner. If the correlation is antiferromagnetic the adjacent spins will be oriented in an antiparallel manner.

In the paramagnetic zone (above T_c) the external magnetic field aligns all spins according to the field, but when the external field is switched off, the spins come back to their natural random position. In the $T < T_c$ zone there is cooperative (long-range) order. If this order is ferromagnetic, all spins are oriented parallel, even without any external magnetic field (Figure 10.26). We may speak about the so-called *molecule-based magnets*.

For many years, one of the principal aims of molecular magnetism has been the synthesis of these molecule-based magnets, i.e., discrete molecules or low-dimensional systems with cooperative phenomena in the crystal net with a higher T_c possible. Experimentally a molecule-based magnet shows the typical characteristics of a classical magnet, such as Fe_3O_4 (magnetite): (i) with decreasing temperature, the magnetization increases sharply to achieve the saturation magnetization, $M_s = Ng\mu_B S$, S being the spin value of the ground state of the molecule. The Brillouin law is not followed. (ii) They exhibit hysteresis: once reaching saturation magnetization, it is not sufficient to cut-off the magnetic field in order to get zero magnetization: it is always necessary to apply an opposite field. There is, thus, a remnant magnetization. To suppress this remnant magnetization it is necessary to apply a field in the opposite direction (coercive field). These concepts can be found in any book on magnetism.

10.2.11.1 Some Examples of Molecule-based Magnets

Many strategies have been tried over the last 25 years to obtain molecule-based magnets:

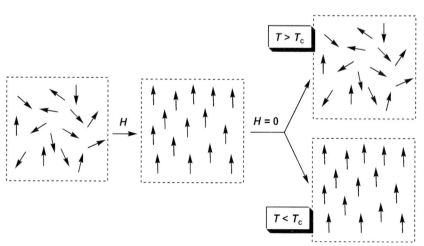

Figure 10.26 Scheme of a paramagnetic and a ferromagnetic phase in the solid state, after removal of the applied magnetic field (*H*). Arrows represent a single molecule (mono or polynuclear).

1. The *metal–radical approach*, in which the interacting magnetic centers are metal ions directly bound to stable organic radicals [20].

2. The so-called *charge-transfer complexes*, in which D + A → D^+ + A^-. It is worth mentioning that the reaction of $[V(CO)_6]$ [or $V(C_6H_6)_2$] and TCNE (tetracyanoethylene) led to the isolation of a molecular-based magnet with a $T_c \approx 400\,K$ [21]. Other similar high-temperature metal–organic magnets have been reported recently [22].

3. The typical *coordination systems*, mainly low-dimensional, tuned or forced to link the "molecular" entities in a ferromagnetic or ferrimagnetic manner in the 3D network. The azide ligand has been one of the best ligands for providing such materials with a 3D *ferromagnetic* order [17]. Oxamato, oxamidato and similar ligands have been, for a long time, the main ligands devoted to the synthesis of molecule-based *ferrimagnetic* magnets with spectacular success [23]. Analogous success has been achieved with $[M(ox)_3]^{3-}$ ($M = Cr^{3+}$ or Fe^{3+}) complexes acting as ligands [24]. The most promising results have been obtained with paramagnetic cyano-complexes as building blocks, giving one- two- and three-dimensional systems many of them showing long-range magnetic order. Let us emphasize the Prussian Blue analogs, because some of them show room temperature T_c values [25].

10.2.11.2 Importance and Perspectives

There are important features that distinguish magnetic materials based on molecules from their analogs in ionic or metallic lattices, such as transparency, low density (possible plastic magnets), solubility in organic solvents allowing several recrystallizations and purification, etc. The synthetic methods are, of course, quite different. Consequently, thin films could be deposited with methods, such as solvent evaporation or others that are familiar from polymer science.

10.2.12
Single-Molecule Magnets [3]

Currently, the most fascinating field in molecular magnetism is that corresponding to the so-called single-molecule magnets (SMMs). These systems are *isolated molecules*, usually with a large S in the ground state, and, mainly, strong anisotropy (preferred direction of the spin). The phenomenon was discovered at the beginning of the 1990s on $[Mn_{12}O_{12}(acetato)_{16}(H_2O)_4]$ (Figure 10.27) (hereafter abbreviated as $Mn_{12}Ac$), comprising a central $[Mn_4^{IV}O_4]^{8+}$ cubane held within a non-planar ring of eight Mn^{3+} ions linked by eight μ_3-O^{2-} ions. Peripheral ligation is provided by sixteen μ-carboxylato groups. This compound has a ground state $S = 10$ and shows relaxation of the magnetization at low temperature (of the order of months at 2 K) [26]. From the magnetization point of view $Mn_{12}Ac$ behaves "like a classical magnet". However, it is still small enough to also show important quantum effects. In fact $Mn_{12}Ac$ (and related molecules, see later) provide the best examples to date of the observation of these effects, such as the quantum tunneling of the magnetization.

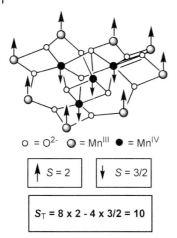

$O = O^{2-}$ $\bigcirc = Mn^{III}$ $\bullet = Mn^{IV}$

| $\uparrow$ $S = 2$ | $\downarrow$ $S = 3/2$ |

$S_T = 8 \times 2 - 4 \times 3/2 = 10$

Figure 10.27 Schematic representation of [Mn$_{12}$Ac] (see text), with the calculation of its S_T ground state.

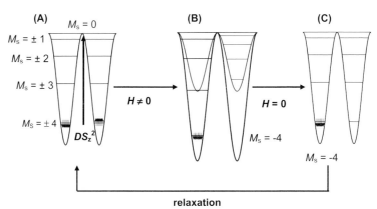

relaxation

Figure 10.28 Scheme of the double-well energy for an $S = 4$ SMM, with negative D value, when applying a magnetic field and after removal of the applied magnetic field (H).

10.2.12.1 Magnetic Relaxation

The Hamiltonian $\hat{H} = DS_z^2$ describes the anisotropy of such systems. An anisotropic $S = 4$ system, taken as an example, is plotted in Figure 10.28A. The states with positive M_s are plotted in one potential well, and those with negative M_s in the other. When no external field is applied all the levels are degenerate pairs (except, logically $M_S = 0$). If D is negative the $M_S = \pm S$ levels will be lowest. When a field is applied parallel to the z axis, the states with a negative M_S, corresponding to a projection of the magnetization parallel to the field, are stabilized, while those with positive M_S, corresponding to a magnetization against the applied field, are destabilized (Figure 10.28B). Thus, if T is low and DS_z^2 large the $M_S = -4$ state will be

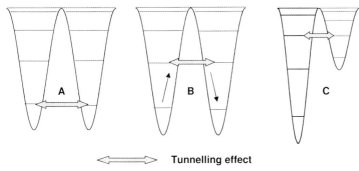

Figure 10.29 Quantum tunneling effect between (A) the ground states and (B) thermally activated states. See text for (C).

the *only one* populated and the magnetization will reach the saturation value. But now a question arises: what happens when the field is removed? (Figure 10.28C). Logically, the system must go back to thermal equilibrium, through a series of steps, with a certain relaxation time. The important feature for these systems is that this thermal relaxation time can be long. The relaxation (the return to equilibrium) can be monitored by measuring the magnetization as a function of time [27].

The thermal relaxation rate follows the Arrhenius law, where the barrier ΔE is $-DS^2$ for integer spin and $-D(S^2 - \frac{1}{4})$ for half-integer spin values. Thus, for a given D value, the higher the S value the longer the relaxation time. The reader must be aware that if $D > 0$ the ground state is $M_S = 0$ and therefore these magnetic phenomena cannot be observed.

10.2.12.2 Quantum-tunneling Effects

Attention will be focused on low temperatures, when few phonons are available, so that the spin can hardly jump over the potential barrier. Assuming the double well of Figure 10.28, if the wavefunction of the left-hand particle extends over the right-hand well, and vice versa, the state of the particle must be described by a superposition of the two states: the probability of observing the left particle in the right well is different from zero: therefore the particle will be in both the right- and left-hand wells. It is as if the particle can pass from left to right without climbing the barrier, but through tunneling. This effect is in fact called *quantum tunneling*, and is one of the most typical manifestations of quantum mechanics. Therefore, the actual possibility of observing tunneling depends on the extent of the interaction of the wavefunctions. The explanation of the strict conditions is beyond the scope of this book (see Ref. [27]).

It is very important to note that tunneling may occur not only between the lowest-lying states $M_S = \pm S$ (Figure 10.29A), but also between pairs of degenerate excited states. This phenomenon is called the *phonon-assisted* (or *thermally-activated*) tunneling mechanism (Figure 10.29B). This mechanism is important at intermediate temperatures. It offers, therefore, a shortcut for the relaxation of the molecules: a molecule does not need to go over the maximum of the barrier, even at relatively high temperatures, since it may find a shortcut and tunnel.

Finally, tunneling can also occur in the presence of magnetic fields. If a magnetic field is applied parallel to the z axis the energies of the M_S levels change rapidly. It is apparent that the pairs of $\pm M_S$ levels will no longer be degenerate, and the condition for tunneling will be lost. However, since the energy of the $+M_S$ level increases and that of the $-M_S + n$ level decreases, they may coincide in energy at some values of the magnetic field and restore the conditions for resonant tunneling (Figure 10.29C).

10.2.12.3 Main Examples of SMMs [28]

Mn_{12}-carboxylato complexes obtained by substitution from $Mn_{12}Ac$ by reaction with other carboxylates have been the most abundant series to date, together with other manganese clusters with higher nuclearity. It has become clear, however, that the high nuclearity is not so important since pentanuclear, tetranuclear, trinuclear and even dinuclear complexes (always with Mn^{3+}) have been reported as SMMs.

The $[Fe_8O_2(OH)_{12}(tacn)_6]^{8+}$ (tacn = 1,4,7-triazacyclononane) was the first properly identified SMM of iron(III) and it was the second molecular species, after the family of $[Mn_{12}]$ clusters. Other SMMs have been reported, including vanadium(III), cobalt(II), nickel(II), and some heterometal SMMs.

The finding of slow magnetization relaxation in lanthanide complexes has opened the possibility of constructing SMMs containing *only* a single metal ion as a magnetic center. Indeed, $[LnPc_2]^-$ (Pc = dianion of phthalocyanine; Ln = Tb, Dy, Ho) show the typical features corresponding to SMMs. Progress made on the synthesis of SMMs up to 2006 is reviewed in Ref. [28].

10.2.12.4 Single Chains Magnets (SCM)

The field of SMMs was further expanded by the discovery in 2001 of a polymeric compound, $[Co(hfac)_2\{NIT(C_6H_4OMe)\}]$ (hfac = hexafluoridoacetylacetonate, NIT = 2-(4′-R)-4,4,5,5-tetramethylimidazoline-1-oxyl-3-oxide), which displays hysteresis behavior above 4 K without undergoing three-dimensional magnetic ordering. The slow dynamics of the magnetization requires a strong Ising anisotropy and a very low ratio of interchain/intrachain interactions. These two requirements are difficult to balance. Progress on the synthesis of SCMs up to 2006 is reviewed in Ref. [29]. A novel dinuclear Mn^{3+} complex, reported in 2007, has behavior intermediate between that of SMMs and SCMs, due to its intermolecular interactions [30].

10.2.12.5 Challenges and Perspectives

Since the anisotropy is a tensorial property it is determined by the tensorial sum of the local anisotropies. It other words, it is not only necessary to choose individual building blocks with the correct anisotropy, but also to achieve an adequate topological arrangement of the local anisotropy tensors: the total anisotropy of the cluster will actually depend on the geometry of the whole structure. On this subject, Gatteschi stated: *"Frankly speaking this still seems to be beyond our control, suggesting that serendipity must play an important role in the development of SMMs"* [27].

Few theoretical studies on how to control the total anisotropy and its relation with the molecular orbitals of the system have been reported. It is worth mentioning the recent DFT work by Ribas-Ariño et al., on Fe^{2+} cubes, which predicts how to enhance the magnetic anisotropy barrier of these systems [31].

The low T_c temperature (a few degrees Kelvin), remains the main obstacle for some of the potential applications of SMMs. Possible applications of magnetic clusters range from novel storage media to quantum computing. Some achievements have been reported in recent years: a direct observation of SMMs organized on gold surfaces showed that the idea of storing information in one molecule has considerable credibility; the immobilization of SMMs in mesoporous silica hosts; the storage of magnetic information on polymers by using patterned SMM (Mn_{12} clusters). Progress on the preparation of novel materials using SMMs up to 2006 is reviewed in Ref. [32].

10.3
Spin Transitions (Spin Crossover (SCO))

10.3.1
Concepts and Mechanisms [33]

The spin crossover centers display labile electronic configurations switchable between the high- and low-spin states. Changes in magnetism, color and structure may be driven by a variation of temperature and/or pressure and by light irradiation. Cooperative spin transitions accompanied by hysteresis (memory effect) may be achieved when the cohesive forces in the solid state propagate the structural changes cooperatively to the whole framework.

Basically, the phenomenon of spin crossover is a property of the isolated complex due to the interplay between the dependence of the ligand field strength and the electron–electron repulsion. However, secondary effects such as substantial deviations from octahedral symmetry, packing effects in the crystal lattices, cooperative interactions, and external perturbation such as pressure or magnetic fields, may influence the physical and photophysical properties of spin crossover compounds to a non-negligible extent.

We have commented in Chapter 8 that octahedral d^n ($n = 4, 5, 6, 7$) complexes can present two different electronic configurations, depending on the ratio between Δ_0 and P. The corresponding ground terms are shown in Figures 8.12–8.15.

When $\Delta \approx P$ the *spin transition* (or *spin crossover*) may occur. Thus LS $\leftrightarrow$ HS may happen: in the vicinity of the crossing point, where Δ and P have similar values, the differences in energy between the HS and LS states is of the order of magnitude of the thermal energy ($k_B T$). In this singular region complexes can adopt both spin states and interconvert in a controlled, detectable and reversible manner, by the action of temperature, pressure or light irradiation.

P changes very little during the spin transition. It is the ligand field strength which changes. It does so in such a way that in the high-spin state $10Dq$ is substantially smaller than *P*, and in the low-spin state $10Dq$ is substantially larger than *P*: $10Dq^{HS} < P < 10Dq^{LS}$.

The SCO phenomenon can be considered to be an intra-ionic electron transfer, where the electrons move between the e_g and t_{2g} orbitals. Given that the e_g subset has an antibonding character its population/depopulation takes place concomitantly with an increase/decrease in the metal-to-ligand bond. This factor contributes to the change of the metal–ligand bond length. It follows that the average metal–ligand bond length is longer in the HS state than in the LS state. This difference is in the range 0.14–0.24 Å for iron(II) complexes ($\Delta S = 2$), 0.11–0.15 Å for iron(III) ($\Delta S = 2$) and 0.09–0.11 Å for cobalt(II) ($\Delta S = 1$). Remarkably, variations for the bond angles are also observed.

The metal-to-ligand bond change affects significantly the constant force of the vibrations (*f*) of the HS and LS forms. An instructive and simple way to represent the SCO center is to consider the totally symmetric breathing mode A_{1g} of the octahedron as the most representative structural change, in a configurational coordinate diagram. The potential energy of the SCO center may be expressed as $E_i = \frac{1}{2}[f_i r_i^2]$, where i = HS, LS and *r* is the average metal–ligand bond length.

In this configurational coordinate diagram the minima of the two potential wells are displaced relative to each other, both vertically and horizontally (Figure 10.30). At very low temperatures the SCO center is in the LS ($n_{vib} = 0$) state. As the tem-

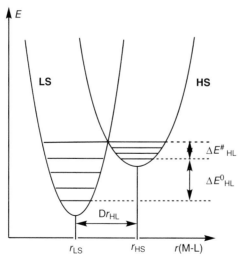

Figure 10.30 Energy curves with their vibrational components for LS and HS states in a spin crossover system (see text for the meaning of the labels).

perature increases, the SCO center transforms the thermal energy into vibrational energy, populating the excited vibrational levels up to the crossing point of the energy curves, where the geometry of the precursor (LS) and the successor (HS) centers are the same. It is at this point that the LS ↔ HS transformation takes place. The energy cost to clear the enthalpy gap between the LS and HS states is compensated by the entropy difference between both states, which favors the population of the HS state at high temperatures. There are basically two contributions to the entropy difference between the two states, namely the electronic contribution due to the greater spin degeneracy of the high-spin state, and a vibrational contribution of the resulting *higher density* of vibrational states in the high-spin state (caused by the population of the e_g^* orbitals, which decreases f and, thus, the particular energy) (see Figure 10.30).

The zero point energy difference between the two states, ΔE_{HL}^0, has to be of the order of thermally accessible energies, $k_B T$. If such is the case, all complexes will be in the low-spin state at very low temperature, whereas at elevated temperatures an entropy-driven, almost quantitative, population of the high-spin state may be observed [34]. In practice, the high-spin state is invariably that of higher temperature.

10.3.1.1 Effect of Pressure
The influence of pressure on the SCO can be understood in terms of the two potential wells. Its main effect is to destabilize the HS state as it is characterized by a larger volume than the LS. Hence, pressure decreases Δr and increases ΔE_{HL}^0. Consequently, pressure decreases the activation energy ($\Delta E_{HL}^\#$, Figure 10.30).

10.3.1.2 Effect of Light
In the 1980s, it was observed that an Fe^{2+} compound in the LS, $(t_{2g})^6$, state was converted by light into the metastable HS, $(t_{2g})^4(e_g)^2$, state and this has a virtually infinite lifetime at sufficiently low temperature [33]. The $[Fe^{II}(ptz)_6](BF_4)_2$ (ptz = 1-propyltetrazole) complex, for example, exhibits a spin transition at 135 K along with a spectacular change of color: from purple in the LS phase to white in the HS phase. Irradiating the sample at 10 K (LS) for a few minutes turns it white and confers to it all the spectroscopic and magnetic properties of the high-temperature phase. The spin state conversion is quantitative and the trapped HS state does not decay within several hours or days if the temperature is maintained around 10 K. Above 50 K the HS state relaxes back to the stable LS state.

This phenomenon became known as the LIESST effect (light-induced excited-spin-state trapping) [35]. The mechanism of the process is not difficult to explain, using the Tanabe–Sugano diagram for a d^6 ion (Chapter 8). Green light (514 nm of an Ar ion laser) is used for the spin-allowed transition $^1A_1 \rightarrow {}^1T_1$ (LS). A fast relaxation cascading over two successive intersystem crossing steps (see Chapter 15), $^1T_1 \rightarrow {}^3T_1 \rightarrow {}^5T_2$, populates the metastable $^5T_2$ state. Radiative relaxation $^5T_2 \rightarrow {}^1A_1$ is forbidden, and decay by thermal tunneling to the ground state $^1A_1$ is slow at low temperatures. Thus, $^5T_2$, which corresponds to HS, remains stable.

10.3.2
Collective Behavior

Although the origin of the SCO phenomenon is purely molecular, the macroscopic manifestation in the solid state is the result of the cooperative interaction of the molecules that constitute the material. The cooperative nature of the SCO has stimulated much interest, given that the first-order transition, often accompanied by hysteresis, confers to the materials a degree of memory. Cooperativity stems from the change in volume of the spin-crossover molecule. Therefore, it has an elasticity leading to long-range interactions. These interactions may be pictured as an internal pressure, which increases with the concentration of the LS species, and interacts with all the molecules in the crystal with the same strength, irrespective of distances.

When there is a great cooperativity between all molecules in the solid, the transition is more abrupt. In principle, when the difference between the radii of the cation in its HS and LS configuration is >10% the transition is usually abrupt. If this difference is <10% the transition is, currently, more smooth. The origin of the hysteresis loop is also a cooperative phenomenon.

10.3.3
Magnetic Measurements

A thermally induced LS $\leftrightarrow$ HS is characterized by an $x = f(T)$ curve, where x is the molar fraction of HS and $(1 - x)$ that of LS molecules. Different magnetic behaviors can occur (Figure 10.31).

1. *Abrupt* transitions that occur within a very small temperature interval, or *smooth* transitions that occur within a very large temperature interval.

2. Complete transitions, both at low temperature and high temperature, or incomplete transitions at high or low temperatures (the incompleteness at low temperature is more often observed).

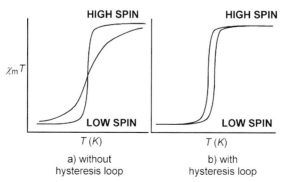

Figure 10.31 Different possible shapes of the $\chi_m T$ vs. T curves for spin-crossover systems (see text for explanation).

3. One-step transition or two-step transitions (not very frequent).

4. Transitions identical when increasing or decreasing the temperature or transitions in which the behavior is different when increasing or decreasing the temperature, showing a *hysteresis effect*. In the latter case the typical temperature $T_c\downarrow$ on cooling is lower than the critical temperature $T_c\uparrow$ upon heating.

10.3.4
Some Examples

The first cases in which the spin transition was observed and studied were in $[Fe^{III}(S_2CNR_2)_3]$, (S_2CNR_2 dithiocarbamate) where the spin crossover occurs at $\Delta_o \approx 16.500\,cm^{-1}$. The existence of *spin equilibrium* depends on very subtle energy terms, and minor changes in the ligand can have a dramatic effect on the position of the equilibrium. The magnetic properties depend on the volume of the dithiocarbamate substituents, R and the solvent. Bulky substituents (such as R = isopropyl groups) tend to give LS complexes even at high temperature or with high critical temperature (T_c) (defined as the temperature for which the molar fraction of LS and HS molecules are both equal to 0.5). Intermediate ligands tend to give spin crossover with T_c around 200 K. Finally, ligands without steric hindrance give HS behavior over the whole temperature range. The spin transition is always very smooth, without detectable hysteresis.

Fe^{2+} (d^6) complexes are by far the most studied [36]. Most of the theoretical studies on this phenomenon take as models two Fe^{2+} complexes, $[Fe(bpy)_2(NCS)_2]$ and $[Fe(NCS)_2(phen)_2]$, which show abrupt transitions (at 212 and 176 K, respectively) with a very narrow hysteresis effect, 0.15 K for the former and 0.40 K for the latter. Several $[Fe^{II}(chelate)_3]$ complexes have also been extensively studied. One of them, the $[Fe(Htrz)_3](ClO_4)_2$, where Htrz is the unsubstituted 1,2,4-1H-triazole, has a one-dimensional structure which favors the cooperativity. The compound, when perfectly dry, shows a rather smooth spin transition around 265 K with a weak thermal hysteresis at about 5 K. When one drop of water is added to ca. 50 mg of $[Fe(Htrz)_3](ClO_4)_2$, the transition becomes very abrupt, both in the warming and cooling mode, with $T_c\downarrow$ = 296 K and $T_c\uparrow$ = 313 K. The thermal hysteresis thus occurs at room temperature. The role of water is most likely to create hydrogen bonds which further favor the cooperativity. This complex is a proof that cooperativity can be considered a measure of the efficiency of transmitting changes, via intermolecular interactions, in the whole molecular crystal.

d^7 (Co^{2+}) complexes with spin transition are also well documented [37]. A unique case with Co^{2+} and cooperativity is that reported by Hauser et al. on $[Co(bpy)_3][LiCr(ox)_3]$. In the three-dimensional oxalate networks $[M^{II}(bpy)_3]$ $[M^IM^{III}(ox)_3]$ the negatively charged oxalate backbone provides perfect cavities for tris-bipyridyl complex cations. The size of the cavity can be adjusted by variation of the metal ions of the oxalate backbone. In $[Co(bpy)_3][NaCr(ox)_3]$, the $[Co(bpy)_3]^{2+}$ complex is in its usual $^4T_{1g}(t_{2g}^5 e_g^2)$ high-spin ground state. Substituting Na^+ by Li^+

reduces the size of the cavity. The resulting chemical pressure destabilizes the high-spin state of $[Co(bpy)_3]^{2+}$ to such an extent that the $^2E_g(t_{2g}^6 e_g^1)$ low-spin state becomes the actual ground state [38].

The influence of the counterions and solvent molecules in spin-crossover systems has been investigated [39]. Novel coordination polymers, with increasing cooperativity, undergoing spin crossover have been reported in recent years [40], as well as a thermal spin transition in a nanaporous iron(II) coordination framework material [41].

10.3.5
Perspectives

The so-called *molecular bistability* is one of the most important challenges in future devices and applications. Indeed, one of the most spectacular examples of molecular bistability is the spin-transition phenomenon. One can envisage that these SCO systems could find applications as sensors, in photonic devices, light modulators and filters, rewritable molecular memory devices and image processing [33, 42]. For example, in the paramagnetic state, the particular system is expected to induce strong magnetic field inhomogeneities, which will alter the NMR signal in spectroscopy as well in imaging, giving the so-called "intelligent contrast agents for magnetic resonance imaging" [43].

References

1 Atkins, P., de Paula, J. (2002) *Physical Chemistry*, 7th ed. Oxford University Press, Oxford.

2 Kahn, O. (1993), *Molecular Magnetism*, VCH, New York.

3 Gatteschi, D., Sessoli, R., Villain, J. (2006) *Molecular Nanomagnets*, Oxford University Press, Oxford.

4 Hatscher, S., Schilder, H., Lueken, H., Urland, W. Practical Guide to Measurements and Interpretation of Magnetic Properties, *Pure Appl. Chem.* 2005, *77*, 497.

5 Mabbs, F.E., Machin, D.J. (1973) *Magnetism and Transition Metal Complexes*, Chapman and Hall, London.

6 Carlin, R.L. (1986) *Magnetochemistry*, Springer-Verlag, New York.

7 Herrera, J.M., Bleuzen, A., Dromzee, Y., Julve, M., Lloret, F., Verdaguer, M. Crystal Structures and Magnetic Properties of Two Octacyanotungstate(IV) and (V)-Cobalt(II) Three-Dimensional Bimetallic Frameworks, *Inorg. Chem.* 2003, *42*, 7052.

8 Curély, J., Barbara, B. *Struct. Bond.*, 2006, *122*, 207.

9 Hay, P.J., Thibeault, J.C., Hoffmann, R. Orbital Interactions in Metal Dimer Complexes, *J. Am. Chem. Soc.* 1975, *97*, 4884.

10 Mealli, C., Proserpio, D.M. CACAO Program (Computer Aided Composition for Molecular Orbital Analysis), PC version 4.0, July 1994. Original reference, MO Theory Made Visible, *J. Chem. Educ.* 1990. *67*, 399.

11 Ruiz, E. Theoretical Study of the Exchange Coupling in Large Polynuclear Transition Metal complexes Using DFT Methods, *Struct. Bond.*, 2004, *113*, 71.

12 Rodríguez-Fortea, A., Alemany, P., Alvarez, S. Ruiz, E. Exchange Coupling in Halo-Bridged Dinuclear Cu(II) Compounds: A Density Functional Study, *Inorg. Chem.* 2002, *41*, 3769.

13 (a) Julve, M., Faus, J., Verdaguer, M., Gleizes, A. Copper(II), a Chemical Janus: Two Different (Oxalato) (bipyridyl)copper(II) Complexes in One Single Crystal. Structure and Magnetic Properties, *J. Am. Chem. Soc.* 1984, *106*, 8306; (b) Soto, L., García, J., Escrivá, E., Legrós, J-P., Tuchages, J-P., Dahan, F., Fuertes, A. Synthesis, characterization and magnetic properties of μ-oxalato- and μ-oxamido-bridged copper(II) dimers. Crystal and molecular structures of [Cu$_2$(mepirizole)$_2$(C$_2$O$_4$)(H$_2$O)$_2$](PF$_6$)$_2$ mepirizole$_3$H$_2$O and [Cu$_2$(mepirizole)$_2$ (C$_2$O$_4$)(NO$_3$)$_2$(H$_2$O)]$_2$[Cu$_2$(mepirizole)$_2$ (C$_2$O$_4$)(NO$_3$)$_2$], *Inorg. Chem.* 1989, *28*, 3378; (c) Veit, R., Girerd, J-J., Kahn, O., Robert, F., Jeannin, Y. Copper(II) and nickel(II) trinuclear species with dithiooxamide derivative ligands: structural, magnetic, spectroscopic, and electrochemical properties, *Inorg. Chem.* 1986, *25*, 4175; (d) Veit, R., Girerd, J.J., Kahn, O., Robert, F., Jeannin, Y., El Murr, N. Amino acid amides of dithiooxalic acid: spectroscopic, electrochemical, and magnetic properties of copper(II) binuclear complexes and crystal structure of [*N,N′*-(1,2-dithioxoethane-1,2-diyl)bis(methyl methio ninato)]bis(bromocopper(II)), *Inorg. Chem.* 1984, *23*, 4448.

14 Verdaguer, M. (1984) Interaction dans les systèmes polymétalliques: du complexe binucléaire a la chaine ferrimagnétique, PhD Thesis. Univ. Paris-Sud (Orsay).

15 Vicente, R., Ribas, J., Alvarez, S., Seguí, A., Solans, X., Verdaguer, M. Synthesis, X-ray diffraction structure, magnetic properties, and MO analysis of a binuclear (μ-tetrathiooxalato)copper(II) complex, (AsPh$_4$)$_2$[(C$_3$OS$_4$)CuC$_2$S$_4$Cu(C$_3$O S$_4$)], *Inorg. Chem.* 1987, *26*, 4004.

16 Kahn, O. Magnetism of the Heteropolymetallic Systems, *Struct. Bond.*, 1987, *68*, 89.

17 Ribas, J., Escuer, A., Monfort, M., Vicente, R., Cortés, R., Lezama, L., Rojo, T. Polynuclear N$_i^{II}$ and M$_n^{II}$ azido bridging complexes. Structural trends and magnetic behavior, *Coord.. Chem. Rev.* 1999, *193–195*, 1027.

18 Ruiz, E., Cano, J., Alvarez, S., Alemany, P. Magnetic Coupling in End-On Azido-Bridged Transition Metal Complexes: A Density Functional Study, *J. Am. Chem. Soc.* 1998, *120*, 11122.

19 Escuer, A., Harding, C.J., Dussart, Y., Nelson, J., McKee, V., Vicente, R. Constrained ferromagnetic coupling in dinuclear μ$_{1,3}$-azido nickel(II) cryptate compounds. Crystal structure and magnetic behaviour of [Ni$_2$(L1)(N$_3$)(H$_2$O)] [CF$_3$SO$_3$]$_3$ · 2H$_2$O · EtOH {L1 = N[(CH$_2$)$_2$NHCH$_2$(C$_6$H$_4$- *m*)CH$_2$NH(CH$_2$)$_2$]$_3$N}, *Dalton Trans.*, 1999, 223.

20 (a) Canneschi, A., Gatteschi, D., Sessoli, R., Rey, P. Toward Molecular Magnets: The Metal-Radical Approach, *Acc. Chem. Res.* 1989, 22, 392; (b) Benelli, C., Gatteschi, D. Magnetism of Lanthanides in Molecular Materials with Transition-Metal Ions and Organic Radicals, *Chem. Rev.* 2002, *102*, 2369.

21 (a) Miller, J.S. Magnetically ordered molecule-based assemblies, *Dalton Trans.* 2006, 2742; (b) Her, J-H., Stephens, P.W., Pokhodnya, K.I., Bonner, M., Miller, J.S. Cross-Linked Layered Structure of Magnetically Ordered [Fe(TCNE)$_2$]-z CH$_2$Cl$_2$ Determined by Rietveld Refinement of Synchrotron Powder Diffraction Data, *Angew. Chem. Int. Ed.* 2007, *46*, 1521.

22 Jain, R., Kabir, K., Gilroy, J.B., Mitchell, K.A.R., Wong, K.C., Hicks, R.G. High-temperature metal-organic magnets, *Nature*, 2007, *445*, 291.

23 Stumpf, H.O., Pei, Y., Kahn, P., Sletten, J., Renard, J.P. Dimensionality of Mn^{II}Cu^{II} Bimetallic Compounds and Design of Molecular-Based Magnets, *J. Am. Chem. Soc.* 1993, *115*, 6738.

24 Coronado, E., Galán-Mascarós, J.R., Gómez-García, C.J., Martínez-Agudo, J.M. Molecule-Based Magnets Formed by Bimetallic Three-Dimensional Oxalate Networks and Chiral Tris(bipyridyl) Complex Cations. The Series [Z^{II}(bpy)$_3$] [ClO$_4$][M^{II}Cr^{III}(ox)$_3$] (Z^{II} = Ru, Fe, Co, and Ni; M^{II} = Mn, Fe, Co, Ni, Cu, and Zn; ox = Oxalate Dianion), *Inorg. Chem.* 2001, *40*, 113.

25 (a) Ohba, M., Okawa, H. Synthesis and magnetism of multi-dimensional cyanide-bridged bimetallic assemblies, *Coord. Chem. Rev.* 2000, *198*, 313; (b) Nelson, K.J., Giles, I.D., Troff, S.A., Arif, A.M., Miller, J.S. Solvent-Enhanced Magnetic Ordering Temperature for Mixed-Valent Chromium Hexacyanovanadate(II), $Cr_{0.5}^{II}Cr^{III}[V^{II}(CN)_6] \cdot z$MeCN, Magnetic Materials, *Inorg. Chem.* 2006, *45*, 8922.

26 (a) Caneschi, A., Gatteschi, D., Sessoli, R., Barra, A.L., Brunel, L.C., Guillot, M. Alternating Current Susceptibility, High Field Magnetization, and Millimeter Band EPR Evidence for a Ground S = 10 State in $[Mn_{12}O_{12}(CH_3COO)_{16}(H_2O)_4]$. $2CH_3COOH.4H_2O$, *J. Am. Chem. Soc.* 1991, *113*, 5873; (b) Sessoli, R., Tsai, H. L., Schake, A.R., Wang, S., Vincent, J.B., Folting, K., Gatteschi, D., Christou, G., Hendrickson, D.N. High-Spin Molecules: $[Mn_{12}O_{12}(O_2CR)_{16}(H_2O)_4]$, *J. Am. Chem. Soc.* 1993, *115*, 1804.

27 Gatteschi, D., Sessoli, R. Quantum Tunneling of Magnetization and Related Phenomena in Molecular Materials, *Angew. Chem. Int. Ed.* 2003, *42*, 268.

28 Aromí, G., Brechin, E.K. *Struct. Bond.*, 2006, *122*, 1.

29 Coulon, C., Miyasaka, H., Clérac, R. *Struct. Bond.*, 2006, *122*, 163.

30 Lecren, L., Wernsdorfer, W., Li, Y-G., Vindigni, A., Miyasaka, H., Clérac, R. One-dimensional Supramolecular Organization of Single-Molecule Magnets, *J. Am. Chem. Soc.* 2007, *129*, 5045.

31 Ribas-Ariño, J., Baruah, T., Pedersen, M.R. Toward the Control of the Magnetic Anisotropy of Fe^{II} Cubes: A DFT Study, *J. Am. Chem. Soc.* 2006, *128*, 9497.

32 Cornia, A., Costantino, A.F., Zobbi, L., Caneschi, A., Gatteschi, D., Mannini, M., Sessoli, R. *Struct. Bond.*, 2006, *122*, 133.

33 (a) Real, J.A., Gaspar, A.B., Muñoz, M.C. Thermal, pressure and light switchable spin-crossover materials, *Dalton Trans.* 2005, 2062; (b) Hauser, A. Ligand Field Theoretical Comriderations, *Adv. Polym. Sci.* 2004, *233*, 49; (c) Gütlich, P., Goodwin, H.A. Spin Crossover – An Overall Perspective, *Top. Curr. Chem.* 2004, *233*, 1.

34 Hauser, A. Intersystem Crossing in Iron (II) Coordination compounds: a model Process Between Classical and Quantum Mechanical Behaviour, *Comments Inorg. Chem.* 1995, *17*, 17.

35 Hauser, A. Light-Induced Spin-Crossover and the High-Spin → Low-Spin Relaxation, *Top. Curr. Chem.* 2004, *234*, 155.

36 Real, J.A., Gaspar, A.B., Niel, V., Muñoz, M.C. Communication between iron(II) building blocks in cooperative spin transition phenomena, *Coord. Chem. Rev.* 2003, *236*, 121.

37 Goodwin, H.A. Spin Crossover in cobalt (II) Systems, *Top. Curr. Chem.* 2004, *234*, 23.

38 Sieber, R., Descurtins, S., Stoeckli-Evans, H., Wilson, C., Yufit, D., Howard, J.A.K., Capelli, S.C., Hauser, A. A Thermal Spin Transition in $[Co(bpy)_3][LiCr(ox)_3]$ (ox=$C_2O_4^{2-}$; bpy=2,2'-bipyridine), *Chem. Eur. J.* 2000, *6*, 361.

39 Galet, A., Gaspar, A.B., Muñoz, M.C., Real, J.A. Influence of the Counterion and the Solvent Molecules in the Spin Crossover System [Co(4-terpyridone)$_2$]$X_p \cdot n$H$_2$O, *Inorg. Chem.* 2006, *45*, 4413.

40 (a) Galet, A., Muñoz, M.C., Real, J.A. Coordination polymers undergoing spin crossover and reversible ligand exchange in the solid, *Chem. Comm.* 2006, 4321; (b) Galet, A., Muñoz, M.C., Real, J.A. {Fe(3CNpy)$_2$[Cu(3CNpy)(μ-CN)$_2$]$_2$}: a One-Dimensional Cyanide-Based Spin-Crossover Coordination Polymer, *Inorg. Chem.* 2006, *45*, 4583.

41 Neville, S.M., Moubaraki, B., Murray, K.S., Kepert, C.J. A Thermal Spin Transition in a Nanoporous iron(II) Coordination Framework Material, *Angew. Chem. Int. Ed.* 2007, *46*, 2059.

42 Nihei, M., Han, L., Oshio, H. Magnetic Bistability and Single-Crystal-to-Single-Crystal Transformation Induced by Guest Desorption, *J.Am. Chem. Soc.* 2007, *129*, 5312.

43 Muller, R.N., Elst, L.V., Laurent, S. Spin Transition Molecular Materials: Intelligent Contrast Agents for Magnetic Resonance Imaging, *J. Am. Chem. Soc.* 2003, *125*, 8405.

11
Electron Paramagnetic Resonance in Coordination Compounds

11.1
Introduction

In the last chapter we commented that on applying an external magnetic field the energy terms are split into their states due to the electronic Zeeman effect. The study of the transitions between two different states, by the application of an electromagnetic field, is the aim of electron paramagnetic resonance (EPR) or electron spin resonance (ESR) spectroscopy.

11.2
Fundamentals of EPR Spectroscopy

Let us recall the electronic Zeeman effect, already studied in the previous chapter (Figure 11.1). The simplified Zeeman Hamiltonian is $\hat{H} = g\mu_B H\hat{S}$. As a consequence, when an external field (H) is applied to an isolated electron, the two M_s components of the $S = 1/2$, are split, according to their energies, into $\pm 1/2 g\mu_B H$. The g value for a free electron is $g_e = 2.0023193$ (≈ 2.00).

For any external magnetic field, there will be an electromagnetic radiation with energy $h\nu$ ($\Delta E = g\mu_B H$) that will create resonance between the two $\pm 1/2$ states. For example, for a free electron subjected to a magnetic field of ca. 3000 G (a very standard field) the resonance occurs at approximately 9 GHz ($\approx 0.3\,\mathrm{cm}^{-1}$). This is the *microwave* region of the electromagnetic spectrum.

The population of the two states ($M_s = \pm 1/2$) is not *exactly* the same owing to Boltzmann distribution. It is just the small difference in occupancy which still persists, proportional to $e^{-E/kT}$, which gives rise to a net absorption of radiation. Therefore the intensity is enhanced at low temperature.

EPR spectra are generally measured at fixed frequency (microwave region) and varying magnetic field. A suitable sample containing unpaired electrons is placed in a static magnetic field. Plane polarized radio waves are sent through the sample with their direction of propagation perpendicular to the static field and with their

Coordination Chemistry. Joan Ribas Gispert
Copyright © 2008 WILEY-VCH Verlag GmbH & Co. KGaA, Weinheim
ISBN: 978-3-527-31802-5

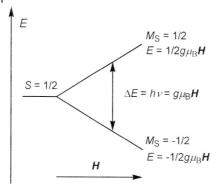

Figure 11.1 Electronic Zeeman effect: fundamentals of EPR spectroscopy.

magnetic vector usually also perpendicular to the static field (*standard procedure*), but occasionally parallel to it. The microwave is directed through a metallic guide to the resonant cavity where the sample is placed. The frequency is tuned to amplify the power of the microwaves in the sample cavity. The external magnetic field that interacts with the sample must be very homogenous to avoid great bandwidth and overlap between different bands. Finally, a detector sensitive to microwaves measures the absorption of microwaves.

In commercial spectrometers two common frequencies are used: the X and Q bands. The X-band corresponds to a frequency of approximately 9.5 GHz (1 GHz = 10^9 Hz, cycles s^{-1}). This frequency corresponds to 0.3 cm^{-1}. The resonant field for a free electron in these conditions is close to 3400 G. The Q-band corresponds to a frequency of 35 GHz (1.16 cm^{-1}). The resonant field for a free electron in these conditions is close to 12 000 G. The Q-band is more sensitive than the X-band (sensitivity increases roughly as v^2). However, there are several limitations when working with Q-band, such as the homogeneity of the magnetic field in the cavity is much more difficult to achieve. For this reason the standard experiments in EPR spectrometry are usually made in X-band.

Note When the magnetic vector of the microwave is perpendicular to the static field (*the standard procedure*), the selection rule is $\Delta Ms = \pm 1$. For some special cases (see below) it is better to work with the microwave magnetic field oriented parallel to the static field, H. Then, the selection rule changes from $\Delta M_s = \pm 1$ to $\Delta M_s = 0$.

11.3
Systems Suitable for Research with EPR

In general, any system with unpaired electrons in its ground or excited state can be studied by EPR spectroscopy:

1. Organic radicals or inorganic species with an unpaired electron (such as NO).
2. Organic biradicals and inorganic molecules with more than one unpaired electron (O_2).
3. Transition (d or f) ions with unpaired electrons. Due to the character of this book, the study of these ions will be the main goal of this chapter.
4. Several point defects in solids.
5. Systems with conducting electrons.

For any of these systems, one of the advantages of EPR spectroscopy is its extreme sensitivity to very small amounts of paramagnetic materials. For example, under favorable conditions a signal of the DPPH radical (diphenylpicrylhydrazil), which is widely used as a standard signal, can be detected if there is 10^{-12} g of the material in the spectrometer. This high sensitivity of EPR measurements has been of practical utility in biological systems (see Chapter 17).

11.4
Recording EPR Spectra and Their Information

The sample tube employed for recording EPR spectra is not trivial. If the signal-to-noise ratio is low, a quartz tube is necessary. Commercial glasses absorb more of the microwave power and also exhibit EPR signals. The EPR spectra can be recorded in different conditions, giving different, and complementary, information. The most usual conditions are:

- Microcrystalline powders.
- Solutions in solvents with low absorption in the microwave zone. Water is not appropriate because it absorbs very strongly. The reader can find a wide list of appropriate solvents or mixtures of them in the classical book of Drago [1].
- "Frozen" solutions, at low temperatures. In this case, it is necessary that the solvent does not give oriented phases when frozen. Thus, solvents able to give glasses are the best.
- Monocrystals (the most difficult but giving the most complete information).
- Doping of paramagnetic ions in diamagnetic hosts (very useful in spectroscopy, but not so useful in coordination chemistry, due to the inherent difficulties).

The EPR spectrum of a paramagnetic entity in fluid solution is usually characterized by a symmetric shape whose main features are the resonant field (and thus the value of g) and the number, relative intensity and spacing of the hyperfine lines (see below). After a rotation of the sample tube in the cavity by a given angle the spectrum remains absolutely the same. If an EPR spectrum is instead run for a single crystal containing a number of identical paramagnetic centers the spectrum also appears quite symmetric but the position of the spectral line and the splitting between the various hyperfine lines (if the system exhibits a hyperfine structure) *vary dramatically with the orientation of the crystals in the external magnetic field*. An intermediate situation is observed for samples that are solid but divided

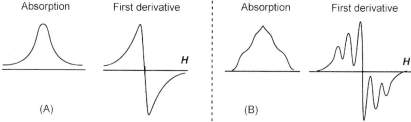

Figure 11.2 Absorption and first derivative plots in EPR measurements.

into a large number of small crystals (polycrystalline materials or powders). In such a case the spectrum is usually less symmetric than the solution spectrum but does not change its profile with the orientation in the external field. The powder spectrum covers an extended magnetic field range, being a sort of convolution of all possible single crystal spectra for the various orientations.

The commercial and conventional spectrometers give directly the first derivative of the absorption. The need for visualizing the spectra as the first derivative is clearly manifested when the spectra are anisotropic, with marked hyperfine coupling or with some bands very close to each other. These features are very common in coordination compounds. As an idealized case, the absorption and its first derivative for two kinds of spectra (simple and more complicated) are shown in Figure 11.2. The current spectrometers, guided by computers, can easily give the absorption or the second derivative. Let us stress that, in organic radicals, a g value of 2.01 can be very significant compared with a value of 2.02. However, in coordination chemistry, the necessary precision is less: a g value of 2.21 or 2.22 is practically the same.

11.5
g Values

Electrons are not isolated but they are on a molecule and, as well as the spin component, they always have an orbital component. Depending on the type of orbital in which an electron is placed, it will resonate with a g value different from g_e. Thus the g value gives us structural information through the orbital component. The difference between g and g_e is very small for free radicals but it can be significant in the case of paramagnetic transition metal ions (see Chapters 8 and 10).

Starting from the formula given in Figure 11.1, $h\nu = g\mu_B H$, the experimental g value is:

$$g = h\nu/\mu_B H \qquad (11.1)$$

$g = (6.625 \times 10^{-27} \, \text{erg s} \times A \times 10^9 \, \text{s}^{-1})/(9.27 \times 10^{-21} \, \text{erg G}^{-1} \times H \, \text{G}) = 714.6 \times (A/B)$ (A in GHz and H in G), hence g is non-dimensional.

11.6
Hyperfine Coupling

The spin Hamiltonian given above is oversimplified. In fact, the spectrum of an isotropic system (nucleus + electron) with spin $S = 1/2$ must be interpreted according to the following spin-Hamiltonian

$$\hat{H} = g\mu_{Be}H\hat{S} - g_{n}\mu_{Bn}H\hat{I} + A\hat{S} \cdot \hat{I} \tag{11.2}$$

where the first and second term account for the electron-Zeeman and nuclear-Zeeman interactions, respectively. The third term $(A\hat{S} \cdot \hat{I})$ describes the coupling of the electron and nuclear spin moments, which classically corresponds to the dot product of these two vectors. μ_{Bn} is the nuclear Bohr magneton (a thousand times smaller than the electronic Bohr-magneton), I is the nuclear spin, S the electronic spin, A the electron–nucleus coupling constant. The Zeeman nuclear Hamiltonian (which is the typical Hamiltonian in NMR spectroscopy) has a different sign to the Zeeman electronic Hamiltonian, owing to the different sign of protons and electrons.

The effect of the second and third Hamiltonians is to modify and split the electronic states created by the electronic Zeeman effect by means of the coupling between nuclei and electrons. If we consider that the unpaired electron is only *on* a nucleus (transition metal ion, for instance) the coupling will be only with this nucleus. However, the reality is much more complex: any electron is delocalized in a molecular orbital in which all the atoms of the ligand can, more or less, participate. This delocalization – due to the covalence – allows any electron to be coupled to several nuclei. This effect should be considered with the logical condition that $I \neq 0$.

There are several types of electron spin–nuclear spin interactions which determine the hyperfine coupling constant. The *isotropic* hyperfine coupling or Fermi contact term is related to the finite probability of finding the unpaired electron on the nucleus and is given by

$$A = A_{iso} = \frac{8\pi}{3}g_{e}\mu_{e}g_{n}\mu_{n}|\Phi_{0}|^{2}$$

where $|\Phi_{0}|^{2}$ is the square of the value of the wavefunction of the unpaired electron evaluated in the volume of the nucleus. It is different from zero for s electrons. Even in system where the unpaired electron is formally in d or f orbitals, a fraction of s character is determined by spin polarization.

If the electron and nuclear magnetic dipoles were to behave classically and an externally applied static field H ($\parallel z$) was present so as to align them, there is the energy of dipole–dipole interaction. Since the electron is not localized at one position of the space, its energy calculation must be averaged over the electron probability distribution function. The energy is averaged out to zero when the electron cloud is spherical (s orbital) and comes to a finite value in the case of axially symmetric orbitals (p, d, etc.). This interaction is anisotropic in that it depends on the orientation of the orbital with respect to the applied magnetic field. This dipolar spin–spin component is averaged to zero in a fluid solution: the *observed* hyperfine coupling thus corresponds to the isotropic part only.

A third contribution is the interaction of the orbital momentum with the nuclei. In many systems this contribution is negligible (angular momentum quenched) but may be important for systems with near-orbital degeneracy, as in transition metals.

Note The interaction energy between the electron spin and magnetic nucleus is characterized by the *hyperfine coupling constant*, A with units of J. A/h and $A/(hc)$ may be reported in MHz and cm^{-1}, respectively.

Hyperfine interaction results in splitting of the lines in an EPR spectrum. The *hyperfine splitting* (a) is measured in units of millitesla, mT, (currently in G). For simple cases, the hyperfine splitting is related to the absolute value of the hyperfine constant, A by $A = ag\mu_B$.

11.6.1
The Simplest Example: the Hydrogen Atom

Let us assume the simplest case: an unpaired electron on a nucleus of $I = 1/2$. In fact this is the case of the hydrogen atom. In the hydrogen atom case, there are four possible wavefunctions:

$$\Phi_1 = |\alpha_e \alpha_n >; \quad \Phi_2 = |\alpha_e \beta_n >; \quad \Phi_3 = |\beta_e \alpha_n > \quad \text{and} \quad \Phi_4 = |\beta_e \beta_n >$$

($\alpha = 1/2$; $\beta = -1/2$). Applying the Hamiltonian reported in Eq. (11.2) it is easy to obtain the energy values for the four functions, Φ_i. The reader can do this calculation, without great difficulty (see Ref. [1]). The four final energies are:

$$\Phi_1 = |\alpha_e \alpha_n > \quad E = 1/2 g\mu_B H - 1/2 g_n \mu_{Bn} H + 1/4 A$$

$$\Phi_2 = |\alpha_e \beta_n > \quad E = 1/2 g\mu_B H + 1/2 g_n \mu_{Bn} H - 1/4 A$$

$$\Phi_3 = |\beta_e \alpha_n > \quad E = -1/2 g\mu_B H - 1/2 g_n \mu_{Bn} H - 1/4 A$$

$$\Phi_4 = |\beta_e \beta_n > \quad E = -1/2 g\mu_B H + 1/2 g_n \mu_{Bn} H + 1/4 A$$

From these energies, the corresponding energy diagram (in an arbitrary scale, since the Zeeman electronic is a thousand times stronger than the Zeeman nuclear), is schematized in Figure 11.3.

Remark Taking into account the spectroscopic selection rules, $\Delta M_S = \pm 1$ and $\Delta M_I = 0$, there will be two allowed transitions at $g\mu_B H \pm A/2$.

Considering, as indicated above, that in EPR spectroscopy the frequency is fixed and the external magnetic field changes, the scheme of the energy splitting for the

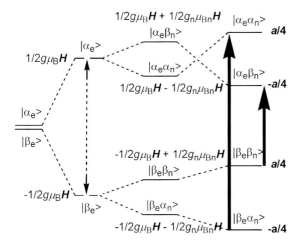

SELECTION RULES: $\Delta M_I = 0$; $\Delta M_S = \pm 1$

Figure 11.3 Origin of the hyperfine splitting in a hydrogen atom ($S = 1/2$ and $I = 1/2$)

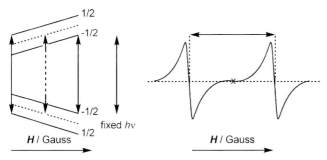

Figure 11.4 Energy scheme (fixed frequency) and simulated EPR spectrum for a hydrogen atom.

hydrogen atom (together with its spectrum) is shown in Figure 11.4. The two solid lines represent the resonant fields for the fixed radiation energy, $h\nu$. The dotted line represents the field for the resonance without hyperfine coupling. Thus, the EPR spectrum will have two equally intense and equidistant signals with respect to the hypothetical g value if the hyperfine coupling is zero. The **H** value, indicated as **x**, (Figure 11.4B) represents the field value that is taken for the calculation of g.

11.6.2
Hyperfine Coupling Created by Equivalent $I = 1/2$ Nuclei

Let us study some more complicated cases than the hydrogen atom: the first will be the case of one unpaired electron coupled to two or three $I = 1/2$ equivalent

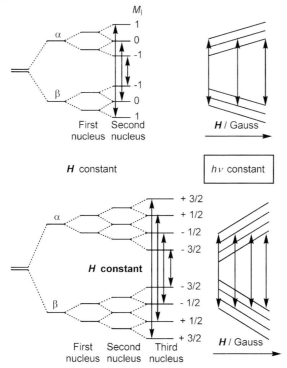

Figure 11.5 Energy scheme for an electron coupled to 2–3 equivalent nuclei with $I = 1/2$.

nuclei, with the same hyperfine coupling, A. The energy diagram is shown in Figure 11.5. In the first case (two equivalent nuclei) the EPR spectrum has three signals of $1:2:1$ relative intensity, because the two $M_I = 0$ states are degenerate. In the second case (three equivalent nuclei) the EPR spectrum will have four signals with $1:3:3:1$ relative intensity, according to its multiplicity. The simulated spectra (assuming $g = 2.00$) are shown in Figure 11.6.

In general, for n equivalent nuclei, the number of signals is $2nI + 1$. If $I = 1/2$, the number of signals is $n + 1$, and the intensity of these signals, owing to their multiplicity, follows the coefficients of the binomial expansion $(a + b)^m$. This result is, logically, exactly the same as that found for the coupling of protons in NMR. This formula is restricted to equivalent protons or other nuclei having $I = 1/2$.

11.6.3
Hyperfine Coupling Originated by $I = 1/2$ Non-equivalent Nuclei

Let us assume two different cases: (i) an electron coupled to two non-equivalent protons, with hyperfine coupling a and a', and (ii) an electron coupled to three protons, two of them equivalent (a) and the third non-equivalent (a'). If $a > a'$, the energy diagram in both cases is shown in Figure 11.7 and the corresponding EPR spectra (assuming $g = 2.00$) in Figure 11.8.

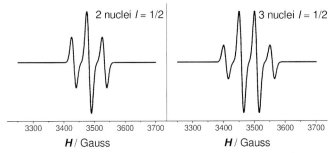

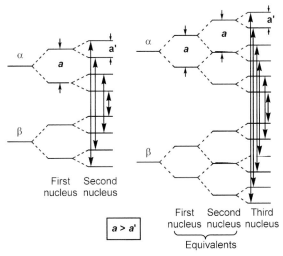

Figure 11.6 Simulated EPR spectra corresponding to two cases of Figure 11.5.

Figure 11.7 (A) Energy scheme for an electron coupled to 2 non-equivalent nuclei with $I = 1/2$, (B) energy scheme for an electron coupled to 2 equivalent and one non-equivalent nuclei with $I = 1/2$.

In the first case, the EPR spectrum shows four signals (doublet of doublets). Each doublet has its own hyperfine coupling ($a > a'$). The second case is a triplet of doublets. The triplet is due to the two equivalent protons (with hyperfine coupling a) and the doublet is due to the coupling to the non-equivalent hydrogen (hyperfine coupling, a'). In general, such as occurs in NMR spectra, the total number of signals due to the hyperfine coupling with non-equivalent nuclei is $(2nI_i + 1)(2mI_j + 1)$, n being the number of equivalent nuclei with spin I_i and m the number of equivalent nuclei with spin I_j.

11.6.4
Hyperfine Coupling Originated by Nuclei with $I > 1/2$

Let us assume an unpaired electron coupled to several nuclei with $I > 1/2$. The energy scheme for coupling to one or two equivalent nuclei with $I = 1$ (nitrogen,

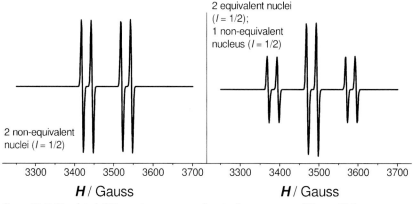

Figure 11.8 Simulated EPR spectra corresponding to the two cases of Figure 11.7.

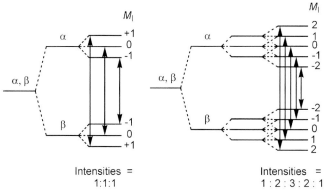

Figure 11.9 (A) Energy scheme for an electron coupled to one nucleus with $I = 1$, (B) energy scheme for an electron coupled to 2 equivalent nuclei with $I = 1$.

for instance) is shown in Figure 11.9. The number of bands is $(2nI + 1) = (2n + 1)$. In the first case (only one nucleus) three bands will appear, with relative intensity $1:1:1$, while in the second case five bands will appear, with relative intensity $1:2:3:2:1$.

The simulated spectra for the two cases are shown in Figure 11.10(a, b), assuming $g = 2.00$. As a real example, in Figure 11.10(c) is given the simulated spectrum of the polynuclear cobalt complex $[Co_3(CO)_9Se]$ ($I_{Co} = 7/2$). This spectrum was reported in 1971 by Strouse [2]. The structure of this cluster is tetrahedral with three equivalent cobalt atoms in three vertices and a selenium atom in the fourth vertex of the tetrahedron. The spectrum suggests the existence of only one unpaired electron coupled to three equivalent cobalt nuclei. Indeed, $2nI + 1 = 2 \times 3 \times \frac{7}{2} + 1 = 22$ signals. Finally, another example for understanding the effect of the hyperfine coupling, was reported in 1996 by Christou et al. concerning the $[V_4O_8(NO_3)(tca)_4]^{2-}$ complex (tca = thiophene-2-carboxylate) (Figure 11.10(d), simulated spectrum) [3].

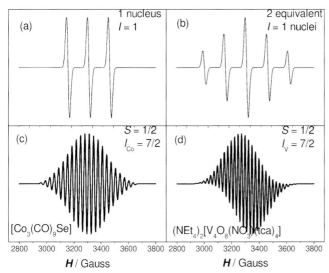

Figure 11.10 Simulated EPR spectra for one electron coupled to: (a) one nucleus with $I = 1$, (b) two equivalent nuclei with $I = 1$, (c) three equivalent nuclei with $I = 7/2$, (d) four equivalent nuclei with $I = 7/2$ (see text).

In this complex there are formally three V^{5+} (d^0) ions and one V^{4+} (d^1) ion. There is, thus, only one unpaired electron, which is equally delocalized on the four vanadium ions. Indeed, $2nI + 1 = 2 \times 4 \times \frac{7}{2} + 1 = 29$ signals. It is worth mentioning the difference in the symmetry of these two spectra: the spectrum of the cobalt complex is more symmetric than the spectrum of the vanadium complex. This is due to the relative bandwidth compared to the hyperfine coupling constants.

In coordination chemistry the ion most studied with the EPR technique is, undoubtedly, Cu^{2+}. Its nuclear spin, I, is $3/2$. Thus, without considering now the typical anisotropy of copper(II) (studied later) the EPR spectrum of a Cu^{2+} complex should show four identical signals. Actually, the spectrum will be more complicated if the unpaired electron is in a molecular orbital in which other ligand atoms with $I \neq 0$ also participate. Let us assume, for instance, a macrocyclic Cu^{2+} complex in which the donor atoms are nitrogen ($I = 1$). Given that the unpaired electron *belongs essentially* to the copper(II) ion, the coupling constant with Cu (A) will be stronger that the coupling constant with the four nitrogen atoms (A'). Therefore, the spectrum will be a quadruplet (due to the copper) split into nine signals, due to the four nitrogen atoms (Figure 11.11). In theory, the more similar are a and A', the more difficult will be the qualitative interpretation of the spectra. In this case it will be necessary to simulate the spectrum with appropriate computer programs.

Note When it is assumed that an unpaired electron belongs to a nucleus, the corresponding coupling with the nuclear spin is calledled the *hyperfine coupling*. When the coupling is created by the delocalization of the unpaired electron on the

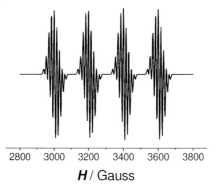

2800   3000   3200   3400   3600   3800

H / Gauss

Figure 11.11 Simulated EPR spectrum for a copper(II) ion
($I = 3/2$) showing the hyperfine coupling with the copper ion
and with four equivalent N atoms ($I = 1$) of the ligand. It is
assumed that the two coupling constant are very different.

ligand nuclei, then the coupling should be called *superhyperfine coupling*. In the literature, these two names are frequently mixed, using hyperfine coupling, to express both kinds of couplings.

11.7
Isotropic Polyelectronic Systems

The systems we have studied so far have only been monoelectronic ($S = 1/2$) systems. In coordination chemistry we must very frequently treat ions with more than one unpaired electron ($S > 1/2$). Assuming an isotropic system, splitting for any M_s state (between 1 and 5/2) is shown in Figure 11.12. In this kind of splitting, all possible transitions between $\Delta M_s = \pm 1$ have the same energy giving a single signal in the EPR spectrum. As a consequence, the EPR spectrum of isotropic ions will show a single signal at $g \neq 2.00$ (see Chapter 8), which will be split –at least theoretically– due to the hyperfine coupling with the central nucleus or with the nuclei of the ligands.

Two simulated examples of spectra of isotropic polyelectronic ions with hyperfine splitting are shown in Figure 11.13. Part (a) corresponds to a typical spectrum for a salt of Mn^{2+}, whose nuclear spin is 5/2. Six bands of equal intensity are expected. Part (b) corresponds to a typical spectrum of a Cr^{3+} compound. Cr^{3+} complexes are quite isotropic, giving only a single signal. However, in nature chromium exists as two different isotopes with very different ratio: the ^{52}Cr ($\approx 90\%$) has $I = 0$ giving, thus, a single signal without hyperfine coupling, while the other isotope, ^{53}Cr ($\approx 10\%$) has $I = 3/2$, giving, thus, four signals of equal intensity. Given the ratio between the two isotopes, the single signal due to the ^{52}Cr isotope will be approximately ten times more intense than the sum of the other four signals of the ^{53}Cr isotope. This feature explains why the four isotopomer lines are not always easy to detect.

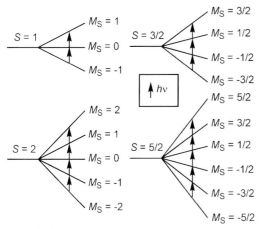

Figure 11.12 Splitting of the energy terms corresponding to $S = 1, 3/2, 2$, and $5/2$, assuming isotropic behavior.

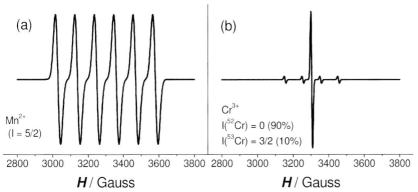

Figure 11.13 Simulated spectra of two different isotropic polyelectronic ions, showing the hyperfine coupling and the isotopomers (right).

11.8
Study of the Anisotropy in Monoelectronic Systems with Orbitally Non-degenerate Ground States

11.8.1
The g Parameter

The use of the effective spin Hamiltonian, $\hat{H} = g\mu_B H\hat{S}$ assumes that g is a symmetric (3×3) matrix. The true Hamiltonian must be written as:

$$\hat{H} = \mu_B [S_x S_y S_z] \begin{bmatrix} g_{xx} & g_{xy} & g_{xz} \\ g_{yx} & g_{yy} & g_{yz} \\ g_{zx} & g_{zy} & g_{zz} \end{bmatrix} \begin{bmatrix} H_x \\ H_y \\ H_z \end{bmatrix} \tag{11.3}$$

The mathematical difficulty in operating with this tensor can be minimized by a diagonalization matrix-process that allows one to obtain the principal values of the tensor. The g matrix becomes simply

$$\begin{pmatrix} g_{xx} & 0 & 0 \\ 0 & g_{yy} & 0 \\ 0 & 0 & g_{zz} \end{pmatrix}$$

This operation is equivalent to a rotation of the Cartesian axes. It is important to underline that this change of axis allows treatment with only three g values: g_{xx}, g_{yy} and g_{zz} (or g_x, g_y and g_z). This general case corresponds to total anisotropy. When the anisotropy is only axial ($x = y \neq z$) then $g_x = g_y$ are called $g_\perp$ and g_z is $g_\parallel$. This last case is the most frequently observed in copper complexes.

Remembering that the g anisotropy depends on the second-order spin–orbit coupling, i.e. on λ and ΔE, the general formula for the g values is: $g = 2.00 - (n\lambda/\Delta E)$ (Chapter 8).

For $S = 1/2$ ions it is possible to demonstrate that the n values of this formula can be calculated through the so-called "magic pentagon", shown in Figure 11.14, which summarizes the results of the evaluation of the matrix elements $<0|L|n>$ where 0 means the ground state and n one of the excited states. In fact the magic pentagon is only a simple mnemonic method to find the n values rigorously calculated by the operators of quantum chemistry.

Let us assume a Cu^{2+} complex the magnetic orbital of which is the $x^2 - y^2$ (square planar, square pyramidal, elongated O_h). This leads to the following g values:

$$g_\parallel = 2.00 - (8\lambda/\Delta E)$$

$$g_\perp = 2.00 - (2\lambda/\Delta E)$$

8 and 2 are deduced from the pentagon. To calculate $g_\parallel$ it is assumed that the magnetic field is parallel to the z axis. The second-order spin–orbit perturbation

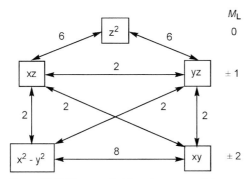

Figure 11.14 The so-called "magic pentagon" (see text for explanation).

will operate through L_z that can only mix orbitals with identical $|M_L>$ values. In this case, these orbitals are xy, $x^2 - y^2$ (linear combinations of $|\pm2>$). The number that links these two orbitals in the pentagon is 8. On the contrary, to calculate the $g_\perp$ value it will be necessary to consider different orbitals whose $|M_L>$ values differ by unity. Indeed, in this case the orbital operator will be $L_{x,y}$ (related to the operators, L_+, L_-). In the present case, the pentagon indicates that the magnetic orbital $x^2 - y^2$ can be mixed with the xz, yz orbitals, with an n value equal to 2.

If the magnetic orbital is the z^2 (trigonal bipyramid, compressed O_h), then:

$$g_\| = 2.00 \quad \text{and} \quad g_\perp = 2.00 - (6\lambda/\Delta E)$$

Looking at the pentagon, L_z cannot mix this orbital with any other, and the g value does not vary with regard to the g value for an isolated (free) electron. However, the perpendicular component (x, y) will experience variation since the z^2 orbital will be able to mix with other orbitals with $M_L = \pm1$ (xz, yz), through the raising and lowering operators. In the pentagon, the n value that relates z^2 to xz, yz is 6. Thus, to summarize:

The unpaired electron in the $x^2 - y^2$ orbital: $g_\| > g_\perp > g_e$

The unpaired electron in the z^2 orbital: $g_\| = 2.00; g_\perp > g_\|$ $\qquad$ (11.4)

Note We are dealing with Cu^{2+}, with 9 electrons; λ is, thus, negative. A detailed study of anisotropic six-coordinated copper complexes with $g_\perp > g_\|$ has been reported by Gatteschi et al. [4].

11.8.2
Shape of the EPR Spectra for Anisotropic Monoelectronic Ions

As previously indicated, the anisotropy of the samples in EPR measurements can only be observed in appropriate conditions:

- *In monocrystals.* This is the ideal condition to study the anisotropy. However, it is not always possible or easy to do.

- *In dilute frozen solution.* This is the standard situation for working with soluble species. In a frozen solution, all molecules are oriented in a random manner, but fixed in the space. For any direction, and assuming axial symmetry, the g value is $g_\|^2 \cos^2 \theta + g_\perp^2 \sin^2\theta$, θ being the angle formed between the magnetic field H and the z direction of the molecule. The actual spectrum will be the result of the sum of the spectra of all the directions, weighed according to their statistical probability. The more intense feature corresponds to crystallites with the magnetic field in the xy plane, and the other to crystallites with the field parallel to z. The different intensity can be understood, in practical terms, taking into account that the probability to have a molecule in a frozen solution oriented in the xy plane is higher than the probability of an orientation along z. Considering that the concentration can be made at sufficient dilution, it is

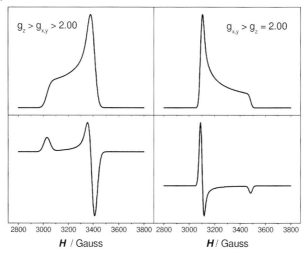

Figure 11.15 Absorption and first derivative plots for anisotropic monoelectronic systems with axial distortion.

possible to avoid in the experimental spectrum any signal due to interactions among the molecules such as usually occurs in polycrystalline samples.

- *In microcrystalline solids.* The major fraction of the EPR spectra recorded in the solid state comes from polycrystalline materials or powders whose spectra are usually called "powder" spectra. As many chemical systems of interest are 1D, 2D, 3D systems, in which the solution would destroy the actual structure, the practical interest in powder spectra in EPR is very high. Powder spectra are recorded for various real chemical systems such as glasses, ceramics, catalysts, minerals, micro- and nano-porous materials and many coordination compounds. In this case, the anisotropy can be hidden or masked by other different phenomena due to the magnetic or dipolar coupling among the molecules in the solid.

Two EPR spectra with axial symmetry are simulated in Figure 11.15, indicating the absorption spectrum and its first-derivative (typical) spectrum. In the case of total anisotropy (rhombic tensor), the EPR spectrum is similar to that shown in Figure 11.16. Three g values will be found, usually indicated as g_x, g_y and g_z or g_1, g_2 and g_3. The g_z value corresponds (in an arbitrary form) to the highest field; g_y is the intermediate and g_x corresponds to the lowest field. Of course, no information is available on the actual orientation of the x, y and z axes in the molecule.

The hyperfine splitting can be seen, mainly, in frozen solution [5]. In most of Cu^{2+} complexes with $g_{\parallel} > g_{\perp} > 2.00$ the hyperfine coupling with the copper nucleus is only seen in the $g_{\parallel}$ component, not in the $g_{\perp}$.

A recent EPR study of a copper (II) complex of N-confused tetraphenylporphyrin, labelled with ^{13}C in the macrocycle, allowed the direct observation of the σ metal–carbon bond, through a detailed study of the carbon hyperfine interaction. Interest-

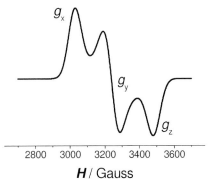

Figure 11.16 Simulated EPR spectrum for a totally anisotropic monoelectronic system.

ingly, the delocalization is approximately 11% larger than that for "normal" CuTPP [6].

11.8.3
Examples of S = 1/2 Systems Other Than Cu²⁺ Complexes

The vast majority of $S = 1/2$ systems reported in the literature correspond to copper(II) complexes. However, there is a significant number of other cases where the $S = 1/2$ comes from different metal ions. Octahedral Ni^{3+} complexes [7], low-spin square-planar or square pyramidal Co^{2+} [8], pentacoordinated oxovanadium(IV) complexes [9], low-spin Fe^{3+} bis-axially ligated porphyrinate complexes [10], have been reported and studied. Stable radical anions of heavier Group 13 elements, $[(^tBu_2MeSi)_3E\cdot^-]$ (E = Al, Ga), have also been reported. The planarity of these complexes in solution is evidenced by the EPR study. The Al complex (^{27}Al, $I = 5/2$, 100%) shows a well-resolved sextet at $g = 2.005$. The Ga case is more complex due to the presence of two isotopes (^{69}Ga; $I = 3/2$, 60.1%; ^{71}Ga; $I = 3/2$; 39.9%.) [11].

11.9
Zero-field Splitting: Anisotropic Polyelectronic Systems (S > 1/2) [12, 13]

The spin Hamiltonian with the **g** and **D** tensors (omitting the hyperfine interactions and other higher order terms) is: $\hat{H} = g\mu_B H\hat{S} + \hat{S}\cdot\mathbf{D}\cdot\hat{S}$. This Hamiltonian can be simplified with appropriate selection of axes, giving:

$$\hat{H} = \mu_B(g_x H_x S_x + g_y H_y S_y + g_z H_z S_z) + D(S_z^2 - \frac{1}{3}S(S+1)) + E(S_x^2 - S_y^2) \qquad (11.5)$$

When D is of the same order of magnitude as the electronic Zeeman term− *the usual case*− the perturbational treatment must be abandoned and the spin Hamiltonian must be diagonalized exactly. Gatteschi et al. have studied these systems, giving the spin Hamiltonian matrix and, when an exact solution is possible, the energy levels and transitions for the various spin systems in orthorhombic symmetry [12].

D is a new anisotropy factor that arises from *electron–electron* spin interaction (dipolar interactions, very important in organic biradicals), the distortion from the regular symmetry and the second-order spin–orbit coupling effect. This important feature has already been mentioned in Chapters 8 and 10. As indicated by Drago, "*In transition metal ions systems, the ZFS term is employed to describe any effect that removes the spin degeneracy, including dipolar interactions and spin–orbit coupling. A low symmetry crystal field often gives rise to a large zero-field effect*" [1].

From both experimental and theoretical viewpoints, it is convenient to distinguish between the two types of anisotropic polyelectronic systems: (i) S integer and (ii) S non-integer.

11.9.1
S Integer (i.e. $S = 1$, Ni^{2+} or $S = 2$, Mn^{3+})

Let us choose, as the simplest example, the Ni^{2+} ($S = 1$) ion. The axial spin Hamiltonian for an $S = 1$ spin system can be derived from Eq. (11.5).

If $D > h\nu$ the energy diagram produced by the ZFS effect and the Zeeman effect is shown in Figure 11.17A. This figure clearly shows that if the energy of the electromagnetic radiation ($h\nu$) is lower than that required to allow excitation from $M_S = 0$ to $M_S = -1$ *no EPR spectrum will be observed*, unless very high fields are available. This is the typical feature for almost all the Ni^{2+} complexes, in which the D value is usually of the order of 5–6 cm^{-1}. These systems have been frequently named EPR-silent (X and Q bands). However, today, it is rather easy to work with non-conventional spectrometers, called *high-field-high-frequency spectrometers*,

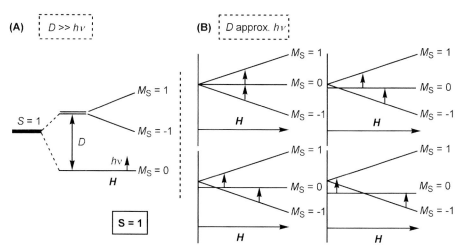

Figure 11.17 Energy diagram and possible transitions for an $S = 1$ system with $D \gg h\nu$ (A) and $D < h\nu$ (B).

which give the possibility of showing resonance at very high fields. In these conditions Ni^{2+} or Mn^{3+} are not EPR silent.

If $D < h\nu$, an EPR spectrum can be observed. The Zeeman splitting for a field parallel to z in the simplest case, $S = 1$, is shown in Figure 11.17B. When $D = 0$ there are two allowed transitions, $|-1> \rightarrow |0>$ and $|0> \rightarrow |1>$, which have the same energy. If we increase the D value, with the condition $D < h\nu$, the single signal is split into two signals, symmetrically placed with regard to the hypothetical g value of $D = 0$. However, the real system is more complicated: *a non-zero D parameter necessarily requires distortion of the regular geometry.* This feature implies anisotropy of the g value. Assuming axial symmetry, the splitting of the $|M_i>$ states will be different in the x and y directions, compared to the z direction. To summarize, four signals will be expected: two for $g_{\parallel}$ and two for $g_{\perp}$. If the anisotropy is total $(x \neq y \neq z)$ we could expect six different signals: two for g_x, two for g_y and two for g_z.

From the corresponding matrix, the energy of the levels can be calculated together with the resonant fields when the static magnetic field is parallel to one of the principal axes. The exact eigenvalues and magnetic functions for $S = 1$ system are given in Table 11.1 [12, 13].

The eigenvalues in the z-direction can be assigned to the quantum numbers 0, +1, −1. However, in the x and y directions, the eigenvalues cannot be associated with quantum numbers at small fields because, in this limit, there is mixing of quantum numbers [14]. This causes the curving of the energy plots and the appearance of new bands that can be associated to the theoretically forbidden $\Delta M_s = 2$ transition. For large enough fields, the M_s become "good" quantum numbers again [14]. The resulting energy diagram for $S = 1$, assuming $D = 0.15\,cm^{-1}$ and $E = 0$ or $D = 0.15\,cm^{-1}$ and $E = 0.03\,cm^{-1}$ are shown in Figures 11.18 and 11.19,

Table 11.1 The exact eigenvalues and magnetic functions for the $S = 1$ system.

Direction	Shifted eigenvalues for $E \neq 0$ (general case) $(G_a = g_a\mu_B H_a)$
z	$\varepsilon_{1z} = D + (E^2 + G_z^2)^{1/2}$
	$\varepsilon_{2z} = 0$
	$\varepsilon_{3z} = D - (E^2 + G_z^2)^{1/2}$
x	$\varepsilon_{1x} = D + E$
	$\varepsilon_{2x} = \frac{1}{2}(D - E) - [1/4(D - E)^2 + G_x^2]^{1/2}$
	$\varepsilon_{3x} = \frac{1}{2}(D - E) + [1/4(D - E)^2 + G_x^2]^{1/2}$
y	$\varepsilon_{1y} = \frac{1}{2}(D + E) + [1/4(D + E)^2 + G_y^2]^{1/2}$
	$\varepsilon_{2y} = \frac{1}{2}(D + E) - [1/4(D + E)^2 + G_y^2]^{1/2}$
	$\varepsilon_{3y} = D - E$

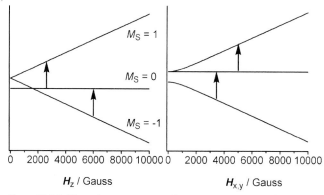

Figure 11.18 Allowed EPR transitions for an $S = 1$ system with axial anisotropy (only D parameter).

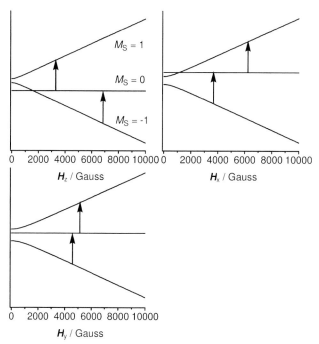

Figure 11.19 Allowed EPR transitions for an $S = 1$ system with total anisotropy (D and E parameters).

respectively. In the first case four bands can be expected, whereas in the second case (with $E \neq 0$) even six bands could be expected.

This situation is easily found in organic biradicals. In coordination compounds, dealing with polynuclear systems of copper(II) (without single-ion D), it is easier to find many spectra with small molecular D values. Let us assume, for instance,

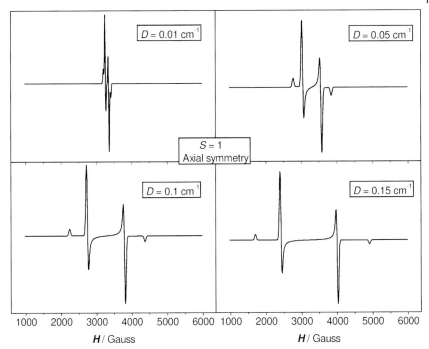

Figure 11.20 Simulated EPR spectrum for an axially anisotropic S = 1 system, with different D values.

a dinuclear copper(II) complex. If the coupling is ferromagnetic, at low temperatures only the triplet ground state will be populated. If the coupling is antiferromagnetic, at high temperatures the triplet will also be populated. Working, thus, at temperatures appropriate for the population of the triplet, we are likely to see very good spectra the pattern of which is undoubtedly due to the ZFS of the $S = 1$ state.

Some simulated spectra for $S = 1$ state and small D values are shown in Figure 11.20. As indicated above, these spectra will show four signals in the most favorable conditions (appropriate D value). However, depending on the D value, some of these signals disappear (negative magnetic field, which is absurd, or too great fields). If the symmetry is less than axial, the E parameter appears in the corresponding Hamiltonian and then, not only four, but up to six bands can appear (Figure 11.21). As indicated, at low field the forbidden $\Delta M_s = 2$ band can also appear (not shown in these two figures).

It can be demonstrated, from the formulae of Table 11.1, that the difference between the two corresponding signals is proportional to the magnitude of the D value. For example, the separation of the two z bands is equal to $2D$.

A real example, with only axial distortion, is shown in Figure 11.22 [15]. The best simulation was made with $D = 0.47\,\text{cm}^{-1}$ and $E = 0\,\text{cm}^{-1}$. In this figure only the two perpendicular signals and one parallel are seen. The other parallel "should

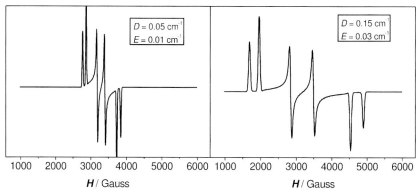

Figure 11.21 Simulated EPR spectrum for a totally anisotropic $S = 1$ system, with different D and E values.

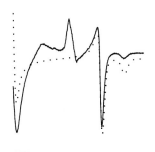

Figure 11.22 Experimental (solid line) and simulated (dotted line) spectrum for a $[Cu_2^{II}(carboxylato)_4]^{2+}$ system ($S = 1$), showing the splitting of the bands due to the ZFS (see text).

appear" at negative field values. It is important to realize that in this (and other) real systems, the spectra is due to the $S = 1$ triplet, the singlet $S = 0$ state being diamagnetic. Thus, any small amount of monomeric impurities will give the typical copper(II) signal centered at g close to 2.00. Less than 1% of impurities produce this signal, which is seen in the middle of the real spectrum in Figure 11.22.

In general, in the two cases ($E = 0$ or $E \neq 0$) the bands occupy an important part of the whole range of the spectrum, from 0 to $10\,000\,G$ (X-Band).

> This kind of structure (pattern) derived from the D (and E) parameters is called *fine* structure.

Two questions arise from these comments:

1. How can we calculate the g values, when the spectra afford such a great number of signals? The calculation of g_z, g_x and g_y is not always an easy task in these systems. For $S = 1$, these formulae can be easily derived from those indicated in Table 11.1. In general, for any S value, the use of computer programs is very useful and necessary to simulate some of these spectra.

2. How can copper(II) complexes present D (and E) parameters? Copper(II) is a d^9 system (only one unpaired electron, thus $D = 0$). However, as previously indicated, two main factors contribute to the value of D, (i) single ion anisotropy and (ii) dipole–dipole interactions. The dipolar component can be treated as the polar interaction between negative charges placed at short distances from each other. These dipolar interactions contribute with very small values to the ground D parameter, depending on the M–M separation.

For the $S = 2$ spin system the orthorhombic spin Hamiltonian is more complicated because it must include fine structure terms up to fourth order in the spin operators, although usually they can be neglected as compared to D. The energy diagrams for the two limiting cases ($D > h\nu$ and $D < h\nu$) are shown in Figure 11.23. The shape of the observed spectra depends strongly on D. If $D \gg h\nu$, only the "forbidden" transition within $|{\pm}2\rangle$ and $|{\pm}1\rangle$ doublets will be observed. The allowed $|{-}1\rangle \to |0\rangle$ transition will be observed depending on the strength of the magnetic field. If $D < h\nu$, many $\Delta M_S = \pm 1$ can be observed. The corresponding formula when applying the external magnetic field according to z, y or x are given in Refs. [12] and [13].

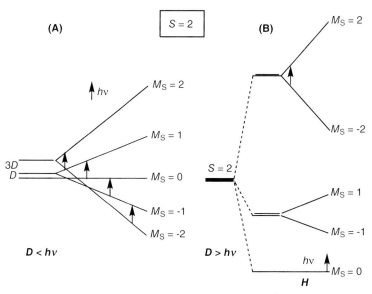

Figure 11.23 ZFS and Zeeman effect for an $S = 2$ system ($H \perp z$).

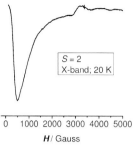

0   1000 2000 3000 4000 5000
***H*/ Gauss**

Figure 11.24 EPR spectrum corresponding to the transition within $M_S = \pm 2$ for an $S = 2$ system (see text for explanation).

Furthermore, there is an important aspect that must be emphasized: in high-spin integer systems with $S \geq 2$ and rhombic symmetry, some nominally spin-forbidden transitions can be detected in certain cases, usually at very low magnetic field, mainly when using a parallel mode detection of the EPR signal [16]. This is the case, for example, for Mn^{3+} complexes: An EPR signal may be observed between the levels of the $|\pm2\rangle$ doublet under certain conditions. g values of 8.7 or 10 have been reported (Figures 11.23B and 11.24). The study of EPR spectra by means of parallel and perpendicular mode detection is named "dual mode".

A study of the anisotropic hyperfine interaction in the $[Mn^{III}(H_2O)_6]^{3+}$ ($S = 2$; $I_{Mn} = 5/2$) ion has been reported recently [17].

11.9.2
S Non-integer (i.e. $S = 3/2$, Cr^{3+} or $S = 5/2$, Mn^{2+})

For $S = 3/2$ the spin Hamiltonian is also derived from Eq. (11.5). The spin Hamiltonian matrix and energy levels can be found in Refs. [12] and [13].

The appearance of the spectra is strongly dependent on the relative order of magnitude of $h\nu$ and the ZFS. If the ZFS is small compared to $h\nu$, three $\Delta M_S = \pm 1$ transitions will be recorded for every principal direction of g (Figure 11.25A). When, on the contrary, the ZFS is larger than $h\nu$, the two Kramers doublets constituting the $S = 3/2$ manifold will be separated in energy and only transitions within each of them will be obtained. In this case the spectra can be described by an effective $S = 1/2$ spin Hamiltonian. The effective g values of the $S = 1/2$ are strongly anisotropic (Figure 11.25B).

The Cr^{3+} ($S = 3/2$) complexes usually show a very small (almost negligible) D parameter. Reference [18] contains many spectra of tetragonally distorted Cr^{3+} complexes and a detailed analysis of them. EPR spectra of anisotropic $S = 3/2$ states can also be detected in other kind of complexes. A good illustrative case is a tetrahedral Co^{2+} ($e^4 t_2^3$) complex. Its ground state is also $S = 3/2$. The second-order spin–orbit coupling between the ground $^4A_2$ term and the excited $^4T_2$ term is greater than in octahedral complexes because $\Delta_T < \Delta_O$. As a consequence, the g values are close to 2.3, larger than the g value for the free electron. The magnitude of g values can be $g_\parallel \approx 2$ and $g_\perp \approx 4$.

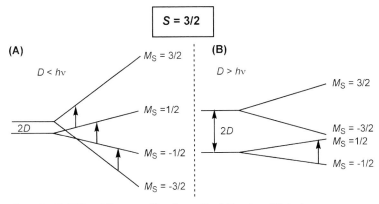

Figure 11.25 ZFS and Zeeman effect for an $S = 3/2$ system ($H \perp z$).

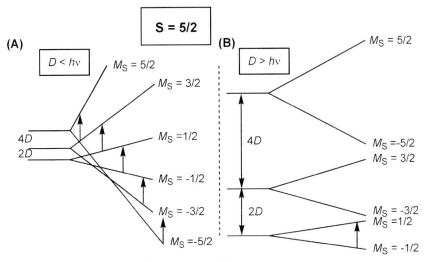

Figure 11.26 ZFS and Zeeman effect for an $S = 5/2$ system ($H \perp z$).

For $S = 5/2$ the spin Hamiltonian needs terms of higher order, although, as in $S = 2$ systems, these terms are smaller and, often, they are not included in the Hamiltonian.

Two different perspectives of the splitting of the ^6S state of an octahedral manganese(II) complex are illustrated in Figure 11.26. The zero-field splitting produces three doubly degenerate spin states, $M_s = \pm 5/2, \pm 3/2$ and $\pm 1/2$ (Kramers' degeneracy). Each of these is split into two singlets by the applied field, producing six levels. As a result of this splitting, considering only the z direction, five transitions ($-5/2 \rightarrow -3/2; -3/2 \rightarrow -1/2; -1/2 \rightarrow 1/2; 1/2 \rightarrow 3/2$ and $3/2 \rightarrow 5/2$) are expected if $D < h\nu$. If $D \gg h\nu$, the situation is shown in Figure 11.26B and, thus, only one very anisotropic transition is seen in the EPR, corresponding to $|-1/2\rangle$

and $|+1/2\rangle$. It is demonstrated that, in this case, $g_\perp = 3g_\parallel(g_e) \approx 6.0$. It must be mentioned that often, together with these lines, intense $\Delta M_S = \pm 2$ forbidden transitions are observed.

11.10
EPR Spectra of Polynuclear Metal Complexes

The reader interested in this very complex subject may consult the book of Gatteschi and Bencini [19]. The full interpretation of the EPR of polynuclear complexes is not always an easy task and it is beyond the scope of this non-specialized book.

In general, two kinds of spectra can be considered: at very low temperature (4.2 K for example) it can be assumed that *only* the ground state is populated. In this case, the spectrum is very similar to that of a mononuclear complex, but with g, A and D values being a linear combination of those corresponding to single ion values. The relations between the spin Hamiltonian parameters of the individual ions and those of the pairs have been extensively studied and verified [19]. Furthermore, the dipolar contribution due to the separated single ions, although very small, is always significant. The spectra at higher temperatures are much more complicated because they show all the resonances corresponding to all states (different S values).

11.10.1
Some Typical Examples

We have already dealt with the coupled Cu–Cu complexes in this chapter. If we assume AF coupling, such as in the [Cu$_2$(carboxylato)$_4$] species, the energy diagram is given in Figure 11.27. For $|J|$ noticeable (strong AF coupling) transition from $S = 0$ to $S = 1$ state are not observed in the EPR. The exchange interaction gives rise to a lower energy state $S = 0$, so the intensity of the signals decreases with decreasing temperature. The temperature dependence of the signals' intensity is a signature of the J value. At high temperature, several transitions with $\Delta M_s = |1|$ are possible and indicated in Figure 11.27. Furthermore, the theoretically forbidden transition with $\Delta M_s = |2|$ is also frequently detected. However, the most important feature of this spectrum is, again, that this splitting is *only* valid for **H** parallel to the z axis. For **H** parallel to the x and y axes the splittings are different (non-linear) and the actual spectrum will be much more complicated, as already commented previously in this chapter.

A possibility to study an anisotropic $S = 3/2$ state consists of recording the spectrum of a ferromagnetically coupled Cu$_3^{II}$ complex [20], or a Ni^{2+}–Cu^{2+}–Ni^{2+} complex with antiferromagnetic coupling between the Ni–Cu ions. The ground state will be 3/2 (Figure 11.28A). If the antiferromagnetic coupling is strong enough, as happens in the complexes derived from the ligand of Figure 11.28A, at 4 K only the ground state, $S = 3/2$, will be populated. In this case this $S = 3/2$

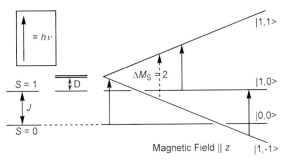

Figure 11.27 Scheme of the EPR transitions for a dinuclear copper(II) complex with AF coupling.

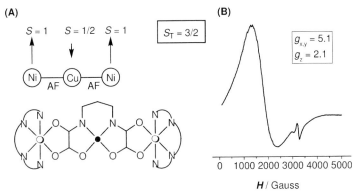

Figure 11.28 (A) Scheme of a Ni^{II}–Cu^{II}–Ni^{II} complex and (B) its EPR spectrum at low temperature.

comes from a Ni^{2+} ion in which the D factor is important. The single-ion zero-field splitting of the Ni^{2+} contributes to the molecular ZFS of the trinuclear entity. Figure 11.28B represents the idealized spectrum of a Ni–Cu–Ni trinuclear system [21]. By changing the blocking amine in the terminal nickel(II) ions, different complexes with very anisotropic EPR spectra ($g_{\perp} \approx 4.5$–6 and $g_{\parallel} \approx 2.1$–2.3) have been studied [21].

A different spin quartet ground state ($S = 3/2$) with a large positive zero-field splitting has been reported and analyzed for a $[Ru_2]^{5+}$ paddlewheel ($\sigma^2\pi^4\delta^2\pi^{*2}\delta^{*1}$ electron configuration) [22].

In some cases the EPR can be very complicated by the appearance of the hyperfine coupling, giving extraordinarily beautiful spectra, such as the example reported by McPherson et al., for a coupled Cu(II)–Mn(II) dimer, with $S_{G.S.} = 2$. In the latter, the three components of g, with values close to 2.00, are present. Moreover, the spectrum is complicated by the ZFS parameters, $D = 0.051\,cm^{-1}$ and $E = 0.013\,cm^{-1}$ [23]. As indicated by the authors, "*the beautifully resolved spectrum leaves little doubt that the resonance arises from a magnetically coupled copper(II)-manganese(II) dimer. The four groups of lines correspond to the fine structure expected of a system having a* *total spin of two. Superimposed on this fine structure is the hyperfine structure from the manganese and the copper nuclei. Each fine structure component is split into six lines by the interaction with the ^{55}Mn nucleus (I = 5/2). Each of these lines is then split into four by the copper nucleus (I = 3/2) to produce the characteristic 24-lines pattern".*

11.11
EPR Spectra of Orbitally Degenerate Terms

All the above-mentioned spectra are typical for non-orbitally degenerate ground terms. In these cases, spectroscopic studies demonstrate that the relaxation time is very appropriate to see the spectra. A very small relaxation time will produce the saturation of the excited state, impeding the observation of the signal. If the relaxation was too rapid, the bandwidth would be too large and the spectra would be impossible to see. The first-order spin–orbit coupling provides a way to relax the electron very rapidly, avoiding the possibility of seeing the signal, at least at room temperature. In these cases, it is necessary to diminish the temperature to "freeze" the relaxation and to have the possibility to see the signal. Thus, currently, the spectra of systems with T terms, should be recorded at liquid helium temperature (4 K).

This spin–orbit coupling mechanism is so efficient concerning the relaxation time, that many systems with tetrahedral geometry having only second-order spin–orbit coupling, have very broad bands and, thus, it is necessary to decrease considerably the temperature to be able to observe the corresponding spectra.

It must be emphasized that the interpretation of the EPR spectra of complexes with a T ground term, is very complicated and is beyond the limits and scope of this book. Thus, only some qualitative aspects will be treated here with regard to the spectrum of a Co^{2+} ion in octahedral symmetry ($^{4}T_{1g}$ term). Due to the first-order spin–orbit coupling the EPR spectrum can be seen only at a very low temperature (usually 4 K). As has been commented in Chapter 8, the ground multiplet after the application of the spin–orbit coupling is a Kramers doublet with an effective $S' = 1/2$ spin. Thus, at very low temperatures, only this Kramers doublet will be populated, having a $<g>$ close to ≈ 4.3 (13/3). However, Co^{2+} systems are often distorted and, thus, anisotropic. The existence of the spin–orbit coupling together with the distortion produces poorly defined spectra, with strong bandwidth that overlap different g values. Furthermore, there is the possible hyperfine coupling with the cobalt nucleus (I = 7/2). All these features make the spectra difficult to interpret. The reader interested in this field can consult the review published by Gatteschi et al. [24].

11.12
High-frequency and High-field EPR Spectroscopy [25]

As indicated previously in this chapter, for transition-metal ions with integer-spin ground states, the magnitude of the axial ZFS parameter, $|D|$ is often larger than

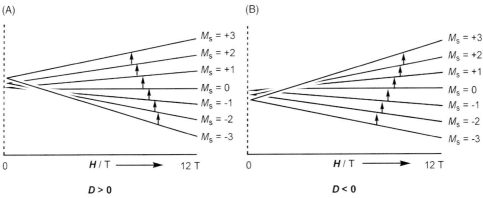

Figure 11.29 Spin levels of an $S = 3$ state in a magnetic field parallel to the unique axis. (A) $D > 0$, (B) $D < 0$. Simulated EPR transitions.

the variable microwave quantum and the system is named "EPR-silent". High-frequency and high-field EPR (HFEPR; $v > 90\,GHz$) has the ability to overcome this difficulty, so that EPR signals are observable even in $S = 1$ systems characterized by large ZFS, whether axial or rhombic symmetry, such as Ni^{2+} (d^8) and V^{3+} (d^2), both with pseudooctahedral symmetry. Most HFEPR measurements have also been made with distorted octahedral Mn^{3+} ($S = 2$) complexes.

The advantages of HFEPR have been recently summarized by Gatteschi et al. [26]: (i) increased resolution, (ii) increased sensitivity, (iii) simplification of the spectra and their assignment, due to the greater separation of signals, (iv) observation of the spectra in EPR silent species, (v) determination of the sign of the ZFS anisotropy. The latter should become apparent with the following example: the energy levels for $S = 3$ in a field parallel to z are shown in Figure 11.29, for $D > 0$ and $D < 0$. It is evident that in the former case the transition involving the lowest energy M levels occurs at high field, while in the latter at low field. At intermediate temperature the intensities of the signals will be weighted according to the Boltzmann population of the M levels. Inspection of the changes in band intensities with the temperature variation provides evidence for the sign of D. See Chapter 10 for the importance of the sign of D in polynuclear complexes acting as SMMs.

11.13
Relaxation Times and Linewidths in EPR

In a possible transition, the lower energy levels are more highly populated. Without spin relaxation, thus, stimulated absorption will be dominant and saturation will inevitably occur. Thus some mechanisms are necessary in order to have the appropriate relaxation of the energy absorbed. This relaxation is the main origin of the linewidth of the EPR spectra. There are two main effects.

11.13.1
Spin–Lattice Relaxation

The spin-lattice relaxation time, T_1, which is temperature dependent, measures the characteristic time for recovery of the magnetization of the paramagnetic system along the applied magnetic field direction after equilibrium has been disturbed.

Let us define a thermodynamic parameter, called *the spin-temperature T_s*, by means of the relation $N_e/N_g = \exp(-\Delta E/kT_s)$, where N_e and N_g are the occupancy numbers of the excited and ground states. If now we suppose that the spin system is subject to a radiation tuned so that the energy photon matches ΔE, the resulting EPR energy absorption by the spin causes a change in the population; that is, the ratio N_e/N_g is altered. Since the spin system has gained energy, it can be considered "hot" compared to its surroundings. The spin system undergoes interactions with the surroundings such that it "cools" and its temperature is eventually restored. The parameter T_1, the relaxation time, is the characteristic time for the energy to flow from the spin system into the surroundings; it reflects the degree to which the spin system is connected to its surroundings. If T_1 is too long, the excited state becomes saturated (no possibility of more excitation) but if T_1 is too short, the relaxation phenomena are too rapid, producing problems in the signal.

This T_1 called spin–lattice relaxation time because it occurs primarily through dynamic interactions with the environment. This can be the surrounding solvent or lattice (if in a crystalline solid – hence the origin of the name). The lattice generally can be thought of as having infinite heat capacity, so that its temperature can be taken as constant throughout the EPR experiment.

The variation in spin–lattice relaxation times in different systems is quite large. For some compounds it is sufficiently long to allow the observation of spectra at room temperature, while for others this is not possible. Since the relaxation time, T_1 usually increases as the temperature decreases, many complexes of the transition metal ions need to be cooled to liquid N_2 or even to liquid He temperature before well-resolved spectra are observed.

11.13.2
Spin–Spin Relaxation

Spin–spin relaxation on the other hand, is concentration dependent and temperature independent. Spin–spin interaction results from the small magnetic fields that exist on neighboring paramagnetic ions. As a result of these fields, the total field at the ions is slightly altered and the energy levels are shifted. This effect varies as $(1/r^3)(1 - \cos^2 \theta)$, where r is the distance between the ions and θ is the angle between the field and the symmetry axis. This kind of broadening will show a marked dependence upon the direction of the field. The effect can be reduced by increasing the distance between paramagnetic ions by diluting the salt with an isomorphous diamagnetic material, or by using a much diluted frozen solution.

It is important to distinguish between *homogeneous* and *inhomogeneous* broadening. The former involves coupling of the spin system to the lattice (T_1) and/or the mutual interactions between all of the paramagnetic centers in a sample (T_2), and the combined effect of T_1 and T_2 processes gives rise to what is called *homogeneous* broadening. Homogeneous line broadening for a set of spins occurs when all these "see" the same net magnetic field and have the same spin-Hamiltonian parameters. This means that the line-shape (i.e. the transition probability as a function of magnetic field) is the same for each dipole.

Inhomogeneous broadening of lines comes from the additional effects of unresolved interactions with a finite number of surrounding nuclei or from a limited set of nearby paramagnetic ions. Generally the unpaired electrons in a sample are not all subjected to exactly the same *H* values. The observed line is then a superposition of a large number of individual components, each slightly shifted from the others. We have, thus, a resultant envelope. The main causes of inhomogeneous broadening are: (i) an inhomogeneous external field, (ii) unresolved hyperfine structure, (iii) anisotropic interactions in randomly oriented systems in the solid state, (iv) dipolar interactions with other fixed paramagnetic centers.

Finally, in magnetically coupled complexes, two opposite parts contributing to the linewidth must be considered: (i) the dipolar contribution commented above, which tends to broaden the line and (ii) the *exchange narrowing*. Indeed, unpaired electrons centered on more than one ion, are more or less influenced, each one depending on the strength of the coupling (ferro or antiferromagnetic). This magnetic coupling creates a sharpening of the bands, referred to as *exchange narrowing*. From a qualitative point of view it is very easy to understand that when several electrons are coupled together, not isolated, the relaxation time is slower. Thus, for slower relaxation, the band becomes narrower.

References

1 Drago, R.S. (1992) *Physical Methods for Chemists*, 2nd edn., Saunders, Orlando, USA.

2 Strouse, C.E., Dahl, L.F. Organometallic Chalcogen Complexes. XXII. Syntheses and Structural Analyses by X-Ray Diffraction and Electron Spin Resonance Single Crystal Methods of $Co_3(CO)_9Se$, $FeCo_2(CO)_9Se$, and $FeCo_2(CO)_9Te$. The Antibonding Metallic Nature of an Unpaired Electron in an Organometallic Cluster System, *J. Am. Chem. Soc.* 1971, *93*, 6032.

3 Karet, G.B., Sun, Z., Heinrich, D.D., McCusker, J.K., Folting, K., Streib, W.E., Huffman, J.C., Hendrickson, D.N., Christou, G. Tetranuclear and Pentanuclear Vanadium(IV/V) Carboxylate Complexes: $[V_4O_8(NO_3)(O_2CR)_4]^{2-}$ and $[V_5O_9X(O_2CR)_4]^{2-}$ (X = Cl⁻, Br⁻) Salts, *Inorg. Chem.* 1996, *35*, 6450.

4 Bertini, I., Gatteschi, D., Scozzafava, A. Six-Coordinate Copper Complexes with $g_{\parallel} < g_{\perp}$ in the Solid State, *Coord. Chem. Rev.* 1979, *29*, 67.

5 Ribas, J., Diaz, C., Monfort, M., Corbella, M., Solans, X., Font-Altaba, M. Crystal structure and spectroscopic properties of N,N'-bis(8-quinolilethylenediamine)monochlorocopper(II) perchlorate, *Inorg. Chim. Acta*, 1986, *117*, 49.

6 Calle, C., Schweiger, A., Mitrikas, G. Continuous-Wave and Pulse EPR Study of the Copper(II) Complex of N-Confused

Tetraphenylporphyrin: Direct Observation of a σMetal-Carbon Bond, *Inorg. Chem.* 2007, *46*, 1847.

7 Goldcamp, M.J., Robison, S.E., Bauer, J.A.K., Baldwin, M.J. Oxygen Reactivity of a Nickel(II)–Polyoximate Complex, *Inorg. Chem.* 2002, *41*, 2307.

8 (a) Urbach, F., Bereman, R.D., Topich, J.A., Hariharan, M., Kalbacher, B.J. Stereochemistry and Electronic Structure of Low-Spin, Square-Planar Cobalt(II) Chelates with Tetradentate Schiff Base Ligands, *J. Am. Chem.Soc*, 1974, *96*, 5063; (b) McGarvey, B.R. Theory of the Spin Hamiltonian Parameters for Low Spin Cobalt(II) Complexes, *Can. J. Chem.* 1975, *53*, 2498.

9 (a) Hanson, G.R., Kabanos, T.A., Keramidas, A.D., Mentzafos, D., Terzi, A. Oxovanadium(IV)-Amide Binding. Synthetic, Structural, and Physical Studies of {N-[2-(4-Oxopent-2-en-2-ylamino)phenyl]pyridine-2-carboxamido} oxovanadium(IV) and {N-[2(-4-Phenyl-4-oxobut-2-en-2-ylamino)phenyl]pyridine-2-carboxamido}oxovanadium(IV), *Inorg. Chem.* 1992, *31*, 2587; (b) Rangel, M., Leite, A., Amorim, M.J., Garribba, E., Micera, G., Lodyga-Chruscinska, E. Spectroscopic and Potentiometric Characterization of Oxovanadium(IV) Complexes Formed by 3-Hydroxy-4-Pyridinones. Rationalization of the Influence of Basicity and Electronic Structure of the Ligand on the Properties of $V^{IV}O$ Species in Aqueous Solution, *Inorg. Chem.* 2006, *45*, 8086.

10 Watson, C.T., Cai, S., Shokhirev, N.V., Walker, F.A. NMR and EPR Studies of Low-Spin Fe(III) Complexes of *meso*-Tetra-(2,6-Disubstituted Phenyl)Porphyrinates Complexed to Imidazoles and Pyridines of Widely Differing Basicities, *Inorg. Chem.* 2005, *44*, 7468.

11 Nakamoto, M., Yamasaki, T., Sekiguchi, A. Stable Mononuclear Radical Anions of Heavier Group 13 Elements: [(^{t}Bu$_2$MeSi)$_3$E$^•{-}$]·[K$^+$(2.2.2-Cryptand)] (E = Al, Ga), *J. Am. Chem. Soc.* 2005, *127*, 6954.

12 Bencini, A., Gatteschi, D. (1982) *Transition Metal Chemistry. A Series of Advances*, ESR Spectra of Metal Complexes of the First Transition Series in Low-Symmetry Environments, Vol. 8, pp. 1–154. Marcel Dekker, Inc., New York.

13 Boca, R. Zero-field splitting in metal complexes, *Coord. Chem. Rev.* 2004, *248*, 757.

14 Ballhausen, C.J. (1979) *Molecular Electronic Structures of Transition Metal Complexes*, McGraw Hill, New York.

15 Huang, Z., Song, H-B., Du, M., Chen, S-T., Bu, X-H., Ribas, J. Coordination Polymers Assembled from Angular Dipyridyl Ligands and Cu^{II}, Cd^{II}, Co^{II} Salts: Crystal Structures and Properties, *Inorg. Chem.* 2004, *43*, 931.

16 Dexheimer, S.L., Gohdes, J.W., Chan, M.K., Hagen, K.S., Armstrong, W.H., Klein, M.P. Detection of EPR Spectra in S = 2 States of Trivalent Manganese Complexes, *J. Am. Chem. Soc.* 1989, *111*, 8923.

17 Krivokapic, I., Noble, C., Klitgaard, S., Tregenna-Piggott, P., Weihe, H., Barra, A-L. Anisotropic Hyperfine Interaction in the Manganese(III) Hexaaqua Ion, *Angew. Chem. Int. Ed.* 2005, *44*, 3613.

18 Pedersen, E., Toftlund, H. Electron Spin Resonance of Tetragonal Chromium(III) Complexes. I. *trans*-[Cr(NH$_3$)$_4$XY]$^{n+}$ and *trans*-[Cr(py)$_4$XY]$^{n+}$ in Frozen Solutions and Powders. A Correlation between Zero-Field Splitting and Ligand Field Parameters via Complete d-Electron Calculation, *Inorg. Chem.* 1974, *13*, 1603.

19 Bencini A. and Gatteschi, D. (1990) *EPR of Exchange Coupled Systems*, Springer-Verlag, Berlin.

20 Glaser, T., Heidemeier, M., Grimme, S., Bill, E. Targeted Ferromagnetic Coupling in a Trinuclear Copper(II) Complex: Analysis of the S_t = 3/2 Spin Ground State, *Inorg. Chem.* 2004, *43*, 5192.

21 Tercero, J., Diaz, C., Ribas, J., Maestro, M., Mahía, J., Stoeckli-Evans, H. Oxamato-Bridged Trinuclear Ni^{II}Cu^{II}Ni^{II} Complexes: A New (Ni^{II}Cu^{II}Ni^{II})$_2$ Hexanuclear Complex and Supramolecular Structures. Characterization and Magnetic Properties, *Inorg. Chem.* 2003, *42*, 3366.

22 Chen, W-Z., Cotton, F.A., Dalai, N.S., Murillo, C.A., Ramsey, C.M., Ren, T., Wang, X. Proof of Large Positive Zero-Field Splitting in a Ru$_2^{5+}$ Paddlewheel, *J. Am. Chem. Soc.* 2005, *127*, 12691.

23 Krost, D.A., McPherson, G.L.
Spectroscopic and Magnetic Properties of
an Exchange Coupled Copper(II)-
Manganese(II) Dimer, *J. Am. Chem. Soc.*,
1978, *100*, 987.

24 Banci, L., Bencini, A., Benelli, C.,
Gatteschi, D., Zanchini, C. Spectral-
Structural Correlations in High-Spin
Cobalt(II) Complexes, *Struct. Bond.*, 1982,
52, 37.

25 Krzystek, J., Ozarowski, A., Telser, J.
Multi-frequency, high-field EPR as a
powerful tool to accurately determine zero-
field splitting in high-spin transition metal
coordination complexes, *Coord. Chem. Rev.*
2006, *250*, 2308.

26 Gatteschi, D., Sessoli, R., Villain, J. (2006)
Molecular Nanomagnets, Oxford University
Press, Oxford.

Bibliography

Weil, J.A., Bolton, J.R., Wertz, J.E. (1994)
Electron Paramagnetic Resonance, John
Wiley, New York.

Pilbrow, J.R. (1990) *Transition Ions Electron
Paramagnetic Resonance*, Oxford Science
Publications, Oxford.

Abragam, A., Bleaney, B. (1970) *Electron
Paramagnetic Resonance of Transition Ions*,
Dover Publications, New York.

Part Three   Electron Transfer

12
Redox Mechanisms

12.1
Historical Introduction

Oxidation–reduction reactions involve changes in the oxidation state. There are two species, the *oxidant*, which receives the electron(s), and the *reductant*, which gives the electron(s). Much of the experimental and theoretical background for the understanding of electron transfer processes in solution was developed between 1950 and 1975. The classical treatments of Marcus and Hush, and their experimental validation dominated that period [1]. In 1952 Libby underlined the relevance of the Franck–Condon principle to thermal electron transfer in solution and its application to the reactions of metal complexes. Quantum mechanical formulations were firmly established by the mid 1970s.

The distinction between *inner-sphere* and *outer-sphere* mechanisms was already widely appreciated at that time [2]. Taube and coworkers showed in 1953 that chloride transfer occurred in the reaction:

$$[CoCl(NH_3)_5]^{2+} + [Cr(H_2O)_6]^{2+} + 5H_3O^+ \rightarrow [Co(H_2O)_6]^{2+} + [CrCl(H_2O)_5]^{2+} + 5NH_4^+$$

implying a chlorido-bridged transition-state [2].

Henry Taube received the Nobel Prize in 1983. Rudolph. A. Marcus received the Nobel Prize in 1992.

Redox reactions are usually (with exceptions) second-order reactions (first-order in the oxidant and first-order in the reductant: $v = k[\text{oxidant}][\text{reductant}]$.

12.2
Mechanisms of Redox Reactions

Two mechanisms are recognized for redox reactions: the *inner-sphere* and the *outer-sphere* mechanisms (Figure 12.1). The inner-sphere mechanism includes atom-transfer processes: the coordination spheres of the reactants share a ligand for a short time and form an intermediate with a bridge. The outer-sphere redox reactions simply involve electron transfer. The reactants are in contact without sharing

Coordination Chemistry. Joan Ribas Gispert
Copyright © 2008 WILEY-VCH Verlag GmbH & Co. KGaA, Weinheim
ISBN: 978-3-527-31802-5

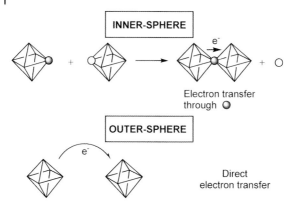

Figure 12.1 General scheme of the electron transfer through outer- and inner-sphere mechanisms.

any bridging ligand. In the outer-sphere mechanism the coordination spheres remain more or less intact as the electron donor–acceptor pair passes through the transition state.

Although the differences between the two mechanisms are clear, it is not always an easy matter to establish which mechanism operates in specific cases. The methodology of proving the outer-sphere mechanism usually relies on providing convincing evidence that the alternative inner-sphere mechanism is not available. Activation enthalpies are generally small, being lowest in outer-sphere reactions. Both types of reaction show negative activation entropies. Thus, standard kinetic parameters are not very useful for distinguishing between the two mechanisms. As underlined by Tobe and Burgess, there is an *"uncomfortably large number of reactions in the grey area in between"* [3].

12.2.1
The Outer-sphere Mechanism

In 1956–1960 Marcus published the first expressions for *outer-sphere* electron transfer rate constants. He assumed that the electronic coupling was large enough for the electron transfer to occur adiabatically (see below) but not so large as to lower the activation barrier significantly. The inner coordination shells of the reactants need to be reorganized and the surrounding solvent to be readjusted prior to electron transfer. Within a 'classical' model, energy conservation requires that the actual electron transfer occurs at the intersection of two energy surfaces, one for the reactants (R) and one for the products (P). Electronic interactions of the reactants give rise to a splitting at the intersection of the surfaces to form first-order surfaces. The splitting was considered to be small.

In this type of reaction nothing is 'broken' and nothing is 're-made'. *It is simply a problem of collisions.* The coordination spheres of the two complexes are unchanged. The great simplicity of this mechanism makes positive proof very hard to obtain. There are a number of inorganic redox reactions in which it is extremely difficult to imagine an inner-sphere mechanism operating. Such reactions are those in which both the oxidant and reductant are substitutionally inert (half-lives with respect to substitution measured in hours) and yet electron transfer is very rapid. A typical example of this mechanism is the reduction of $[Fe(phen)_3]^{3+}$ by $[Fe(CN)_6]^{4-}$. Both complexes have inert coordination spheres, with no possibility of substitution of a ligand. The only possible mechanism is a transfer between the complex ions, through an outer-sphere contact.

Any outer-sphere reaction can be divided into a sequence of four elementary steps. The first step is the diffusion of the oxidant and reductant to form an assembly that is a precursor to electron transfer. This is followed by the electron transfer itself and the subsequent, generally rapid, dissociation of the successor complex:

$$A + B \leftrightarrow [A, B]$$

$$[A, B] \leftrightarrow [A, B]^{\ddagger}$$

$$[A, B]^{\ddagger} \leftrightarrow [A^-, B^+]$$

$$[A^-, B^+] \leftrightarrow A^- + B^+$$

In fact these steps can be reduced to two:

1. A fast pre-equilibrium association of the reagents, $(M^M + N^N \leftrightarrow M^M, N^N)$

followed by

2. The act of electron transfer, $(M^M, N^N \rightarrow M^{M-1} + N^{N+1})$

In principle, the second step can be seen as a composite, with the electron transfer followed by the separation of the two components of the 'successor' complex. In practice, this separation to give the final products is always fast, as there is no bond-breaking involved as in the case of the inner-sphere successor complex.

In order to understand these electron transfer processes, it is of great importance to separate the formation of the 'electron transfer precursor' from k_{et} (the rate of the electron transfer).

12.2.1.1 The Electron Transfer Precursor

The pre-association of the reactants brings them sufficiently close together for the transfer of the electron to occur with reasonable probability and speed. In fact, the interactions between the ions form an assembly of ion pairs, in which the intimate interactions differ. Usually, in outer-sphere reactions, the ion association constant is small and formation of the ion pair is rapid, generally close to the limit at which the ions can diffuse through solution, since the energy barrier in this step of the reaction is low. However, in suitable circumstances, where electron transfer is slow and the electrostatic and other interactions between the reactants are strongly favorable, it is possible to 'detect' the formation of ion pairs through the deviation from the second-order behavior caused in the rate law.

Table 12.1 K_{ip} values (ion pairing equilibrium constant) and k_{et} values for a range of outer-sphere redox reactions between transition metal complexes, according to z^+z^-

Reactants		K_{ip}	k_{et} (s^{-1})
Charge product −12			
$[Co(NH_3)_5(H_2O)]^{3+}$	$[Fe(CN)_6]^{4-}$	1500	0.19
$[Co(NH_3)_5(py)]^{3+}$	$[Fe(CN)_6]^{4-}$	2400	0.015
$[Co(NH_3)_5(3\text{-cyanopyridine})]^{3+}$	$[Fe(CN)_6]^{4-}$	1300	0.346
Charge product −9			
$[Co(dmso)(NH_3)_5]^{3+}$	$[Fe(CN)_5(imidazole)]^{3-}$	450	2.6
$[Co(dmso)(NH_3)_5]^{3+}$	$[Fe(CN)_5(pyridine)]^{3-}$	490	0.15
$[Co(NH_3)_5(pyridine)]^{3+}$	$[Fe(CN)_5(pyridine)]^{3-}$	860	0.0068
$[Co(NH_3)_5(pyridine)]^{3+}$	$[Fe(4,4'\text{-bipyridine})(CN)_5]^{3-}$	1860	0.0021
$[Co(phen)_3]^{3+}$	$[Co(ox)_3]^{3-}$	650	0.24
Charge products −8			
$[Co(acetate)(NH_3)_5]^{2+}$	$[Fe(CN)_6]^{4-}$	300	0.00037
$[Co(benzoate)(NH_3)_5]^{2+}$	$[Fe(CN)_6]^{4-}$	240	0.00062
$[CoCl(NH_3)_5]^{2+}$	$[Fe(CN)_6]^{4-}$	38	0.027
$[CoN_3(NH_3)_5]^{2+}$	$[Fe(CN)_6]^{4-}$	49	0.00062

In general, reactions in which precursor complexes are detected are rare. The first convincing demonstrations of pre-association (ion-pairing) before electron transfer reported in 1971 came from a series of $[Co^{III}(NH_3)_5(H_2O)]^{3+}$ and $[Fe^{II}(CN)_6]^{4-}$ reactions in the presence of disodium dihydrogenethylenediaminetetracetate [4]. The half-life for electron transfer from Fe(II) to Co(III) within the binuclear outer-sphere complex is ~4 s. Here both metals are in the low-spin d^6 configuration. This ensures substitution-inertness prior to electron transfer. In this case, a large charge (z^+z^-) encourages ion-pairing, and electron transfer is slow enough to permit observation of the pre-association of the reactants. However, the effect of (z^+z^-) should not be overestimated because, in some cases, solvent, hydrogen bonds, weak interaction effects, etc., are more important and the precursor could exist even with very small electric charges. Some illustrative K_{ip} (ion-pairing) values are gathered in Table 12.1.

K_{ip} depends on the charge product, z^+z^-, and the internuclear distance, which can be estimated from the magnitude of K_{ip}.

12.2.1.2 The Electron Transfer Step

The two most characteristic outer-sphere redox reactions are: (i) *Symmetrical systems* such as the exchange reaction with the $[Fe(H_2O)_6]^{2+}$ and $[Fe(H_2O)_6]^{3+}$ ions at an appropriate separation distance (one of the Fe ions is isotopically labelled(*)

$$[Fe(H_2O)_6]^{2+}/[Fe^*(H_2O)_6]^{3+} \rightarrow [Fe(H_2O)_6]^{3+}/[Fe^*(H_2O)_6]^{2+}$$

From a chemical point of view, this is not a 'true' redox reaction, but is very useful for understanding the redox mechanisms. (ii) *Non-symmetrical (true redox) systems* such as the following exchange reaction:

$$[Fe(H_2O)_6]^{2+}/[Co(H_2O)_6]^{3+} \rightarrow [Fe(H_2O)_6]^{3+}/[Co(H_2O)_6]^{2+}$$

Marcus made use of potential energy surfaces and statistical mechanics to provide a detailed description of the electron transfer process. The potential energies of the initial and final states of an electron transfer reaction (the reactants plus surroundings medium and the products plus surrounding medium) can be represented by *harmonic free energy curves* that are a function of a single reaction coordinate. These energy surfaces will have minima corresponding to the more stable nuclear configurations of the reactants and products and will intersect where the reactants and products have the same configurations and energies. Within this framework, for an electron transfer reaction with zero standard free energy change, free energy parabolas with identical force constants describe the distortion of the reactants and products from their equilibrium configuration. A symmetrical system, such as $Fe^{3+/2+}$, just simplified in *one-dimension*, is shown in Figure 12.2.

These parabolas can be written (in a simplified manner) as:

$$E = \lambda Q \pm f Q^2 \tag{12.1}$$

Here, Q is the *reaction coordinate*, which can be related to the elongation of the bonds (vibrations). Indeed, in most cases of electron transfer the nuclear

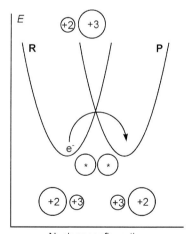

Nuclear configuration
(Reaction Coordinate)

Figure 12.2 Plot of the potential energy of the reactants (R, precursor complex) and products (P, successor complex) as a function of the nuclear configuration for an electron-exchange reaction.

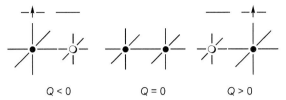

$Q < 0$ $\qquad\qquad$ $Q = 0$ $\qquad\qquad$ $Q > 0$

Figure 12.3 Coordination spheres of the reactants and products as a function of the vibrational coordinate, Q ($Q_A - Q_B$; see text).

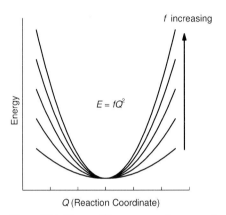

Figure 12.4 Shape of the energy surfaces as a function of the reduced force constant (f).

coordinate for the reorganization corresponds to a normal coordinate associated with the harmonic stretch of the metal–ligand bond (Figure 12.3). Simplifying, if we call Q_A and Q_B the normal coordinates of the reactants A and B (Fe^{2+} and Fe^{3+} in symmetrical reactions), we can assume that electronic transfer *matches* the 'out-of-phase' vibration (Q_A–Q_B). This $Q_A - Q_B$ antisymmetric vibration will be the Q value, known as the *reaction coordinate*. When $Q = 0$, the coordination spheres of A and B are identical; when $Q < 0$ there is an increase in the M–L distances of A and a decrease in B; when $Q > 0$, the situation is inverted (Figure 12.3).

f is the reduced force constant (which gives the shape of the corresponding parabola) (Figure 12.4).

λ is defined as the vertical difference between the free energies of the reactants and products at the reactants' equilibrium configuration (minimum) *for an electron transfer reaction with zero standard free energy change* (Figure 12.5).

λ must be divided into two components, λ_{in} and λ_{out} ($\lambda = \lambda_{in} + \lambda_{out}$). λ_{in} (inner-shell, intramolecular and vibrational) is related to the magnitude of the distortion of products and reactants in their equilibrium geometry: it is thus proportional to the difference in M–L distances in the two oxidation states (Figure 12.6). It depends on the reorganization of the bond lengths and angles within the *precursor complex* assembly.

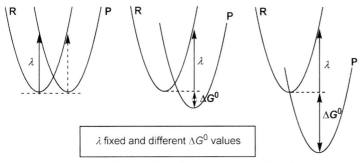

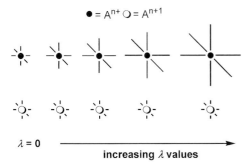

Figure 12.5 Graphical definition of λ and its comparison with ΔG^0.

Figure 12.6 Concept of the λ parameter, as a function of Δd (oxidized and reduced form of the ion).

The major changes in structure are generally found in metal–ligand bond length, which is generally shorter at higher oxidation. Metal–ligand vibrational frequencies can give an idea of bond strengths and thus of the work which may be required to stretch or compress bonds so as to attain the appropriate energy for electron transfer. For example, the Co—N stretch in $[Co(NH_3)_6]^{2+}$ is $357\,cm^{-1}$ and in $[Co(NH_3)_6]^{3+}$ is $494\,cm^{-1}$. Table 12.2 gives some correlations of bond length changes with electron transfer rate constants.

In general, the lower Δd, the greater k_{et} becomes. When λ_{in} is close to 0 the rate of the electron transfer process will be largely dependent on λ_{out}.

λ_{out} (the outer-shell, solvent reorganization free energy) depends on the properties of the solvent (dielectric properties, for example), on the distance separating the donor and acceptor sites and, for a given separation, on the shape of the reactants. The values for the outer-sphere term, λ_{out}, can be estimated from a classical electrostatic treatment for the movement of charge in a medium of continuous dielectric constant.

Summary All previous features are illustrated in Figure 12.7A for a symmetrical system and in Figure 12.7B for a non-symmetrical (true redox, $\Delta G^0 \neq 0$) system. The free energy of the close-contact reactants and surrounding medium (Curve R)

Table 12.2 Relationships between Δd and k_{et} in outer-sphere redox reactions.

Reaction	Δd (Å)	k_{et} (M^{-1} s^{-1})	Reaction	Δd (Å)	k_{et} (M^{-1} s^{-1})
$[Fe(CN)_6]^{3-/4-}$	0.03	2×10^4	$[Fe(H_2O)_6]^{3+/2+}$	0.13	1.1
$[Ru(H_2O)_6]^{3+/2+}$	0.09	20	$[Cr(H_2O)_6]^{3+/2+}$	0.20	1.9×10^{-5}
$[Co(bpy)_3]^{3+/2+}$	0.19	5.7	$[Co(NH_3)_6]^{3+/2+}$	0.22	2×10^{-8}

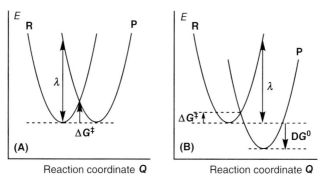

Figure 12.7 Concept of λ and $\Delta G^{\#}$ for redox reactions with $\Delta G^0 = 0$ (A) and $\Delta G^0 < 0$ (B).

and the free energy of the close-contact products and surrounding medium (Curve P) are plotted against the reaction coordinate. λ and $\Delta G^{\ddagger}$ (activation energy) are indicated. The activation energy is the difference between the free energies of the reactants in their transition state configuration and in their equilibrium configuration.

Factors That Determine Electron Transfer   First, it is necessary to imagine that both species are vibrationally rearranged to become "practically equal" (Figure 12.8).

In this vibrational excitation, *only* the totally symmetric normal modes of vibration take part. Electron transfer is very easy when the nuclei of the two complexes have positions which ensure that the electron has the same energy in all these positions. From this hypothesis, it can be deduced that the rate of the electron transfer and the activation energy of the process are governed by the capacity of the nuclei to adopt positions that can render the energies equal. Thus, the main factors that determine the electron transfer rate are:

1. *The shape of the energy surfaces.* If the parabola has a great slope (this depends on f), there will be a rapid increase in energy when the reaction coordinate, Q, increases. As the intersecting point is high, the activation energy is very great. On the contrary, plate (flattened) energy surfaces indicate low activation energy values (Figures 12.9A and B, respectively).

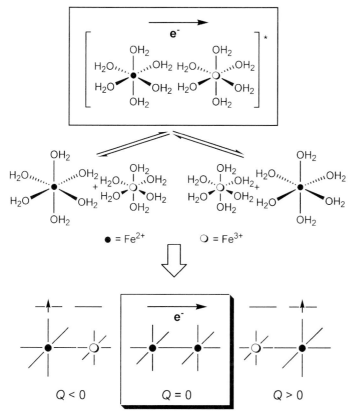

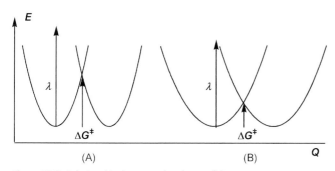

Figure 12.8 Molecular rearrangement (vibrational) in an outer-sphere mechanism.

Figure 12.9 Relationship between the shape of the energy
surfaces (depending on f) and λ and $\Delta G^{\#}$ (see text).

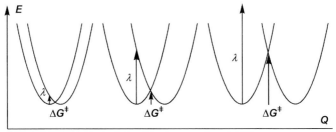

Figure 12.10 Graphical relationships between λ and $\Delta G^{\#}$.

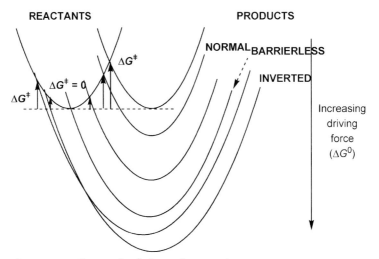

Figure 12.11 Influence of ΔG^0 (driving force) on the energy curves (normal, barrierless and inverted regions).

2. *Large variations in the internuclear distances in the equilibrium* (λ) mean that the equilibrium points are very far away and that the intersecting point will be reached only with great distortion (Figure 12.10).

3. *The third factor is* ΔG^0 *of the reaction* ($\Delta G^0 = G_P - G_R$, at the equilibrium configuration). Starting from $\Delta G^0 = 0$, when ΔG^0 increases, $\Delta G^{\ddagger}$ decreases. At the limit of the exoenergetic reaction (when the curve of the products cuts the curve of the reactants at its minimum), the so-called *barrierless point* is reached ($\Delta G^{\ddagger} = 0$; k_{et} close to 10^{10-12}). After this point, when ΔG^0 increases, $\Delta G^{\ddagger}$ increases and the rates of the reaction diminish (this is the 'inverted region' in Marcus' terminology) (Figure 12.11).

12.2.1.3 Zero-order and First-order Potential Energy Surfaces

All previous considerations correspond to the so-called zero-order energy surfaces. The surfaces intersect in regions where the precursor and successor complexes

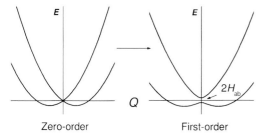

Figure 12.12 From the zero-order to first-order redox reactions: influence of H_{AB}.

have the same nuclear configuration and energies. However, once the system reaches the intersection of the two curves the probability of going from the R to the P surface depends on an important and new factor: *the extent of coupling of the electronic orbitals of the two reactants. Electronic coupling of the reactants is necessary if the system is to pass from the initial to the final state.*

Electronic coupling removes the degeneracy at the intersection of the zero-order surfaces and leads to the formation of two new surfaces, the first-order or adiabatic states of the system (Figure 12.12). The splitting at the intersection is $2H_{AB}$, H_{AB} being $<\psi_A|\hat{H}|\psi_B>$, in which ψ_A and ψ_B are the wavefunctions for the donor and acceptor orbitals, respectively. Thus, the progress of the electron transfer reaction can also be described in terms of the mixing of the wavefunctions for these states.

The magnitude of H_{AB} determines the behavior of the system on reaching the intersection region. If H_{AB} is not very small, then the system will remain on the lower potential energy surface on passing through the intersection region. Since the system remains in the same electronic state throughout the course of the reaction, such reactions are called *adiabatic,* distinguishing them from those in the zero-order class, which are called *non-adiabatic.*

12.2.1.4 Theoretical Treatments of the ET Transfer: From Classical to Quantum Theories

The process by which the reactants become products depends on precisely how the barrier is passed over, tunnelled through or otherwise avoided. As indicated by Barbara et al., *'This is the topic of electron transfer kinetics and rate theory'* [5].

The time scales for ET can be very slow or very fast. The simple barrier picture of previous figures suggests that control of the barrier height is a major factor in controlling the time scale. Indeed, achieving such control by a combination of barrier height and the extent of electronic coupling has been the aim of many ET studies. In all theories for understanding ET processes, the basic parabola model explained above is very useful.

Classical Theories: Marcus Approach   Most of the features indicated so far correspond to what is known as *'classical theory'* (in nuclear terms; in electronic terms the treatment method is *always* quantum mechanics). ET needs, simply, to cross the barrier between the reactants and products. This barrier can be calculated by straightforward algebra. *Nuclear tunneling is not important.*

In the classical theory [1] the rate constant for electron transfer within the precursor complex formed by the reactants in a bimolecular reaction is given by:

$$k_{et} = \kappa_{el} \nu_n \exp(-\Delta G^{\ddagger}/kT) \tag{12.2}$$

κ_{el} is the electronic factor (or electronic transmission coefficient) for the reaction. There is no classical approximation of this parameter. Its calculation requires quantum mechanical treatment. In general, the electronic factor is related to the extent of the overlap between donor and acceptor orbitals, and can be estimated by calculation of the electronic matrix coupling element $H_{AB} = <\psi_A|\hat{H}|\psi_B>$. The magnitude of this coupling depends greatly on the relative orientation of the complexes. In quantum terms, it depends on the overlap integral, S. If $H_{AB} = 0$, $\kappa_{el} = 0$; if H_{AB} is important, $\kappa_{el} = 1$ (in the classical limit). Provided that the electronic interaction of the reactants (H_{ab}) is 'sufficiently' strong, then $\kappa_{el} \approx 1$ and the electron transfer will occur with near unit probability in the intersection region: the electron transfer reaction is adiabatic with the system remaining on the lower energy surface on passing through the intersection region. Under these conditions,

$$k_{et} = \nu_n \exp(-\Delta G^{\ddagger}/kT) \tag{12.3}$$

As a consequence of κ_{el}, one of the most important factors in the rate of electron transfer processes is the nature, π^* or σ^*, of the donor and acceptor MOs. The electron transfer is easier when both MOs are π^* since these electrons are more exposed and more sensitive to their surroundings than the σ^* electrons, which are directed towards the vertices of the octahedron. Thus, a $\pi^* \rightarrow \pi^*$ transfer is more rapid than a $\sigma^* \rightarrow \sigma^*$ one (Figure 12.13). One of the characteristic features of these reactions arises when comparing transfer data for $[Co(NH_3)_6]^{2+}/[Co(NH_3)_6]^{3+}$ and transfer data for Fe^{2+}/Fe^{3+} and Ru^{2+}/Ru^{3+} amminated complexes. For cobalt ions the rate constant is close to 10^{-9}, whereas it is positive for Fe and Ru analogs. In this latter case (Fe and Ru), the starting configuration is d^6 and the final one is d^5, with the orbitals of both configurations being of the π^* type. However, in the cobalt case, the initial configuration is d^7 (high spin) and the final is d^6 (low spin) which is, thus, a change from π^* to σ^* orbitals. For this reason, the Co—L bond distance changes by $0.178\,\text{Å}$, four times the variation for similar Fe or Ru ions.

The vibration frequency, ν_n, with which the activated complex approaches the transition state, i.e. the vibration frequency that takes the system through the intersection region is generally in the range 10^{12} to $10^{14}\,\text{s}^{-1}$. Where λ is large, internal rearrangement dominates the activation process and the effective nuclear frequency corresponds to some combination of internal breathing modes of the complexes, typically around $400\,\text{cm}^{-1}$ for a M—N bond stretch, corresponding to $\nu_{eff} = 10^{13}\,\text{s}^{-1}$. Where internal rearrangements are less important, reorientation of solvent dipoles dominates the energy of the process.

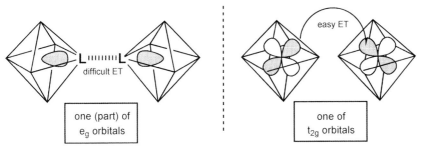

Figure 12.13 Relationship between outer-sphere electron transfer and symmetry of the atomic orbitals.

The activation barrier ($\Delta G^{\ddagger}$) and its relationship with λ and H_{ab} From the potential energy curves given in previous figures it is easy to demonstrate mathematically that, when H_{ab} is no longer negligible, the free energy of activation for a self-exchange reaction is given by:

For symmetrical systems, $\Delta G^{0} = 0$

$$\Delta G^{\ddagger} = \lambda/4 - H_{ab} + H_{ab}^{2}/\lambda = \lambda(1 - 2H_{ab}/\lambda)^{2}/4 \qquad (12.4)$$

For unsymmetrical systems, $\Delta G^{0} \neq 0$

$$\Delta G^{\ddagger} = \frac{\lambda}{4} + \frac{\Delta G^{0}}{2} + \frac{(\Delta G^{0})^{2}}{4(\lambda - 2H_{ab})} - H_{ab} + \frac{H_{ab}^{2}}{(\lambda + \Delta G^{0})} \qquad (12.5)$$

These equations only apply in the '*normal region*' (Figure 12.11).

Kinetic parameters and free energy regions (i) The normal free energy region is characterized by $|\Delta G^{0}| < \lambda$. In this region $\Delta G^{\ddagger}$ decreases as λ or ΔG^{0} decreases (with increased driving force). (ii) When $|\Delta G^{0}| = \lambda$, $\Delta G^{\ddagger} = 0$, the reaction is barrier-less. (iii) The inverted free energy region is characterized by $|\Delta G^{0}| > \lambda$. In this region $\Delta G^{\ddagger}$ increases as λ decreases or ΔG^{0} becomes more negative (with increased driving force). These cases are illustrated in Figure 12.14.

The rate of the electron transfer (k_{et}) increases from the normal region to the barrierless region and then decreases in the inverted region with increasing driving force.

The 'inverted region' was already predicted by Marcus in the early 1960s. However, the first examples were reported in 1984 in organic reactions [6] and, later, many examples from coordination systems [7] and several from bioinorganic chemistry have also been reported [8].

All the equation and comments above are only valid for symmetrical/unsymmetrical systems as long as the system is described by a double-well potential. However, if the electronic coupling is very strong, the double well disappears and the equations are no longer valid. This does not happen in outer-sphere redox

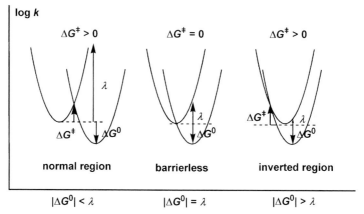

Figure 12.14 Relationships between $\Delta G^{\neq}$, ΔG^0 and λ in normal, barrierless and inverted regions.

reactions, but it will be found in mixed-valence complexes, class III (see next Chapter).

Empirical Marcus Law for a "Cross-reaction" The full theoretical Marcus treatment requires a number of assumptions and estimates, but several of the complications disappear comparing rate constants for a given reaction with rate constants for self-exchange for the component redox couples, for example if one compares k_{12} for the cross-reaction (with equilibrium constant K) with k_{11} and k_{12}:

$$[Ru(NH_3)_6]^{3+} + [Ru(NH_3)_6]^{2+} - k_{11} \rightarrow [Ru(NH_3)_6]^{2+} + [Ru(NH_3)_6]^{3+}$$

$$[Co(phen)_3]^{3+} + [Co(phen)_3]^{2+} - k_{22} \rightarrow [Co(phen)_3]^{2+} + [Co(phen)_3]^{2+}$$

$$[Co(phen)_3]^{3+} + [Ru(NH_3)_6]^{2+} - k_{12} \rightarrow [Co(phen)_3]^{2+} + [Ru(NH_3)_6]^{3+}$$

Marcus deduced his equation of the rate constant for an outer-sphere reaction from the self-exchange rate constants (k_{11} and k_{22}) and the equilibrium constant of the overall reaction (K):

$$k_{12}^2 = f k_{11} k_{22} K \quad \text{or} \quad k_{12} = (k_{11} k_{22} K)^{1/2} w \tag{12.6}$$

$$\text{or} \quad 2\ln k_{12} = \ln k_{11} + \ln k_{22} + \ln K + \ln(f) \tag{12.7}$$

f (or w) is a work term representing the algebraic sum of the energies expended or liberated in bringing the reactants together in the self-exchange and cross-reaction. Clearly w is related to the association constants, especially for a cross-reaction involving charges of opposite sign, and may be much affected by the solvent. In fact, the f factor is a very complex parameter that in approximate calculations is taken as close to unity.

K can sometimes be measured directly, for example using UV–visible spectroscopy, but is usually obtained from redox potential data ($\ln K = \Delta E^0/0.059$). The rate constant (k_{12}) is generally measured directly, by conventional or fast-reaction techniques. The determination of the rates of electron self-exchange processes is an important element in testing Marcus' theory. The rate k_{11} (k_{22}) associated with dynamic equilibrium is independent of the free energy change. It involves no apparent reaction, which creates certain difficulties for measurements. These studies were conducted by isotopic labelling, structural substitutions or spectroscopic measurements (NMR, EPR). Thus, k_{11} and k_{22} (self-exchange rate constants) are sometimes available by direct measurement. In other systems they are obtained by means of a Marcus analysis.

The importance of the empirical Marcus' equation is that it connects thermodynamic and kinetic aspects (K and k). If K increases, the rate constant k also increases, at least in the normal region. Consequently the more thermodynamically favorable reactions are also the faster ones. The Marcus–Hush theory has proved extremely valuable in correlating, rationalizing, explaining and predicting many inorganic systems.

Semi-classical and Quantum Theories In parallel with the development of the classical electron transfer theories, quantum-mechanical descriptions were developed [9]. They include a quantum mechanical treatment of the vibrational modes. Furthermore, the intersection of the potential energy surfaces is de-emphasized and, instead, nuclear tunneling from the initial to the final state is emphasized. At high temperature, the occupation of the upper vibrational levels is sufficient to ensure that most of the reaction proceeds classically. The electron transfer occurs, thus, from the lowest vibrational level of the initial state to the jth vibrational level of the final state: i.e. only $0 \rightarrow j$ vibronic transitions are considered. However, at low temperature the nuclear tunneling contribution becomes increasingly important because of the depopulation of the upper vibrational levels.

The quantum effect is made clearer by inserting the harmonic oscillator vibrational energy levels within the potential curves, as shown schematically in Figure 12.15.

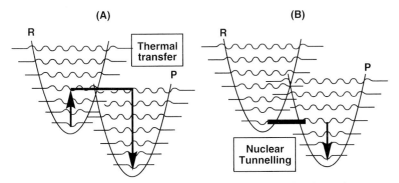

Figure 12.15 Quantum treatment of the energy surfaces: thermal and tunneling processes.

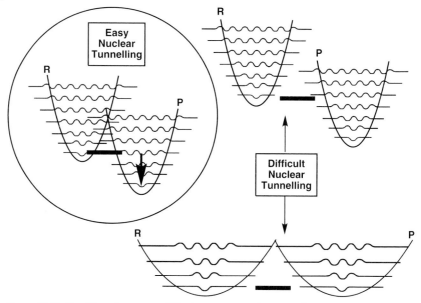

Figure 12.16 Conditions for easy or difficult quantum nuclear tunneling

As well as the thermal transfer, tunneling occurs optimally between the ground level in the reactant potential curve and one vibrational level in the product curve. Tunneling, thus, must be taken into account in describing the rate constant, though its importance depends on the extent of the vibrational overlap between the initial and final states. To achieve a great tunneling effect there are two main possibilities, concerning the nuclear position: (i) a small energy barrier and (ii) a sharp barrier. These features are shown in Figure 12.16.

Comparison of the Models   The difference between classical, semi-classical and quantum-mechanical treatment concerns *only nuclear movement*. In the classical treatment, Marcus et al. worked with the Newton, Hamilton and Lagrange formulations. In the semi-classical treatment, as indicated below, nuclear quantum tunneling is added to the prior treatment. Finally, in quantum treatment, the nuclear movement is treated through the Schrödinger nuclear equations, including the vibrational degrees of freedom, etc. Nuclear tunneling enters naturally into quantum-mechanical treatments. Such treatments are required to interpret low-temperature rates and/or reactions in the inverted region when high-frequency vibrational modes are involved. In the absence of such modes, classical reorganization expressions are generally adequate. The expressions derived from the classical model can be considered as the high-temperature limits of quantum mechanical expressions. At lower temperatures they must be corrected for quantum mechanical effects.

Table 12.3 Rate constants and activation parameters for some
self-exchange Co^{3+}/Co^{2+} complexes.

	$^{298}k_{11}$	$\Delta H^{\ddagger}$ $(kJ\,mol^{-1})$	$\Delta S^{\ddagger}(J\,K^{-1}\,mol^{-1})$	$\Delta V^{\ddagger}(cm^3\,mol^{-1})$
$[Co(phen)_3]^{3+/2+}$	12	21	−156	
$[Co(sep)]^{3+/2+}$	5	41	−95	−6
$[Co(N\text{-macrocycles})]^{3+/2+}$	10^{-2} to 10^{-5}			
$[Co(en)_3]^{3+/2+}$	7.7×10^{-5}	59	−132	−20
$[Co(tmen)_3]^{3+/2+}$	8.5×10^{-8}			

12.2.1.5 Some Examples of Outer-sphere Reactions

Self-exchange Reactions Significant data for Co^{3+}/Co^{2+} complexes in aqueous
solution are given in Table 12.3.

The most reliable bond distances (Co—N) indicate a change of 0.20–0.21 Å and
thus a considerable reorganization barrier. The change in the spin involved in this
electron transfer contributes significantly to the overall activation barrier, but is
probably rather less than the reorganization energy associated with change in the
cobalt–nitrogen distance.

The substitution lability of cobalt(II) complexes causes problems in obtaining
reliable self-exchange rate constants. The best way to circumvent this problem is
to encapsulate the cobalt atom, so as to make its escape from the surrounding
donor atoms very difficult. Kinetic parameters have been determined for a number
of self-exchange reactions of encapsulated cobalt complexes (sep, for example) (see
Chapter 2). Rate constants range from 2.4×10^{-2} to $2.8 \times 10^{4}\,dm^3\,mol^{-1}\,s^{-1}$.

In nitrogen macrocycles, there are a great variety of values: triazamacrocyclic
ligands permit faster electron exchange than tetrazamacrocycles. Substitution of
hydrogen by four methyl groups in each ethylenediamine ligand reduces the rate
constant by 10^3. This marked decrease can be attributed to steric factors: the oxidized and reduced forms have to be linked up carefully, keeping various methyl
groups out of each other's way, in order for the cobalt centers to be in reasonable
proximity.

Non-symmetrical ('True' Redox) Reactions Selected values of k_{et} for this type of
reaction have already been given in Table 12.1. The internuclear distance is important in relation to k_{et} and can allow some speculative data. This effect has
been fully studied in the electron transfer process between $[Co(NH_3)_5B]^{3+}$ and
$[Fe(CN)_5A]^{3-}$ (A and B pyridine derivatives, Table 12.1, charge products −9). The
shorter metal–metal distance determined from kinetic measurements supports an
outer sphere mechanism, with the metal centers as close to each other as possible,
suggesting that the transfer pathway could involve either solvent molecules or

Table 12.4 Comparison of calculated and observed rate constants for some outer-sphere cross-reactions [11].

Reaction	Observed	Calculated
$[IrCl_6]^{2-} + [W(CN)_8]^{4-}$	6.1×10^7	6.1×10^7
$[IrCl_6]^{2-} + [Fe(CN)_6]^{4-}$	3.8×10^5	7.0×10^5
$[Mo(CN)_8]^{3-} + [W(CN)_8]^{4-}$	5.0×10^6	4.8×10^6
$[Ru(NH_3)_6]^{3+} + [V(H_2O)_6]^{2+}$	1.5×10^3	4.2×10^3
$[Fe(H_2O)_6]^{3+} + [Ru(en)_3]^{2+}$	8.4×10^4	4.2×10^5
$[Mn(H_2O)_6]^{3+} + [Fe(H_2O)_6]^{2+}$	1.5×10^4	3.0×10^4

direct metal–metal overlap, rather than the ligand-mediated pathway (*inner-sphere*) [10].

Finally, Table 12.4 shows the utility of the Marcus' approach for modelling outer-sphere electron transfer cross-reactions.

12.2.2
The Inner-sphere Mechanism [2]

The formalisms explained above are exclusively for outer-sphere systems. However, there have also been attempts to apply them to inner-sphere reactions, provided that the coupling of the donor–acceptor sites is not too large. These attempts have been unsuccessful. In principle, solvent reorganization and electronic coupling in inner-sphere systems should be less difficult to treat theoretically because the distance separating the donor and acceptor sites and the relative orientation of the two sites can be much better defined than for outer-sphere systems. *In practice*, however, this is not true. Only in mixed-valence complexes (see Chapter 13) is the inner-sphere pathway perfectly known. Furthermore, it differs from outer-sphere transfer in an important aspect, namely, that the electronic coupling of the acceptor and donor sites in these bridged systems can become very large. In these cases, a completely new treatment is necessary (see Chapter 13).

In the inner-sphere mechanism, bonds break and form at the same time as electron transfer. The formation of an intermediate is required: this intermediate is a bridging complex made by one of the ligands of the complex. Once this precursor is formed, the electron transfer follows and, finally, the breaking of the bridge occurs. Any one of these steps can be the determinant process of the reaction. The bridging ligand is retained in the most inert complex.

12.2.2.1   The Experiments of Taube
The first time that an inner-sphere mechanism was confirmed was in the reduction of the $[Co^{III}Cl(NH_3)_5]^{2+}$ complex by the action of $[Cr^{II}(H_2O)_6]^{2+}$:

$$[CoCl(NH_3)_5]^{2+} + [Cr(H_2O)_6]^{2+} + 5H_3O^+ \rightarrow [Co(H_2O)_6]^{2+} + [CrCl(H_2O)_5]^{2+} + 5NH_4^+$$

All evidence indicates that the chloride has been transferred from the coordination sphere of the first complex to the second one. The redox reaction has a half-life of less than 1 ms under normal stopped-flow operating conditions. However, aquation of the $[CoCl(NH_3)_5]^{2+}$ cation has a half-life for the loss of chloride ion from the cobalt of more than four days. Therefore, virtually every cobalt atom enters the redox transition state with its chloride still attached. In a complementary way, the half-life for the formation of $[CrCl(H_2O)_5]^{2+}$ from $[Cr(H_2O)_6]^{3+}$ and Cl^- is a number of days, whereas the characterization of the $[CrCl(H_2O)_5]^{2+}$ product was achieved in a matter of minutes. Thus it follows that the chloride must have been bound to the chromium before the redox transition state separated into its product components and must have been bound to the cobalt as the transition state was formed. In other words, the chloride must be bound to both metals in the electron transfer transition state.

The efficiency of inner-sphere electron transfer may be illustrated by comparing the rate constant for $[Cr^{II}(H_2O)_6]^{2+}$ reduction of $[Co^{III}Cl(NH_3)_5]^{2+}$, $k = 6 \times 10^5 \, mol^{-1} \, s^{-1}$, with that for $[Cr^{II}(H_2O)_6]^{2+}$ reduction of $[Co^{III}(NH_3)_6]^{3+}$, where there is no possibility of bridge formation and k is only $9 \times 10^{-5} \, mol^{-1} \, s^{-1}$.

12.2.2.2 Mechanism

Any inner-sphere reaction has to be considered as a three-step mechanism:

1. formation of the precursor bridging complex

$$ML_6 + XM'L_5' \rightarrow L_5M-X-M'L_5' + L$$

2. electron transfer within the precursor

$$L_5M^*-X-M'L_5' \rightarrow L_5M-X-M^{*'}L_5' \ (\text{*oxidized})$$

3. decomposition of the bridged complex giving the products of the reaction.

The rate-determining step of the overall redox reaction can be any one of these three steps though the one currently considered most likely is step 2. Like outer-sphere reactions, inner-sphere reactions most commonly – but not always – conform to a simple second-order rate.

The rates of these inner-sphere reactions vary within a large interval (several powers of ten) and are strongly affected by the nature of the metals, the bridging ligand and the distance of the metal centers. Finally, rate constants for electron transfer may be strongly influenced by the nature of the other, non-bridging, ligands.

Inner-sphere reactions show considerable complexity and the development of a theoretical framework for discussion of mechanistic details has been much slower than for outer-sphere reactions. On the positive side, there is a detailed picture of the transition state for the reaction, which is not available in outer-sphere reactions.

12.2.2.3 Criteria and Factors Influencing the Inner-sphere Mechanism

The most essential requirement for inner-sphere electron transfer is a *suitable bridging ligand*. The ligands that can easily act as bridges are those with electrons which can be shared by two metallic centers. Thus, in $[Co(NH_3)_5L]^{3+}$, L = pyridine is not a potential bridging ligand, but ligands such as 4-cyanopyridine, pyrazine, piperazine and 4,4′-bipyridyl are effective bridging ligands for inner-sphere electron transfer. The role of the bridging ligand is dual: to bring up the metallic ions (*thermodynamic contribution*) and to mediate in their electronic transfer (*kinetic consideration*).

Electronic Structure of the Metal Centers A major factor influencing k_{et} is the nature of the two metal centers. Significant comparisons can be made, for example between the '*precursor*' binuclear cations $[(NH_3)_5M^{III}–pyrazine–Ru^{II}(NH_3)_5]^{5+}$ with M = Co or Ru, where $k_{et} = 0.055\,s^{-1}$ and $2.7 \times 10^{10}\,s^{-1}$, respectively. Thus, the forming of a bridge between the two ions is necessary, but not sufficient. It is also necessary to consider the symmetry of the orbitals of the reductant and the oxidant. This effect is very marked when the two orbitals are of σ^* type such as in Ru^{III}/Ru^{II} ions (just the opposite to what happens in the outer-sphere mechanisms) (Figure 12.17). This indicates that the electron transfer is effected through the bridging ligand, with its orbital directed to the σ^* orbitals of the metals.

In general, the capacity to carry out electronic transfer can be assigned to the symmetry of the orbitals of the starting reactants, the bridge and the final products.

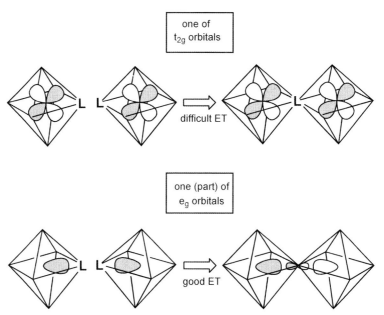

Figure 12.17 Relationship between inner-sphere electron transfer and symmetry of the atomic orbitals.

They must belong to the same symmetry species. If the reducing complex yields an e_g electron to an e_g orbital (the oxidant), then a ligand bridge strongly σ (Cl⁻) is better than a bridge with π orbitals, such as acetato or azido. However, t_{2g} electrons are better adapted to interact with acetato and azido bridges, owing to the good overlap between the π systems of these ligands. In inner-sphere reactions, the overlap of the metal-centered orbitals involved in the electron transfer is modified by the orbitals of the bridge, and the extent of this interaction will depend on the nature of the reaction partner. Where there is a mismatch in symmetry, the inner-sphere pathway is put at a disadvantage and outer-sphere pathways become important.

Thus, when the symmetry is favorable, the electron is transferred to the ligand to give an organic radical that, in the following step, transfers its electron to the second ion. However, most frequently, the bridge is only a mere electron transmitter, without the existence of the radical.

Other Factors   In general, the effects of bridging ligands are very complicated, and important questions are still unanswered at present. The possibility exists of facile electron transfer from a remote site through a conjugated ligand. It should be noted that, in some cases, electron transfer still takes place reasonably rapidly over a separation of more than 20 Å.

It is clear that pyrazine, for example, is a particularly good ligand for mediating electron transfer, that 4,4′-bpy is much less effective, and that 1,2-bis(4-pyridyl)ethene is less effective still. Although full delocalization across the bridge is maintained in this series, the actual distance to be traversed by the electron increases steadily. There is a linear correlation between $\log k_{et}$ and the reciprocal of the distances between the two metal ions.

Several experiments demonstrate that putting a rigid non-conducting barrier into a bridge causes a large decrease in k_{et}. The effect of inserting saturated blocks, such as —CH₂— or —CH₂CH₂—, between the two pyridyl rings in a 4,4′-bipyridyl bridge is important. In some cases, the rather small slow-down produced by saturated linkages may indicate that there are contributions from 'through-space' electron transfer. For ligands that are rigid, electron transfer presumably occurs through the ligand. However, for more flexible ligands electron transfer is probably 'outer-sphere', as the flexibility of the hydrocarbon chain that connects the two pyridine rings allows for close approach between the metal centers. The importance of the ability of flexible ligands to 'circle round' is clearly shown by the values of k_{et} for electron transfer from ruthenium(II) to cobalt(III) in the series of cations indicated in Figure 12.18, with $n = 0$–4. From $n = 0$ to $n = 2$, k_{et} decreases and the ruthenium–cobalt distance increases, but as n increases further to 3 and 4, k_{et} starts to increase again as direct outer-sphere electron transfer begins to make an important contribution [12].

12.2.2.4   Formation of Precursor Complexes
Definitive evidence for an inner-sphere transfer is provided by the detection of an 'unstable' bridged precursor and/or successor complex. In the Taube reaction, as

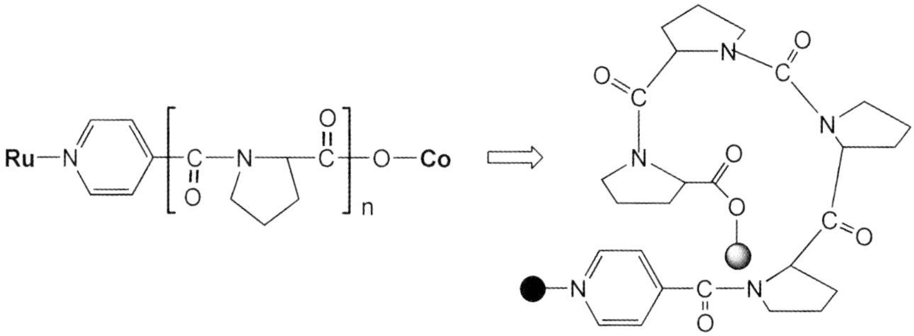

Figure 12.18 From inner-sphere to outer-sphere mechanism according to the flexibility of the bridge.

indeed in the great majority of inner-sphere redox reactions, the formation of the association and precursor complexes is rapid, and usually reversible, preceding rate-limiting ET. The first kinetic demonstration of the intermediacy of a precursor complex, with a lifetime of about 10 s, was in the reduction of $[Co^{III}(NH_3)_5(nta)]$ (nta = nitrilotriacetate, $N(CH_2COO)_3^{3-}$) by $[Fe(H_2O)_6]^{2+}$: nitrilotriacetic acid forms a cobalt(III) complex in which tertiary nitrogen and two carboxyl groups of the nta remain free to coordinate a second metal ion (Figure 12.19) [13].

The reduction of $[Co(NC\text{-}acac)(NH_3)_5]^{2+}$ (NC-acac = 3-cyano-2,4-pentadionate) by $[Cr(H_2O)_6]^{2+}$ is another important example where the precursor, the binuclear intermediate $[(NH_3)_5Co^{III}\text{-}NCacac\text{-}Cr^{II}(H_2O)_4]^{4+}$, has been detected. [14].

12.2.2.5 An Organic Radical?

The inner-sphere mechanism involves linking the two metal centers through the atoms and orbitals of the bridging ligands. The electron, therefore, traverses the bridge in the course of the net redox process. If it remains on the ligand for an appropriate time, then there is an intermediate in which the metal centers are transiently bridged by a radical ligand. This is sometimes called the *chemical, radical or two step mechanism*. However, in most inner-sphere reactions, electron transfer through the bridging ligand is too fast for such an intermediate to be chemically meaningful.

The first kinetic indications came from $[Cr(H_2O)_6]^{2+}$ reductions of similar carboxylato complexes, $[Co(NH_3)_5(O_2CR)]^{2+}$. Second-order rate constants spanned only a remarkably short range despite marked variation of the groups R. This kinetic evidence is seemingly incompatible with 'direct' electron transfer. *It will be noted that almost all the radical bridges reported involve cobalt(III) as the oxidizing center*. Transfer of the electron from the bridging radical to cobalt(III) is slow, permitting the detection of the intermediate, because the electron has to move from an orbital of π-symmetry of the bridging carboxylato ligand to the e_g orbital of σ-symmetry on the metal in order to reduce the $Co^{3+}(t_{2g}^6)$ to $Co^{2+}(t_{2g}^6e_g^1)$. In contrast, reduction of $Ru^{3+}(t_{2g}^5)$ to $Ru^{2+}(t_{2g}^6)$ is rapid since the t_{2g} orbital on the Ru^{3+} to which the electron passes has the same π-symmetry as that on the bridging ligand.

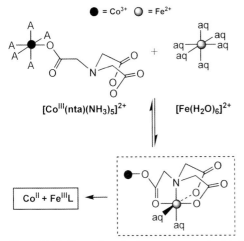

Figure 12.19 Formation of a "precursor complex" in the inner-sphere redox reaction shown in the figure (see text for explanation) (H$_3$nta is not necessarily fully deprononated).

Hence, no evidence has yet been obtained for the *chemical mechanism* in Ru^{3+}/Ru^{2+} systems.

A significant example is the reduction of [Co^{III}(NH$_3$)$_5$(pyrazine-2-carboxylate)]$^{2+}$ by [Cr(H$_2$O)$_6$]$^{2+}$. The reaction occurs via the formation of a radical-bridged species that has been characterized by rapid-flow EPR spectroscopy (a signal characteristic of an organic radical at g = 2.003 and hyperfine coupling to two non-equivalent nitrogen atoms) [15]. The intrinsic reducibility of the aromatic, heterocyclic carboxylic acid appears to be the driving force for the rapid formation of relatively stable radical species.

12.2.2.6 Post-electron Transfer Intermediates

Post-electron transfer intermediate complexes usually contain either one or two labile metal centers and thus dissociate at rates that are much faster than in the rate-limiting step discussed above. Final products are therefore normally obtained without detecting any intermediate post-electron transfer complex.

However, if both M^{m+1} and N^{n-1} are inert, then the binuclear post-electron transfer complex will be slow to dissociate to the final mononuclear products. Not only should the post-electron transfer complex be detectable, but it should be possible to measure k_{diss} and to demonstrate transfer of the bridging ligand. This case is most likely to arise from combination of d^6 or d^3 and d^6 centers. The very few good examples of post-electron transfer complexes include:

1. [Cr(H$_2$O)$_6$]$^{2+}$ reduction of Ru^{3+} or Ir^{4+} complexes (with Cr^{3+}, d^3; and Ru^{2+} or Ir^{3+}, d^6 post-electron transfer). For example, the reaction of the [Ir^{IV}Cl$_6$]$^{2-}$ ion with [Cr(H$_2$O)$_6$]$^{2+}$ was the original example (1954) of an inner-sphere electron transfer

with the identification of the post-electron intermediate [2]. The transient deep-blue color observed was attributed to the post-electron transfer complex, $[Cl_5Ir^{III}-Cl-Cr^{III}(H_2O)_5]$. This contains two substitution-inert centers (d^6, d^3) and so is reluctant to dissociate. The greater inertness of the third-row, d^6, iridium(III) ensures that the bridging chloride is *not only* transferred to the chromium.

2. Oxidation of Co^{2+} complexes by Fe^{3+} or Ru^{3+} complexes (d^6, d^6 post-electron transfer). For example, $[Fe^{III}(CN)_6]^{3-}$ and $[Co^{II}(CN)_5]^{3-}$ react to give the d^6–d^6 complex $[(NC)_5Fe^{II}-CN-Co^{III}(CN)_5]^{6-}$, which can be isolated as its potassium or barium salt and is robust enough to be oxidized to $[(NC)_5Fe^{III}-CN-Co^{III}(CN)_5]^{5-}$ [3].

12.2.3
Inner-sphere vs. Outer-sphere Reactions

Where there is incontrovertible evidence for simultaneous electron and ligand transfer, it is clear that the mechanism is inner-sphere. In cases when electron transfer is very much faster than the substitution required for inner-sphere electron transfer, the outer-sphere mechanism must be operative. Fortunately, the mechanism of a large number of electron transfer reactions involving transition metal complexes can be deduced with confidence from these simple criteria. However, in several cases, it has proved very difficult to establish the mechanism with any confidence, even by assembling evidence from a number of different mechanistic probes. For inner-sphere reactions, the relative orientation of the metal centers is firmly established and the effects of electronic and geometric factors can be ascertained. However, for outer-sphere reactions, because of the lability of the pairs, the relative orientation of cation and anion, when at least one of these does not possess octahedral symmetry, is unknown. Under these circumstances, indirect criteria must be invoked to obtain information about the detailed pathway for the electron transfer.

In the absence of direct evidence, the operation of an inner-sphere mechanism can sometimes be inferred from reactivity patterns. For example, an inner-sphere hydroxido-bridged pathway is indicated when the hydroxido complex is much more reactive than the aqua complex. Electron transfer reactions between couples of aqua complexes, frequently exhibit a rate law consisting of an acid-independent (considered to arise from an outer-sphere pathway) or an inverse acid term (commonly interpreted in terms of an inner-sphere hydroxido-bridged pathway).

In an inner-sphere pathway, a *'wrong-bonded'* isomer may be formed which may undergo spontaneous isomerization. *Inner-sphere reactions in which this occurs tend to proceed relatively slowly.* This is the case for SCN^- bridges. Such a relatively unfavorable change cannot occur when the closely related azido is the bridging ligand. Consequently, azido/thiocyanato rate comparisons serve as a diagnosis of the electron transfer mechanism. Comparison of rate constants for analogous azido and thiocyanato complexes has proved useful. Indeed, since azido is a more efficient bridging group the differences in inner-sphere electron transfer are

Table 12.5 The use of reactivity ratios for azido and thiocyanato complexes, in the form of complexes $[CoX(NH_3)_5]^{2+}$, in the assignment of an electron transfer mechanism [3].

Reductant	$k(X = N_3^-)/k(X = NCS^-)$	Mechanism
$[TiOH]^{2+}$ aq	400 000	inner-sphere
U^{3+} aq	40 000	inner-sphere
Cr^{2+} aq	20 000	inner-sphere
Fe^{2+} aq	3000	probably inner-sphere
Eu^{2+} aq	300	probably inner-sphere
V^{2+} aq	40	probably outer-sphere
$[Cr(bpy)]^{2+}$	4	outer-sphere
$[Ru(NH_3)_6]^{2+}$	1.5	outer-sphere

large. However, there is little difference for outer-sphere electron transfer (Table 12.5).

References

1 (a) Sutin, N. Nuclear, Electronic, and Frequency Factors in Electron transfer Reactions, *Acc. Chem. Res.* 1982, *15*, 275; (b) Sutin, N. Theory of Electron Transfer Reactions: Insights and Hindsights, *Prog. Inorg. Chem.* 1983, *30*, 441; (c) Marcus, R.A., Sutin, N. Electron Transfers in Chemistry and Biology, *Biochim. Biophys. Acta*, 1985, *811*, 265; (d) Sutin, N. Electron Transfer Reactions in Solution: A Historical Perspective, *Adv. Chem. Phys.* 1999, *106*, 7.

2 Haim, A. Mechanisms of Electron Transfer Reactions: The Bridged Activated Complex, *Prog. Inorg. Chem.*, 1983, *30*, 273.

3 Tobe, M. and Burgess, J. (1999) *Inorganic Reaction Mechanisms*, Longman, New York.

4 Gaswick, D., Haim, A. Direct Measurement of a First-Order Rate Constant for an Elementary Electron Transfer Step, *J. Am. Chem. Soc.* 1971, *93*, 7347.

5 Barbara, P.F., Meyer, T.J., Ratner, M.A. Contemporary Issues in Electron Transfer Research, *J. Phys. Chem.* 1996, *100*, 13148.

6 (a) Miller, J.R., Calcaterra, L.T., Closs, G.L. Intramolecular Long-Distance Electron Transfer in Radical Anions. The effect of Free Energy and Solvent on the Reaction Rates, *J. Am. Chem. Soc.* 1984, *106*, 3047; (b) Gould, I.R., Farid, S. Dynamics of Bimolecular Photoinduced Electron Transfer Reactions, *Acc. Chem. Res.* 1996, *29*, 522.

7 see for example, Katz, N.E., Mecklenburg, S.L., Graff, D.K., Chen, P., Meyer, T.J. Calculation of Electron Transfer Rate Constants in the Inverted Region from Absorption Spectra, *J. Phys. Chem.* 1994, *98*, 8959.

8 see for example, Mines, G.A., Bjerrum, M.J., Hill, M.G., Casimiro, D.R., Chang, I-J., Winkler, J.R., Gray, H.B. Rates of Heme Oxidation and Reduction in Ru(His33)cytochrome *c* at Very High

Driving Forces *J. Am. Chem. Soc.* 1996, *118*, 1961.

 9 see, for example, Kestner, N.R., Logan, J., Jortner, J. Thermal Electron Transfer in Polar Solvents, *J. Phys. Chem.*, 1974, *78*, 2148.

10 see, for example, Miralles, A.J., Armstrong, R.E., Haim, A. The Outer-Sphere Reduction of Pyridinepentaammi necobalt(III) and Pyridinepentaammineru thenium(III) by Hexacyanoferrate(II), *J. Am. Chem. Soc.* 1977, *99*, 1416.

11 Atwood,J.D. (1977) *Inorganic and Organometallic Reaction Mechanism*, 2nd edn., VCH Publishers, New York.

12 Isied, S.S., Vassilian, A. Electron Transfer Across Polypeptides. 3. Oligoproline Bridging Ligands, *J. Am. Chem. Soc.*, 1984, *106*, 1732.

13 (a) Cannon, R.D., Gardiner, J. A Binuclear Intermediate Preceding the Cobalt(III)-Iron(II) Electron Transfer Process, *J. Am. Chem. Soc.*, 1970, *92*, 3800; (b) Cannon, R.D., Gardiner, J. A Long-Lived Intermediate in a Cobalt(III)–Iron(III) Electron Transfer Reaction, *Inorg. Chem.*, 1974, *13*, 390.

14 Balahura, R.J., Johnston, A.J. Kinetics of the reaction of Chromium(II) with (3-cyano-2,4-pentanedionato-*N*)pentaammine cobalt(III): Intramolecular Electron Transfer between Cobalt(III) and Chromium(II), *Inorg. Chem.*, 1983, *22*, 3309.

15 Spiecker, H., Wieghardt, K. Kinetic and Electron Spin Resonance Spectroscopic Evidence for a Chemical Mechanism in the Chromium(II) Reduction of two (Pyrazinecarboxylato)amminecobalt(III) Complexes, *Inorg. Chem.* 1977, *16*, 1290.

Bibliography

Lappin, A.G. (1994) *Redox Mechanisms in Inorganic Complexes*, Ellis Horwood, Chichester.

Wilkins, R.G. (1991) *Kinetics and Mechanism of Reactions of Transition Metal Complexes*, 2nd edn., VCH, Weinheim.

Bolton, J.R., Mataga, N., McLendon G. (eds.) (1991) *Electron Transfer in Inorganic, Organic and Biological Systems*, Vol. 228 of *Advances in Chemistry Series*, American Chemical Society, Washington D.C. See especially the article of N. Sutin, *Nuclear and Electronic Factors in Electron Transfer: Distance Dependence of Electron Transfer Rates*, p. 25.

Brunschwig, B.S., Sutin, N. *Energy Surfaces, reorganization energies, and coupling elements in electron transfer*, Coord. Chem. Rev. 1999, *187*, 233.

13
Mixed-valence Compounds

13.1
Introduction

Chemists have long been interested in certain compounds, found mainly in geology and inorganic chemistry that are characterized by their strong colors and their electrical conductivity. It was rapidly seen that all these systems could be assembled together in a family, called *mixed-valence compounds*. Their typical features do not depend only on the nature of the elements but also on their capacity for having different oxidation states and exchanging their valence electrons. The three most studied features of these mixed-valence compounds were their colors, the deep changes in their electrical conductivity and their new magnetic properties.

13.2
Experimental Features

13.2.1
Color

Alfred Werner was the first to report the colors that appear when one chemical element has two different oxidation states in the same crystal. In 1896 he prepared a series of potassium bis(oxalato)platinate(II) which was a yellow-lemon color when it had $K_2[Pt(C_2O_4)_2]\cdot 2H_2O$ stoichiometry. It was oxidized quickly giving materials with metallic brightness. In 1918 Watson carried out the following experiment: when the mineral vivianite has the stoichiometric composition $Fe_3^{II}(PO_4)_2\cdot 8H_2O$ it is colorless; but most of the natural samples are pale-blue due to the partial oxidation of Fe^{2+} to Fe^{3+}. When the mineral is ground, it becomes darker because of the increased Fe^{3+}, as demonstrated by chemical analysis. Following the oxidation process, the mineral turned blue–green, then yellow–green and finally, yellow–lemon when all the iron ions had been oxidized to Fe^{3+}. However, when the sample was ground in a CO_2 atmosphere, no color changes were observed.

Coordination Chemistry. Joan Ribas Gispert
Copyright © 2008 WILEY-VCH Verlag GmbH & Co. KGaA, Weinheim
ISBN: 978-3-527-31802-5

The idea that light absorption could produce electron transfer phenomena between different atoms of the same element or different elements was developed in the following 25 years. The expression *"constitution color"* was used as early as 1915 by Hoffmann and Hoschele, to describe those solids whose color is not just the addition of their constituent colors. In 1922, Wells assumed that the atoms of a metal in two different oxidation states, instead of having fixed valences, could continuously exchange their electrons. These researchers stated that the most intense colors were due not only to the different oxidation states of the atoms but also to their similar/dissimilar geometrical environment.

13.2.2
Electrical Conductivity

On studying certain ionic solids, such as transition metal oxides, it was found that when their composition was other than stoichiometric, the electrical conductivity increased considerably. For example, NiO (green color, insulating) can be oxidized to give a product with only 0.5% oxygen excess. Though this quantity seems ridiculously small, the conductivity of NiO has increased by ten orders of magnitude. An analogous effect can be obtained by heating the stoichiometric NiO with Li_2CO_3 at 800 °C. Under these conditions, some Ni^{2+} ions are replaced by Li^+ ions in the "NaCl" net, and an equivalent number of Ni^{2+} ions are oxidized to Ni^{3+}. These type of systems are known as *"controlled-valence semiconducting"* solids.

13.2.3
Magnetic Properties

Mixed-valence in solids is the origin of unexpected magnetic effects. The best known case is that of the "magnet" characteristics of Fe_3O_4 (magnetite). Magnetite is a double-oxide, $FeO + Fe_2O_3$, with an *inverted spinel* structure in which some of the iron ions have the oxidation state 2+ and the rest 3+. Neutron diffraction has shown that there is a weak antiferromagnetic coupling between Fe^{2+} ions in a tetrahedral environment, but between Fe^{2+} and Fe^{3+} in octahedral environments a new ferromagnetic and dominant interaction is created. This behavior occurs because one of the valence electrons of the Fe^{2+} (d^6) does not remain motionless, but moves to another Fe^{3+} neighbor (d^5). This electron transfer occurs 10^8 times a second. Mössbauer measurements corroborate that electron transfer only takes place between the octahedral holes, as the electron transfer possible between the octahedral and tetrahedral holes is inefficient.

13.3
Definition, Stability and Electronic Delocalization

Mixed-valence compounds contain identical (or different) ions in two distinct oxidation states, which allow electron transfer. The archetypal example is the Creutz–

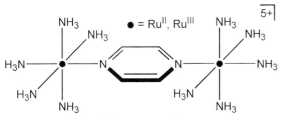

Figure 13.1 Scheme of the Creutz–Taube ion.

Taube ion, $[(NH_3)_5Ru^{II}\text{-pyrazine-}Ru^{III}(NH_3)_5]^{5+}$ with d^6–d^5 electronic configuration (Figure 13.1).

The concept is more general: a Fe^{2+}–Co^{3+} system (d^6–d^6) can stand in equilibrium with the Fe^{3+}–Co^{2+} system (d^5–d^7); a Re^+–Ru^{3+}, with d^6–d^5 electronic configuration, can stand in equilibrium with the Re^{2+}–Ru^{2+} system (d^5–d^6 configuration). Thus, the most important requirement is the capacity of the electrons to be at both sites. Mixed-valence complexes may exist with or without a bridging ligand. In thermodynamic terms, the mixed-valence complex must be stable for the two parent integer compounds. For example, in the Creutz–Taube ion:

$$[II, II] + [III, III] \Leftrightarrow 2[II, III]$$

The statistical factor of K_c (*comproportion*) in this reaction is 4. K_c varies from the statistical value of 4 for localized complexes to as high as 10^{24} with completely delocalized systems.

It is highly appropriate to consider the two ions in a certain oxidation state, but with an *additional electron* either trapped in one of them or delocalized over the two centers. All experimental data seem to indicate that the optical, electric and magnetic properties depend on the degree of delocalization. Thus, the extent of delocalization was the first measure by which mixed-valence compounds were classified: the *Robin and Day* classification [1]. In Class I complexes, interaction is very weak and the properties are those of the discrete, unperturbed metal centers. In these systems the extra-electron is not delocalized giving two perfectly isolated oxidation states. In Class III complexes, interaction is very strong and the metal centers are identical. The electron can jump from one ion to another much more quickly than any possible control by physical methods. In between these two extremes are Class II systems, in which the metal centers are *almost* identical and the unpaired electron is localized but weakly coupled.

Mixed-valence compounds are found not only in molecular systems but also in one-, two-, or three-dimensional nets. Tungsten bronzes, for example, are structurally derived from the ReO_3 structure. The starting product, WO_3, is colorless and insulating. However, Na^+ ions can be introduced into this neutral network. When the number of Na^+ ions increases, the solid becomes bronze-colored and conducting. This is because each Na^+ gives a supplementary electron to the WO_3 net, so reducing one W^{VI} to W^V, without great net distortion (the sodium ions are simply intercalated in the net). At the final stoichiometric limit we obtain the perovskite $NaWO_3$, in which all W ions are W^V and the color and conducting character are lost.

13.4
Electrons Jumping from One Center to Another

Electrical conductivity is not found in a solid that contains metallic ions in only one oxidation state. In this case, if the electron could "jump over" from one center to another, the energy necessary for the electron transfer would be very great. Fe_2O_3 is, for example, insulating. This situation can be drastically changed if we have Fe^{2+} and Fe^{3+} ions in the net: then the electron transfer needs very little energy and the new systems become conducting. This is the case of the mixed-oxide Fe_3O_4, already mentioned above. However, if we try to extrapolate this feature to other mixed-oxides, such as Co_3O_4 or Mn_3O_4, the conclusions would be wrong. These two mixed-valence systems show very low conductivity, clearly lower than for Fe_3O_4. The three oxides belong to the group called spinels, with general formula $(A^{II})(B^{III})_2O_4$. In this net there are two different cationic positions, tetrahedral and octahedral, which are occupied in a different manner according to the kind of spinel. In the *normal spinel*, A and B ions occupy the tetrahedral and octahedral holes, respectively. This is the case for Co_3O_4 and Mn_3O_4. However, Fe_3O_4 is an *inverted spinel*: A ions occupy octahedral holes, while B ions occupy half the octahedral holes and half the corresponding tetrahedral holes. The great electrical conductivity of Fe_3O_4 is simply due to the fact that the iron ions, *in different oxidation state (+2 and +3), have the same octahedral environment in the net*. This feature facilitates the electron transfer.

For a better understanding of this phenomenon we can reason as follows, assuming only two positions, A and B, with two possible configurations: A^+B^- and A^-B^+ (+ and − indicate greater or lower oxidation state, such as Fe^{2+} and Fe^{3+}). If the environment (symmetry) of A and B are different, the two configurations cannot mix. This is because valence electrons can be described as wavefunctions with singular symmetry properties. The zero-order ground-state wavefunction is simply $\phi_A^{II}\phi_B^{III}$, but an excited state $\phi_A^{III}\phi_B^{II}$, in which one electron has been exchanged into the two sites, can also exist. When the symmetry properties are the same these two functions can easily mix, and so attain a new ground state that is a linear combination of the stated wavefunctions

$$(1-\alpha^2)^{1/2}(\phi_A^{II}\phi_B^{III}) + \alpha(\phi_A^{III}\phi_B^{II})$$

The mixing coefficient (α) is a measure of the degree of similarity of the A and B environments. When the symmetry of A and B sites is very different ($\alpha = 0$), the possibility of function-mixing is almost zero and the electron remains "trapped" in one ion (Class I). In Class II the mixing coefficient is small but not zero. The environment of the two ions is similar but not identical, and the two entities are distinguishable. Finally, Class III systems have the maximum value of ($\alpha = 1/\sqrt{2}$) and the positions of A and B are indistinguishable: the electron will be mobile within the two completely equal sites.

13.5
The Robin–Day Classification

Robin and Day were the pioneers in differentiating and classifying mixed-valence compounds, on the basis of the relationship between electronic delocalization and similarity of cationic sites [1]. The properties and main features of this classification are given in Table 13.1.

Some significant examples of complexes according to the Robin–Day classification are shown in Table 13.2.

13.6
Theory of Mixed-valence Compounds

13.6.1
Introduction

Following the general treatment of the electron transfer mechanism (Chapter 12), the theoretical study of mixed-valence systems started around 1967–1970, with the

Table 13.1 Characteristics of the Robin–Day classification.

	Optical properties	*Electrical properties*	*Magnetic properties*
Class I (Trapped electrons in ions of different symmetry)	No intervalence transfer band in the electronic spectrum. Color due to the isolated ions	Insulating	Properties of each ion
Class II (Ions of almost identical symmetry)	Intervalence transfer band in the visible or near/medium infrared. Deeper color than in isolated ions	Semiconducting (for a 1D, 2D or 3D solid)	Magnetically diluted. F or AF at low temperature
Class III (Delocalized electrons)			
Clusters (equivalent and indistinguishable ions)	Intervalence transfer band in the visible or near/medium infrared. Strong color.	Insulating	Magnetically diluted
Infinite Net (equivalent and indistinguishable ions)	Absorption threshold close to the IR. Opaque. Dark color. Metallic brightness	Metallic conductivity	Possible long-range ferromagnetic order at high T_C

Table 13.2 Examples of mixed-valence systems of different classes.

Class	Examples	Geometry of A	Oxid. State	Geometry of B	Oxid. State
I	$[Cu(en)_2][CuBr_2]_2$	D_{4h}	2	linear	1
	$[Co(NH_3)_6]_2[CoCl_4]_3$	O_h	3	T_d	2
	$Ga[GaCl_4]$	dodecahedral	1	T_d	3
II	$[M_3O(carboxilato)_6L_3](M = Mn,$ Fe, Ru)	O_h	2	O_h	3
	$Fe_4[Fe(CN)_6]_3 \cdot 4H_2O$	O_h	2	O_h	3
	$[Pt(etn)_4][PtCl_2(etn)_4]Cl_4$	D_{4h}	2	O_h	4
	$(NH_4)_2[SbBr_6]$	O_h	3	O_h	5
III-A	$[Nb_6Cl_{12}]Cl_2$	O_h	2.33		
(clusters)	$[Fe_4S_4(SCH_2Ph)_4]^{2-}$	T_d	2.5		
	$[(NH_3)_5Os{-}N_2{-}Os(NH_3)_5]^{5+}$	O_h	2.5		
	$[Re_2Cl_8]^{3-}$	C_{4v}	2.5		
	Creutz–Taube ion	O_h	2.5		
III-B	$K_{1.75}[Pt(CN)_4] \cdot 1.5H_2O$	D_{4h} (1D)	2.25		
(1, 2, 3D)	$K_2[Pt(CN)_4]Br_{0.30} \cdot 3H_2O$	D_{4h} (1D)	2.30		
	Na_xTiO_2 $(0 < x < 1)$	O_h (3D)	(3–4)		
	Na_xWO_3 $(0.4 < x < 0.9)$	O_h (3D)	(5–6)		

research of Robin–Day and Allen–Hush [1, 2]. The theory is, in fact, the same as that reported for the general outer-sphere mechanism. It was stated in Chapter 12 that for the inner-sphere mechanism this theory does not work. However, for mixed-valence complexes, it does. The bridging ligand is not always necessary in mixed-valence systems. Indeed, we will study later in this chapter some complexes with metal–metal bonds, giving very strong delocalization.

A new electronic band, called intervalence transfer (IT), metal–metal charge transfer (MMCT) or intervalence charge transfer (IVCT), which cannot be attributed to each ion separately, appears and must be related to the electron transfer. Many theoretical studies have tried to relate the energy, shape and amplitude of this new IT to the degree of delocalization. Once again, these theories can be divided into *classical* or *quantum*. All theories focus on the factors developed in the previous chapter. The study of the potential energy of the surfaces (parabolas) can give more or less importance to electron–phonon coupling (vibronic coupling) and to the tunneling effect. When these two factors are important, we are dealing with quantum theories.

All models are derived from very simple systems, generally dinuclear complexes with simple bridges, such as Cl^-, O^{2-}, CN^-, pyrazine, etc. Let us assume a general case: a dinuclear complex AB, in which the electron (schematized as *) can move freely from A to B and vice versa.

$$A * {-}B \leftrightarrow A - B *$$

As indicated in the previous chapter, there is always a certain barrier for electron transfer or activation energy ($\Delta G^{\ddagger}$) due to changes in the coordination sphere of A and B and their solvation effects. This feature can be correlated with the "out-of-phase" (Q_A–Q_B) vibration of the molecule considered as a whole. We will call this new antisymmetric vibration simply Q. Q will again be the abscissa in the energy curves.

As the electron transfer is much more rapid than nuclear rearrangement, if we assume that the electron jumps from one ruthenium to another ca. 10^{16} times each second in the Creutz–Taube ion (Figure 13.1), then the [2+, 3+] ground "starting" configuration is converted to an excited configuration [3+, 2+], in which the 3+ ions have the geometrical environment corresponding to 2+ ions and vice versa. As the differences in Ru—N are not excessive (Ru^{II}—N = 2.14 Å and Ru^{III}—N = 2.10 Å), a vibrational relaxation will be enough to return to the ground state [2+, 3+]. This vibrational relaxation will be the vibrational mode Q (Q_A–Q_B).

13.6.2
Classical Theory

This theory was developed by Hush and Sutin [2–4] who examined the vibrational modes and the need for electronic coupling (H_{AB}) that occurs by mixing metal-based donor and acceptor orbitals with orbitals of appropriate symmetry in the bridge. As indicated in the previous chapter, $H_{AB} = <\psi_A|\hat{H}|\psi_B>$. The free energy of the two states is represented by two parabolas that depend on Q, λ and H_{AB} (Figure 13.2) (see previous Chapter).

Hush provided an analysis of IT band shapes based on parameters that define the electron transfer barrier and the orbital mixing of the starting and final states. Localization or delocalization depends on λ and H_{AB}. The greater λ, the larger the localization becomes. The relationship between the intervalence band and λ is shown in Figure 13.3. When $\lambda = 0$, the two curves coincide and the intervalence band does not appear. When λ increases, the energy of the IT band increases and often changes from the near-infrared to the visible.

If $H_{AB} = 0$, we are dealing with Class I; if $H_{AB} \neq 0$, Class II is feasible; if H_{AB} is very large, Class III is predominant. In this last case, the interaction is very strong

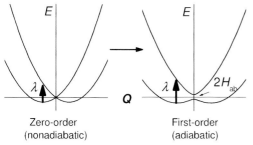

Zero-order
(nonadiabatic)

First-order
(adiabatic)

Figure 13.2 Intervalence band in a mixed-valence complex (see text for the meaning of λ, Q and H_{ab}).

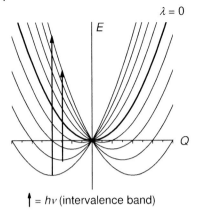

$$\uparrow = h\nu \text{ (intervalence band)}$$

Figure 13.3 Graphical relationship between the intervalence band (IT) and the λ parameter.

and *other theories are required*. Indeed, for weakly-coupled Class II systems, $2H_{AB} \ll \lambda$. If $2H_{AB} \gg \lambda$, the thermal barrier to intramolecular electron transfer vanishes and the ground state adiabatic surface exhibits a single minimum (see below).

For class II a number of important relationships between optical (specifically their energy (E_{IT}, E_{op}), intensity (ε) and bandwidth at half-height ($\Delta \bar{\nu}_{1/2}$)), and thermal electron transfer can be derived from the Hush theory [3, 4]. The most important are shown below, with $\bar{\nu}$ in cm^{-1} (see Chapter 12).

$$E_{IT} = \bar{\nu}_{max} = \lambda + \Delta G^0 \qquad\qquad E_{IT} = 4\Delta G^\ddagger$$

$$\Delta G^+ = (\bar{\nu}_{max})^2 / [4(\bar{\nu}_{max} - \Delta G^0)] \quad \alpha^2 = [H_{AB}^2 / \bar{\nu}_{max}^2]$$

$$\Delta \bar{\nu}_{1/2} \approx [2130 \times E_{IT}]^{1/2} \qquad H_{AB} = (1/r)(2.06 \times 10^{-2})(\varepsilon_{max} \bar{\nu}_{max} \Delta \bar{\nu}_{1/2})^{1/2}$$

with α being what is called the delocalization parameter. The upper limit for Class II mixed-valence compounds is $\alpha = 0.25$. Less than 0.25 indicates that the metal centers can be considered as essentially trapped.

13.6.3
Quantum Theory: the Piepho–Krausz–Schatz (PKS) Model

The quantum theory developed by Piepho, Krausz and Schatz (known as the PKS theory) [5, 6] uses equations that also depend on H_{AB}. In this case, these equations describe electron transfer as occurring through a series of vibrational channels from initial vibrational level j to final level j'. The electron transfer occurs from the lowest vibrational level of the initial state to the jth vibrational level of the final state; that is, only $0 \rightarrow j_{vibronic}$ transitions are considered (Figure 13.4).

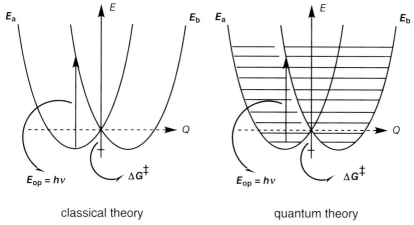

classical theory   quantum theory

Figure 13.4 Schematic representation of the energy surfaces for a classical and quantum model of mixed-valence complexes.

In this treatment, solvent reorganization is minimized because, in fact, vibronic coupling can provide a dissipative channel for electron transfer in the complete absence of the solvent. When the solvent is introduced it is usually treated classically.

> The main principle of the PKS quantum model consists of determining the energy of all vibrational states of the complex and then calculating the intensities of the transitions allowed between the vibrational levels of the ground state and the vibrational levels of the excited state. Thus, the *quantum theory of the vibrational modes* is the main factor in this theory (unlike in the classical ones).

For a symmetric case, with certain simplifications and similar to the classical theories

$$\begin{cases} E_a = \lambda Q + 1/2\,kQ^2 \\ E_b = -\lambda Q + 1/2\,kQ^2 \end{cases} \tag{13.1}$$

with λ, k and Q having the usual meaning (Chapter 12).

The main difference between this quantum theory and the classical theory is that now we have to take the quantum vibrational states into consideration (Figure 13.4). Thus, the zero-point energy of the curves is not necessarily the minimum of the curves (classical theories), but the ground vibrational mode ($n = 0$) that can be more or less separate from the mathematical minimum. *This gap may be important in large coordination compounds in which there are a lot of normal vibrational modes (3N − 6).*

The IT band (E_{op}) corresponds, thus, to the vertical transition between the ground vibrational state ($n = 0$) of the left parabola to one of the excited vibrational states of the right parabola (Figure 13.4). As in classical theories, this hypothesis is not realistic unless the interaction between the two ions is included. The parameter called the resonance integral β corresponds to H_{AB} in classical theory. The new Hamiltonian operator is then

$$E = 1/2\,kQ^2 \pm [\beta^2 + (\lambda Q)^2]^{1/2} \tag{13.2}$$

In summary, the properties of the mixed-valence compounds in the PKS model can be interpreted by means of two parameters: β (the measurement of the interaction between the mononuclear centers) and λ, the measurement of the vibronic coupling, which is directly related to the difference between the equilibrium values of the totally symmetric normal modes, a_{1g}, of the two oxidation states.

Some drawings of the corresponding energies in the space Q for different β and λ values are shown in Figure 13.5. If β and λ are zero, the potential surfaces of the two parabolas coincide (Figure 13.3). If $\lambda \neq 0$ but $\beta = 0$, the two potential surfaces become horizontally separated (Figures 13.3 and 13.5a). The two curves (E_- and E_+) are symmetric in the Q space. E_- corresponds to the ground state and E_+ to the excited state. The two Q values for which the energy has a minimum correspond to $dE/dQ = 0$. These values are $Q^2 = \lambda^2/k^2 - \beta^2/\lambda^2$. As a consequence, if $|\beta| < \lambda^2/k$ there will be two energy minima (one for each potential surface). The electron can jump from one curve to another, overcoming the energy barrier, which depends on β. When the electronic coupling increases, the potential surfaces are modified; and when $|\beta| = \lambda^2/k$, the two minima disappear. Then $Q = 0$ and there is only one minimum instead of two. Finally, if β is much greater than λ (mathematically, it is like assuming $\lambda = 0$, for any value of β) the two surfaces are identical only with a vertical gap of energy $2|\beta|$. *This value corresponds, then, to the intervalence band in Class III systems.*

The curves (a)–(d) of Figure 13.5 represent mixed-valence complexes with more or less trapped electrons, whilst the last two curves (e) and (f) represent fully delocalized mixed-valence systems. In all these curves, to simplify the figures, the vibrational levels are not drawn. The intervalence band is related to $\Delta G^{\ddagger}$ ($E_{op} = 4\Delta G^{\ddagger}$) in Class II and to β ($E_{op} = 2\beta$) in Class III compounds. For asymmetric complexes a new parameter of asymmetry, W, has to be introduced because $\Delta G^0 \neq 0$

$$E = 1/2\,kQ^2 \pm [\beta^2 + (\lambda Q + W)^2]^{1/2} \tag{13.3}$$

The potential surfaces are similar to those corresponding to symmetric systems ($W = 0$) but are now asymmetric.

13.6.4
Classification of Mixed-valence Systems According to the Classical or PKS Theories

Class I, II and III mixed-valence complexes correspond to $|\beta| \ll (\lambda^2 + W)$; $|\beta| \leq (\lambda^2 + W)$; $|\beta| > (\lambda^2 + W)$. Class I systems have no electronic interaction (β)

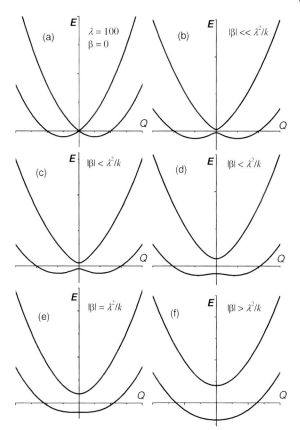

Figure 13.5 Schematic representation of the energy surfaces of a mixed-valence complex varying the β parameter from 0 to a very large value.

owing to the long M–M distances and the different environments of the two ions which cause total electronic localization. Class III systems have a strong interaction parameter (β) between the two metallic centers. The properties of the two isolated ions "disappear" and new properties, typical of the dinuclear complex, "appear". Class II compounds have intermediate properties: a certain degree of delocalization, but with oxidation states still distinguishable ("trapped valence"). Thus, their properties are a mixture of the isolated ions and the molecular entity. In Class II systems there is an energy barrier between the two minima, corresponding to distorted geometries. The electron is actually "trapped" in one minimum but there is a not-zero probability of thermal transfer from one center to another.

When going from a strongly localized system to a strongly delocalized system we are dealing with a different quantum basis (Φ_a, Φ_b) (valence bonding approach) going to ($\Phi_\pm$) (molecular orbital approach). $\Phi_\pm$ *are simply bonding or antibonding molecular orbitals.* In other words, the formation of the molecular orbital that contains the electrons of the two centers is the final cause of delocalization in mixed-valence complexes. This feature allows us to predict that the formation of a metal–metal bond will favor complete electronic delocalization, whereas the formation of metal–bridge(s)–metal bonds will favor electronic localization. For symmetrical Class III systems, while the terms "IT transition" and "mixed-valence" are retained, the transitions do not involve net charge transfer and the systems are more accurately defined as "averaged valence".

13.6.4.1 Characteristics of the Intervalence Bands

The characteristics of the intervalence band are one of the best ways of classifying mixed-valence systems:

For Class II (localized systems), the intervalence band is broad, symmetric, low-intensity, temperature dependent and *solvent dependent*.

For Class III (delocalized systems), the intervalence band is intense ($\varepsilon_{max} \gg 5000\,M^{-1}cm^{-1}$), narrow ($\Delta\bar{v}_{1/2} < 2000\,cm^{-1}$), asymmetric, temperature independent and *solvent independent*. This band is narrow because it may be associated with the *vertical separation* between the lower and upper surfaces. However, in reality, it is broadened by quantum and solvent effects.

For the intermediate Class II–III, the intervalence band is narrow, solvent independent but with localized oxidation states.

In general, if λ is dominant over β (Class II), when λ and/or W increase, the intervalence band is shifted to greater energies, its intensity decreases and its width increases. If β is the dominant factor (class III), when β increases, the energy of the intervalence band also increases (2β), its intensity increases and its width decreases.

13.6.4.2 Dynamics Considerations and the Class II–III Classification

The elucidation of the factors which govern the shift between localization and delocalization is complicated by the interplay between the timescales for intramolecular electron transfer and the coupled nuclear and solvent vibrations. Meyer and coworkers have addressed these issues, by defining mixed-valence systems in the "localized-to-delocalized" regime as Class II–III [7]. In Class II the solvent and exchanging electron are localized; in Class II–III the solvent modes are averaged (delocalized) but the exchanging electron is localized. The solvent, once averaged, no longer contributes to the dynamics of the barrier crossing. They are *valence-localized / solvent-decoupled*. In Class III, the exchanging electron is delocalized and the solvent and vibrational modes are averaged.

If thermal electron transfer between donor and acceptor metal ions is greater than the rate of solvent relaxation $10^{12}\,s^{-1}$, the mixed-valence properties of the complex will be solvent independent even though on the electronic time scale, $10^{15}\,s^{-1}$, the odd electron is trapped on one metal ion. Such systems, on the borderline of class II and III behavior, have properties normally associated with both localization and delocalization:

- The appearance of narrow, *solvent-independent* bands at low energy which can reasonably be assigned to an IT transition.

- Direct experimental evidence for localization from a crystal structure or the appearance of an oxidation state marker (see below).

- The appearance of a symmetric bridging ligand vibration such as $v(pz)$ or $v(N_2)$. The intensity of these bands should be related at some level to the extent of delocalization.

Class II–III answers many of the questions relating to the Creutz–Taube ion (see below).

13.7
Degree of Delocalization: Factors Favoring Localization or Delocalization

When we are dealing with M–L–M complexes, there are some general trends:

1. There is a correlation between the number of atoms in the ligand and the degree of interaction. The interaction decreases exponentially as the size of the bridge increases.

2. Strong interactions only take place with small ligands that have a marked π-donor (O^{2-}) or π-acceptor (N_2, pyrazine) character.

3. Saturated bonds in the ligand diminish the coupling, whereas conjugated bridges favor the coupling.

Research into mixed-valence systems reveals that cation symmetry and packing effects (or similar effects due to the ligands) are, in addition to the bridging ligand, important factors in controlling mixed-valence states. Effectively, the lowest-energy electronic states are vibronic and, as a result, the complexes are very sensitive to their environment. Let us study some examples of these two effects.

13.7.1
Influence of the Symmetry / Asymmetry Created by Ligands

13.7.1.1   Cu^{II}–Cu^{I} Complexes with Binucleating Macrocyclic Ligands
One of the first cases of the Cu^{II}–Cu^{I} type was studied by Hendrickson and coworkers in 1983 by means of EPR measurements (Figure 13.6) [8]. Complexes I–IV are symmetric, whereas the other three are asymmetric. Although the bridge that links the two copper ions is the same in all cases, the localization–delocalization of the mixed-valence electron is very different. EPR spectra and their temperature variation are very significant: the solution spectra of symmetric complexes (I, II, III) at room temperature show an isotropic band with seven hyperfine signals (Figure 13.7A), which indicate delocalization over the two copper ions, in the time scale of the technique. However, asymmetric complexes (except V) show hyperfine splitting with only four signals, typical of delocalization over one copper ion (Figure 13.7B). When the temperature is decreased, symmetric complexes start to coalesce from seven to four bands close to 230 K. Logically, asymmetric complexes continue

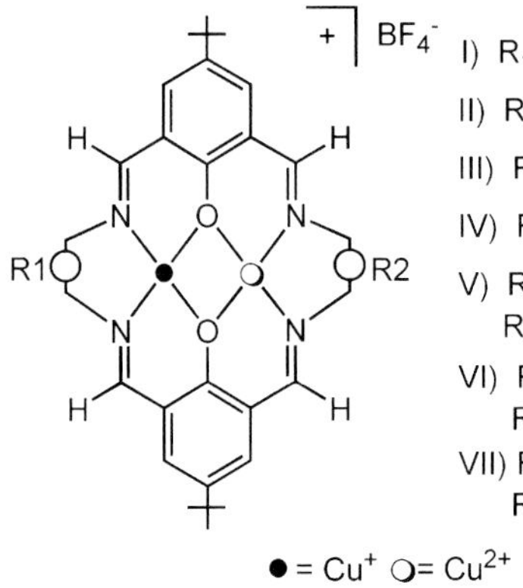

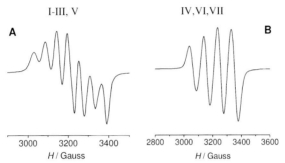

I) $R_1 = R_2$ = propylene

II) $R_1 = R_2$ = 2,2'-dimethylpropylene

III) $R_1 = R_2$ = butylene

IV) $R_1 = R_2$ = 2,2'-biphenylene

V) R_1 = propylene
  R_2 = 2,2'-dimethylpropylene

VI) R_1 = propylene
  R_2 = 2,2'-biphenylene

VII) R_1 = propylene
  R_2 = butylene

● = Cu$^+$ ○ = Cu^{2+}

Figure 13.6 Schematic representation of a dinuclear copper(I,II) complex, varying R1 and R2 groups (see text for the explanation of their properties).

I-III, V   IV,VI,VII

A   B

3000 3200 3400   2800 3000 3200 3400 3600

H / Gauss H / Gauss

Figure 13.7 X-band EPR spectra of the complexes shown in Figure 13.6 (see text for explanation).

with the electron fully localized. In frozen solution, all complexes (I to VII) show identical electron-localized spectra.

It can be deduced from the EPR and electronic spectra that in this series of complexes, the β parameter scarcely varies: λ and W are the dominant parameters. With mixed ligands, the greater asymmetry favors the electrons being more trapped. The calculated $\Delta G^{\ddagger}$ vary from 15.1 kJ mol^{-1} for I to > 31.1 kJ mol^{-1} for VI and VII.

Similar features have also been reported for the Cu^{II}–Cu^{I} complex drawn in Figure 13.8A [9]. The solution EPR spectrum consists of seven lines, in accord with the interaction of the odd electron with the two copper centers ($I = 3/2$). In contrast, a frozen solution (in CH$_2$Cl$_2$ or CH$_3$CN) at liquid nitrogen temperature shows four-line anisotropic spectra consistent with localization of the odd electron on a single copper center. Solutions of this Cu^{II}–Cu^{I} complex show a very broad,

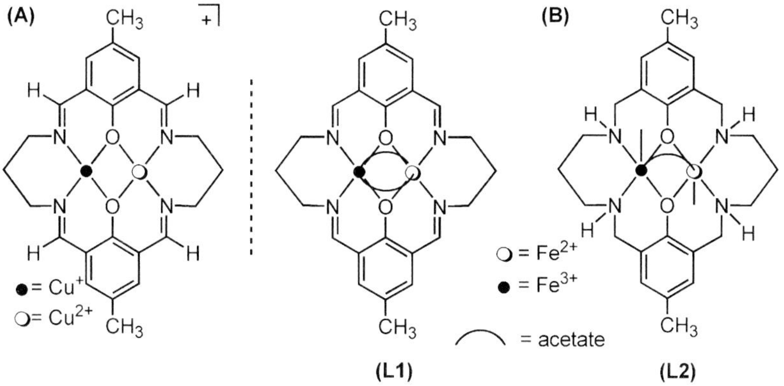

Figure 13.8 (A) A mixed-valence Cu(I,II) complex; (B) similar mixed-valence Fe(II,III) complexes (see text for explanation of their properties).

solvent-dependent band in the near-IR region: it appears at 1200 nm in non-coordinating CH_2Cl_2 and at 900–1000 nm in weakly coordinating solvents such as CH_3OH, acetone or DMF. No such band is apparent in CH_3CN, which has a strong affinity for Cu^+. The reaction with CO localizes the electron on the Cu^+ site at both high and low temperature. As a consequence, there is no band in the visible–near-IR region of the electronic spectrum attributable to an IT transition. Thus, this CO derivative behaves as a Class I species.

13.7.1.2 Fe^{II}–Fe^{III} Complexes with Binucleating Macrocyclic Ligands

The Fe^{II}–Fe^{III} complexes, $[Fe_2L^1(\mu\text{-}AcO)_2](ClO_4)$ and $[Fe_2L^2(\mu\text{-}AcO)(AcO)(H_2O)]$ (ClO_4), being L^1 and L^2 binucleating macrocycles ligands (Figure 13.8B) are paradigmatic cases [10]. L^1 keeps imino groups, while in L^2 these groups are reduced to amino groups. The main difference between the two complexes lies in their symmetry. The two Fe ions are hexacoordinated. With L^1 hexacoordination is achieved by two acetato bridges, giving a highly symmetric molecule, whereas with L^2 hexacoordination is reached by one acetato group and two monodentate ligands (acetato and aqua) so giving a much more asymmetric molecule. The symmetric complex shows an intense intervalence band at 1060 nm ($\varepsilon = 1250\,M^{-1}\,cm^{-1}$), which is solvent independent. The asymmetric complex shows an intervalence band between 1110 and 1160 nm, with lower intensity ($\varepsilon = 560$–$640\,M^{-1}\,cm^{-1}$) and is solvent dependent. Mössbauer spectra and their temperature variation also provide clarification: the symmetric complex gives at room temperature an isomeric shift (δ) and quadrupole splitting values (ΔE_Q) intermediate between the values for analogous Fe^{III}–Fe^{III} and Fe^{II}–Fe^{II} complexes with a similar environment. By lowering the temperature the bandwidth increases. The asymmetric complex shows, at 293 K, the signals for Fe^{II} and Fe^{III} ions. All these features indicate that the symmetric complex has great delocalization, whilst the asymmetric one has more trapped (localized) valence.

13.7.1.3 Cyanido-bridged Complexes

Cyanide is a very good bridge for giving Class II mixed-valence complexes, because of its ability to electronically mediate the transfer process and because the energy of the $\nu(CN)$ bridge depends on the metals to which it is attached, the nature of the linkage, and the oxidation state of the metals. For example, the binuclear $[(CN)_5Ru^{II}-CN-Ru^{III}(NH_3)_5]^-$ has a maximum at 675 nm in H_2O (dark blue solution, $\varepsilon = 2.84 \times 10^3\,M^{-1}\,cm^{-1}$) attributed to an intervalence transition [11a]. The Raman spectrum under near-resonance conditions with the $Ru^{II}-Ru^{III}$ charge-transfer band shows enhancement of the bridging cyanide stretching [11b]. Electronic spectra of these complexes in different solvents give values of $H_{AB} = 2000\,cm^{-1}$ and α^2 between 0.06 and 0.12 which indicates a Class II species. A review of molecular mixed-valence cyanide bridged $Co^{III}-Fe^{II}$ complexes has been published recently [11c].

13.7.2
Influence of the Asymmetry of the Ions

13.7.2.1 Manganese Dinuclear Complexes

Although most of the $[Mn^{II}-Mn^{III}]$ and $[Mn^{III}-Mn^{IV}]$ manganese complexes belong to Class I, one of the most typical Class II mixed-valence manganese complexes corresponds to the series $[(L)_2Mn^{III}(\mu\text{-}O)_2Mn^{IV}(L)_2]^{3+}$ ion, where L = bpy or phen (Figure 13.9) [12]. The electronic spectrum shows broad bands in the near-infrared, assigned to the mixed-valence bands, that do not appear in the $Mn^{III}-Mn^{III}$ or $Mn^{IV}-Mn^{IV}$ parent complexes. The relatively low energy of this band indicates a moderate electronic delocalization. This feature is confirmed by the different environment of the two manganese ions (Mn^{IV}, d^3 and Mn^{III}, d^4-low-spin with strong Jahn–Teller effect). Thus, parameters λ and W are rather large, but the parameter β is small.

These dinuclear complexes are, currently, antiferromagnetically coupled: $S_{G.S.} = (2-3/2) = 1/2$, corresponding to one unpaired electron. If the antiferromagnetic coupling is strong enough, at low temperature only the $S = 1/2$ ground state will be populated and, thus, the EPR will correspond to one unpaired electron

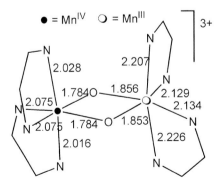

Figure 13.9 Scheme of a mixed-valence Mn(III,IV) oxido-bridged complex, showing the main Mn–ligand bond lengths.

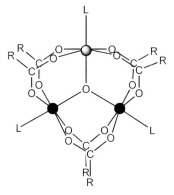

[M$_3$O(carboxylato)$_6$L$_3$]

○ = M^{II} ● = M^{III}

Figure 13.10 Scheme of the [M$_3$O(carboxylato)$_6$L$_3$] mixed-valence (II,III) complexes.

delocalized on the two manganese ions in a non-equivalent manner. Indeed, the spectra at low temperature show a great number of signals (currently 16) attributed to the hyperfine coupling with the two non-equivalent manganese ions. The degree of delocalization is thus intermediate between the total equivalence (11 lines) and the total non-equivalence of the two ions (36 lines). Similar features have been reported for other analogous mixed-valence [Mn^{II}–Mn^{III}] and [Mn^{III}–Mn^{IV}] complexes. The interpretation of the single-crystal EPR spectrum and its simulation has been done by Yano and coworkers [13].

13.7.2.2 [M$_3$O] Complexes
The influence of the asymmetry caused by the ions is clearly shown is the [M$_2^{III,III}$M^{II}O]$^{6+}$ units (Figure 13.10), with M = Mn, Fe, Ru.
    The following features are worth pointing out:

M = Mn, Fe   A rapid electron transfer is predicted, in some of them, in view of the D_{3h} symmetry. It occurs, for example, in [Mn$_3$O(acetate)$_6$(py)$_3$]·py at 223 K, indicating that the three Mn centers are crystallographically equivalent. This compound has a first-order phase transition at T_c = 184.65 K, which has been attributed to the conversion of the Mn$_3$O complex from valence detrapped to valence trapped and to the fact that pyridine solvate molecules become dynamic above T_c. Below T_c the compound loses its C_3 symmetry axis. This behavior is characteristic of mixed-valence Class II systems. Other Mn$_3$O complexes, such as [Mn$_3$O(X-benzoate)$_6$(py)$_2$(H$_2$O)]·0.5MeCN (X = H, Cl, Br), have an important difference: the Mn^{2+} ion has a H$_2$O ligand, whereas the two Mn^{3+} ions each have a pyridine ligand. These compounds appear to have an almost valence-trapped structure with a very low rate of electron transfer.
    The mixed-valence [Fe$_3$O(carboxylato)$_6$L$_3$]·S complexes, where L = water or pyridine, have been extensively studied by X-ray data, Mössbauer and infrared

spectroscopy, optical spectroscopy and susceptibility methods [14]. In most cases, Fe—O distances are identical at room temperature, assuming a delocalized unpaired electron. However, multi-temperature X-ray data (when reported) has revealed that at very low temperatures (10 K) the systems become valence-trapped. The Mössbauer spectrum of $[Fe_3O(acetato)_6(H_2O)_3]$ is markedly temperature dependent: at 17 K, signals due to distinct Fe(II) and Fe(III) sites are observed, whereas at 300 K a single absorption is seen. This feature is characteristic of all these $Fe_2^{III}Fe^{II}$ molecules, indicating Class II type. In the solid state, the valence-trapped to valence-detrapped phase transition occurs together with an onset of motion of the solvate molecule (S).

[Ru₃O] Trinuclear $[Ru_2^{III}Ru^{II}O(RCOO)_6L_3]$ where R = H, CH_3, C_2H_5 etc. and L = H_2O, PPh_3, CO, N-heterocycles, etc. have been extensively investigated [15]. Most of the complexes have their three metal atoms arranged equilaterally, which implies strong electronic delocalization within the Ru_3O core. An electronic pH-dependent band, close to 900 nm, is characteristic of these $Ru_3^{III,III,II}O$ complexes. These features are different when the clusters contain CO or isocyanide ligands that preferentially stabilize the Ru^{2+} site: these complexes are better described as isosceles triangles.

Two Ru_3O derivatives can be linked by pyrazine, 4,4′-bpy, 1,2-bis(4-pyridyl)ethene and 1,2-bis(4-pyridyl)ethane (BL), forming dimers such as $[(Ac)_6(py)_2Ru_3O$-BL-$Ru_3O(Ac)_6(py)_2]^{2+}$ (Figure 13.11A) [15], which can be oxidized or reduced (3+, 1+) giving "mixed-valence" dimers. Intercluster electronic interaction depends mainly on the bridging ligand. It seems particularly strong in the case of the pyrazine

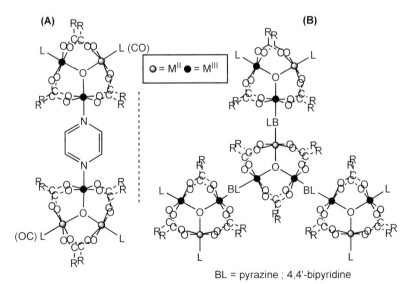

BL = pyrazine ; 4,4′-bipyridine

Figure 13.11 Linkage of several $[Ru_3O(carboxylato)_6L_3]$ complexes with different bridging ligands (BL).

ligand, because of the favorable overlap between the Ru$_3$O cluster d-orbitals and bridging π^*-orbitals. In the "mixed-valence" species [Ru$_3^{III,III,II}$O–pz–Ru$_3^{III,II,II}$O], an electronic band observed in the range of 800–925 nm has been ascribed to an inter-cluster charge-transfer transition. Estimates of H_{AB}, supports delocalized Class III behavior.

The rate of intra- and inter-molecular electron transfer in this type of pyrazine-bridged dimeric mixed-valence cluster has been investigated through the analysis of vibrational peak profiles. CO has been used as a suitable peripheral ligand in the cluster moiety, acting as an infrared vibrational probe [16,17]. In [{Ru$_2^{III}$Ru^{II}O(CH$_3$COO)$_6$(CO)(L)}$_2$(μ-pz)] (Figure 13.11A), each isosceles Ru$_3$O unit can be understood as resulting from trapped valence (as mentioned before, the CO ligand stabilizes the Ru^{II}). The one-electron reduction can be generated electrochemically, giving an inter-cluster mixed valence state: (Ru$_2^{III}$Ru^{II}–py–Ru^{III}Ru$_2^{II}$). If L is different in each triangle, this results in a change in the pattern of CO stretching vibrations from two discrete bands to a single broad averaged band at 1907 cm^{-1}. The degree of "coalescence" of the ν(CO) IR band in the inter-cluster mixed-valence state depends on the degree of electronic coupling between the pyrazine-linked Ru$_3$ clusters, modified by L.

The Raman spectra for these mixed-valence oxidation states are resonant with the transition assigned as the IT band [17a].

Ligand-bridged trimeric, tetrameric and polymeric clusters have also been reported [15]. They can be considered "*electron sponges*" in terms of the extent of their multiple electron transfer behavior. The characterization of the symmetric tetrameric clusters was facilitated by their high D_{3h} symmetry, which can be readily checked by using ^{1}H- and ^{13}C-NMR techniques (Figure 13.11B).

13.7.3
Non-discrete Systems

13.7.3.1 One-dimensional Complexes
The ensemble of one-dimensional Pt complexes with a halido bridge, such as [Pt^{II}L$_4$][Pt^{IV}L$_4$X$_2$](ClO$_4$)$_2$, has been exhaustively studied [18]. All of them have the same basic structure (Figure 13.12) and show one very peculiar and unique feature: their deep coloration (when the external ligand is ethylamine, they are called *red-Wolfram salts*). This color cannot be attributed to the isolated Pt^{2+} + Pt^{4+} colors but to the intervalence band. In all of them, conductivity through the chain is ca. 300 times greater than in the plane perpendicular to the chain. These two features are indicative of a certain degree of delocalization between the two oxidation states of platinum. The extent of this delocalization changes as a function of the halido bridge (Cl < Br < I), confirming the possibility of orbital overlap between the d orbitals of the metals and the appropriate orbitals of the ligands. As a logical consequence, the complexes with an iodido bridge show the greatest conductivity.

The Raman spectrum at the wavelength corresponding to the intervalence band (*resonance Raman effect*) is very peculiar in these complexes. The resonance is attributed to the stretching modes X–Pt^{IV}–X, which enable up 17 overtones to be

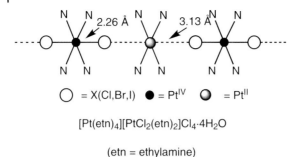

$$[Pt(etn)_4][PtCl_2(etn)_2]Cl_4\cdot 4H_2O$$

(etn = ethylamine)

Figure 13.12 Scheme of the 1D Wolfram red salt (and similar systems).

seen. Such an intense resonance Raman effect in the intervalence band implies that the Pt^{IV}—X distances have to change substantially during excitation, confirming the process postulated for the conductivity along the chain:

Ground state: $\cdots Pt^{II}\cdots X–Pt^{IV}–X\cdots Pt^{II}\cdots X–Pt^{IV}–X\cdots$

Excited state: $\cdots Pt^{III}\cdots X\cdots Pt^{III}\cdots X\cdots Pt^{III}\cdots X\cdots$

More than 200 mixed-valence platinum 1D compounds have been synthesized and characterized so far. All of them belong to Class II in the Robin–Day classification. The mixed-valence properties can be tuned by varying the metal ion (Pd instead of Pt, for example), bridging halogen atoms, in-plane ligands, and counteranions. The effect of the counteranion, for example, has been recently studied in $[Pd^{II}(en)_2][Pd^{IV}Br_2(en)](C_nH_{2n+1}SO_3)_4$, with $n = 7, 8, 9,$ and 10. With increasing alkyl-chain length (n) of the counter anion, the $Pd^{II}\cdots Pd^{IV}$ distance decreases, indicating that the oxidation states of the Pd atoms are closer to the Pd^{III} state (X-ray and optical conductivity measurements) [18b].

Halogen-bridged Pt^{II}/Pt^{IV} mixed-valence ladder compounds have been recently reported [19]. These novel ladder-type compounds have strong advantages for studying the boundary region between 1D and 2D electronic systems.

The use of building blocks of discrete $[Pt_2(L–L)_4]$ ($L–L = H_2P_2O_5^{2-}$ = diphosphonate, pop) or $CH_3CS_2^-$ (dithioacetate) has recently been demonstrated to lead to linear-chain complexes of platinum (Figure 13.13), that is $[Pt_2(dta)_4I]$ and $R_4[Pt_2(pop)_4I]\cdot nH_2O$ (R = alkylammonium). These chain compounds can take four electronic structures such as one average valence (AV, all $Pt^{2.5+}$) and another three more localized (2+,3+ in different alternations) [20]. Temperature and R are important for the degree of delocalization. The AV systems form golden or bronze metallic crystals with a broad band around 600 nm, completely absent in the aqueous solution spectra. The XPS spectra indicate that the platinum atoms are equivalent. The conductivity of these complexes ranged between 10^{-3} and $10^{-4}\,\Omega^{-1}\,cm^{-1}$, indicating that they are better semiconductors than those "classical" ones mentioned above, for which the conductivity range is between 10^{-7} and $10^{-10}\,\Omega^{-1}\,cm^{-1}$.

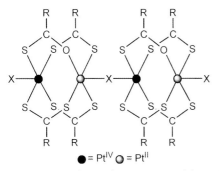

● = Pt^{IV} ○ = Pt^{II}

Figure 13.13 Scheme of a 1D compound formed by dinuclear Pt(II,IV) unities.

13.7.3.2 Three-dimensional Complexes

Prussian Blues are the most characteristic examples of these complexes. The intervalence band in Prussian Blues and its dependence on temperature has been extensively studied. Bandwidth can be adjusted to a vibrational mode of $430\,cm^{-1}$, in accord with the stretching vibrations Fe—CN. Its semiconducting character corroborates that Prussian Blue is class II.

13.7.4
The Creutz–Taube Ion: [(NH$_3$)$_5$Ru–(μ-pyrazine)–Ru(NH$_3$)$_5$]$^{+5}$ and Similar: Class II–III?

As indicated by Meyer, "*The first designed mixed-valence complex has proven to be the most difficult to understand*" [7]. Ru(II,III) complexes analogous to the Creutz–Taube ion were first classified as Class II systems [21], although many authors claimed that they should be considered as class III. A low-energy absorption band appears in the near-IR at 1560 nm (6410 cm^{-1}) in D$_2$O. This band is narrow ($\Delta\bar{v}_{1/2} = 1480\,cm^{-1}$) and "almost" solvent independent. Raman at 1320 nm (close to the IT transition) is strongly resonance-enhanced. EPR data show that the odd electron occupies an orbital that lies along the Ru—pyz—Ru axis. Vibrations of pye, NH$_3$, Ru—NH$_3$ are averaged compared to those in the Ru^{II}–Ru^{II} and Ru^{III}–Ru^{III} forms. A consensus appears to have been reached that the Creutz–Taube ion is delocalized, but evidence for its localization exists. It has been concluded that the Creutz–Taube ion belongs to the class II–III [7].

The main features (position and intensity of their intervalence band) of five Creutz–Taube-like ions, are given in Figure 13.14. The immediate environment of the ruthenium ions is very similar in the five complexes, which indicates that the λ parameter is very similar in them. E_{op} depends on λ, which in turn depends on the distance between the two active centers. The most important variation lies in the bridging ligands that considerably affect the parameter β (which influences the intensity of the band, ε). Thus, it is logical that the greatest electronic delocalization is found in the first complex (the Creutz–Taube ion) because delocalization occurs through a *short aromatic* bridge. The second and third compounds are intermediate, with the third having lower delocalization due to the double bond in the center of the bridge. With the methyl derivative in the pyridine ring, owing

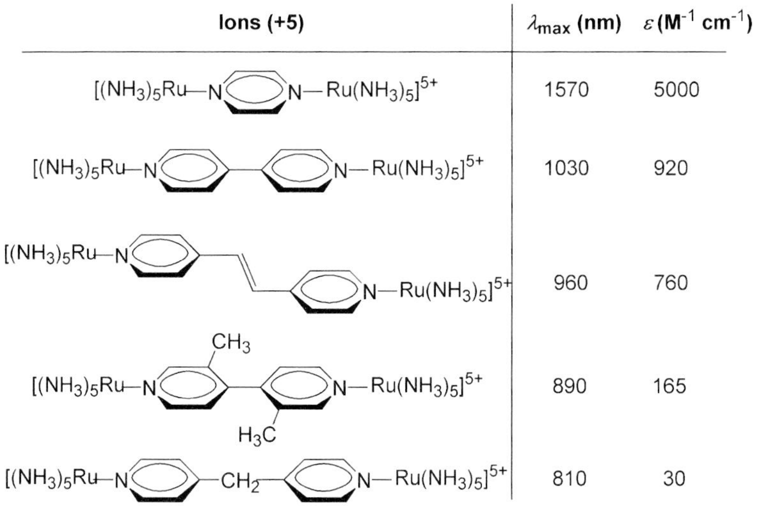

Ions (+5)	λ_{max} (nm)	$\varepsilon\,(M^{-1}\,cm^{-1})$
$[(NH_3)_5Ru\!-\!N\bigcirc N\!-\!Ru(NH_3)_5]^{5+}$	1570	5000
$[(NH_3)_5Ru\!-\!N\bigcirc\!-\!\bigcirc N\!-\!Ru(NH_3)_5]^{5+}$	1030	920
$[(NH_3)_5Ru\!-\!N\bigcirc\,...\,\bigcirc N\!-\!Ru(NH_3)_5]^{5+}$	960	760
$[(NH_3)_5Ru\!-\!N\bigcirc\!-\!\bigcirc N\!-\!Ru(NH_3)_5]^{5+}$ (CH$_3$, H$_3$C)	890	165
$[(NH_3)_5Ru\!-\!N\bigcirc\!-\!CH_2\!-\!\bigcirc N\!-\!Ru(NH_3)_5]^{5+}$	810	30

Figure 13.14 Some Creutz–Taube ion derivatives, with the main features of their IT bands.

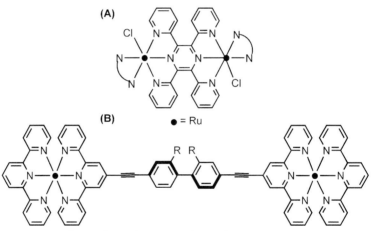

Figure 13.15 Scheme of several Creutz–Taube ion derivatives.

to the donor-electron effect of the methyl groups, delocalization through the π-orbitals of the ligand decreases. In the last compound, delocalization is interrupted by the presence of the CH$_2$ group between the pyridine rings. In the last three complexes, as the Ru–Ru distance is not very different, E_{op} is similar, but H_{AB} changes dramatically changed: the ε values are completely different.

Similar mixed-valence Ru^{II}–Ru^{III} have also been studied with other different bridging ligands. With the tppz (2,3,5,6-tetrakis(2-pyridil)pyrazine) ligand (Figure 13.15A), the Ru^{II}–Ru^{III} species has a rather narrow intervalence charge transfer band at $1700\,cm^{-1}$ ($\varepsilon = 2250\,M^{-1}\,cm^{-1}$) suggesting a class II–III mixed-valence state [22]. Its electronic coupling constant (H_{AB}) is calculated as $2940\,cm^{-1}$. The width at half height ($\Delta \bar{v}_{1/2}$) is 1390 cm^{-1}, whereas applying the "localized" Hush theory for a class II mixed-valence system, it would be 3680 cm^{-1}. Therefore, the tppz ligand increases the electronic delocalization of Creutz–Taube-like ions.

It has been possible to establish how the molecular architecture helps to control through-bond electronic coupling in these and similar series. Apart from solvent effects and the distance dependence, it has been shown that the electronic coupling matrix element between metal centers separated by a 4,4'-bipyridyl spacer decreases when different groups are substituted at the 2,2'-positions (torsion effect) (Figure 13.15B). Indeed, H_{AB} depends critically on the torsion of the bpy bridge [23].

Many asymmetric Creutz–Taube-like complexes with terpy, bpy and NH$_3$, using pyrazine as well as CN$^-$, 4-CNpy, 4,4'-bpy and 1,2-bis(4-pyridyl)ethylene as bridging ligands have been reported [24]. The calculated values of α^2 enabled the authors to describe all these systems as class II, even with pyrazine as bridging ligand. Their asymmetry does not favor delocalization.

Readers can find in a recent review on mixed-valence ruthenium complexes, the effect on the delocalization of many different ligands [25].

13.7.5
Class III-A (Isolated Complexes)

These complexes are fully delocalized (solvent and exchange electron transfer). They can be divided, in principle, into complexes without a metal–metal bond and clusters with a metal–metal bond.

13.7.5.1 Clusters Without a Direct Metal–Metal bond (or at Least Total Evidence of It)

Cu^I–Cu^{II} Systems. There are two classes of copper-containing metalloproteins responsible for rapid intra- and inter-molecular electron transfer in biological systems: those containing mononuclear blue copper sites and those containing the binuclear Cu$_A$ center. The Cu$_A$ center found in nitrous oxide reductase and cytochrome *c* oxidase (see Chapter 17) is a completely delocalized (i.e. class III) mixed-valence binuclear center with two Cu$^{1.5+}$ ions about 2.4–2.5 Å apart and bridged by two cysteine ligands. The Cu$_A$ center can perform rapid long-range electron transfer at rates of the order of 10^3–10^5 s^{-1}.

Parallel studies of mixed-valence model complexes help explain this behavior. For example, Tolman's complexes, [{LCu$^{1.5}$}$_2$]$^+$, in which the copper ions have a symmetric environment with the thiolate-bridged [Cu$_2$(μ-SR)$_2$]$^+$ core structure, are comparable to Cu$_A$ electron-transfer sites [26]. The Cu$_2$S$_2$ unit is planar, with an average Cu···Cu distance of 2.92 Å that is slightly longer than that suggested for Cu$_A$ (~2.5 Å). The geometry of each of the metal ions can be considered intermediate between geometries favorable for Cu$^+$ and Cu^{2+} oxidation states, presumably in order to facilitate redox reactions and to stabilize a Cu$^{1.5}$Cu$^{1.5}$ form. These complexes feature an EPR signal consistent with a fully delocalized class III complex:

a solution at 4.2 K gives a nearly axial signal with clearly recognizable seven-line hyperfine splitting patterns in the two components.

Molybdenum Complexes: Soluble Molybdenum Blues (MB) [27]   Solutions of MB are obtained almost instantaneously by the reduction of Mo^{VI}-type species in acid solution (pH < 3) by a great variety of reducing agents (see a detailed description of the structures in Chapter 5). The intense blue color consists of IT (Mo^V–Mo^{VI}) bands with the ε-value correlating linearly with the number of Mo^V centers. The resonance Raman spectrum is also indicative of class III mixed-valence compounds.

N_2-bridge Osmium Complexes [28]   The N_2 bridging ligand stabilizes mixed-valence Os^{II}–Os^{III} complexes, such as $[(NH_3)_5Os-N_2-Os(NH_3)_5]^{5+}$, $[(H_2O)(NH_3)_4Os-N_2-Os(NH_3)_5]^{5+}$, $[Cl(NH_3)_4Os-N_2-Os(NH_3)_5]^{4+}$ and $[Cl(NH_3)_4Os-N_2-OsCl(NH_3)_4]^{3+}$. K_c is very large in all cases, ca. 7×10^{12}. Its main feature is that the $\nu(N_2)$ practically does not appear in the IR spectra, indicating equivalence of the two Os ions on the time scale of the lifetime of a vibrational state. XPS measurements reveal only one set of Os 5d binding energies consistent with delocalization. The electronic spectra are very similar, showing an intervalence band close to $14000\,cm^{-1}$ (710–720 nm) with $\varepsilon = 4 \times 10^3\,M^{-1}\,cm^{-1}$.

By replacing ammonia with aromatic terminal ligands, such as terpy or bpy, the mixed-valence complexes are more localized (class II or II–III). In these cases, $\nu(N_2)$ clearly appears close to $2010\,cm^{-1}$ (electronic localization), but the IT transitions at higher energy are narrow and solvent independent (delocalization).

13.7.5.2 Metallic Clusters with Metal–Metal Bond

Dinuclear complexes   The simplest dinuclear clusters are $[Ru_2(carboxylate)_4]Cl$ ($\sigma^2\pi^4\delta^2\delta^{*2}\pi^{*1}$), $K_3[Mo_2(SO_4)_4](\sigma^2\pi^4\delta^1)$, $(NH_4)_3[Tc_2Cl_8](\sigma^2\pi^4\delta^2\delta^{*1})$, $[Re_2Cl_5(dppm)_2]$ ($\sigma^2\pi^4\delta^2\delta^{*1}$) and some other analogous systems. In all these cases MO theory is required to understand the origin of the unpaired electron (see Chapter 6, Figure 6.2). All metallic clusters are characterized by their intense coloration. In some cases, their EPR spectra are characteristic, although in other cases serious difficulties arise due to the number of different isotopes in the metals [29].

Mixed-valence complexes having a single metal–metal bond, have also been studied. One of the most fascinating cases is the anion $[CH_3N(PF_2)_2]_3[Co_2(CO)_2]^-$, obtained electrochemically in solution from the neutral $[CH_3N(PF_2)_2]_3[Co_2(CO)_2]$, whose structure is known [30]. All spectroscopic studies of the reduced solution indicate that the structure of the oxidized parent complex is retained. The EPR spectra of this Co^0–Co^{-1} (d^9–d^{10}) anion are indicative of class III: there is complete delocalization over the two cobalt sites of the radical anion, even at 4 K. This anion is, therefore, a class III mixed-valence compound.

Localization and delocalization in mixed-valence Cu^+–Cu^{2+} helicates, $[Cu_2(L)_2]^{3+}$ has been recently reported [31]. X-ray crystal analysis provides structural evidence for the presence of an internuclear Cu–Cu bond with an even distribution of spin density across the two Cu centers. Room–temperature UV–vis spectroscopy is consistent with this finding; however, frozen-glass EPR spectroscopy suggests solvatochromic behavior at 110 K, with the $[Cu_2]^{3+}$ core varying from localized to delocalized, depending on the solvent polarity: a highly polar solvent such as $MeNO_2$ results in the trapping of a valence-localized state (on the EPR time scale). On introducing toluene (much less polar) a system crossover from spin-localized to spin-delocalized begins to occur as the glass is no longer sufficiently polar to fully stabilize the large dipole moment associated with a valence-localized state.

Trinuclear Complexes Some trinuclear structures containing metal–metal bonds are shown in Figure 13.16. A-type complexes are characterized by their six ligands on each metal, giving a distorted octahedral geometry. Each metal only has three d orbitals available to form the metal–metal bonds. There are, thus, nine possible MOs, three bonding, *one non-bonding* and five antibonding. Several complexes with six, seven and eight electrons with these MOs have been synthesized and reported. With six electrons, no mixed-valence properties are shown; but with seven and/or eight valence electrons, the corresponding complexes are mixed-valence systems. For example, the $[Nb_3X_8]$ halido clusters have seven electrons in these MOs. Thus, the formal oxidation state of each Nb ion is 2.67. It is important to point out that athough the overall structure is two-dimensional, formally the clusters can be considered as discrete. B-type complexes, such as $[Mo_3(OEt)_2(carboxilato)_6(H_2O)_3]^{2+}$ have metallic ions linked to seven ligands. Thus, each metal has only two d orbitals available for the formation of metal–metal bonds. All six atomic orbitals produce three bonding MOs and three antibonding MOs. The molybdenum complex shown in Figure 13.16 has eight valence electrons: the bond order is reduced to 2/3. The oxidation state of each molybdenum ion is 3.33. It is curious that these complexes are similar to the manganese ones previously reported in this chapter, belonging to class I–II. Why do molybdenum complexes belong to class III? Once again, the difference lies in the presence of metal–metal bonds, which exist in molybdenum complexes but not in manganese complexes. Thus, the similarity is *only* geometric.

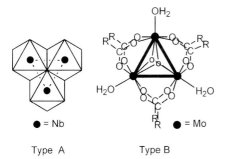

Type A Type B

Figure 13.16 Schematic representation of two mixed-valence trinuclear complexes: (A) $[Nb_3X_8]$ and B) $[Mo_3(OEt)_2(carboxilato)_6(H_2O)_3]^{2+}$.

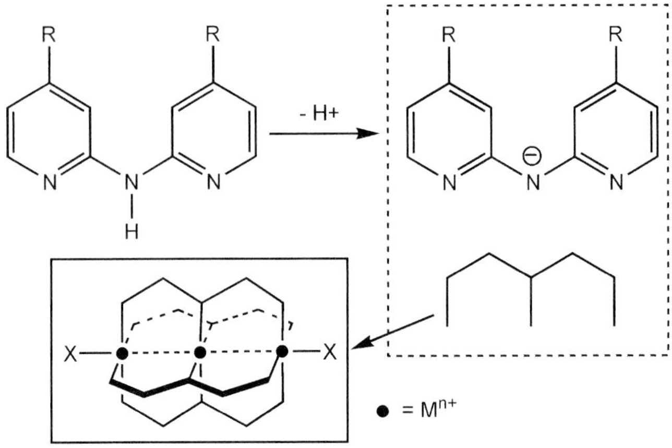

Figure 13.17 Drawing of the mixed-valence $[M_3(dpa)_4X_2]^+$ ions
(M = Co^{2+}, Ni^{2+}, Cu^{2+}, Ru^{2+} and Rh^{2+}) (see text for the
differences between them).

Cotton et al. [32]. reported some novel mixed-valence trinuclear complexes starting from the ligand di(2-pyridyl)amine (Hdpa), which stabilizes linear trinuclear coordination complexes of Cr^{2+}, Co^{2+}, Ni^{2+}, Cu^{2+}, Ru^{2+} and Rh^{2+}, of formula $[M_3^{II}(dpa)_4Cl_2]$. These trinuclear complexes are the simplest prototypical examples of compounds with extended metal atom chains (see Chapter 6). One-electron oxidation allows the synthesis and characterization of very new and peculiar systems, $[M_3(dpa)_4X_2]^+$ (Figure 13.17). When M = Cr and Cu, the oxidized systems are electron-localized and when M = Co or Ni they are electron-delocalized. The M—M length in the solid complexes is clearly indicative of localization (different lengths) or delocalization (same lengths). As $Cr_2^{II}Cr^{III}$ are asymmetric, there is a localized system with a quadruple bound $[Cr_2]^{4+}$ unit and a Cr^{3+} atom. The copper complex is also localized, but without the existence of any metal–metal bond. $[Co_3(dpa)_4Cl_2]^+$ and $[Ni_3(dpa)_4Cl_2]^+$ are symmetric, proving complete electronic delocalization of the unpaired electron.

Hexanuclear Complexes   Many hexanuclear mixed-valence complexes have also been reported. This group of clusters is dominated by two series, $[M_6X_{12}]^{n+}$ (M = Nb, Ta, W; X = F, Cl, Br, I) and $[M_6X_8]^{n+}$ (M = Nb, Mo, W; X = Cl, Br, I) (see Chapter 6, Figure 6.6) (Tables 13.3 and 13.4). MO calculations demonstrate that the first series contains up to 16 electrons which are delocalized in the M—M bonds. The second series has up to 24 electrons. These features allow multiple reversible redox reactions without important structural changes. They are, thus, class III mixed-valence clusters.

Multiple metal–metal bond complexes as building-blocks   Studies of electronic coupling and mixed valence in metal–metal quadruply bonded complexes linked by

Table 13.3 Examples of hexanuclear clusters (type $[M_6X_{12}]^{n+}$).

Compound	Electrons cluster	d (M—M), Å	Bond order
$[M_6^{2.33}Cl_{12}]^{2+}$ (M = Nb, Ta)	16	2.90–2.91	0.667
$[M_6^{2.50}Cl_{12}]^{3+}$ (M = Nb, Ta)	15	2.80–2.96	0.625
$[M_6^{2.67}Cl_{12}]^{4+}$ (M = Nb, Ta)	14	2.96–3.01	0.583
$[Zr_6^{2.5}Cl_{12}]Cl_3$	9	3.20	0.375

The Nb and Ta compounds are deeply colored with an intervalence band between 350 and 1000 nm.

Table 13.4 Examples of hexanuclear clusters (type $[M_6X_8]^{n+}$).

Compound	Electrons cluster	d (M—M), Å	Bond order
$[Nb_6^{1.83}I_8]I_3$	19	2.85	0.792
$Cs[Nb_6^{1.66}I_8]I_3$	20	2.826	0.833
$[W_6^{2.33}Br_8]Br_6$	22	2.64	0.917

different bridges have been reported. A unique complex with four Cl$^-$ as bridging ligands is the $[\{cis\text{-}Mo_2(L)_2\}_2(\mu\text{-}Cl)_4](PF_6)$ salt, whose core is shown in Figure 13.18A (L = N,N′-di-p-anisylformamidinate) [33]. It consists of one $[Mo_2]^{4+}$ and one $[Mo_2]^{5+}$ unit linked by four Cl bridges: thus, there are Mo—Mo and Mo—Cl—Mo bonds. It has one unpaired electron and a fully delocalized structure; $K_c = 1.3 \times 10^9$ is three orders of magnitude larger than that of the Creutz–Taube ion. For this reason it can be assigned to class III. The Mo—Mo distances increase from 2.119 Å in the reduced $[\{cis\text{-}Mo_2(L)_2\}_2(\mu\text{-}Cl)_4]^0$ species to 2.145 Å in the oxidized one, $[\{cis\text{-}Mo_2(L)_2\}_2(\mu\text{-}Cl)_4]^+$.

The use of poly-carboxylates and closely related ligands has also been reported (Figure 13.18B) [34]. Chemical oxidation results in the formation of mixed-valence

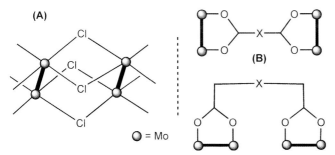

Figure 13.18 Two representative examples of linkage between mixed-valence metal–metal multiple bond complexes.

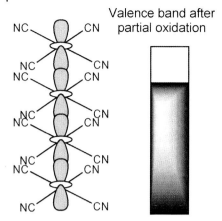

Valence band after partial oxidation

Figure 13.19 Scheme of the valence band of the partially oxidized $[Pt^{II}(CN)_4]^{2-}$ compound

species that are particularly well-suited for the study of the Class II/III border. The extent of electronic coupling has been determined by a variety of spectroscopic techniques, including EPR and electronic absorption spectroscopy. In some cases the importance of the conformation of the linker in mediating electronic communication between metal-containing units has been demonstrated [35].

13.7.6
Class III-B: Extended Systems

13.7.6.1 One-dimensional Wires
The most widely studied class III-B compounds are the partially oxidized tetracyano-platinate salts, commonly known as Krogmann salts, $K_{1.75}[Pt(CN)_4] \cdot 1.5H_2O$ or $K_2[Pt(CN)_4]Br_{0.3} \cdot 3H_2O$ (KCP). KCP is obtained by partial oxidation of $K_2[Pt(CN)_4] \cdot aq$ with Br_2. Its electronic structure is drawn in Figure 13.19, in which the d_{z^2} orbitals for each Pt are indicated. These orbitals are responsible for the conducting properties when one-dimensional systems are partially oxidized.

$K_2[Pt(CN)_4] \cdot aq$ is a totally white complex, without any important electric property. When partial oxidation occurs, the d_{z^2} orbital becomes partially empty, creating a conducting band that allows strongly anisotropic metallic conductivity. Moreover, a new, very strong electronic intervalence band is formed, giving the salts their characteristic brightness and dark-metallic color. Two main factors need to be remembered:

1. Similar Pt species can be obtained by replacing CN^- by oxalate or other anions.

2. d^8 square-planar complexes of the third transition row (such as Au^{3+} or Ir^+) also give, by partial oxidation, new one-dimensional conducting mixed-valence systems. However, the elements of the second row, such as Pd^{2+}, do not give

these mixed-valence systems so easily, owing to the lower diffuse character of the d_{z^2} orbitals, which impedes a good overlap. The first row elements, such as Ni^{2+}, do not give any mixed-valence compound of this kind.

The comparison of these one-dimensional conducting compounds (class III) and Wolfram red salt (class II, see Section 13.7.3.1) with less conductivity, is new evidence of the importance of the metal–metal bond in the classification and properties of mixed-valence systems. KCP and its derivatives have a direct Pt–Pt bond whereas, in Wolfram salt, chloride is the bridge between the two platinum ions.

Starting from discrete dinuclear complexes with a metal–metal bond as precursors, new promising "molecular wires" have been reported. For example, reduction of $[Rh_2(CH_3CN)_{10}](BF_4)_4$ (a Rh^{II}–Rh^{II} complex) allows the synthesis of a new mixed-valence 1D rhodium system, $[\{Rh(CH_3CN)_4\}(BF_4)_{1.5}]_\infty$, formed by mixed-valence $Rh_2^{I,II}$ species that consist of alternating Rh–Rh interactions of 2.844 and 2.928 Å. Although the charge transport experiments indicate that this new 1D system is only a semiconductor, theoretical calculations predict that the d_{z^2} orbitals interact quite strongly along the chain [36]. New theoretical considerations and practical developments are opened up by this new class of metal–metal mixed-valence systems.

13.7.6.2 Three-dimensional Systems

Fe_3O_4, magnetite, is the archetype of these systems, as discussed in the first part of this chapter. The main features of magnetite are its spontaneous magnetization at room temperature and its high electrical conductivity. A mobile electron jumps very easily from one site to another, polarizing the rest of the electrons and so creating ferromagnetic ordering.

13.8
Conclusions

Is it possible to tune the extent and magnitude of electronic delocalization and the electron-transfer rate? By the end of this chapter, the conclusion is affirmative.

1. By playing with the distortion of the ligands or their own metal ions, the extent of electronic delocalization can be enhanced or diminished. However, this possibility is still too serendipitous.

2. By playing with the bridging ligand, it is also possible to enhance or diminish the electronic interaction. We have already commented that small ligands, such as N_2, or CN^-, are better delocalizers than pyrazine or larger ligands. Clearly, if these ligands allow good overlap between their orbital and the corresponding orbitals of the two metal centers, delocalization will be almost as important as in the metal–metal bond mixed-valence complexes.

3. Playing with terminal ligands also permits modification of the extent of electronic delocalization. For example, π-acid ligands, such as CO, nitriles,

aromatic rings rather than non-aromatic ones, tend to stabilize the smallest oxidation state, which gives a more localized system (passing from class III to class II, for example). Some representative examples have been given throughout this chapter.

References

1 Robin, M.B., Day, P. Mixed Valence Chemistry. A Survey and Classification, *Adv. Inorg. Chem. Radiochem.*, 1967, *10*, 247.

2 Allen, G.C., Hush, N.S. Intervalence-Transfer Absorption. Part I. Qualitative Evidence for Intervalence-Transfer Absorption in Inorganic Systems in Solution and in Solid State, *Prog. Inorg. Chem.* 1967, *8*, 357.

3 Hush, N.S. Intervalence-Transfer Absorption. Part II. Theoretical Considerations and Spectroscopic Data, *Prog. Inorg. Chem.* 1967, *8*, 391.

4 Brunschwig, B.S., Creutz, C., Sutin, N. Optical transitions of symmetrical mixed-valence systems in the Class II–III transition regime, *Chem. Soc. Rev.* 2002, *31*, 168.

5 (a) Piepho, S.B., Krausz, E.R., Schatz, P.N. Vibronic Coupling Model for Calculation of Mixed Valence Absorption Profiles, *J. Am. Chem. Soc.* 1978, *100*, 2996; (b) Wong, K.Y., Schatz, P.N., Piepho, S.B., Vibronic Coupling Model for Mixed-Valence Compounds. Comparisons and Predictions, *J. Am. Chem. Soc.* 1979, *101*, 2793.

6 Wong, K.Y., Schatz, P.N. A Dynamic Model for Mixed-Valence Compounds, *Prog. Inorg. Chem.* 1981, *28*, 369.

7 Demadis, K.D., Hartshorn, C.M., Meyer, T.J. The Localized-to-Delocalized Transition in Mixed-Valence Chemistry, *Chem. Rev.* 2001, *101*, 2655.

8 Long, R.C., Hendrickson, D.N. Intramolecular Electron Transfer in a Series of Mixed-Valence Copper(II)–Copper(I) Complexes, *J. Am. Chem. Soc.* 1983, *105*, 1513.

9 (a) Gagné, R.R., Koval, C.A., Smith, T.J. Binuclear Complexes of Macrocyclic Ligands. A Mixed-Valence Copper(II)–

Copper(I) Complex which Exhibits Unusual Temperature-Dependent Behavior *J. Am. Chem. Soc.* 1977, *99*, 8367: (b) Gagné, R.R., Koval, C.A., Smith, T.J., Cimolino, M.C. Binuclear Complexes of Macrocyclic Ligands. Electrochemical and Spectral Properties of Homobinuclear Cu^{II}Cu^{II}, Cu^{II}Cu^{I}, and Cu^{I}Cu^{I} Species Including an Estimated Intramolecular Electron Transfer Rate *J. Am. Chem. Soc.* 1979, *101*, 4571.

10 Dutta, S.K., Ensling, J., Werner, R., Flörke, U., Haase, W., Gütlich, P., Nag, K. Valence-Delocalized and Valence-Trapped Fe(II)–Fe(III) Complexes: Drastic Influence of the Ligands, *Angew. Chem. Int. Ed. Eng.* 1997, *36*, 152.

11 (a) Siddiqui, S., Henderson, W.W., Shepherd, R.E. Pentacyanoruthenate(II)-Pentaammineruthenium(II/III) Binuclear Complexes Bridged By Cyanogen, Cyanide, and 4,4′-Bipyridine, *Inorg. Chem.*, 1987, *26*, 3101; (b) Doorn, S.K., Hupp, J.T. Intervalence Enhanced Raman Scattering from [(NC)$_5$Ru-CN-Ru(NH$_3$)$_5$]$^-$. A Mode-by-Mode Assessment of the Franck-Condon Barrier to Intramolecular Electron Transfer. *J. Am. Chem. Soc.*, 1998, *111*, 1142. (c) Bernhardt, P.V., Bozoglian, F., Macpherson, B.P., Martínez, M. Molecular mixed-valence cyanide bridged Co(III)–Fe(II) complexes, *Coord. Chem. Rev.* 2005, *249*, 1902.

12 (a) Cooper, S.R., Calvin, M. Mixed Valence Interactions in Di-μ-oxo Bridged Manganese Complexes, *J. Am. Chem. Soc.* 1977, *99*, 6623; (b) Cooper, S.R., Dismukes, G.C., Klein, M.P., Calvin, M. Mixed Valence Interactions in Di-μ-oxo Bridged Manganese Complexes. Electron Paramagnetic Resonance and Magnetic Susceptibility Studies, *J. Am. Chem. Soc.* 1978, *100*, 7248; (c) Mabad, B., Tuchagues,

J.P., Hwang, Y.T., Hendrickson, D.N. Mixed-Valence Md[II]Mn[III] Complexes: Models for the Manganese Site in the Photosynthetic Electron-Transport Chain, *J. Am. Chem. Soc.* 1985, *107*, 2801.

13 Yano, J., Sauer, K., Girerd, J-J., Yachandra, V.K. Single Crystal X- and Q-Band EPR Spectroscopy of a Binuclear Mn$_2$(III,IV) Complex Relevant to the Oxygen-Evolving Complex of Photosystem II, *J. Am. Chem. Soc.* 2004, *126*, 7486.

14 see, for example, Wilson, C., Iversen, B.B., Overgaard, J., Larsen, F.K., Wu, G., Palii, S.P., Timco, G.A., Gerbeleu, N.V. Multi-Temperature Crystallographic Studies of Mixed-Valence Polynuclear Complexes; Valence Trapping Process in the Trinuclear Oxo-Bridged Iron Compound, [Fe$_3$O(O$_2$CC(CH$_3$)$_3$)$_6$(C$_5$H$_5$N)$_3$], *J. Am. Chem. Soc.* 2000, *122*, 11370.

15 Toma, H.E., Araki, K., Alexiou, A.D.P., Nikolaou, S., Dovidauskas, S. Monomeric and extended oxo-centered triruthenium clusters, *Coord. Chem. Rev.* 2001, *219–221*, 187.

16 Yamaguchi, T., Imai, N., Ito, T., Kubiak, C.P. A Strongly Coupled Mixed Valence State Between Ru$_3$ Clusters. Intramolecular Electron Transfer on the Infrared Vibrational Time Scale in a Pyrazine (pz) Bridged Dimer of Triruthenium Clusters, [{Ru$_3$(μ_3-O) (μ-CH$_3$CO$_2$)$_6$(CO)(abco)}$_2$(μ(abco-pz)] (abco = 1-azabicyclo[2,2,2]octane), *Bull. Chem. Soc Jpn.*, 2000, *73*, 1205.

17 (a) Londergan, C.H., Rocha, R.C., Brown, M.G., Shreve, A.P., Kubiak, C.P. Intervalence Involvement of Bridging Ligand Vibrations in Hexaruthenium Mixed-Valence Clusters Probed by Resonance Raman Spectroscopy, *J. Am. Chem. Soc.* 2003, *125*, 13912; (b) Crutchley, R.J. Charge-Transfer Isomers and Mixed-Valence Properties, *Angew. Chem. Int. Ed.* 2005, *44*, 6452.

18 (a) Interrante, L.V., Browall, K.W., Bundy, F.P. Studies of Intermolecular Interactions in Transition Metal Complexes. IV. A HighPressure Study of Some Mixed-Valence Platinum and Palladium Complexes *Inorg. Chem.* 1974, *13*, 1158 and Studies of Intermolecular

Interactions in Transition Metal Complexes. V. Electrical Properties of Some Mixed-Valence Platinum and Palladium Complexes, *Inorg Chem.* 1974, *13*, 1162; (b) Arakawa, H., Kawakami, D., Takaishi, S., Kajiwara, T., Miyasaka, H., Sugiura, K., Yamashita, M., Kishida, H., Okamoto, H. Tuning of Electronic Structures of Quasi-One-Dimensional Bromo-Bridged Pd[II]–Pd[IV] Mixed-Valence Complexes by Substituting Counter Anions, *Bull. Chem. Soc. Jpn*, 2007, *80*, 189.

19 Kawakami, D., Yamashita, M., Matsunaga, S., Takaishi, S., Kajiwara, T., Miyasaka, H., Sugiura, K., Matsuzaki, H., Okamoto, H., Wakabayashi, Y., Sava, H. Halogen-Bridged Pt(II)/Pt(IV) Mixed-Valence Ladder Compounds, *Angew. Chem. Int. Ed.* 2006, *45*, 7214.

20 Yamashita, M., Takaishi, S., Kobayashi, A., Kitagawa, H., Matsuzaki, H., Okamoto, H. Tuning of electronic structures of quasi one-dimensional iodide-bridged dinuclear platinum mixed-valence complexes, *Coord. Chem. Rev.* 2006, *250*, 2335.

21 (a) Creutz, C., Taube, H. A Direct Approach to Measuring the Franck-Condon Barrier to Electron Transfer between Metal Ions *J. Am. Chem. Soc.* 1969, *91*, 3988; (b) Creutz, C., Taube, H. Binuclear Complexes of Ruthenium Ammines, *J. Am. Chem. Soc.* 1973, *95*, 1086.

22 Chanda, N., Sarkar, B., Fiedler, J., Kaim, W., Lahiri, G.K. Synthesis and mixed valence aspects of [{(L)ClRu}$_2$(μ-tppz)]$^{n+}$ incorporating 2,2'-dipyridylamine (L) as ancillary and 2,3,5,6-tetrakis(2-pyridyl)pyrazine (tppz) as bridging ligand, *Dalton Trans.* 2003, 3550.

23 Benniston, A.C., Harriman, A., Li, P., Sams, C.A., Ward, M.D. Orientational Control of Electronic Coupling in Mixed-Valence, Binuclear Ruthenium(II)-Bis(2,2':6',2''-Terpyridine) Complexes, *J. Am. Chem. Soc.* 2004, *126*, 13630.

24 (a) Altabef, A.B., Ribotta de Gallo, S.B., Folquer, M.E., Katz, N.E. Syntheses and characterization of new mononuclear and dinuclear complexes derived from ruthenium polypyridines, *Inorg. Chim. Acta*, 1991, *188*, 67; (b) Fagalde, F., Katz,

N.E. Distance Depedence of Intramolecular Electron-Transfer Parameters in Mixed-Valence Asymmetric Complexes of Ruthenium, *Polyhedron*, 1995, *14*, 1213; (c) Mellace, M.G., Fagalde, F., Katz, N.E., Crivelli, I.G., Delgadillo, A., Leiva, A.M., Loeb, B., Garland, M.T., Baggio, R. Dinuclear Asymmetric Ruthenium Complexes with 5-Cyano-1,10-phenanthroline as a Bridging Ligand, *Inorg. Chem.* 2004, *43*, 1100.

25 Kaim, W., Sarkar, B. Mixed valency in ruthenium complexes – Coordinative aspects, *Coord. Chem. Rev.* 2007, *251*, 584

26 Houser, R.P., Young Jr, V.G., Tolman, W.B. A Thiolate-Bridged, Fully Delocalized Mixed-Valence Dicopper(I,II) Complex That Models the Cu$_A$ Biological Electron-Transfer Site, *J. Am. Chem Soc.*, 1996, *118*, 2101.

27 Muller, A., Serain, C., Soluble Molybdenum Blues-"des Pudels Kern", *Acc. Chem. Res.* 2000, *33*, 2.

28 (a) Magnuson, R.H., Taube, H. Mixed Oxidation States in Osmium Ammine Dinitrogen Complexes, *J. Am. Chem. Soc.* 1972, *94*, 7213; (b) Richardson, D.E., Sen, J.P., Buhr, J.D., Taube, H. Preparation and Properties of Mixed-Valence (μ-Dinitrogen) bis(pentaammine) Complexes of Osmium and Ruthenium, *Inorg. Chem.* 1982, *21*, 3136; (c) Demadis, K.D., El-Samanody, E-S., Coia, G.M., Meyer, T.J. Os^{III}(N$_2$)Os^{II} Complexes at the Localized-to-Delocalized, Mixed-Valence Transition, *J. Am. Chem. Soc.* 1999, *121*, 535.

29 Cotton, F.A., Murillo, C.A., Walton, R.A. (2005) *Multiple Bonds between Metal Atoms*, 3rd edn., Springer, Berlin.

30 Babonneau, F., Henry, M., King, R.B., El Murr, N. Spectroscopic Properties of a Mixed-Valence Binuclear Cobalt Complex: [CH$_3$N(PF$_2$)$_2$]$_3$CO$_2$(CO)$_2$$^-$, *Inorg. Chem.* 1985, *24*, 1946.

31 Jeffery, J.C., Riis-Johannessen, T., Anderson, C.J., Adams, C.J., Robinson, A., Argent, S.P., Ward, M.D., Rice, C.R. Localization and Delocalization in a Mixed-Valence Dicopper Helicate, *Inorg. Chem.* 2007, *46*, 2417.

32 Berry, J.F., Cotton, F.A., Daniels, L.M., Murillo, C.A., Wang, X. Oxidation of Ni$_3$(dpa)$_4$Cl$_2$ and Cu$_3$(dpa)$_4$Cl$_2$: Nickel–Nickel Bonding Interaction, but No Copper–Copper Bonds, *Inorg. Chem.* 2003, *42*, 2418.

33 Cotton, F.A., Liu, C.Y., Murillo C.A., Wang, X. A mixed-valence compound with one unpaired electron delocalized over four molybdenum atoms in a cyclic tetranuclear ion, *Chem. Commun.*, 2003, 2190.

34 (a) Chisholm, M.H., Patmore, N.J. Studies of Electronic Coupling and Mixed Valency in Metal-Metal Quadruply Bonded Complexes Linked by Dicarboxylate and Closely Related Ligands, *Acc. Chem. Res.* 2007, *40*, 19; (b) Cotton, F.A., Dalal, N.S., Liu. C.Y., Murillo, C.A., North, J.M., Wang, X. Fully Localized Mixed-Valence Oxidation Products of Molecules Containing Two Linked Dimolybdenum Units: An Effective Structural Criterion, *J. Am. Chem. Soc.* 2003, *125*, 12945.

35 Cotton, F.A., Liu, C.Y., Murillo, C.A., Zhao, Q. Electronic Localization versus Delocalization Determined by the Binding of the Linker in an Isomer Pair, *Inorg. Chem.* 2007, *46*, 2604.

36 Ber, J.K., Dunbar, K.R. Chain Compounds Based on Transition Metal Backbones: New Life for an Old Topic, *Angew. Chem. Int. Ed.* 2002, *41*, 4453.

Bibliography

Brown, D.B. (ed.) (1979) *Mixed-Valence Compounds*, NATO ASI Series, C. Reidel, Dordrecht (The Netherlands).

Prassides, K. (ed.) (1990) *Mixed Valence Systems: Applications in Chemistry, Physics and Biology*. NATO ASI Series, Vol. 343, Kluwer, Dordrecht (Holland)

Creutz, C. Mixed Valence Complexes of d^5-d^6 Metal Centres, *Prog. Inorg. Chem.* 1983, *30*, 1.

Clark, R.J.H. The Chemistry and Spectroscopy of Mixed-Valence Complexes, *Chem. Soc. Rev.* 1984, 219.

Richardson, D.E., Taube, H. Mixed-Valence Molecules: Electronic Delocalization and Stabilization, *Coord. Chem. Rev.* 1984, *60*, 107.

Young, C.G. Mixed-Valence Compounds of the Early Transition Metals, *Coord. Chem. Rev.* 1989, *96*, 89.

Crutchley, R.J. Intervalence Charge Transfer and Electron Exchange Studies of Dinuclear Ruthenium Complexes, *Adv. Inorg. Chem.* 1994, *41*, 273.

Kaim, W., Klien, A., Glöckle, M. Exploration of Mixed-Valence Chemistry: Inventing New Analogues of the Creutz-Taube Ion, *Acc. Chem. Res.* 2000, *33*, 755.

-D'Alessandro, D.M., Richard Keene, F. Current Trends and Future Challenges in the Experimental, Theoretical and Computational Analysis of Intervalence Charge Transfer (IVCT) Transitions, *Chem. Soc. Rev.* 2006, *35*, 424.

D'Alessandro, D.M., Richard Keene, F. Intervalence Charge Transfer (IVCT) in Trinuclear and Tetranuclear Complexes of Iron, Ruthenium, and Osmium, *Chem. Rev.* 2006, *106*, 2270.

Kaim, W., Lahiri, G.K. Unconventional Mixed-Valent Complexes of Ruthenium and Osmium, *Angew. Chem. Int. Ed.* 2007, *46*, 1778.

Part Four   New Trends in Modern Coordination Chemistry

14
Supramolecular Chemistry, Metallosupramolecular Chemistry and Molecular Architecture

14.1
Supramolecular Chemistry: Definitions

In the book of Balzani and coworkers, entitled "Molecular Devices and Machines: a Journey into the Nanoworld" [1], the authors give a glossary with these two definitions:

"a- classical definition by J.M. Lehn: the chemistry beyond the molecule, bearing on organized entities of higher complexity that result from the association of two or more chemical species held together by intermolecular forces …

b- more general definition used in this book: the chemistry of species made of two or more molecular components."

The textbook "Supramolecular Chemistry" [2] emphasizes this difference: *"The rapid expansion in supramolecular chemistry over the past 15 years has resulted in an enormous diversity of chemical systems … which may lay some claim, either in concept, origin or nature, to being supramolecular. In particular, workers in the field of supramolecular photochemistry have chosen to adopt a rather different definition of a supramolecular compound as a group of molecular components that contribute properties that each component possesses individually to the whole assembly (covalent or non-covalent). Thus an entirely covalent molecule comprising, for example, a chromophore, spacer and redox centre might be thought of as supramolecular."*

In this chapter we are going to follow and develop the idea of Lehn, applied to coordination chemistry entities.

14.1.1
Lehn's Idea [3]

The concept and the term supramolecular chemistry were introduced in 1978 by Lehn: *"Just as there is a field of molecular chemistry based on the covalent bond, there is a field of supramolecular chemistry, the chemistry of molecular assemblies and of the*

Coordination Chemistry. Joan Ribas Gispert
Copyright © 2008 WILEY-VCH Verlag GmbH & Co. KGaA, Weinheim
ISBN: 978-3-527-31802-5

intermolecular bond ... covering the structure and functions of the entities formed by the association of two or more chemical species".

It has been reformulated, as *"chemistry beyond the molecule"* relating to the organized entities of higher complexity that result from the association of two or more chemical species held together by intermolecular forces ... *Supramolecular entities are by nature constitutionally dynamic by virtue of the lability of non-covalent interactions* [4].

It is important to emphasize this latter aspect: *the intrinsic dynamic character of any supramolecular process.* Whereas the supramolecular entities are *dynamic by nature*, the molecular entities are dynamic by intent [4].

The *dynamic character* has been forgotten many times in the literature, which has classified as supramolecular entities many species with interesting properties, but which cannot be included in this category. This problem has been extremely pronounced in coordination chemistry, which is also widely used in supramolecular chemistry. The geometry requirements of metal ions, combined with the design of specific ligands have permitted the construction of complex molecular topologies. However, many times the dynamic character of these species has been completely omitted. It is worth noting that Constable coined the term *"metallosupramolecular chemistry"* to describe metal-directed assembly processes [5].

Finally, a very interesting comparison – initiated by Lehn – has been developed by Stoddart et al.: [6] *"Chemists are just starting to write their own 'words'. They know how to produce the 'words'. Now, they are learning how to write the 'sentences'. The 'grammar' they will use will be dictated by the nature of the noncovalent bond. The 'modern languages' are about to evolve. Materials science and the life sciences will be beneficiaries."*

14.1.2
Intermolecular Bonds

Concerning the nature of the intermolecular bonds that hold the components together, various types of interactions may be distinguished, that present different degrees of strength and directionality: *electrostatic forces, hydrogen bonding, van der Waals interactions, donor–acceptor interactions, π–π stacking interactions, metal ion coordination, etc.* Their strengths range from weak or moderate, as in hydrogen bonds, to stronger, for metal ion coordination. Intermolecular forces are in general weaker than covalent bonds, so that supramolecular species are thermodynamically less stable, kinetically more labile and dynamically more flexible than molecules. The bond energy of a typical single covalent bond is around $350\,kJ\,mol^{-1}$ ($942\,kJ\,mol^{-1}$ for the triple bond in N_2). The strengths of many of the non-covalent interactions used by supramolecular chemists are generally much weaker ranging from $2\,kJ\,mol^{-1}$ for dispersion forces, through $20\,kJ\,mol^{-1}$ for a hydrogen bond to $250\,kJ\,mol^{-1}$ for an ion–ion interaction.

14.1.3
Concepts and Perspectives [3]

14.1.3.1 Host–Guest Chemistry

The supramolecular field started with the selective binding of alkali metal cations by natural as well as by synthetic macrocyclic and macropolycyclic ligands, the crown ethers and the cryptands. The outlook broadened, leading to the identification of *molecular recognition* as a novel domain of chemical research. The chemistry of molecular recognition is at the core of *"host–guest chemistry or receptor–substrate selective binding"*.

The selective binding of a specific substrate to its receptor yields the supramolecular species and involves a molecular recognition process. If, in addition to binding sites, the receptor also bears reactive function it may effect a *chemical transformation* on the bound substrate thus behaving as a supramolecular reagent or catalyst. Thus, molecular recognition and transformation represent the basic functions of supramolecular species.

14.1.3.2 Self-assembly Processes

The second milestone of the field is the concept of *self-assembly*. Behind the term "self-assembly" lies the design of programmed systems that self-organize through explicit manipulation of molecular recognition features. As we will see later, the dynamic character necessary for supramolecularity has often been ignored, creating a great confusion between "molecular self-assembly" and "supramolecular self-assembly".

14.1.3.3 Perspectives

The interest in supramolecular chemistry is, in part, inspired by the challenge to mimic the complex and cooperative functions of natural systems. Biological systems routinely store information in the shape, size, and electronic properties of molecules. This information is read by the way the molecules recognize and interact with each other. For example DNA is the central carrier of information in cells. As chemists learn to manipulate molecules with increasing dexterity, synthetic systems may not only mimic biological systems, but they may surpass them in both biological and abiotic applications.

14.1.3.4 The "Philosophy" of This Chapter

From a personal point of view, I prefer to consider the limitations (and precisions) indicated at the beginning of this chapter, i.e. following the idea of Lehn and not the too general idea given in photochemistry. However, as this is a general book on Coordination Chemistry, I would like to give a rather broad viewpoint, emphasizing by means of different examples whether each system is better described as molecular rather than supramolecular or vice versa. The two main aspects of supramolecular chemistry, molecular recognition and self-assembly processes, will be discussed in this chapter.

14.2
Molecular Recognition

In the late 1960s, Pedersen, Lehn and Cram published the synthesis of macrocyclic molecules (crown-ethers, cryptands, spherands, and so forth) that are able to selectively bind ions or small organic molecules via non covalent interactions. In this sense, Lehn defined molecular recognition as the energy and the information involved in the binding and selection of substrate(s) by a given receptor molecule [3]. Recognition implies *geometrical preorganization* and *interaction complementarity* between the associated partners:

- The concept behind *preorganization* is the construction of a host that exactly matches, both sterically and electronically, the requirements of the guest before the guest is bound.

- Matching host and guests are said to be *complementary*. Hosts must have binding sites which cooperatively contact and attract binding sites of guests without generating strong nonbonded repulsions. The individual interactions between hosts and guests are all relatively weak – much weaker than a covalent bond – and hence it is only through the complementary interaction of multiple pairs of binding sites that strong, selective complexation can take place.

> The generalized complementarity principle extending over energetic features as well as over geometrical ones is represented by the *"lock and key"*, steric fit concept of Fisher.

As indicated by Lehn, "*Biological molecular recognition represents the most complex expression of molecular recognition leading to highly selective binding, reaction, transport, regulation etc. It provides study cases ... for the design of model systems as well as of abiotic receptors*" [3].

14.2.1
Molecular Receptors (Hosts)

Molecular receptors represent *generalized coordination chemistry*; they are not limited to transition-metal ions but extend to all types of substrates: cationic, anionic, or neutral species of organic, inorganic, or biological nature. Many receptors of extremely varied structural types have been conceived and investigated. For the purpose of this book, the focus will be on cations and other inorganic species.

14.2.1.1   Cation-binding Hosts
The coordination chemistry of s and p cations has developed over the last 40 years with the discovery of many types of more or less powerful and selective cyclic or

Table 14.1 Comparison of the diameter of different crown ethers with the ionic diameter of various metal cations.

Cation	Diameter (Å)	Crown ether	Cavity diameter (Å)
Li$^+$	1.36	[12]crown-4	1.20–1.50
Na$^+$	1.90	[15]crown-5	1.70–2.20
K$^+$	2.66	[18]crown-6	2.60–3.20
Cs$^+$	3.38	[21]crown-7	3.40–4.30

acyclic ligands. Many of the receptors bind cations via electrostatic ion–dipole interactions, and in some cases are enhanced by the formation of hydrogen bonds. Some special classes may be distinguished:

Crown Ethers   Crown Ethers (see Chapter 2) are ubiquitous in supramolecular chemistry as hosts for both cations and neutral molecules. Depending on the size of the cycle they show great selectivity towards a given cation. Pedersen discovered that dibenzo-18-crown-6 shows great selectivity toward K$^+$ as compared to the Na$^+$ ion; in contrast, benzo-15-crown-5 has a special affinity for Na$^+$ due to the size of its cavity. Furthermore, Pedersen recognized that the increased solubility of the macrocycle in hydroxylic solvents in the presence of Na$^+$ was due to the crown binding a sodium cation. A relationship exists between the cavity size, cationic radius and stability of the resulting complex (Table 14.1).

The better the fit of the cation into the crown, the stronger the complex formed. This concept is referred to as *optimal spatial fit*. For example, [18]crown-6 forms the most stable complexes with K$^+$ whereas [21]crown-7, a large crown, binds Cs$^+$ more strongly than K$^+$. Furthermore, K$^+$ cation sits perfectly in the middle of its macrocycle. For smaller cations, such as sodium, the crown ether distorts, wrapping itself around the metal in an attempt to maximize the electrostatic interactions, but at the same time increasing the strain of the ligand. Larger cations such as cesium have to perch above one face of the macrocycle because they are too large to fit into the cavity. Indeed, although coordination generally takes place at the center of the macrocycle, there are also metal derivatives that show other types of coordination, for example, the "sandwich" type in K$^+$-(benzo[15]crown-5)$_2$. The inverse situation also occurs, with large crowns being capable of binding more than one cation at once; dibenzo[24]crown-8, for example, can encapsulate two sodium cations simultaneously.

Crown ethers can be functionalized with pendant arms containing additional coordinating groups: the *lariat* crown ethers (see Chapter 2). The effect of a lariat ether stabilizing the binding of K$^+$ cation, not in the center of the ether, is shown in Figure 14.1. For this reason lariat ethers are designed not only for complexing but mainly as carrier species to transport cations across lipophilic membranes because they possess higher binding constants than regular crowns, but remain kinetically labile (i.e. the rates of complexation and decomplexation are fast—a requirement for efficient transport) due to the flexibility of the arm.

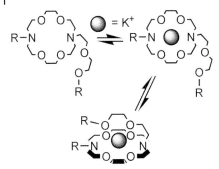

Figure 14.1 Complexation of K⁺ cation in a lariat ether.

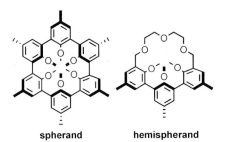

spherand   hemispherand

Figure 14.2 Scheme of a spherand and a hemispherand.

Heterocrowns The incorporation of softer donor atoms (e.g. N, S, P) results in noticeable changes in both the binding ability of the ligand and the geometry of the resulting complex. For example, vastly enhanced binding of the soft Ag⁺ (and corresponding drop in affinity of K⁺) occurs as soon as hard oxygen donors are exchanged for S or N. Polyaza- and polythia-ligands have been used in the recognition of transition metal ions. Their selectivity allows strong complexation of toxic heavy metals such as cadmium, lead and mercury.

Silicon analogs of crown-ethers and cryptands have been reported recently. The authors gave their work the title *"A New Chapter in Host–Guest Chemistry?"* [7].

Spherands Since the beginning of the history of crown ethers, new similar species but with better defined cavities have been synthesized: *spherands* and the hybrid *hemispherands* (Figure 14.2). While crown ethers are relatively flexible in solution, Cram realized that if a rigid host could be designed, which had donor sites that were forced to converge on a central binding pocket even before the addition of a metal cation, then strong binding and excellent selectivity between cations should be observed. For example, the spherand shown in Figure 14.2 forms extraordinarily stable Li⁺ complexes. All other cations are excluded (except, to a certain extent, Na⁺) because they are simply too big to fit between the binding pocket. They differ in their affinities for alkali metal cations by a factor of up to 10^{10}. Consequently, selected spherands can be used to remove Li⁺ and Na⁺ impurities to obtain ultrapure samples of potassium salts. *The spheransd are preorganized whereas the crown ethers are conformationally mobile.* The complexes when formed are termed *spheraplexes.*

Cryptands   As already commented in Chapter 2, whereas macrocycles define a two-dimensional, circular hole, macrobicycles define a three-dimensional, spheroidal cavity, particularly well suited for binding alkali and alkaline-earth cations. Indeed, macrobicyclic cryptand ligands form cryptates [$M^{n+} \subset$ cryptand] by inclusion of a metal cation inside the molecule. [2.2.1]cryptand is very selective for Na^+. The binding of K^+ by [2.2.2]cryptand in methanol is some 10^4 times stronger than its crown analog. The cation exchange kinetics of the cryptates is several orders of magnitude slower than those of the macrocyclic complexes and usually decreases as the stability of the complex increases.

They show pronounced selectivity as a function of the size complementarity between the cation and the intramolecular cavity, a feature termed *spherical recognition*. Selectivity depends on the length of the bridges. Plateau selectivity is observed for flexible cryptands, which contain longer chains and therefore larger, more adjustable cavities.

Calixarenes   As indicated in Chapter 2, the *calixarenes* are extremely versatile hosts and, depending on their degree of functionalization, may act as host for cations (mainly alkali and alkaline earth), anions and neutral molecules. The aromatic cavities of calixarenes are capable of coordinating to guest species. For example, a Cs^+ cation is held in the cavity of *p-tert*-butylcalix[4]arene via π–cation interactions (a molecule of acetonitrile is also coordinated to the metal cation). These π–cation interactions arise from favorable electrostatic interactions between the electron deficient cation and the electron-rich aromatic ring.

Different functional groups can be appended to the lower rim or, after removal of the *tert*-butyl groups, to the upper rim of the calixarene. Calixarene groups have been functionalized with crown ethers or spherands giving rise to complexes of great selectivity for certain alkali ions and the possibility of studying their mobility as membrane models. The 1,3-calix[4]arene-crown-5-ether receptor (Figure 14.3A), shows higher K^+/Na^+ selectivity than the natural ionophore valinomycin. The 1,3-alternate calix[4]arene-bis(crown-6-ether) (Figure 14.3B) and functionalized derivatives exhibit remarkable selectivity towards Cs^+ vs. Na^+ and to a lesser extent vs. K^+ [8a]. They have been extensively used for solvent extraction from nuclear waste solutions. The hexaester of calix[6]arene has been developed as a blood sensor for potassium [8b].

Functionalization of a calixarene with the same calixarene, i.e. a bis-calix[4]arene scaffold (known as *calix[n]tubes*) (Figure 14.3C) leads to extraordinary K^+ selectivity. Molecular modeling studies confirm that the potassium metal cation is complexed via the axial route, passing through the calix[4]arene annulus. This selectivity is accompanied by modifications of the $O-CH_2-CH_2-O$ chains of the calix[4]tube, the K^+ ion being placed at the center of a distorted cube [9]. For ions smaller than K^+, e.g. Na^+ and Li^+, the ion proceeds more readily to the center of the tube, while

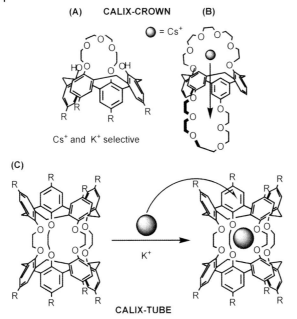

Figure 14.3 Recognition of alkali cations in calixarene derivatives (see text).

for ions larger than K⁺, e.g. Tl⁺, Rb⁺, Cs⁺, the ion remains in the intermediate position close to the aromatic rings.

14.2.1.2 Selectivity of Cation Complexation

The thermodynamic selectivity of a given host for a particular cation represents the ratio between the host's affinity for a given metal and other guest cations. Selectivity is governed by a large number of factors, such as: (i) size match between cation and host cavity, (ii) electrostatic charge, (iii) solvent (polarity, hydrogen bonding and coordination stability), (iv) degree of host preorganization, (v) enthalpic and entropic contributions to the cation–host interaction, (vi) cation and host free energies of solvation, (vii) nature of the counter-anion and its interactions with the solvent and the cation, (viii) cation binding kinetics and (ix) chelate ring size.

Crown ethers exhibit *plateau selectivity*, indicating that they have a similar affinity for several metal ions. They have poor discrimination between Na⁺ and K⁺ or between Rb⁺ and Cs⁺. This may be attributed to the two-dimensional nature of the complexation, which enables the crown ligands to accommodate a wide variety of metal cations, irrespective, to some extent, of ionic radius (Table 14.1). This selectivity may be improved in the *lariat ethers*, due to the presence of the pivoting, podand arm designed to bring some tridimensionality of cation complexation.

In contrast to the crown ethers, the three-dimensional cryptands display *peak selectivity*. The cryptand cavity is much more preorganized for cation binding, meaning that there is less entropically, and often enthalpically, unfavorable conformational rearrangement that must take place in order to adopt the optimum complex geometry. For [2.2.2]cryptand the K^+/Na^+ selectivity is about 10^3, whereas for [18]crown-6 under the same conditions the discrimination is less than 10^2. The small [2.1.1]cryptand is ineffective at binding any of the divalent ions Mg^{2+}– Ba^{2+}. Increasing the size, [2.2.1]cryptand has a peak selectivity for Sr^{2+} (but with similar stability for Ca^{2+} and Mg^{2+}). [2.2.2]cryptand is an extremely effective ligand for Ba^{2+}.

14.2.1.3 Anionic Recognition [10]

Anions play essential roles in many processes, both chemical and biological, and this makes their selective recognition an area of intense interest. The first halide complexants, were termed *katapinands* (Greek, καταπινω = swallow up) (Figure 14.4A).

Anions are relatively large and therefore require receptors of considerable greater size than cations. For example, one of the smallest anions, F^-, is comparable in ionic radius to K^+ (1.33 Å vs. 1.38 Å). The geometries for anion complexes are strikingly similar to those observed in transition-metal coordination chemistry. Indeed, simple inorganic anions occur in a range of shapes and geometries, e.g. spherical (halides), linear (SCN^-, N_3^-), planar (NO_3^-, CO_3^{2-}), tetrahedral (SO_4^{2-}, PO_4^{3-}), octahedral (PF_6^-), as well as more complicated examples.

The negative charge of the anions suggests that both neutral, and especially positively charged, hosts will bind anions. Some neutral macrocyclic and macro-polycyclic amide-based ligands have been investigated for their anion recognition capability. A significant effect of the ring size on the stability constants of anion complexes has been demonstrated: the 20-membered macrocyclic tetramide is a better anion receptor than both its 18- and 24-membered analogs (Figure 14.4B). It forms the most stable complexes with all anions, irrespective of their size. Polyamide-based cryptands give rise to similar effects. These organic hosts are

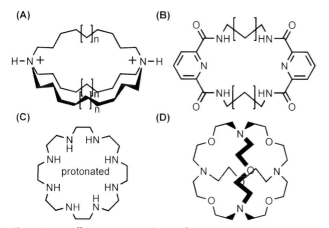

Figure 14.4 Different receptors (hosts) for anionic recognition.

promising *oxoanion extractants*. Calixarene, calix[*n*]pyrroles, and other similar systems have also been studied as anion receptors.

Positively charged hosts, such as those depicted in Figure 14.4C (and other larger ones) have been extensively studied for their capability of coordination to large inorganic anions such as $[PdCl_4]^{2-}$ and $[Fe(CN)_6]^{4-}$ [2]. The term "*supracomplexes*" was adopted for these complexes of metal complexes.

14.2.1.4 From Anionic Recognition to Cationic Recognition: A Simple Change in pH

The macrotricyclic cryptand shown in Figure 14.4D has been termed a "*soccer ball*" molecule because of its nearly perfect spherical shape. The presence of the four nitrogen bridgeheads on this molecule makes it a highly versatile example of a tetrahedral receptor. In addition to its Lewis basic properties, which enable it to bind strongly to cations, in its tetraprotonated form it is also capable of binding anions such as Cl^-, $[Cl^- \subset ligand, 4H^+]$, acting via the formation of four hydrogen bonds supported by electrostatic interactions with the ether oxygen atoms. Thus, depending on the pH of the medium, the neutral host may bind cations, especially ammonium; in its diprotonated state, it binds neutral molecules such as water, $[H_2O \subset ligand, 2H^+]$; and in its tetraprotonated state, it binds anions such as Cl^-. This macrotricyclic ligand behaves, according to the expression of Lehn, "*like a sort of molecular chameleon responding to the pH changes in solution*" [3].

14.2.1.5 Neutral Molecules Recognition: Binding of Neutral Molecules

In spite of its great interest, this kind of *molecular* recognition is not a typical field within coordination chemistry, because the included molecules are usually organic molecules. The interested reader can consult the already mentioned textbook on Supramolecular Chemistry [2]. We will give only a limited outlook on this subject.

In general, individual host molecules possessing an intrinsic cavity that is present in both the solid state and in solution are termed *cavitands*. A cavitand is, thus, a molecular container with an enforced concave surface. The molecular cavity is necessarily open at least at one end. Inclusion of guest species within a cavitand results in a *cavitate* (or *caviplex*). Calixarenes (see above), resorcinarenes, cyclodextrins, cucurbiturils, etc., are examples of synthetically accessible cavitands.

The *cyclodextrins* were the first receptor molecules whose binding properties were recognized and extensively studied. Cyclodextrins (CDs) are important host compounds, with a wide variety of industrial uses in the food, cosmetics and pharmaceutical sectors, generally as slow-release and component-delivery agents. They are cyclic oligosaccharides comprising (usually) six to eight D-glucopyranoside units (α-, β- and γ-cyclodextrins), linked by a 1,4-glycosidic bond (Figure 14.5). The shape of a cyclodextrin is often represented as a tapering torus or truncated funnel with two different faces. The toroidal structure of CDs has a hydrophilic surface resulting from the 2-, 3-, and 6-position hydroxyl groups (faces), and the cavity is composed of glucoside oxygens and methylene hydrogens, giving it a

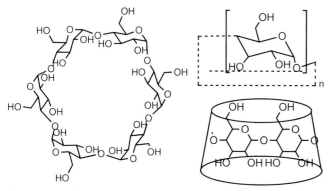

Figure 14.5 Different schematic views of cyclodextrines.

hydrophobic character. As a consequence, CDs can include apolar molecules of appropriate dimensions inside their cavity, making them more water-soluble. As a typical coordination example, the ferrocene (neutral) complex gives an inclusion compound more easily than the ferrocinium cation (positively charged, polar), in β-CD [11a]. Metallated cyclodextrins are able to fully encapsulate medium-sized inorganic anions [11b]. A review of cyclodextrins as supramolecular hosts for organometallic complexes has been published recently [11c].

Carcerands are closed molecular containers or capsules without portals of significant size. Guest species are permanently *"incarcerated"*. Cram was the first to be able to couple the upper rims of two resorcinarene bowls to obtain the first carcerands. The carcerand–guest complex is called a *carceplex*. Carceplexes are comprised of host and guest components that cannot separate from one another without the breaking of a covalent bond. *Hemicarcerands* describe closed molecular containers from which the guest can enter and exit with a measurable activation barrier, forming *hemicarceplexes* with the guest. These receptors then possess larger entry and exit portals around the recognition site and allow slow exchange of guest molecules. Usually heat is required to force the guest out of the cavity entropically.

The *template effect* of the guests in forming the corresponding carceplexes or hemicarceplexes has been underlined since the beginning of this chemistry, when Cram proposed the idea of a carceplex in 1983. Indeed, such cages will often only form in the presence of a guest, which acts as a template.

In addition to intrinsically curved species, there exists a wide range of hosts that may be broadly termed cyclophane hosts: *cyclophane* means any organic host molecule containing a bridged aromatic ring. Cyclophane hosts commonly bind both neutral molecules and organic cations. A striking case, from the inorganic viewpoint, is the crownophane schematized in Figure 14.6, with two hydroxy and two amide groups around the cavity. It is able to include carbonic acid formed from both carbon dioxide and water molecules to give a stable 1:1 adduct at room temperature [12]. It should be noticed that the amide group plays a crucial role in

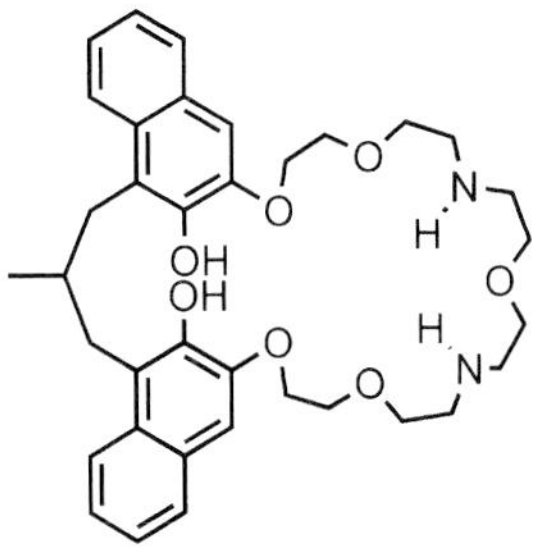

Figure 14.6 Crownophane able to include carbonic acid (see text).

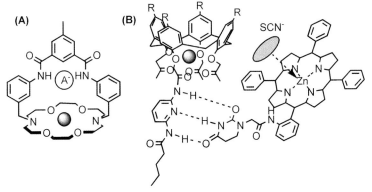

Figure 14.7 Scheme of two cases of multiple recognition (cation and anion) (see text).

fixing carbonic acid at room temperature. The same ring with only O atoms does not absorb CO_2 molecules. *Cryptophanes* are capsular hosts with bridged aromatic rings. They are like the cryptands of the cyclophanes.

14.2.1.6 Molecular Coreceptors and Multiple Recognition

The following step is to consider polytopic coreceptor molecules containing several discrete binding subunits which may cooperate for the simultaneous complexation of several substrates. Polytopic receptors are *homotopic* or *heterotopic*, depending on whether they contain identical or different subunits, respectively. *Heterotopic coreceptors* present, usually, *haptoselectivity*, i.e., preferential fixation of a given substrate at a given subunit, resulting from local complementarity of size, shape and binding interactions. Some typical examples follow.

The ditopic, macrobicyclic receptor of Figure 14.7A has adjacent anion and cation binding sites. Thus, it is able to extract a range of monovalent salts, such as KAcO, $LiNO_3$, $NaNO_3$, KNO_3, $NaNO_2$, into chloroform solution [13].

A self-assembled bifunctional receptor is shown in Figure 14.7B. The strategy applied is based on the assembly of both receptors through hydrogen bonding. The cation receptor calix[4]arene complexes Na^+ with high selectivity. The SCN^- anion is very selective in the Zn-porphyrin [14]. The significant enhancement of the hydrogen bonding as well as the anion complexation points to a cooperativity in the individual interactions. The assembly can be "switched on" by complexation of the Na^+ ion and "switched off" by addition of MeOH, which is able to destroy hydrogen bonds.

14.3
Supramolecular Dynamics: Reactivity, Catalysis and Transport Processes

A receptor–substrate supramolecule is characterized by its *internal dynamics*. Indeed, molecular receptors bearing appropriate reactive groups in addition to binding sites may complex a substrate, react with it (with given rate, selectivity, and turnover) and release the products, thus regenerating the reagent for a new cycle.

Supramolecular reactivity thus involves two main steps: *binding*, which selects the substrate, and *transformation* of the bound species into products within the supramolecule formed. Catalysis additionally comprises a third step, the *release* of the substrate.

Furthermore, *transport* represents one of the basic functional features of supra-molecular species, together with recognition and catalysis. The so-called *carrier-mediated transport* consists of a transfer of a substrate across a membrane, facilitated by a carrier molecule located in the membrane, giving a cyclic process. In fact, selective alkali cation transport was one of the initial objectives of Lehn's work on cryptates. Natural acyclic and macrocyclic ligands (such as valynomicin and others) were found early on to act as selective ion carriers, *ionophores*, and have been extensively studied.

14.4
The Self-assembly Concept and Its Application in Molecular and Supramolecular Chemistry [2, 3, 15]

According to Lehn "*Self-assembly can be taken to designate the evolution towards spatial confinement through spontaneous connection of a few/many components, resulting in the formation of discrete/extended entities at either the molecular (covalent) or the supramolecular (non-covalent) level … The name tecton (from τεκτων: builder) has been proposed for designation of the components that undergo self-assembly*" [3].

Self-assembly of a supramolecular architecture is a multi-step process allowing the spontaneous but controlled generation of complex organic or inorganic archi-tecture on the basis of the molecular information stored in the components. Such processes connect input components with output entity(ies), with a fidelity/reliability depending on the robustness of the program, that is, its ability to resist interference from factors other than the directing/dominant coding interactions.

Supramolecular self-assembly requires *reversibility* of the connecting events, i.e. kinetic lability and rather weak bonding (compared with covalent bonds). Kinetically labile metal–ligand interactions are, by nature, dynamic, as the coordinative bonding may, in principle, dissociate and associate under the influence of the chemical and physical conditions of the environment.

> Self-assembly can be molecular or supramolecular, according to the strength and lability of the bonds.

Many other authors have tried to contribute with their ideas to a best development and final clarification of this concept. In 2002, Whitesides and Grzybowski published an article entitled "Self-assembly at all Scales" [16]. They state: "*Self-assembly is the autonomous organization of components into patterns or structures without human intervention ... Here, we limit the term to processes that involve pre-existing components (separate or distinct parts of a disordered structure), are reversible, and can be controlled by proper design of the components ...*". They add, "*Self-assembly is not a formalized subject, and definitions of the term 'self-assembly' seem to be limitlessly elastic. As a result, the term has been overused to the point of cliché.*"

Stoddart creates the term "strict self-assembly" [6]: "*This term applies to pathways that produce a final product directly and spontaneously when the correct components are mixed under appropriate conditions. The pathway must be completely reversible and the product should be stable at thermodynamic equilibrium. In addition, the constituent components of the final structure must contain all the information necessary for correct assembly to occur. The self-assembly of the DNA double helix from two complementary oligonucleotides is perhaps the most well-known and intensively studied biological paradigm of strict self-assembly ... The non-strict self-assembly assumes covalent modifications, either in natural or unnatural systems*".

Summary The terms molecular and supramolecular self-assembly have similar meanings, which are intimately related, however, they are different. Molecular self-assembly is static (strong bonds) and can promote the synthesis of very complicated and exciting inorganic architectures. Supramolecular self-assembly is dynamic and reversible (weak/labile bonds): on mixing different appropriately designed components in solution, the intermolecular forces that exist between them control their orientation, leading to the *reversible* assembly of a specific "supramolecule".

14.5
Metallosupramolecular Chemistry: Different Strategies and Types

The self-assembly of inorganic architectures was named by Constable "Metallosupramolecular Chemistry" [5]. In metallosupramolecular chemistry the non-covalent interactions are coordinative bonds. In fact, this kind of bond cannot really be considered as non-covalent due to a high covalency of the interaction. However, *in the case of* ligand–metal ion systems, in which thermodynamically stable but *kinetically labile* complex units are involved, the reversibility of the complex formation enables "self-assembly".

The emphasis lies in the design of the ligands and the choice of the metal ions in order to produce defined architectures in a controlled fashion from multiple subunits. Metal ions have properties of special interest as components of (supra)molecular systems. They provide (i) a set of coordination geometries, (ii) a range of binding strengths, from weak to very strong, (iii) a range of formation and dissociation kinetics, from labile to inert.

Chemists have developed a number of strategies for generating metallo[supra]molecular assemblies. We will give a brief explanation of the main types:

14.5.1
Helicates [17]

The term helicate was introduced by Lehn and coworkers in 1987, for the description of a polymetallic helical double-stranded complex. Figure 14.8 represents a typical helication process in which a double helical ligand array is assembled from the interaction of two ligands with two metal ions. In a helicate it is necessary to emphasize the importance of the two components: the metal ion and the organic ligands which favor the self-assembly process. *A one-dimensional helicoidal complex is not a helicate.*

The ligands should possess (i) several binding units along the strand allowing the recognition and coordination of the various metal ions; (ii) judicious spacers between the binding units which are rigid enough to prevent the coordination of several binding units of one strand to the same metal ion, but flexible enough to undergo helication and to wrap around the metal ions to produce stable polynuclear complexes.

As pointed out by Albrecht et al. [18], in principle, helicates are Werner-type coordination compounds with two (or more) complex moieties connected by

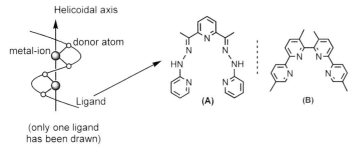

Figure 14.8 Scheme of helicates. Two ligands that give helicates (see text).

(A)

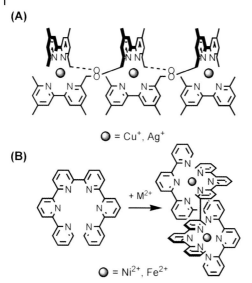

= Cu$^+$, Ag$^+$

(B)

+ M^{2+}

= Ni^{2+}, Fe^{2+}

Figure 14.9 Two ligands than can give helicates with T_d or O_h coordination environment.

spacers. For example, double-stranded helicates are formed from linear ligand strands with bidentate chelating units and a tetrahedrally coordinated metal ion. However, due to the simplicity of helicates, they developed over the years to a kind of metallosupramolecular chemists "drosophila": they allow the study of fundamental mechanistic principles of self-assembly processes.

The ligand strands and the metal ions are the basic point of the classification of helicates [17]. First, the number of coordinated strands is associated with *single- double-, triple*-stranded helicates, possessing, respectively, one, two or three coordination strands wrapped around the metal ions. Double-stranded helicates with tetrahedral coordination were the first to be synthesized, in a serendipitous manner, from the ligand shown in Figure 14.8A. The first attempts with a predetermined strategy were carried out by Lehn, Sauvage and Ziessel, from the tetrapyridine ligand shown in Figure 14.8B, obtaining the new systems [Cu$_2$L$_2$] and [Ag$_2$L$_2$].

In general, these kinds of helicates with tetrahedral coordination are easily formed by self-assembly of cations with a tendency to T_d geometry and bpy derivative ligands. A typical scheme of these complexes is given in Figure 14.9A. Six-coordination is also very common if the bpy units are substituted by tridentate groups. With this strategy many helicates of general formula [M$_2$(L)$_2$]$^{4+}$ (M=Cd^{2+}, Mn^{2+}, Fe^{2+}, Ni^{2+}, Cu^{2+}) have been reported (Figure 14.9B). Most of the triple-stranded helicates present hexacoordinated metal ions (Figure 14.10).

It is important to note that many helicates need the template effect of cations for their formation. In the absence of templating cations only mixtures of oligomeric compounds are observed [18].

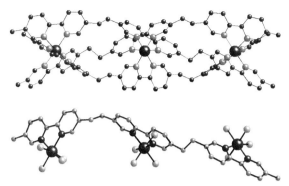

Figure 14.10 (A) Crystal structure of a triple stranded helicate with c.n. 6 for the cation; (B) part of the structure showing only one strand.

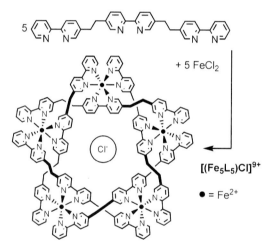

Figure 14.11 Cyclic helicate which needs the template effect of the Cl⁻ anion (see text).

14.5.1.1 Cyclic Helicates

A polydentate ligand strand combined with metal ions does not systematically give polynuclear double(triple)-stranded helicates but, in some cases, cyclic helicates are obtained. Two extraordinary and beautiful cyclic helicates have been reported by Lehn (Figures 14.11 and 14.12). In the first, $[Cl \subset Fe_5^{II}(L)_5]^{9+}$, the Cl⁻ anion is *necessary* for the self-assembling process [19]. With SO_4^{2-} a different $[Fe_6^{II}(L)_6]^{12+}$ entity is formed, with sulfate as counteranion. The Ni^{2+} with the same ligand forms a triple-stranded helicate. The second, $[Cu_{12}(L)_4]^{12+}$, is made up from four ligand molecules and twelve metal ions. It has nanometric dimensions with an external diameter of 28 Å; the central cavity has a diameter of 11 Å and contains four PF_6^- anions as well as solvent molecules [20].

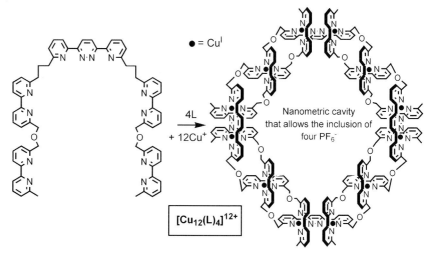

Figure 14.12 Cyclic helicate with a nanometric cavity in the centre (see text).

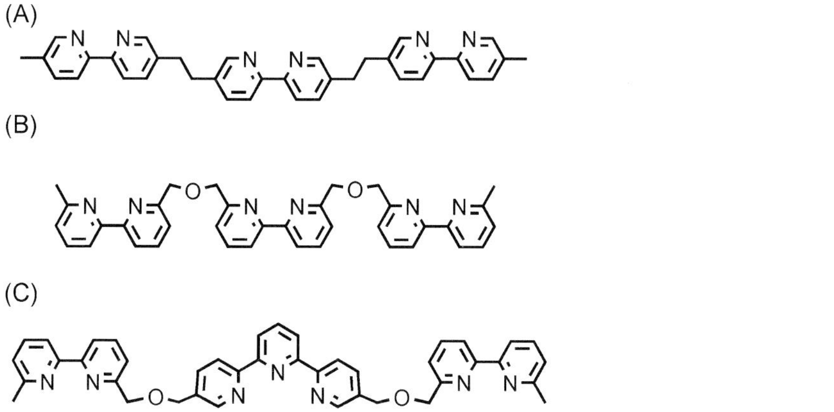

(A)

(B)

(C)

Figure 14.13 Different ligands that allow recognition of different cations (see text).

Remark Cyclic helicates are currently in equilibrium with non-cyclic ones or even with other metallosupramolecular architectures. This feature is essential for attributing the *supramolecular nature* to most of them.

14.5.1.2 Self-recognition in Helicates [21]

Self-assembly and molecular recognition are complementary. When a mixture of the two tris-bipyridine ligands shown in Figure 14.13A,B is allowed to react simultaneously with copper(I) and nickel(II), only the double helicate with copper(I) and A (less flexible) $[Cu_3(A)_2]^{3+}$ (c.n. = 4) and the triple helicate of nickel(II) and B (more flexible), $[Ni_3(B)_3]^{6+}$ (c.n. = 6) are formed. Bidentate diimines are coded for tetrahedral metal ions, and tridentate terimines are coded for octahedral metal ions (Figure 14.13C). Lehn and coworkers have recently published an article discussing

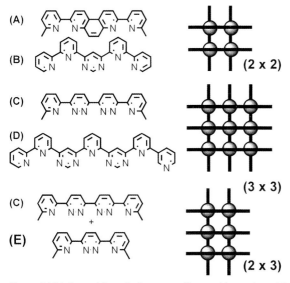

Figure 14.14 Several ligands that can self-assemble to give grids. Scheme of different grids.

the self-assembly of the helicates as well as the selective generation of other entities [22].

14.5.2
Grid-type Metal Ion Architectures

The design of grid-like metal ion arrays (Figure 14.14) requires perpendicular arrangement of the ligand planes at each metal center. Ligands containing either bidentate or tridentate binding subunits in combination with metal ions possessing tetrahedral or octahedral coordination are required [23]. Some examples are: Ligand A (Figure 14.14) can self-assemble with 4 Cu^+ or Ag^+ metal ions, giving a grid molecule [2 × 2], in which metallic ions occupy tetrahedral holes, whereas ligand B will do the same, but with occupancy of octahedral holes (Co^{2+}). [3 × 3] and [3 × 2] grids are shown in the same figure. Many other different types of ligands have been reported for self-assembly of grids [24].

There have been extensive redox and magnetic studies on these systems, opening new perspectives in photochemical properties, redox, and molecular information storage and processing.

Solvent-modulated *reversible* conversion of a [2 × 2]-grid into a pincer-like mononuclear complex of Co(II) has been reported recently, implementing three general features: (i) adaptation to external/medium effects, (ii) dual ligand plasticity and (iii) effector-induced facilitation of structural interconversion [25]. Such processes are of much interest because they represent an adaptive response of a metallosupramolecular architecture to environmental changes.

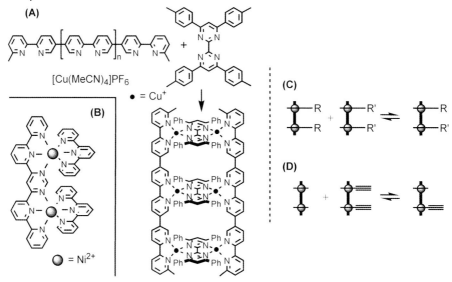

Figure 14.15 (A, B) Scheme of a ladder and a rack (see text); (C, D) scheme of two supramolecular racks, showing chemical reversibility.

14.5.3
Ladder and Racks Architectures

The self-assembly process is, in these cases, multicomponent. Typical examples of ladders and racks are shown in Figure 14.15A and B, respectively. Schmittel and coworkers have demonstrated a very important feature in some multicomponent racks based on copper phenantroline and terminal alkines, i.e., the ligand exchange that allows the reversible nature of the aggregates (Figure 14.15C, D) [26]. Now we are truly entering into the realm of supramolecular self-assembly ...!

14.5.4
Discrete Nanostructures Mediated by Transition Metals.
Molecular Paneling via Coordination [27]

The terms "*molecular architecture*" or "*molecular paneling*" indicate the synthesis of structures specifically designed to exhibit novel specific shapes in two or three dimensions. Since the first report in 1990 [28], of the polynuclear square [Pd(en)(4,4'-bpy)]$_4$(NO$_3$)$_8$ (Figure 14.16) the design of self-assembled molecular architectures has received considerable attention.

Thus, through directional coordinative bonding, two- and three-dimensional self-assemblies are readily available by the *spontaneous* combination of metal centers with appropriate organic ligands (Figure 14.17). Each unit has, in principle,

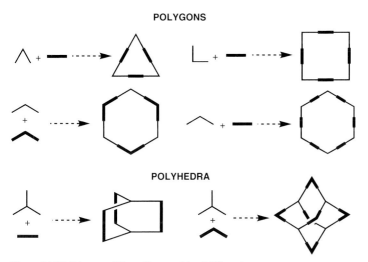

Figure 14.16 The first self-assembled Fujita's polygon.

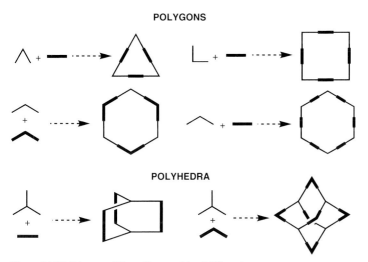

Figure 14.17 Scheme of the self-assembly of different fragments to give polygons or polyhedra.

the programmed information to get a predicted structure (size, shape, etc.). The requirements for a self-assembly of a supramolecular metallocycle (or polyhedron) are: (i) the coordination bonds must be kinetically labile so as to allow self-correction and (ii) the desired assembly must be thermodynamically more favorable than any competing species. Some representative examples are given in the following paragraphs.

14.5.4.1 Polygons (or Metallocycles)

The most numerous self-assembled polygons are, for entropic reasons, triangles and squares (or rectangles). The scheme of the self-assembly of a square metallo-cycle is shown in Figure 14.16 [28]. The square complexes have a cavity similar to those of cyclodextrins. In water, they offer a hydrophobic cavity, with the ability to recognize neutral aromatic compounds. The versatility of this approach has been studied [29]. With a different perspective, chiral molecular squares with metallo-corners for enantioselective sensing have been reported [30].

It is noteworthy (from a supramolecular point of view) that the assumption that "90° + linear ligand = a square" is not always true. The self-assembly of a square lies in a delicate thermodynamic balance between enthalpy and entropy effects and is often accompanied by the formation of a triangle, which is favored in terms of entropy because it is formed from fewer components than a square. The square/triangle ratio depends on the length and rigidity of the ligand and other factors. The spacer is very important in this dynamic equilibrium. A dynamic equilibrium between metallosupramolecular squares and triangles with Pd(II) and Pt(II) and a new rigid linker 1,4-bis(4-pyridyl)tetrafluorobenzene, has been demonstrated by Ferrer and coworkers [31]. The square/triangle ratio depends on several factors, such as the nature of the metal corners, the concentration, and the solvent. Stang et al. have proved that the selective crystallization of either of the two species can be accomplished by the appropriate choice of solvents and the ratio of anions present in the system [32].

A new equilibrium of Pd(en)-based molecular triangle/square compounds, with 4,4'-bpy as bridging ligands, has been reported recently [33]. The equilibrium largely depends on the solvents and, in this particular case, on the presence of diverse polyoxometalates (POMs), which stabilize the squares. The molecular squares encapsulate some small POMs.

A *reversible* interconversion of copper(II) homochiral triangular macrocycles and helical coordination polymers has been reported recently [34]. The major difference between the connecting metal centers in these two structures is the weakly coordinating axial ligand to the copper ion. In the case of the polymer, the axial ligand is methanol, while in the triangle it is pyridine. The axial ligand is, thus, the "cause" of the reversibility.

Larger polygonal structures are scarcer, as a direct consequence of their entropic disadvantage. A "rational approach" to molecular pentagons can be found in Ref. [35].

14.5.4.2 Polyhedra: Molecular Paneling

Five years after the first report on the square complex, Fujita and coworkers extended the two-dimensional structure into a three-dimensional octahedron. Substituting the 4,4'-bpy linear ligand by an exo-tridentate triangular ligand, 2,4,6-tris(4-pyridyl)-1,3,5-triazine, the nanometer-scale adamantane-like $[M_6L_4]$ symmetric framework of Figure 14.18 was assembled [36]. Later, they synthesized other different tridentate triangular ligands (*molecular panels*) similar to the former, studying their complexation with $[Pd(en)]^{2+}$ block [37].

Self-assembly of many polyhedral solids has been achieved through Pd-L and Pt-L. From a chemical point of view palladium or platinum may be employed without distinction, to obtain identical architectures. However, platinum has the advantage (or disadvantage, according to the desired properties) of giving more

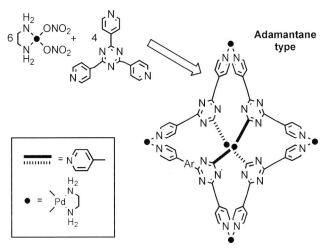

Figure 14.18 Self- assembly process giving an adamantane type polyhedron.

kinetically stable systems. In this sense palladium(II) systems are, per se, "more supramolecular".

Tetrahedral structures, [M_4L_6], have also been reported [38]. The four metal ions act as the vertices and the six ligands act as the edges of the tetrahedron. The {$NH_4 \subset [Fe_4L_6]$} system, with T symmetry, is shown in Figure 14.19A. A triple helicate [M_2L_3] has the same ligand to metal ratio. Thus, it is not surprising than in certain cases a *supramolecular equilibrium* between these two self-assembled species exists. For example, [Ti_4L_6] (Figure 14.19B), is formed in presence of an alkylammonium guest. However, in the absence of this guest, only the [Ti_2L_3] triple helicate is formed [38]. The same ratio (2:3) is found in the truncated tetrahedral chiral [$M_{12}L_{18}$]$^{24+}$ cage (M=Co^{2+}, Zn^{2+}), in which 18 bridging ligands, L, span an edge of the polyhedron [39]. The cube [$Ni_8(tab)_{12}$]$^{16+}$ (tab = 1,2,3,4-tetraaminobutane) also has the same M_2L_3 ratio [39], as well as the [$M_{16}L_{24}$] cuboctahedron or tetracapped truncated tetrahedron [40].

The assembly of [M_2L_4] tetragonal cages remains quite rare, although a few such d- and f-block metal complexes have appeared in the literature recently, such as the [Ln_2L_4] lanthanide cages [41]. The interest in these "molecular lanterns" lies in the [$Ln(H_2O)_8$]$^{3+}$ guests within the cages, which are mobile and are readily replaced reversibly by other Ln^{3+} analogs with the preservation of the original cage structure.

In [M_4L_4] tetrahedral clusters the metal ions occupy the four vertices and the ligands occupy each of the four faces of the tetrahedron (Figure 14.20) [42].

With triangular panels, [M_6L_8] octahedral [43], or truncated octahedral structures[44], have been reported recently. A nanometer-sized hexahedral coordination capsule assembled from 24 components, [$Pd_{18}L_6$]$^{36-}$ (L = triangular assembly unit)

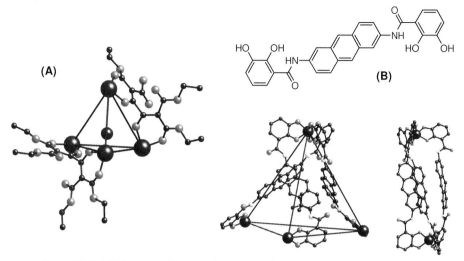

Figure 14.19 (A) Fragment of the crystal structure of {NH$_4$ ⊂ [Fe$_4$L$_6$]} (small black ball = NH$_4^+$; big black balls = iron); (B) supramolecular equilibrium in [Ti$_4$L$_6$], showing the ligand L (see text).

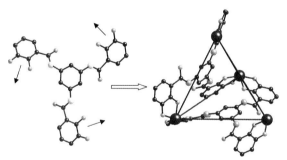

Figure 14.20 Crystal structure of a [M$_4$L$_4$] tetrahedral cluster (black balls = metal ions). View of the corresponding ligand (left).

with an internal volume of 900 Å^3, fully closed but with very small molecules or anions inside the cavity, such as NO$_3^-$, has also been reported [45]. Indeed, in many cases, a different number of cations as well as solvent molecules are observed in the internal cavities.

In addition to triangles, different ligands as molecular panels able to generate prisms, tubes (nanotubes) and box structures have also been reported [46].

As well as molecular polygons, molecular polyhedra are often in equilibrium when their thermodynamic stabilities are comparable. The equilibrium ratio can usually be controlled by changing the conditions (concentration, temperature, solvent, etc.) or by adding a template molecule that selectively binds to and stabilizes one particular assembly [47]. A recent example of solvate-control has been reported in the assembly of $[Pd_3L_6]$ (with MeCN) and $[Pd_4L_8]$ (with DMSO) coordination boxes (L = 1,2-bis[2-(pyridin-4-yl)ethynyl]benzene). It has been demonstrated that the two complexes can be interconverted by simply adding or removing a solvent [48].

Remark It is worth mentioning that in the articles of Fujita and coworkers, they speak of "*molecular self-assembly*" or "*metal-directed-self-assembly*". They do not use the terms "*supramolecular self-assembly*" or "*supramolecular structures*".

14.5.4.3 Cationic and Anionic Control

In several self-assembly processes, the template effect of cations or anions is absolutely necessary for the synthesis of the polyhedra. Paradigmatic examples can be found in Refs. [49] (cationic templation) and [50] (anionic templation: the synthesis of a "tennis ball").

14.6
Encapsulated Guests in Metallo-nanostructures

The construction of large containers from the self-assembly process is mainly focused on the manipulation of chemical reactivity through encapsulation. In general, the reactivity and properties of the guest molecules are different from those in the bulk phases. Some of the $[Ga_4L_6]^{12-}$ cage compounds, for example, encapsulate guest molecules ranging in size from tetramethylammonium to decamethylcobalticinium. Encapsulation can modify or stabilize, in certain cases, the reactivity of guest species: the tropylium cation forms a 1:1 host–guest complex with the $[Ga_4L_6]^{12-}$ assembly, greatly slowing its decomposition [51]. Furthermore, these tetrahedral cages show important catalytic effects in organic reactions [52, 53]. Recently the mechanism of guest exchange of NR_4^+ and PR_4^+ has been demonstrated, indicating that guest exchange does not involve partial dissociation or rupture of the host structure. Guests squeeze through apertures in the host structure and not through larger portals created by partial assembly dissociation [54, 55].

14.7
Supramolecular Assistance in the Synthesis of Molecular (and Supramolecular) Structures

In certain cases of synthesis, many researchers prefer to distinguish between "supramolecular synthesis" and "supramolecular assistance to molecular synthesis" [3, 56], i.e. the synthesis of discrete molecular entities – not supramolecular – held together using covalent bonds, aided by intermolecular, non-covalent interactions. The best examples refer to the "*interlocked molecules*", i.e. rotaxanes, catenanes and knots.

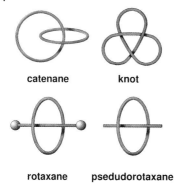

| catenane | knot |

| rotaxane | pseudorotaxane |

Figure 14.21 Schemes of a catenane, knot, rotaxane and peudorotaxane.

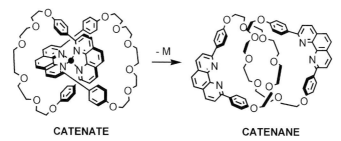

CATENATE   CATENANE

Figure 14.22 Schemes of a Sauvage's catenane and catenate.

14.7.1
Rotaxanes, Catenanes and Knots

Catenanes, rotaxanes and knots are structures with interlocked groups and are shown schematically in Figure 14.21. The introduction of copper(I)-template strategies by Sauvage and colleagues in the mid-1980s marked the first practical synthetic routes to molecular catenanes, rotaxanes and knots [57].

14.7.1.1 Catenanes, Catenates

Catenane motifs involve the interlocking of two cyclic structures, as shown in Figure 14.21. If two such cycles are simultaneously bound to a single metal ion, the resulting motif is termed catenate. Multiple interlocked complexes are also known.

Sauvage's Catenates The catenane of Figure 14.22 was synthesized by Sauvage and coworkers, thanks to the template effect of the copper(I) [57]. The same group has published many variations on the same architecture [58]. They form organized structures in their bulk states over a wide range of temperatures, being particularly promising in relation to molecular machines.

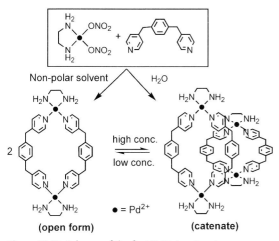

Figure 14.23 Scheme of the first Fujita's catenate.

Fujita's Catenates Fujita et al. reported, in 1994, the palladium ring shown in Figure 14.23 [59a]. The Pd(II)—N bond is weak and its formation is reversible. There is an equilibrium between the catenane and the metallo-ring: in this case, therefore, the catenane is certainly supramolecular. There is a remarkable medium effect on the equilibrium. Less polar media push the equilibrium toward monomer because hydrophobic interaction is reduced in the less polar media. In contrast, this equilibrium is strongly pushed toward catenane by employing a more polar medium which enhances the hydrophobic interaction.

The analogous Pt^{2+} system, once assembled is completely stable, because the Pt(II)—N bond is very strong. It has created a *"molecular lock"*. This bond can be likened to a lock because it is irreversible ("locked") under ordinary conditions but becomes reversible ("released") in a highly polar medium at high temperature. The lock is then released by adding a salt ($NaNO_3$ for example) and heating at 100 °C, allowing the self-assembly of a catenane framework. Finally, this framework is locked by removing the salt and cooling. However, the salt-assisted thermal switching (locking/unlocking) is not a clean and facile process. Fujita has recently demonstrated that this system is a clean photoswitchable molecular lock, through reversible photolabilization of the Pt—N bond. The irradiation in water produces the catenation; irradiation of the catenate in MeCN/MeOH releases the two metallo-rings of the catenate [59b]. Other similar catenates can be found in Ref. [59c].

Puddenphatt's Catenates Puddenphatt and coworkers have developed a new series of supramolecular [2]catenane with gold(I) macrocycles [60]. Indeed, gold(I) is a versatile building-block because of the low steric effect associated with its tendency to linear coordination and the potential to form Au···Au aurophilic attractions that can enhance association between gold(I) centers. Gold(I) is a labile metal center and so easy rearrangements can occur to give thermodynamic control

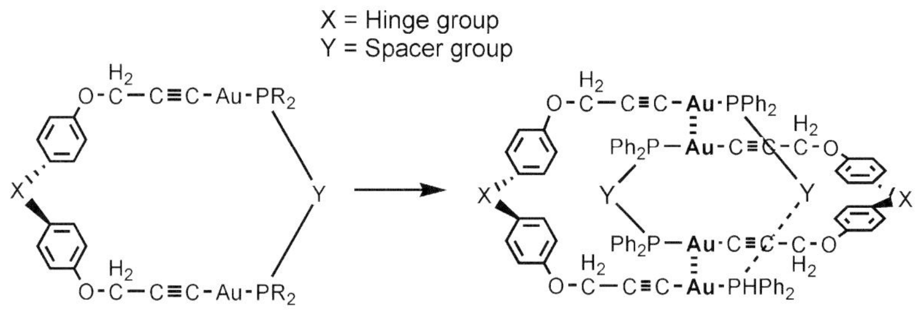

X = Hinge group
Y = Spacer group

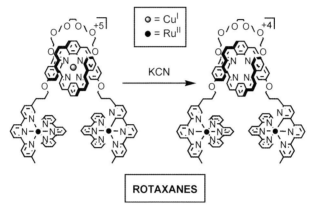

Figure 14.24 Scheme of a Puddenphatt's catenate, showing the aurophilic interactions.

ROTAXANES

Figure 14.25 Scheme of a rotaxane (see text).

of the self-assembly process. One of these catenates is shown in Figure 14.24. Both the "hinge group" (X) and the "spacer group" (Y) determine whether either [2]catenane or macrocyclic complexes are formed and the nature of the equilibrium between them.

14.7.1.2 Rotaxanes

A rotaxane is an interlaced moiety in which a filamentous species, stoppered at each end, is threaded through a cyclic one. The stoppers can be diverse, such as [Ru(terpy)$_2$] units (Figure 14.25), fullerenes, metalloporphyrins, etc. [61]. Pseudorotaxanes are not sterically trapped in their interlace state by the presence of bulky stoppers.

14.7.1.3 Knots

Knots are species in which a single strand alternately passes over and under itself several times in a continuous loop [62]. Knots are named according to the number of times the strand crosses itself. A typical knot is shown in Figure 14.26. Perhaps the most exquisite example of the application of metal–ligand interactions in self-assembly is the synthesis by Sauvage et al. of a molecular trefoil knot [57]. All known knots involve self-assembly processes but with post-modifications, which

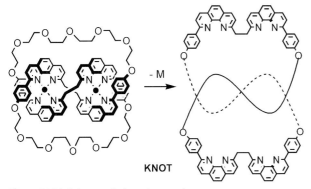

Figure 14.26 Scheme of a knot (see text).

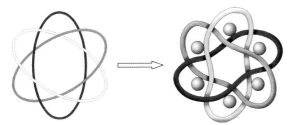

Figure 14.27 Two schemes of a Borromean ring.

are typically a reaction of low statistical probability. Indeed, yields of knot complexes can be low (<10%). Fully or partially demetallated knotates/knotanes can again complex metal ions, thus restoring their conformational rigidity.

Borromean Rings The link known as the Borromean rings has long been regarded as one of the most ambitious targets in this field. Of ancient provenance, this symbol comprises three interlocked rings in an inseparable union, but cut any one of the rings and the whole assembly unravels into three separated pieces (Figure 14.27). In the first Borromean ring a polypyridine ligand has been self-assembled by $Zn(AcO)_2$ [63].

Solomon's Knot This is described mathematically as a "link", since it contains four crossings and comprises two components. A review of the scarce Solomon's links reported in the literature as well as a comparison with Borromean rings has been recently published [64].

14.7.1.4 Necklaces
Multiring catenanes are molecules in which several small rings are threaded onto a single large ring. The first self-assembled "*molecular necklace*" was reported by Kim and coworkers, synthesizing a molecule from nine species, including three molecular "beads" (cucurbituril), three "strings" (*N,N'*-bis(4-pyridylmethyl)-1,4-

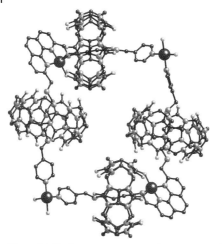

Figure 14.28 The X-ray structure of a Cu(I)-cucurbituril necklace. (large black spheres = copper ions).

diaminobutane) and three "angle connectors" [Pt(en)(NO$_3$)$_2$] [65]. The X-ray structure of a Cu(I)-cucurbituril necklace is shown in Figure 14.28. Other examples can be found in Ref. [66].

14.8
(Supra)molecular Devices and Machines [1]

"*A supramolecular device may be defined as a complex system made up of molecular components with definite individual properties … The interaction energy between the components of the supramolecular device must be small compared with other energy parameters relevant to the system. It does not matter, therefore, how the components are connected together in the device (covalently, hydrogen-bonded, coordination interaction, etc.); all that matters is that each component should contribute something unique and identifiable with that component alone, within the system*" [2]. However, other authors prefer to speak simply of "*molecular machines*" [67]. Common components within molecular devices are *photochemically, redox,* or *chemically* active molecules. Photochemically active devices will be studied in Chapter 15.

Most of these devices use the markedly different stereoelectronic requirements of copper(I) and copper(II). Whereas a coordination number of 4, usually with a roughly tetrahedral arrangement, corresponds to stable Cu(I) systems, copper(II) requires a coordination number 5 (square pyramidal or trigonal bipyramidal geometry) or 6 (octahedral arrangement, with Jahn–Teller distortion). Thus, by switching alternately from copper(I) to copper(II) one should be able to induce changes in the molecule so as to afford a coordination situation favorable to the corresponding oxidation state. This can be made by synthesizing molecular strings

Oxidation

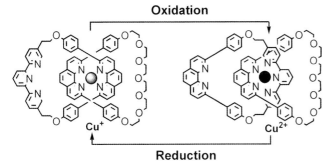

Reduction

Figure 14.29 The first reported case of the movement called "pirouetting of the ring" (redox input).

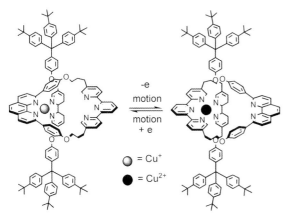

Figure 14.30 Scheme of a different "pirouetting of the ring" (redox input).

containing simultaneously *two* different sites, such as *bi*dentate (bpy-like group) and *ter*dentate (terpy-like group) coordinating units.

14.8.1
Some Different Motions

Pirouetting of the ring: The first molecular motor elaborated and studied in Sauvage's group was a *catenate* containing two different interlocking rings (Figure 14.29). This copper-catenane will allow movement of the functional ligand, through redox reaction of the copper [67]. New bistable rotaxanes, consisting of a 2,2-bpy or phen-containing thread and a ring incorporating both a bidentate chelate and a tridentate fragment undergo an electrochemically driven pirouetting motion of the ring around the axis. The strong influence of the axle structure on the rate of motion has been studied [68]. For example, the movement of the complex in Figure 14.30 take place on the millisecond timescale [69].

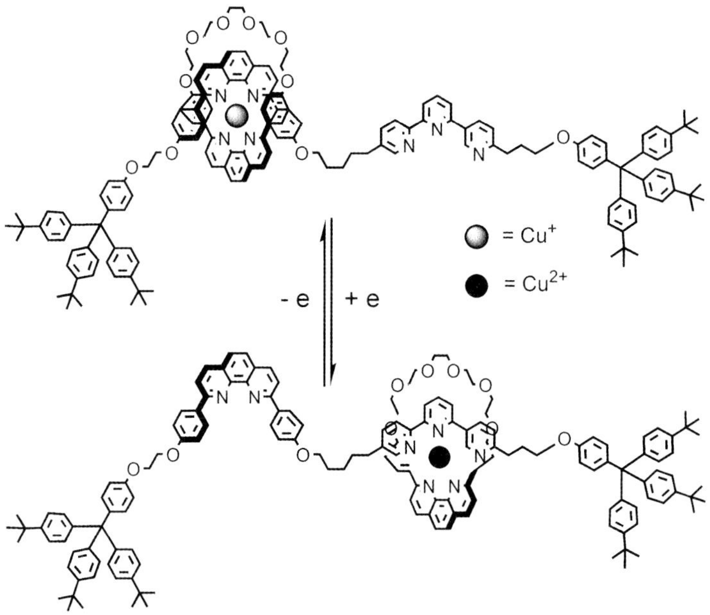

Figure 14.31 Scheme of a translation in rotaxanes, with redox input.

Translation in rotaxanes: A self-assembled [1]pseudorotaxane has been reported in which the filamentous ligand could be electrochemically induced to translate through the cyclic one in a controlled manner. The scheme of the system is indicated in Figure 14.31 [70]. A study of the influence of the ligands on the electrochemically induced translation of the ring has been reported recently [71]. Simple changes can vary the rate from hours or days to milliseconds or seconds.

Contracting–stretching processes: In relation to artificial muscles, one-dimensional molecular assemblies able to undergo stretching and contracting motions represent an exciting target. A fascinating example has been reported by Sauvage and coworkers in a muscle-like rotaxane dimer [72], illustrated in Figure 14.32.

A similar process has been reported recently by Lehn and coworkers, working with different polycondensated helical ligand strands [73].

Molecular gates. The design of a molecular gate is based on a hinge bearing a coordination site oriented divergently from its center and a rotatable handle equipped with a coordination site oriented towards the hinge. In the absence of a metal ion to "lock" the coordination sites together, the handle should rotate freely about the hinge. Thus, without the metal ion, the gate would be "open", but in its presence it would be "closed". In such a system the energy required for the closing of the gate is furnished by the favorable binding of the metal cation. A novel molecular gate based on a Sn-porphyrin and a silver lock has been reported recently (Figure 14.33) [74].

An interesting recent review on transition metal-complexes as molecular machine prototypes can be found in Ref. [75].

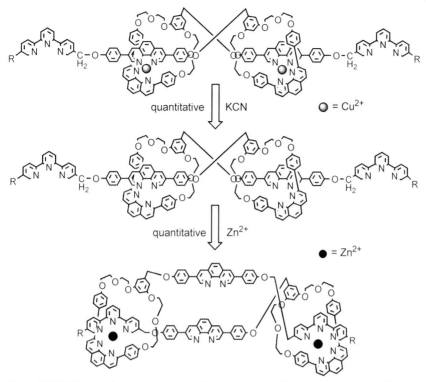

Figure 14.32 Scheme of a contracting–stretching process, in relation to "artificial muscles".

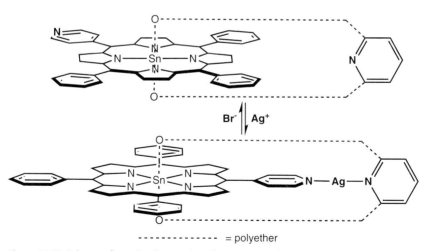

Figure 14.33 Scheme of a molecular gate based on a Sn-porphyrin and a silver lock.

References

1 Balzani, V., Credit, A., Venturi, M. (2004) *Molecular Devices and Machines. A Journey into the Nanoworld*, Wiley-VCH, Weinheim.

2 Steed, W., Atwood, J.L. (2000) *Supramolecular Chemistry*, John Wiley&Sons, New York.

3 Lehn, J-M. (1995) *Supramolecular Chemistry: Concepts and Perspective*, VCH, Weinheim.

4 Lehn, J-M. From supramolecular chemistry towards constitutional dynamic chemistry and adaptive chemistry, *Chem. Soc. Rev.* 2007, *36*, 151.

5 Constable, E.C. Higher Oligopyridines as a Structural Motif in Metallosupramolecular Chemistry, *Prog. Inorg. Chem.* 1994, *42*, 67.

6 Philp, D., Fraser Stoddart, J. Self-Assembly in Natural and Unnatural Systems, *Angew. Chem. Int. Ed.* 1996, *35*, 1154.

7 Ritch, J.S., Chivers, T. Silicon Analogues of Crown Ethers and Cryptands: A New Chapter in Host-Guest Chemistry? *Angew. Chem. Int. Ed.* 2007, *46*, 4610.

8 (a) Souchon, V., Leray, I., Valeur, B. Selective detection of cesium by a water-soluble fluorescent molecular sensor based on a calix[4]arene-bis(crown-6-ether), *Chem. Comm.* 2006, 4224; (b) Chan, W.H., Lee, A.W.M., Tam, W.L., Wang, K.M. Potassium ion-selective optodes based on the calix[6]arene hexaester and application in human serum assay, *Analyst* 1996, *121*, 531.

9 Matthews, S.E., Schmitt, P., Felix, V., Drew, M.G.B., Beer, P.D. Calix[4]tubes: A New Class of Potassium-Selective Ionophore, *J. Am. Chem. Soc.* 2002, *124*, 1341.

10 (a) Bowman-James, K. Alfred Werner Revisited: The Coordination Chemistry of Anions, *Acc. Chem. Res.* 2005, *38*, 671; (b) *Coord. Chem. Rev.* 2006, *250* (the whole volume).

11 (a) Osella, D., Carretta, A., Nervi, C., Ravera, M., Gobetto, R. Inclusion Complexes of Ferrocenes and β-Cyclodextrins. Critical Appraisal of the Electrochemical Evaluation of Formation Constants, *Organometallics*, 2000, *19*, 2791; (b) Poorters, L., Armspach, D., Matt, D., Toupet, L., Jones, P.G. A Metallocavitand Functioning as a Container for Anions: Formation of Noncovalent Linear Assemblies Mediated by a Cyclodextrin-Entrapped NO_3- Ion, *Angew. Chem. Int. Ed.*, 2007, *46*, 2663; (c) Hapiot, F., Tilloy, S., Monflier, E. Cyclodextrins as Supramolecular Hosts for Organometallic Complexes, *Chem. Rev*, 2006, *106*, 767.

12 Hiratani, K., Sakamoto, N., Kameta, N., Karikomi, M., Nagawa, Y. A novel crownophane trapping CO_2 as carbonic acid at room temperature, *Chem. Comm.* 2004, 1474.

13 Mahoney, J.M., Stucker, K.A., Jiang, H., Carmichael, I., Brinkmann, N.R., Beatty, A.M., Noll, B.C., Smith, B.D. Molecular Recognition of Trigonal Oxyanions Using a Ditopic Salt Receptor: Evidence for Anisotropic Shielding Surface around Nitrate Anion, *J. Am. Chem. Soc.* 2005, *127*, 2922.

14 Rudkevich, D.M., Shivanyuk, A.N., Brzozka, Z., Verboom, W., Reinhoudt, D. N. A Self-Assembled Bifunctional Receptor, *Angew. Chem. Int. Ed.* 1995, *34*, 2124.

15 Swiegers, G.F., Malefetse, T.J. New Self-Assembled Structural Motifs in Coordination Chemistry, *Chem. Rev.* 2000, *100*, 3483.

16 Whitesides, G.M., Grzybowski, B. Self-assembly at all scales, *Science*, 2002, *295*, 2418.

17 Albrecht, M. "Let's Twist Again"-Double-Stranded, Triple-Stranded, and Circular Helicates, *Chem. Rev.* 2001, *101*, 3457.

18 Albrecht, M., Fröhlich, R. Symmetry Driven Self-Assembly of Metallo-Supramolecular Architectures, *Bull. Chem. Soc. Jpn.* 2007, *80*, 797.

19 Hasenknopf, B., Lehn, J-M., Kneisel, B.O., Baum, G., Fenske, D. Self-Assembly of a Circular Double Helicate, *Angew. Chem. Int. Ed.* 1996, *35*, 1838.

20 Funeriu, D.P., Lehn, J-M., Baum, G., Fenske, D. Double Subroutine Self-Assembly; Spontaneous Generation of a Nanocyclic Dodecanuclear Cu^I Inorganic Architecture, *Chem. Eur. J.* 1997, *3*, 99.

21 Piguet, C., Borkovec, M., Hamacek, J., Zeckert, K. Strict self-assembly of polymetallic helicates: the concepts behind the semantics, *Coord. Chem. Rev.* 2005, *249*, 705.

22 Giuseppone, N., Schmitt, J.-L., Lehn, J.-M. Driven Evolution of a Constitutional Dynamic Library of Molecular Helices Toward the Selective Generation of [2 × 2] Gridlike Arrays under the Pressure of Metal Ion Coordination, *J. Am. Chem. Soc.* 2006, *128*, 16748.

23 Ruben, M., Rojo, J., Romero-Salguero, F.J., Uppadine, L.H., Lehn, J.-M. Grid-Type Metal Ion Architectures: Functional Metallosupramolecular Arrays, *Angew. Chem. Int. Ed.* 2004, *43*, 3644.

24 as an example see, Barboiu, A., Ruben, M., Blasen, G., Kyritsakas, N., Chacko, E., Dutta, M., Radekovich, O., Lenton, K., Brook, D.J.R., Lehn, J.-M. Self-Assembly, Structure and Solution Dynamics of Tetranuclear Zn^{2+} Hydrazone [2 × 2] Grid-Type Complexes, *Eur. J. Inorg. Chem.* 2006, 784.

25 Ramirez, J., Stadler, A-M., Kyritsakas, N., Lehn, J-M. Solvent-modulated reversible conversion of a [2 × 2]-grid into a pincer-like complex, *Chem. Comm.* 2007, 237.

26 (a) Kalsani, V., Bodenstedt, H., Fenske, D., Schmittel, M. Supramolecular Copper Phenanthroline Racks: Structures, Mechanistic Insight and Dynamic Nature, *Eur. J. Inorg. Chem.* 2005, 1841. (b) Schmittel, M., Kalsani, V., Bats, J.W. Metal-Driven and Covalent Synthesis of Supramolecular Grids from Racks: A Convergent Approach to Heterometallic and Heteroleptic Nanostructures, *Inorg. Chem.* 2005, *44*, 4115.

27 (a) Fujita, M. *Struct. Bond.*, Molecular Poneling through Metal-Directed Self-Assembly 2000, *96*, 177; (b) Fujita, M., Umemoto, K., Yoshizawa, M., Fujita, N., Kusukawa, T., Biradha, K. Molecular paneling via coordination, *Chem. Commun.* 2001, 509.

28 Fujita, M., Yazaki, J., Ogura, K. Preparation of a Macrocyclic Polynuclear Complex, [(en)Pd(4,4′-bpy)]₄(NO₃)₈, Which Recognizes an Organic Molecule in Aqueous Media, *J. Am. Chem. Soc.* 1990, *112*, 5645.

29 Dong, Y-B., Zhang, Q., Liu, L-L., Ma, J-P., Tang, B., Huang, R-Q. [Cu(C₂₄H₂₂N₄O₃)]·CH₂Cl₂: A Discrete Breathing Metallamacrocycle Showing Selective and Reversible Guest Adsorption with Retention of Single Crystallinity, *J. Am. Chem. Soc.* 2007, *129*, 1514.

30 Lee, S.J., Lin, W. A Chiral Molecular Square with Metallo-Corners for Enantioselective Sensing, *J. Am Chem. Soc.* 2002, *124*, 4554.

31 Ferrer, M., Mounir, M., Rossell, O., Ruiz, E., Maestro, M.A. Equilibria between Metallosupramolecular Squares and Triangles with the New Rigid Linker 1,4-Bis(4-pyridyl)tetrafluorobenzene. Experimental and Theoretical Study of the Structural Dependence of NMR Data, *Inorg. Chem.* 2003, *42*, 5890.

32 Schweiger, M., Seidel, S.R., Arif, A.M., Stang, P.J. Solution and Solid State Studies of a Triangle-Square Equilibrium: Anion-Induced Selective Crystallization in Supramolecular Self-Assembly, *Inorg. Chem.* 2002, *41*, 2556.

33 Uehara, K., Kasai, K., Mizuno, N. Syntheses and Characterizations of Palladium-Based Molecular Triangle/ Square Compounds and Hybrid Composites with Polyoxometalates, *Inorg. Chem.* 2007, *46*, 2563.

34 Heo, J., Jeon, Y-M., Mirkin, C.A. Reversible Interconversion of Homochiral Triangular Macrocycles and Helical Coordination Polymers, *J. Am. Chem. Soc.* 2007, *129*, 7712.

35 Campos-Fernández, C.S., Clérac, R., Koomen, J.M., Russell, D.H., Dunbar, K.R. Fine-Tuning the Ring-Size of Metallacyclophanes: A Rational Approach to Molecular Pentagons, *J. Am. Chem. Soc.* 2001, *123*, 773.

36 Fujita, M., Oguro, D., Miyazawa, M., Oka, H., Yamaguchi, K., Ogura, K. Self-Assembly of 10 Molecules Into Nanometer-Sized Organic Host Frameworks, *Nature*, 1995, *378*, 469.

37 (a) Fujita, M., Nagao, S., Ogura, K. Guest-Induced Organization of a Three-Dimensional Palladium(II) Cagelike Complex. A Prototype for "Induced-Fit" Molecular Recognition, *J. Am. Chem. Soc.* 1995, *117*, 1649; (b) Fujita, M., Yu, S-Y., Kusukawa, T., Funaki, H., Ogura, K.,

Yamaguchi, K. Self-Assembly of Nanometer-Sized Macrotricyclic Complexes from Ten Small Component Molecules, *Angew. Chem. Int. Ed.* 1998, *37*, 2082.

38 Caulder, D.L., Raymond, K.N. The rational design of high symmetry coordination clusters, *J. Chem. Soc., Dalton Trans.* 1999, 1185.

39 Bell, Z.R., Jeffery, J.C., McCleverty, J.A., Ward, M.D. Assembly of a Truncated-Tetrahedral Chiral $[M_{12}(\mu\text{-}L)_{18}]^{24+}$ Cage, *Angew. Chem. Int. Ed.* 2002, *41*, 2515.

40 Argent, S.P., Adams, H., Riis-Johannessen, T., Jeffery, J.C., Harding, L.P., Ward, M.D. High-nuclearity Homoleptic and Heteroleptic Coordination Cages Based On Tetra-Capped Truncated Tetrahedral and Cuboctahedral Metal Frameworks, *J. Am. Chem. Soc.* 2006, *128*, 72.

41 Dong, Y.-B., Wang, P., Ma, J-P., Zhao, X.-X., Wang, H-Y., Tang, B., Huang, R-Q. Coordination-Driven Nanosized Lanthanide "Molecular Lantern" with Tunable Luminescent Properties, *J. Am. Chem. Soc.* 2007, *129*, 4872.

42 Yeh, R.M., Xu, J., Seeber, G., Raymond, K.N. Large M_4L_4 (M = Al(III), Ga(III), In(III), Ti(IV)) Tetrahedral Coordination Cages: an Extension of Symmetry-Based Design, *Inorg. Chem.* 2005, *44*, 6228

43 Hiraoka, S., Harano, K., Shiro, M., Ozawa, Y., Yasuda, N., Toriumi, K., Shionoya, M. Isostructural Coordination Capsules for a Series of 10 Different d^5–d^{10} Transition-Metal Ions, *Angew. Chem. Int. Ed.* 2006, *45*, 6488.

44 Moon, D., Kang, S., Park, J., Lee, K., John, R.P., Won, H., Seong, G.H., Kim, Y.S., Kim, G.H., Rhee, H., Lah, M.S. Face-Driven Corner-Linked Octahedral Nanocages: M_6L_8 Cages Formed by C_3-Symmetric Triangular Facial Ligands Linked via C_4-Symmetric Square Tetratopic Pd^{II} Ions at Truncated Octahedron Corners, *J. Am. Chem. Soc.* 2006, *128*, 3530;

45 Takeda, N., Umemoto, K., Yamaguchi, K., Fujita, M. A nanometre-sized hexahedral coordination capsule assembled from 24 components, *Nature*, 1999, *398*, 794.

46 (a) Aoyagi, M., Biradha, K., Fujita, M. Quantitative Formation of Coordination Nanotubes Templated by Rodlike Guests, *J. Am. Chem. Soc.* 1999, *121*, 7457; (b) Tominaga, M., Tashiro, S., Aoyagi, M., Fujita, M. Dynamic aspects in host–guest complexation by coordination nanotubes, *Chem. Commun.* 2002, 2038.

47 Albrecht, M., Janser, I., Fröhlich, R. Catechol imine ligands: from helicates to supramolecular tetrahedra, *Chem. Comm.* 2005, 157.

48 Suzuki, K., Kawano, M., Fujita, M. Solvato-Controlled Assembly of Pd$_3L_6$ and Pd$_4L_8$ Coordination "Boxes", *Angew. Chem. Int. Ed.* 2007, *46*, 2819.

49 Du, M., Bu, X-H., Guo, Y-M., Ribas, J. Ligand Design for Alkali-Metal-Templated Self-Assembly of Unique High-Nuclearity Cu^{II} Aggregates with Diverse Coordination Cage Units: Crystal Structures and Properties, *Chem. Eur. J.* 2004, *10*, 1345.

50 (a) Vilar, R., Mingos, D.M.P., White, A.J.P., Williams, D.J. Anion Control in the Self-Assembly of a Cage Coordination Complex, *Angew., Chem. Int. Ed.* 1998, *37*, 1258; (b) Kim, K.M., Park, J.S., Kim, Y-S., Jun, Y.J., Kang, T.Y., Sohn, Y.S., Jun, M-J. The First Inorganic "Tennis Ball" Encapsulating an Anion, *Angew., Chem. Int. Ed.* 2001, *40*, 2458.

51 Brumaghim, J.L., Michels, M., Pagliero, D., Raymond, K.N. Encapsulation and Stabilization of Reactive Aromatic Diazonium Ions and the Tropylium Ion Within a Supramolecular Host, *Eur. J. Org. Chem.* 2004, 5115.

52 Fiedler, D., Leung, D.H., Bergman, R.G., Raymond, K.H. Selective Molecular Recognition, C-H Bond Activation, and Catalysis in Nanoscale Reaction Vessels, *Acc. Chem. Res.* 2005, *38*, 351.

53 Lützen, A. Self-Assembled Molecular Capsules – Even More Than Nano-Sized Reaction Vessels, *Angew. Chem. Int. Ed.* 2005, *44*, 1000.

54 Davis, A.V., Fiedler, D., Seeber, G., Zahl, A., van Eldik, R., Raymond, K.N. Guest Exchange Dynamics in an M_4L_6 Tetrahedral Host, *J. Am. Chem. Soc*, 2006, *128*, 1324.

55 Pluth, M.D., Raymond, K.N. Reversible guest exchange mechanisms in supramolecular host–guest assemblies, *Chem. Soc. Rev.* 2007, *36*, 161.

56 Fyfe, M.C.T., Fraser Stoddart, J. Synthetic Supramolecular Chemistry, *Acc. Chem. Res.* 1997, *30*, 393.

57 Sauvage, J-P. Interlacing Molecular Threads on Transition Metals: Catenands, Catenates, and Knots, *Acc. Chem. Res.* 1990, *23*, 319.

58 Baranoff, E.D., Voignier, J., Yasuda, T., Heitz, V., Sauvage, J-P., Kato, T. A Liquid-Crystalline [2]Catenane and Its Copper(I) Complex, *Angew. Chem. Int. Ed.* 2007, *46*, 4680.

59 (a) Fujita, M., Ibukuro, F., Hagihara, H., Ogura, K. Quantitative Self-Assembly Of A [2]Catenane From two Preformed Molecular Rings, *Nature*, 1994, *367*, 720; (b) Yamashita, K-I., Kawano, M., Fujita, M. Photoswitchable Molecular Lock. One-Way Catenation of a Pt(II)-Linked Coordination Ring via the Photolabilization of a Pt(II)-Pyridine Bond, *J. Am. Chem. Soc.* 2007, *129*, 1850; (c) Fujita, M. Self-Assembly of [2]Catenanes Containing Metals in Their Backbones, *Acc. Chem. Res.* 1999, *32*, 53.

60 Habermehl, N.C., Mohr, F., Eisler, D.J., Jennings, M.C., Puddephatt, R.J. Organogold(I) macrocycles and [2]catenanes containing pyridyl and bipyridyl substituents – Organometallic catenanes as ligands, *Can. J. Chem.* 2006, *84*, 111.

61 (a) Pomeranc, D., Jouvenot, D., Chambron, J-C., Collin, J-P., Heitz, V., Sauvage, J-P. Templated Synthesis of a Rotaxane with a [Ru(diimine)₃]²⁺ Core, *Chem. Eur. J.* 2003, *9*, 4247; (b) Li, K., Bracher, P.J., Guldi, D.M., Herranz, M. A., Echegoyen, L., Schuster, D.I. [60]Fullerene-Stoppered Porphyrinorotaxanes: Pronounced Elongation of Charge-Separated-State Lifetimes, *J. Am. Chem. Soc.* 2004, *126*, 9156.

62 Lukin, O., Vögtle, F. Knotting and Threading of Molecules: Chemistry and Chirality of Molecular Knots and Their Assemblies, *Angew. Chem. Int. Ed.* 2005, *44*, 1456.

63 (a) Cantrill, S.J., Chichak, K.S., Peters, A.J., Stoddart, J.F. Nanoscale Borromean Rings, *Acc. Chem. Res.* 2005, *38*, 1; (b) Pentecost, C.D., Peters, A.J., Chichak, K.S., Cave, G.W.V., Cantrill, S.J., Stoddart, J.F. Chiral Borromeates, *Angew. Chem. Int. Ed.* 2006, *45*, 4099.

64 Pentecost, C.D., Chichak, K.S., Peters, A.J., Cave, G.W.V., Cantrill, S.J., Fraser Stoddart, J. A Molecular Solomon Link, *Angew. Chem. Int. Ed.* 2007, *46*, 218.

65 Whang, D., Park, K.M., Heo, J., Ashton, P., Kim, K. Molecular Necklace: Quantitative Self-Assembly of a Cyclic Oligorotaxane from Nine Molecules, *J. Am. Chem. Soc.* 1998, *120*, 4899.

66 Kim, K. Mechanically interlocked molecules incorporating cucurbituril and their supramolecular assemblies, *Chem. Soc. Rev.* 2002, *31*, 96.

67 Sauvage, J-P. Transition Metal-ContainIng Rotaxanes and Catenanes in Motion: Toward Molecular Machines and Motors, *Acc. Chem. Res.* 1998, *31*, 611.

68 Collin, J-P., Durola, F., Mobian, P., Sauvage, J-P. Pirouetting Copper(I)-Assembled Pseudo-Rotaxanes: Strong Influence of the Axle Structure on the Motion Rate, *Eur. J. Inorg. Chem.* 2007, 2420.

69 Poleschak, I., Kern, J-M., Sauvage, J-P. A copper-complexed rotaxane in motion: pirouetting of the ring on the millisecond timescale, *Chem. Comm.* 2004, 474.

70 Collin, J-P., Gaviña, P., Sauvage, J-P. Electrochemically induced molecular motions in a copper(I) complex pseudorotaxane, *Chem. Commun.* 1996, 2005.

71 Durola, F., Sauvage, J-P. Fast Electrochemically Induced Translation of the Ring in a Copper-Complexed [2]Rotaxane: The Biisoquinoline Effect, *Angew. Chem. Int. Ed.* 2007, *46*, 3537.

72 Jiménez-Molero, M.C., Dietrich-Buchecker, C., Sauvage, J-P. Chemically Induced Contraction and Stretching of a Linear Rotaxane Dimer, *Chem. Eur. J.* 2002, *8*, 1456.

73 Stadler, A-M., Kyritsakas, N., Graff, R., Lehn, J-M. Formation of Rack- and Grid-Type Metallosupramolecular Architectures and Generation of Molecular Motion by Reversible Uncoiling of Helical Ligand Strands, *Chem. Eur. J.* 2006, *12*, 4503.

74 Guenet, A., Graf, E., Kyritsakas, N., Allouche, L., Hosseini, M.W. A molecular gate based on a porphyrin and a silver lock, *Chem. Comm.* 2007, 2935.

75 Champin, B., Mobian, P., Sauvage, J-P. Transition metal complexes as molecular machine prototypes, *Chem. Soc. Rev.* 2007, *36*, 358.

15
Photochemistry and Photophysics in Coordination Compounds

15.1
Fundamentals

15.1.1
Introduction

It was in the early 1970s that Balzani and Carassiti wrote the first monograph on the photochemistry of transition metal complexes, entitled *Photochemistry of Coordination Compounds* [1], which encouraged the development of coordination photochemistry and photophysics in the following years.

Photochemical reactions differ from thermal reactions in that they are initiated by the absorption of light rather than by heat. Photochemical reactions can, therefore, occur if the compounds to be activated have a chromophore that corresponds to the wavelengths of the exciting radiation. Photochemical excitation is very different to thermal excitation, the other kind of chemical excitation normally used to overcome the activation barrier. It is not possible to excite thermally to 600 K without first passing by 500 K and all other intermediate temperatures. However, the photochemical excitation is much more precise. With monochromatic light of a specific wavelength, *all molecules* can be excited with a pre-determined energy hv. As indicated by Porterfield, "*It is thus possible to have a hot molecule without ever having had a warm molecule*" [2].

15.1.2
Electronic Spectra and Photochemistry

In Chapter 9 we dealt with the absorption of light, but without considering the photophysical processes or photochemical reactions that follow the excitation. The nature of any process resulting from the absorption of photons is strongly determined by the type of electronic transition involved. Figure 15.1 shows a generic molecular orbital diagram for an ideal octahedral complex: there are various kinds of electronic transitions that may occur between the orbitals. These are summarized as: (i) intraligand (IL) bands, (ii) ligand-field (LF), d–d bands or metal centered (MC) bands, (iii) ligand-to-metal-charge-transfer (LMCT) bands, (iv)

Coordination Chemistry. Joan Ribas Gispert
Copyright © 2008 WILEY-VCH Verlag GmbH & Co. KGaA, Weinheim
ISBN: 978-3-527-31802-5

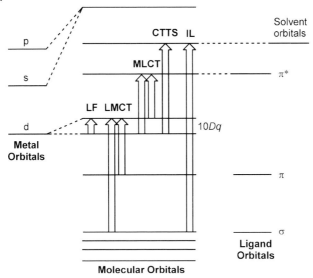

Figure 15.1 A general molecular orbital diagram for an ideal octahedral complex. The electronic transitions that may occur between the orbitals are shown (see text for explanation).

metal-to-ligand-charge transfer (MLCT) bands. When coordination compounds interact with solvent molecules charge transfer to the solvent (CTTS) may occur.

In this chapter we are particularly interested in charge-transfer bands, because of their tendency to produce charge separations which, in turn, may result in photo-redox reactions. Neither IL nor the typical LF transitions normally result in intramolecular photo-induced electron transfer (PET) processes (the most important in photochemistry).

In a general manner, we can state that all chemical compounds are, in principle, capable of electronic transitions in which an electron is promoted from an occupied orbital to an empty orbital at higher energy, but in the vast majority of cases the excited state collapses very quickly back to the ground state, with evolution of heat as the electronic energy is converted to increased vibrational motion in the molecule.

15.1.3
Photochemical Principles for Absorption and Emission

15.1.3.1 Introduction

In a photochemical process a molecule absorbs radiation ($h\nu$), becomes excited and decays by light emission (luminescence), radiationless deactivation and/or by a chemical reaction.

Electronic transitions are much more rapid than molecular vibrations. In a configurational coordinate diagram, electronic transitions are therefore *vertical*

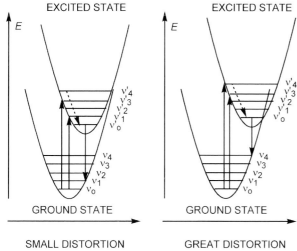

Figure 15.2 Light absorption; vibrational decay and light emission between ground and excited states.

transitions. In a vertical transition an excited vibrational state, v_i', of the electronically excited state is achieved, followed by rapid vibrational relaxation (through collisions) such that the electronically excited state usually ends up in its vibrational ground state (v_0'). All these features are shown in Figure 15.2: light absorption occurs between v_0 of the electronic ground state and v_i' of the electronic excited state. v_i' decays to v_0' and, later, there is either light emission to v_i of the ground state, radiationless deactivation to the ground state or a chemical reaction. In order for a chemical reaction to occur, the excited state must, in general, have a lifetime sufficiently long to be considered as a new and observable chemical species.

15.1.3.2 The Excited State

The initial event in any photoprocess is, thus, the absorption of a photon by an atom, molecule or complex ion, resulting in an electronically excited state. If A and A* represent the ground and excited states, respectively, the process can be written as: A + hv → A*

The excited state can be considered as *a different molecule* compared to that of the ground state. Indeed, distances and angles are different, distortions are different, rotational and vibrational energies may be greater in the excited state, redox potentials are different in the excited state, etc.

It is also possible to achieve the excited state by other procedures, such as through a *sensitizer* previously excited photochemically (A + S* → A* + S); through a chemical reaction (less frequent) or through a *triboluminescent procedure* (by pressure).

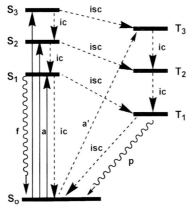

Figure 15.3 A general Jablonski diagram. a and a' stand for absorption of light; f for fluorescence, p for phosphorescence; ic for internal conversion, isc for intersystem crossing. S stands for singlet and T for triplet.

15.1.3.3 Transitions Between Energy States. Decay Processes

A general Jablonski diagram (Figure 15.3) is useful to illustrate the various transitions which connect the energy states of a molecule. Transitions can be classified as 'radiative' or 'non-radiative'. The former involve a 'vertical' change in energy. Non-radiative transitions reorganize internally the energy of the molecule and eventually degrade it into heat in the surroundings. These are *ic* and *isc* in Figure 15.3. We are going to study all these processes in the next section.

Generalities of Decay Processes Once the excited state is achieved by the absorption of photons (*h*ν), it undergoes many processes, either photophysically back to its initial ground state or, frequently, photochemically to a different ground state. Figure 15.4 gives an outline of the many choices of pathways through which the excited state can dissipate the energy gained in absorbing a photon. Many of these processes occur simultaneously and compete with each other such that one usually observes a mixture of outcomes, each with its own quantum yield.

Photophysical Pathways Photophysical pathways can be radiative or non-radiative.

Radiative Decay (Luminescence) After the vibrational relaxation of the excited electronic state, the system may return to the ground state by spontaneously emitting a photon. Two different luminescent processes can be found:

- *Phosphorescence* is a deactivation through light emission between terms of *different spin multiplicity*. It is formally forbidden.
- *Fluorescence* is a deactivation through light emission between terms of the *same spin multiplicity*. It is formally allowed and rapid.

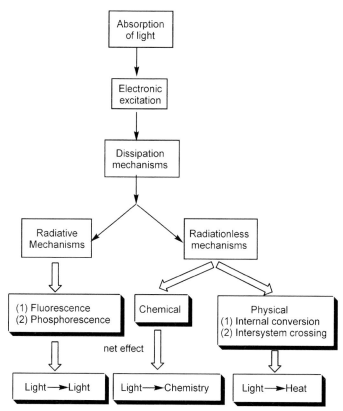

Figure 15.4 Outline of the many pathways through which the excited state can dissipate the energy gained in absorbing a photon.

Radiationless Decay Radiationless decay from excited electronic states may occur if there is sufficient overlap between the vibrational wavefunctions (ground and excited states). There are two types of nonradiative decay:

- *Internal conversion* occurs between electronic terms of the same spin multiplicity.
- *Intersystem crossing* occurs between electronic terms of different spin multiplicity.

All these processes for an octahedral complex of Cr^{3+}, are shown in a Jablonski diagram in Figure 15.5.

The Importance of Spin–Orbit Coupling From a theoretical viewpoint, any transition with different spin multiplicity has a low probability. Thus the excited state has a rather long lifetime. For example, phosphorescence is characterized by lifetimes as long as seconds, minutes or even hours after the absorption of light

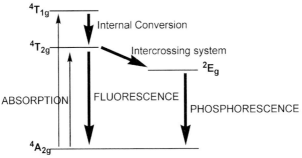

Figure 15.5 Jablonski diagram for an octahedral complex of Cr^{3+}.

(organic compounds). However, in coordination compounds there may be many kinds of radiative transitions that do not fall neatly into these categories. The spin–orbit coupling, not important in organic compounds, but very important in coordination compounds, is the main cause. For this reason, in coordination systems all types of emission are now usually referred to as simply *luminescence*. *Intersystem crossing* is usually forbidden in organic systems but is very frequent (maybe the most frequent) in coordination compounds. These processes are so rapid that they are often indistinguishable from *internal conversion* (same spin multiplicity). Phosphorescence phenomena are also comparatively rapid in coordination compounds.

Photochemical Pathways   The extent to which photochemical products are formed will depend on how well the photochemical pathways compete with the various photophysical decay processes described earlier. Nearly all types of reactions can be classified as either *unimolecular* or *bimolecular*.

Unimolecular (or Intramolecular) Processes   A molecule absorbs a photon, goes to an excited state and produces a reaction:

1. When the absorption is centered in the d–d region of the transition metal ion, the reactions induced are centered in the metal at the center of the complex, resulting for instance in isomerization, dissociation, and substitution reactions.

2. When the absorption is centered in the *transfer charge* region, the most frequent processes are photo-redox processes, in which the oxidation state of the complex is modified.

Bimolecular (or Intermolecular) Processes   In this case, a molecule absorbs a photon and becomes excited, but now the first molecule transfers its energy to another molecule or reacts with it:

1. Through an *electron transfer* process (redox) according to : $M^* + A \rightarrow M^+ + A^-$

2. Through an *energy transfer* process. If the quencher molecule Q has an excited state Q* lower than M*, the excitation energy can be transferred according to

M* + Q → M + Q*. There are two major energy transfer processes: (i) *the Dexter mechanism*. This mechanism requires orbital overlap between donor and acceptor, either directly or mediated by a bridge (through-bond), so that the molecules M* and Q must be in close contact (van der Waals or hard sphere contact). There is an exponential decrease of the orbital overlap with the distance; (ii) *Coulombic mechanism: the Förster mechanism* (also called *resonance, dipole–dipole interaction* or *through-space*). It is a long-range mechanism that does not require physical contact between donor and acceptor. It can be shown that the most important term within the Coulombic interaction is the dipole–dipole term. There is a distance dependence of r^{-6}. Dipole–dipole interaction can be operative at distance of some 100 Å!

3. Through *collisional deactivation*. The excitation is simply dissipated by collisional deactivation according to: M* + A → M + A + heat

15.1.3.4 Quantum Yield

It is important to define an *experimental* parameter, which is the *quantum yield* (ϕ) of a photochemical or photophysical process. The quantum yield of a photochemical or photophysical process is given by the number of molecules undergoing the process per photon absorbed.

$$\phi = \frac{\text{number of molecules undergoing a given process}}{\text{number of photons absorbed}}$$

If there are a number of parallel pathways resulting in different products, then each pathway will have its own quantum yield. "*The absorption of light by a molecule is a one-quantum process, such that the sum of the primary process yields, ϕ, must be unity, i.e. $\Sigma\phi_i = 1$, where ϕ_i is the quantum yield of each of the primary processes*" [3].

15.1.3.5 Main Principles: Kasha's Rule and Stokes Shift

Kasha's rule states that emission occurs almost exclusively from the lowest excited state of a given spin multiplicity. In transition metal complexes spin–orbit coupling is responsible for often very efficient intersystem crossing (isc) and thus in transition metal complexes luminescence originates from the lowest excited state only (Figure 15.2).

The most important consequence is that the *emitted light has always lower frequency than the light absorbed by the same transition*. This difference is called Stokes shift. When the geometrical differences between the ground and excited state are great, that is, if the two energy curves are much shifted relative to each other, then the Stokes shift is noticeable. When the energy curves are little shifted, the Stokes shift is small (Figure 15.2). Let us study the Stokes shift for two electronic transitions in the complex $[Cr^{III}L_6]^{3+}$. In order to analyze this phenomenon it is convenient to use the simplified Tanabe–Sugano diagram for a d^3 ion (Figure 15.6). The spin allowed $^4A_{2g} \rightarrow {}^4T_{2g}$ transition involves a noticeable geometrical change due to the promotion of an electron from a t_{2g} to a e_g orbital ($t_{2g}^3 \rightarrow t_{2g}^2 e_g^1$). Thus, the

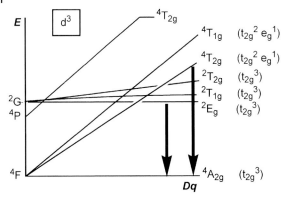

Figure 15.6 Simplified Tanabe–Sugano diagram for a d^3 ion.

Stokes shift will be great. However, the spin forbidden (weak) transition $^4A_{2g} \rightarrow$ ^2E_g, will give a very small Stokes shift, because the two terms originate from the same electronic configuration.

This interpretation is only valid if the symmetry of the excited state does is the same as that of the ground state, as occurs in octahedral Werner-type complexes of Cr^{3+} or Co^{3+}. The explanation can be very different in other kind of complexes. For instance, monomeric $[Au^IL_3]^+$ complexes have visible luminescence with very large Stokes shifts which suggests a significant excited state distortion. Omary et al. have reported a DFT study on the origin of this strong Stokes shift, demonstrating that the triplet excited state has a T-shape structure, in which the two *trans* Au–P bond distances increased significantly [4].

15.1.4
Identification of the Excited State

There are several ways to identify the excited state in a photochemical process, the two most frequent being the use of either a *sensitizer* or a *quenching* agent. The two methods can be schematized as follows:

$$A^* + Q \rightarrow A + Q^* \quad \text{(Quenching of A* by Q)}$$

$$A + S^* \rightarrow A^* + S \quad \text{(Sensitization of A by S*)}$$

Let us discuss which is the excited state of the following reaction:

$$[Cr(CN)_6]^{3-} + DMF \xrightarrow{h\nu} [Cr(CN)_5DMF]^{2-} + CN^- \text{ (DMF = dimethylformamide)}$$

In DMF two phenomena are observed: solvation and phosphorescence. The excited state for the solvation reaction is the $^4T_{2g}$ term of Cr^{3+}, whereas the excited state for phosphorescence is the ^2E_g term (Figure 15.7). Let us comment on how these two different excited states have been deduced.

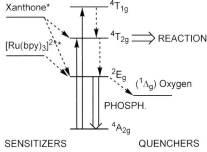

Figure 15.7 Example of identification of the excited state (see text).

1. If a high energy sensitizer (such as phochemically excited xanthone) is added to the reaction, the sensitization with xanthone* may populate the two terms ($^4E_{2g}$ and ^2E_g) given their relative energies (Figure 15.7). Experimentally it has been demonstrated that excited xanthone* induces both processes (phosphorescence and solvation).

2. If a medium-energy sensitizer, such as $[Ru^*(bpy)_3]^{2+}$, is added to the reaction, the corresponding sensitization can only populate the ^2E_g state. Experimentally $[Ru^*(bpy)_3]^{2+}$ induces only phosphorescence but not solvation. Thus, solvation is due to the excited $^4T_{2g}$ term.

3. Finally, if a quencher is added to the reaction, such as the molecule of O_2 (spin triplet which may be excited to spin singlet), the ^2E_g excited term is quenched, impeding the phosphorescence but not the solvation. Thus, we can conclude that phosphorescence is due to the excited ^2E_g term.

15.2
Examples of Main Photochemical Processes

15.2.1
Non-redox Processes

Non-redox photochemical processes are induced by the absorption of light of low energy, corresponding generally to the d–d region of the spectrum. Main examples are:

15.2.1.1 Photoisomerization
In general, light can produce any kind of isomerization. *Linkage isomerization*, for example, has been observed in certain transition metal complexes containing NO^+, NO_2^-, N_2, NCS^-, SO_2 and dimethyl sulfoxide (dmso). Some of these processes have been known from the beginning of coordination chemistry, being in general dissociative such as:

$$[Co(SCN)(NH_3)_5]^{2+} \xrightarrow{hv} [Co(NCS)(NH_3)_5]^{2+}$$

$$[Co(NO_2)(NH_3)_5]^{2+} \xrightarrow{hv} [Co(ONO)(NH_3)_5]^{2+}$$

The single and the first double-linkage isomerism in a six-coordinate iron porphyrin containing nitrosyl and nitro ligands, $[Fe(NO)(NO_2)(TPP)]$ with different hv, have been reported [5].

Phototriggered Ru–S → Ru–O and thermal Ru–O → Ru–S intramolecular linkage isomerizations in both *cis* and *trans*-$[Ru(bpy)_2(dmso)_2]^{2+}$ and $[Ru(L)(dmso)(terpy)]^{2+}$ (L = bpy, tmen, pic, Mepic) complexes have been reported [6]. The yellow *cis*-complex features only S-bonded sulfoxides (*cis*-[S,S]), whereas the orange *trans*-isomer is characterized by S- and O-bonded dmso ligands, *trans*-[S,O]. Following irradiation of the MLCT band, the dmso solutions become deep-red or purple, consistent with the presence of O-bound linkage isomers. In *cis*-$[Ru(dmso)_2(pic)_2]$ there are two photoisomerization processes with very different quantum yields: $\Phi_{SS \rightarrow SO} = 0.46$ and $\Phi_{SO \rightarrow OO} = 0.036$. The isomerization quantum yield for each dmso is, thus, dependent on the ligand that is *trans* to the dmso. The isomerization of a chelating sulfoxide in $[Ru(bpy)_2(OSO)]^+$ (OSO = methylsulfonylbenzoate) has also been reported [6b].

Another example which undergoes *photochromic isomerization* is $[M_2(S_2C_2R_2)_2(\mu$-$S_2)(\mu$-$S_2C_2R_2)_2]$ (M = Mo or W; R = Ph, Me) (Figure 15.8) [7]. Irradiation of **1** (pale gray in CH_2Cl_2) with visible light resulted in the formation of the isomer **1′** (red). In the dark, **1′** returns to **1** in 6 h at room temperature.

15.2.1.2 Photodissociation

Photodissociation is one of the most studied reactions in complexes of Co^{3+} with NH_3. In these complexes the photoliberation of ammonia is very easy. This process can be understood considering the ground state of a Co^{3+} complex ($^1A_{1g}$) that corresponds to the $(t_{2g})^6$ configuration The photochemical excitation produces the $(t_{2g})^5(e_g)^1$ configuration ($^1T_{1g}$ term). This feature produces a weakening of the Co–NH_3 bond which can be easily broken. Other complexes in which photodissocia-

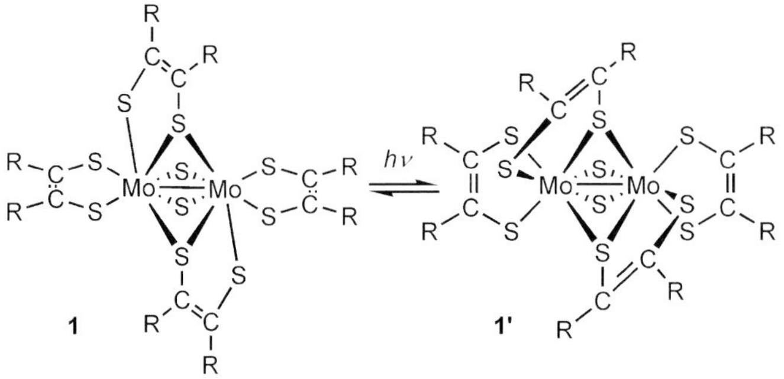

Figure 15.8 Example of linkage photoisomerization (see text).

tion has been extensively studied are the oxalato-complexes (which liberate CO_2), some hydride-complexes (which liberate H_2) and many metal-carbonyl complexes that are easily dissociated giving an activated species with $< 18\,e^-$ by elimination of a CO ligand.

15.2.1.3 Photosubstitution

Photosubstitution is the logical following step after photodissociation. It is very frequent in d^3 and d^6 ions (Cr^{3+}, Co^{3+}, Rh^{3+}, Ir^{3+}). Cr^{3+} is the most appropriate cation for photosubstitution reactions, since the two Cr^{2+} and Cr^{4+} oxidation states are much less stable than Cr^{3+} being, thus, difficult, to take part in redox reactions. Some very simple processes have been extensively studied. Let us mention, for example, the reaction of $[CrX(NH_3)_5]^{2+}$ (X = halido or pseuhalido) with H_2O, giving substitution of the halido or of one NH_3 molecule. By the action of light the evolved NH_3 has the greater yield, while by thermal reaction the evolved X^- has the greater yield.

A particular and important case of photosubstitution is photoaquation. It occurs mainly in Cr^{3+}, Co^{3+}, Rh^{3+}, Ir^{3+}, Ru^{2+} complexes. For example:

$$[Co(CN)_6]^{3-} + H_2O \xrightarrow{h\nu} [Co(CN)_5(H_2O)]^{2-} + CN^-$$

$$[Co(NH_3)_6]^{3+} + H_2O \xrightarrow{h\nu} [Co(H_2O)(NH_3)_5]^{3+} + NH_3$$

It must be emphasized that this aquation reaction in Rh^{3+} and Ir^{3+} complexes is very difficult to achieve, due to their thermal stability. It does not occur thermally but only photochemically. Once again, the photochemical process is due to the resulting instability when the ground state ^1A_g (t_{2g}^6) is excited to the $^1T_{1g}$ ($t_{2g}^5\,e_g^1$) state.

15.2.2
Photo-redox Processes

15.2.2.1 General Aspects

Photochemical redox processes are induced by the absorption of light of high energy, corresponding generally to the MLCT or LMCT region. The following reaction is a typical example:

$$[Rh(N_3)(NH_3)_5]^{2+} \xrightarrow{h\nu} \begin{cases} [Rh(NH)(NH_3)_5]^+ + N_2 & (1) \\ [Rh(H_2O)(NH_3)_5]^{3+} + N_3^{\cdot} & (2) \end{cases}$$

Irradiating in the d–d zone, the quantum yield for the substitution reaction (2) is a maximum at the absorption maximum wavelength, but it is zero at greater $h\nu$. However, for the photoredox reaction (1) it is just the opposite: $h\nu$ must be in the charge-transfer region. As a consequence, we can separate redox or non-redox reactions, enhancing the yields of reactions (1) or (2), according to the wavelength of the light employed. However, these processes are not always so "clean".

15.2.2.2 Mechanism of the Redox Processes

Let us assume the reaction $A + B \rightarrow A^+ + B^-$ is exergonic but with a comparatively high activation energy. This reaction is favored from a thermodynamic viewpoint, but kinetically it will be slow, the corresponding rate constant effectively depending on the activation energy. Light absorption can enhance the reaction rate, because an excited state $A^* + B$ or $A + B^*$ with an energy greater that the ground state activation energy may be formed. A typical example is given in Figure 15.9.

15.2.2.3 Variation in Redox Potentials

In general, *the photochemically excited state is more oxidizing and more reducing than the ground state from which it has been generated.* The scheme of the variation of the redox potentials for an octahedral d^6 ion is showed in Figure 15.10.

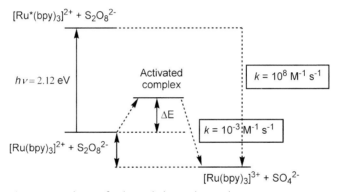

Figure 15.9 Scheme of redox and photoredox mechanisms.

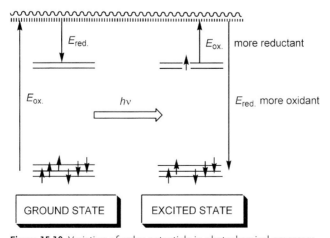

Figure 15.10 Variation of redox potentials in photochemical processes.

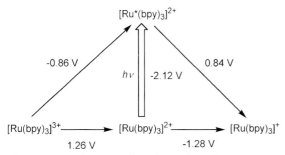

Figure 15.11 Redox potentials in the $[Ru(bpy)_3]^{2+}$ system.

Taking into consideration that $[Ru(bpy)_3]^{2+}$ is the most studied complex during the past decades (see below), let us indicate its redox potentials, with and without photochemical excitation (Figure 15.11). The energy difference between the long-lived ^{3}MLCT state and the ground state of 2.12 eV explains the greatly enhanced reactivity of the excited complex.

15.2.3
Generalities of Some Photochemically Active Series of Compounds

15.2.3.1 $[Ru(bpy)_3]^{2+}$ and Related Complexes [8]

$[Ru^{II}(bpy)_3]^{2+}$ is a d^6 complex with pseudo-octahedral symmetry (actually D_3). The ground state is $^1A_1$ (singlet) with the filled $(t_{2g})^6$ configuration in a strong ligand field. The lowest excited state originates from the excitation of an electron from a metal t_{2g} orbital to a π^*-antibonding orbital of the bpy ligand resulting in a both spin and symmetry allowed $d \rightarrow \pi^*$ (^{1}MLCT) transition at 452 nm ($\varepsilon = 14\,450\,M^{-1}\,cm^{-1}$). At ambient temperature, solutions of $[Ru(bpy)_3]^{2+}$ readily show an orange–yellow emission in the form of a broad band centered around 600 nm. It has been well established that the luminescence is of the spin-forbidden charge transfer type, corresponding to the ^{3}MLCT transition. The fact that the emission occurs from the triplet state irrespective of the irradiation wavelength indicates that there is a fast intersystem crossing to the triplet. The rate constant (k_{isc}) for this intersystem crossing is close to unity, the overall quantum yield for the luminescence in aqueous solution is 0.042. Both the quantum yield and lifetime indicate that the excited state can be formed in sufficient quantities in aqueous solution in order for $[Ru^*(bpy)_3]^{2+}$ to be able to participate in a wide range of bimolecular reactions with added substrates.

Energy and Electron Transfer Reactions Photoprocesses of $[Ru(bpy)_3]^{2+*}$ can be divided into:

1. *Energy transfer reactions.* The reaction $[Ru(bpy)_3]^{2+*} + Q \rightarrow [Ru(bpy)_3]^{2+} + Q^*$ (Q = quencher) is a typical example. A particular case is $[Ru(bpy)_3]^{2+*} + [Cr(CN)_6]^{3-} \rightarrow [Ru(bpy)_3]^{2+} + [Cr(CN)_6]^{3-*}$ or in the double salt $Na[Ru^{II}(bpy)_3][Cr(ox)_3]$ where the phosphorescence from the ^2E_g state of the chromium(III) complex is observed after the excitation.

   This quencher unit (Q) can be covalently connected to a $[Ru(bpy)_3]^{2+}$. A paradigmatic example is the photoinduced electronic energy transfer in modular, conjugated, dinuclear Ru(II)/Os(II) complexes, in which a fast and efficient photoinduced energy transfer takes places from the excited ruthenium moiety to the osmium-based component.

2. *Electron transfer reactions.* The transfer of an electron from the filled $(t_{2g})^6$ orbitals of the metal into the empty π^* orbital of the bpy ligand creates an electron hole at the metal center, while placing an electron on one of the bpy ligands. Thus, the excited state can act as both an oxidizing and a reducing agent:

$$[Ru(bpy)_3]^{2+*} + \text{Reductant} \rightarrow [Ru(bpy)_3]^+ + R^+$$

$$[Ru(bpy)_3]^{2+*} + \text{Oxidant} \rightarrow [Ru(bpy)_3]^{3+} + O^-$$

The *oxidative quenching* has been largely studied with viologens (bipyridinium salts) and many inorganic complexes and salts. The *reductive quenching* is produced mainly by amines, Eu^{2+} ion, ferrocene, etc.

Tuning of the Excited State Reactivity $[Ru(bpy)_3]^{2+}$ has a typical lifetime of ca. 1 μs in deoxygenated solution at room temperature. This lifetime is still too short for many purposes and considerable effort has been spent trying to prolong this lifetime. In fact, the photophysics of ruthenium polypyridyl complexes are governed by the energy gap, ΔE, between the luminescent level of MLCT electronic configuration, and the higher lying metal centered d–d state $^3T_1(t_{2g}^5 \, e_g^1)$, which may provide a thermally activated pathway for fast radiationless decay of the luminescent state. Thus one tries to increase ΔE. Let us explore some possible ways of enhancing the photochemical properties of the $[Ru(bpy)_3]^{2+}$ cation and analogs.

- Substitution of bpy by similar polydentate ligands, such as phen or terpy.

- Exploiting the substituent effects on the bpy (or similar) ligands. Deep effects have been induced by adding macrocycles to the periphery of the bpy ligands.

- Synthesis of multimetallics systems such as $[NC-Ru(bpy)_2-CN-Ru(bpy)_2-NC-Ru(bpy)_2\text{-}CN]^{2+}$ complex, which can be considered to act as an efficient *antenna* (see below).

- Using dendrimers; the good results obtained with many polynuclear ruthenium complexes have encouraged researchers to increase the nuclearity such as in the dendrimer-like complexes already discussed in Chapter 5.

- Using mixed-ligand complexes. The Ru^{II}–C bond increases markedly the electronic coupling with respect to the typical compounds containing only ruthenium–nitrogen bonds [9]. As an example, the cyclometallated complex $[Ru^{II}(bpy)_2(ppH)]^+$ (ppH = 2-phenylpyridine) is very reactive and able to give long binuclear complexes through alkine plus aryl bridges [9a].

- Employing diimine complexes with other metals, such as octahedral d^6 complexes of Re(I), Os(II), Ir(III), and Pt(IV) ions. However, the properties of these diimine complexes are much less interesting than those of their Ru(II) analogs.

15.2.3.2 Porphyrins and Related Complexes

The group of porphyrin compounds is one of the most interesting groups of molecules for life. Chlorophyll plays a key role in photosynthesis. Iron porphyrin in heme is essential to the transport of molecular oxygen by hemoglobin in blood and myoglobin in muscular systems. Electron transport in the respiration cycle is governed by porphyrins in cytochromes.

Free-base porphyrins and their metalloporphyrins generally have an intense π–π^* absorption band in the visible region. The absorption bands in the visible region are explained by transitions between the two highest occupied π orbitals of a_{1u} and a_{2u} symmetry and the two degenerate lowest unoccupied π orbitals of e_g symmetry. The excited energy state diagrams of porphyrins with transition metals are rather complicated when the d orbitals of the metal intervene between the π and π^* orbitals of the porphyrin ring. The $d \rightarrow d$, $\pi \rightarrow d$ and $d \rightarrow \pi^*$ states overlap with the normal porphyrin $\pi \rightarrow \pi^*$ state.

15.2.3.3 Photophysics of Lanthanides

d-Block transition metal complexes often exhibit room-temperature luminescence over the microsecond time scale, while certain lanthanide complexes emit on the millisecond time scale. The lanthanide emission is thus often called '*delayed luminescence*'.

In lanthanide ions the luminescence results from transitions within the partially filled 4f shell of the ions which, in principle, are spin-forbidden transitions. For f-block ions, it may be more appropriate to consider the first coordination sphere as a perturbing environment of the 'bare' metal ion. Thus, the first coordination sphere of an f-metal ion plays a role similar to that played by the second coordination sphere of the d-block metal complexes. Since the photochemistry of the f-block elements is poor, we will only deal with their photophysical properties.

Luminescence The unique electronic properties of lanthanide ions (long-lived luminescence of the order of milliseconds and sharp–narrow emission spectra) make them particularly suitable for the development of diagnostic tools in medical analysis, for sensor development, as luminescent probes and in fluorescence imaging because of their ability to discriminate between background fluorescence and the target signal. Most of the photophysics research on lanthanides has been carried out with the Eu^{3+} and Tb^{3+} ions with emissions at 620 nm (red) and 550 nm (green), respectively. The other ions have all very low quantum yield in aqueous solution and appear to be less useful with respect to similar applications. The three ions Nd^{3+}, Er^{3+} and Yb^{3+} have recently gained in popularity because technical developments have occurred which facilitate the detection of weak NIR (near-infrared) emission which is desirable for many *in vivo* applications. A very good review on luminescent lanthanide ions has been published by Bünzli and Piguet [10].

There are three important problems in lanthanide luminescence which need to be addressed:

1. The bands observed in lanthanide(III) absorbance spectra are normally very weak. Fortunately, some ligands can incorporate a chromophore (sometimes referred to as an *antenna*, see below), which absorbs strongly at a suitable wavelength and transfers its excitation energy to the metal. If the antenna has a high absorption coefficient and the energy transfer process is efficient, then the 'effective' molar absorption coefficient of the metal is vastly increased and intense luminescence may result following excitation by conventional light sources. A review of transition-metal sensitized NIR luminescence from lanthanides in d–f heteronuclear arrays has been published recently [11].

2. The deactivation of the metal emissive states by vibrational energy transfer (vibronic coupling) often primarily involving the solvent water molecules. Even for the best ions in the best conditions, the deactivation effect of water molecules leads to a reduction of an order of magnitude in the intensity of the emission. Thus, it is fundamental to shield the metal as efficiently as possible from the solvent water molecule.

3. The stability of the complex with respect to metal–ligand dissociation. Carefully designed polydentate ligands lead to increased stability of the lanthanide chelate in solution and allow the metal center to be well protected from water molecules. However, the tendency of lanthanide ions to adopt high coordination numbers and their lack of stereochemical preferences make the design of such ligands very challenging. The classical approach was to use linear polydentate and multifunctional ligands, such as polyaminocarboxylates (edta, for example), β-diketonates (investigated as early as 1897), polydentate oligopyridyne amine ligands, etc. Lanthanide phthalocyanines and porphyrins, crown ethers, cryptands and calixarene derivatives have been successfully studied.

There are a number of applications of lanthanide luminescence in solution [12]:

- *Immunoassays.* An immunoassay is a biochemical test that measures the concentration of a substance in a biological liquid using the reaction of an antibody to its antigen, with very specific binding. The presence of both antigen or antibodies can be measured. Detecting the quantity of antibody or antigen can be achieved by a variety of methods. One of the most common is to label them with radioisotopes or *fluorescent* substances. The dissociation enhanced lanthanide fluoroimmunoassay (DELFIA) techniques based on poly-aminocarboxylate chelates of Eu^{3+} have held a prominent place in biological assay systems, with very sensitive results.

  Furthermore, the background fluorescence of biological materials is almost invariably short-lived compared to the long luminescence lifetimes, which may be observed for Eu^{3+} and Tb^{3+} (the latter are of the order of milliseconds). In addition, since the overall experiment is completed in a few milliseconds, it can be repeated several thousands times during a minute, leading to a very high sensitivity within a few minutes.

  Fascinating developments are taking place to overcome some of the limitations inherent in these luminescent probes, namely that (i) the visible emitting

Ln^{3+} ions usually require UV excitation, which compromises *in vivo* applications and (ii) if generated inside a biomaterial, visible light may also be substantially absorbed, reducing the signal from the analyte. One solution is to turn to NIR emission since biological tissues are rather transparent in this spectral range.

- *Magnetic resonance imaging techniques.* The high magnetic moment ($S = 7/2$) and the slow electronic relaxation of gadolinium make it ideal for the design of magnetic resonance imaging (MRI) relaxation agents. All current Gd^{III}-based commercial contrast agents are low-molecular-weight complexes of octadentate poly(aminocarboxylate) ligands such as the macrocycle H_4dota (1,4,7,10-tetraazacyclododecane-*N,N',N'',N'''*-tetraacetic acid), and the acyclic H_5dtpa (diethylenetriamine-*N,N',N''*-pentaacetic acid) (see Chapter 17).

Lanthanide ions have also been used in the solid state. They are widely employed for photonic applications such as light-emitting diodes, tuneable lasers, and optical storage. They are incorporated into glasses of oxides or oxoanions (borate, silicate, aluminate, tungstate), which have many industrial applications such as in television screens, yttrium gadolinium garnet lasers, etc. A range of room-temperature photoluminescent solids in the silicate system $K_3[M_{1-a}Ln_a\,Si_3O_8(OH)_8]$ (M = Y^{3+} and Tb^{3+}; Ln = Eu^{3+}, Tb^{3+} and other) have been prepared recently. These materials exhibit remarkable photoluminescence properties, which may be tuned by carefully choosing the Ln^{3+} ions in the occupancy of the layer and interlayer metal sites.

Fabrication of chemically stable structured nanoparticles with Ln^{3+} in Ln_2O_3, etc. or in nanoporous xerogel and sol–gel SiO_2 glasses, with strong luminescence, has been achieved. Lanthanide-based NIR luminescence is attracting considerable interest not only in medical applications, but also both in the fields of light emitting diodes, and telecommunications. Al_2O_3 doped with Er^{3+} and Yb^{3+} are good devices in this field, such as the Er^{3+}-doped silica fiber amplifiers.

15.3
Photo-molecular Devices and Machines

A *device* is something invented and constructed for a special purpose and a *machine* is any combination of mechanisms for utilizing, modifying, applying or transmitting energy, whether simple or complex. A *molecular-level machine* is a particular type of molecular-level device in which the relative positions of the component parts can change as a result of some external stimuli. If the origin of the excitation is light, then we are dealing with *photo-molecular devices* or *machines*.

In their book on Molecular Devices and Machines [13], Balzani et al., divide devices into (i) wires and related systems, (ii) switching electron- and energy-transfer processes, (iii) light-harvesting antennae and (iv) photoinduced charge separation and solar energy conversion. We will only discuss here cases (ii)–(iv).

15.3.1
Switching Electron-transfer and Energy-transfer Processes

In a general sense, *"molecular-level switch"* describes any molecular-level system that can be reversibly interconverted between two (or more) different states by some external stimulus [13]. Switching requires an external stimulus which, at the molecular level, causes both electronic and nuclear rearrangements. When switching involves large nuclear movements, the mechanical aspect can become more interesting than the switching function itself. Systems of this type are called molecular-scale machines, and will be treated later in this chapter.

The three most important types of stimulus that can be used to switch a chemical compound are (i) light energy (photons), (ii) electrical energy and (iii) chemical energy (in the form of protons, metal ions, specific molecules, etc.). In this chapter we are interested in *photochemical stimulation*.

15.3.1.1 Bistable Systems

Photochromic Systems   The term *'photochromic'* is applied to molecules that can be reversibly interconverted between two forms with different absorption spectra, with at least one of the reactions *being induced by light excitation*. After photochemical conversion, a spontaneous back reaction is expected to occur. Sometimes the photoproduct may be kinetically inert and the process can be reversed only by use of a second light stimulus. Depending on the thermal stability of the photogenerated isomers, photochromic systems can thus be classified into two categories: T-type (thermally reversible types) and P-type (photochemically reversible types).

The families of organic compounds most extensively used in photochromic coordination systems are diarylethenes, spiropyranes and azobenzenes (Figure 15.12). We will only treat here in some detail the diarylethenes. The photochromic performance of diarylethenes can be maintained even after more than 10^4 cycles. π-Electrons are localized in the thiophene rings in the open-ring isomer, whereas they are delocalized throughout the molecule in the closed-ring isomer. As a consequence, the absorption bands occur at much longer wavelengths (up to 600–700 nm) for the closed ring isomers than for the open-ring isomers (no absorption in the visible region). The open form is therefore indicated as the OFF state, and the closed form as the ON state.

When the X and Y substituents in Figure 15.12, are two Ru(II) bipyridine complexes, efficient photocyclization occurs [14]. Upon photoexcitation to the emitting ^{3}MLCT state of Ru-bpy, photoreactive ^{3}IL states are populated by an efficient *energy-transfer process*. The involvement of these ^{3}IL states explains the extremely high quantum yield of the photocyclization, which is independent of the excitation wavelength but decreases strongly in the presence of oxygen. This behavior differs substantially from the photocyclization of the free ligand, which occurs from the lowest ^{1}IL state on a picosecond time scale and is insensitive to oxygen quenching.

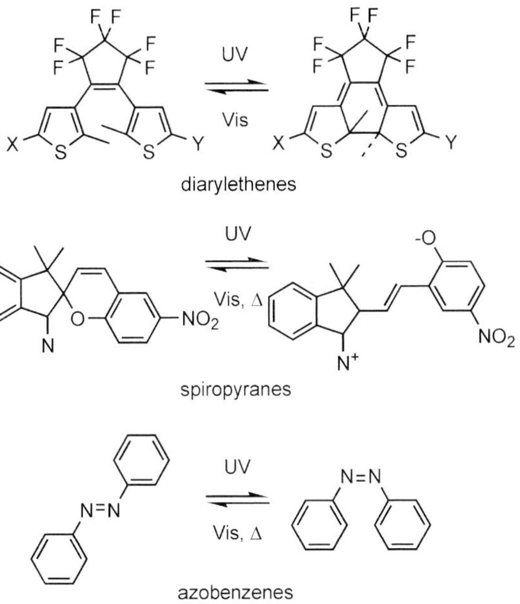

Figure 15.12 Three families of organic compounds used in photochromic coordination systems.

Fluorescent (Luminescent) Switches Some metal complexes, e.g. those of Ni(II), undergo high-spin low-spin interconversion when the temperature is changed. This property has been exploited to obtain a *'fluorescent molecular thermometer'*. When dissolved in a polar medium (water, MeCN) $[Ni^{II}(cyclam)]^{2+}$ exists in two forms at equilibrium: O_h and D_{4h} complexes. The O_h form predominates at low temperature, and the D_{4h} at higher temperatures. The macrocycle can be linked to an aromatic fluorophore, naphthalene, for example: it is known that the octahedral high-spin complex has a pronounced quenching effect. It was observed that in D_{4h} complexes the naphthalene emission is distinctly higher than for O_h complexes. In fact, a solution in MeCN at room temperature is poorly fluorescent; on increasing the temperature, the naphthalene emission increases sharply [15].

The ^{3}MLCT luminescence of $[Ru(bpy)(CN)_4]^{2-}$ may be varied over a wide range by interaction of the externally directed cyanide ligands with additional metal cations, both in the solid state and in solution. The luminescence intensity and lifetime of these salts is highly dependent on the nature of the cations, with Cs^+ affording the weakest luminescence and Ba^{2+} the strongest [16].

15.3.1.2 Multiple Chemical Inputs

Many systems have been constructed that switch under the action of multiple chemical inputs. Let us briefly explain some of them, emphasizing the second input because the *first one is always a photon* (for our purposes).

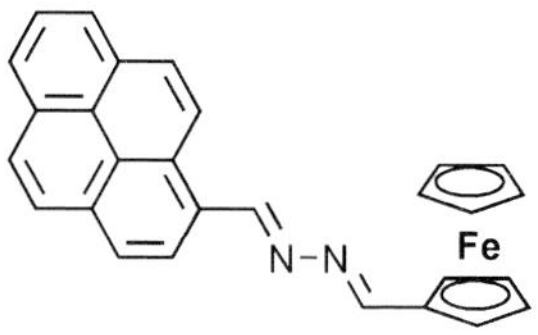

Figure 15.13 A simple and reversible redox-fluorescence molecular switch (see text).

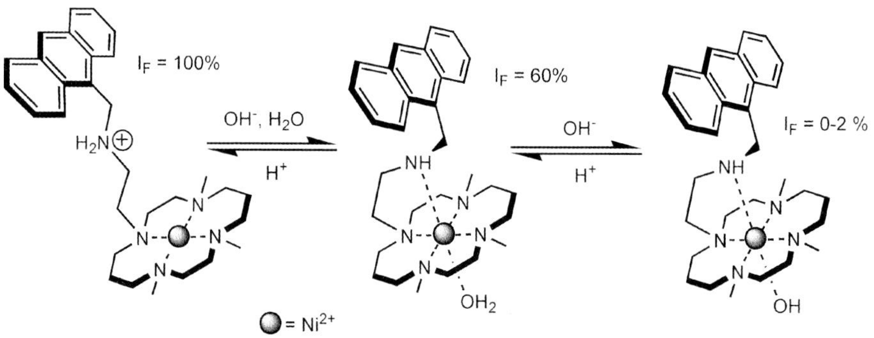

Figure 15.14 Acid–base input: pH dependence of the association of the molecular movement of the anthracene unit with the generation of a luminescent signal.

Redox Inputs A reversible redox-fluorescence molecular switch based on a 1,4-disubstituted azine with ferrocene and pyrene units is shown in Figure 15.13 [17]. This dyad (see Section 15.3.3.1) shows a *fast and reversible* redox-switchable fluorescence emission. Neutral dyad shows only a weak fluorescence because the ferrocene unit quenches the fluorescence of the pyrene subunit. After oxidation of neutral dyad (simply with copper(II) triflate), the electron-donating ability of the ferrocene subunit is reduced and, as a result, the electron transfer is arrested, leading to a fluorescent enhancement of the pyrene. As the ferrocene/ferrocinium pair transformation can be carried out reversibly, there is the possibility of obtaining a redox-fluorescence switch.

Acid–Base Inputs The association of the molecular movement of a 'scorpionate-like' complex with the generation of a luminescent signal has been investigated together with how the emission of the corresponding Ni^{2+} complex is affected by a change in pH [18]. The global effect is shown in Figure 15.14. The highest emission is due to the complex in which the photoexcited anthracene fragment, An^*, is at the furthest distance from Ni^{2+}. This happens in very acid solutions. At pH > 3 the pendant arm goes to occupy one of the axial positions of the coordination octahedron, bringing the An fragment much closer to the metal. At this shorter distance, a PET process takes place between An^* and Ni^{2+} which reduces the emission quantum yield by 60%. The complete quenching of fluorescence which takes place at pH > 9 should not be assigned to any molecular movement, but simply to the release of a proton from the axially bound water molecule of the complex. It has been suggested that the decrease in the overall electric charge of the complex makes the Ni^{2+} to Ni^{3+} oxidation process easier, thus making the occurrence of a PET process from Ni^{2+} to An* possible.

Cation Coordination Input   Lehn and coworkers [19], have reported the ionic modulation of photoluminescence properties in a motional process involving reversible switching between a highly luminescent ligand **L** in a W-shaped state and its poorly luminescent U-shaped complex [**LZn**], triggered by ion complexation–decomplexation reactions (Figure 15.15). The interconversion of **L** to [**ZnL**] and of [**ZnL**] to **L** is possible by adding to the solution the *tren* ligand, which forms a strong, pH-dependent complex $[Zn(tren)]^{2+}$ with Zn^{2+} ions. The transformation **L** → [**ZnL**] is characterized by the almost complete quenching of the fluorescence of **L**. On addition of *tren*, the typical luminescence of **L** is restored. Lowering the pH induces protonation of the terminal nitrogen atoms of *tren* causing the release of the complex's Zn^{2+} ions, which then bind to **L** leading to [**ZnL**] again. Repetitive conversion–interconversion of **L** and [**ZnL**] (luminescence vs. non-luminescence) can thus be pursued by successive additions of acid and base. The two polybenzenes of Figure 15.15 can be substituted by porphyrins (free or metallated) [20] with similar results. An interesting example of simultaneous cation and temperature inputs can be found in Ref. [21].

Anion-binding-controlled Switches   An interesting and particular case is the light-emitting devices based on the $[Zn^{II}(tren)]$ fragment. The tripodal ligand occupies four positions of the coordination polyhedra, leaving an axial site available for coordination of either a solvent molecule or an anion (the *input*). The $[Zn^{II}L]^{2+}$ complexes display the typical blue fluorescence of the anthracene fragment. However, the fluorescence is quenched when the Zn complex reacts with carboxylates due to an intramolecular electron transfer process (Figure 15.16).

15.3.2
Light-harvesting Antennae

Construction of artificial light-harvesting systems by metal coordination and the following step, the *charge separation*, will we treated in the next two paragraphs.

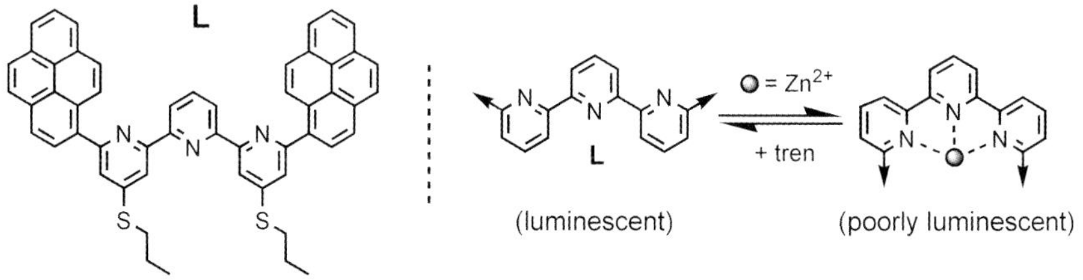

Figure 15.15 Effect of the cation coordination input on luminescent properties (see text).

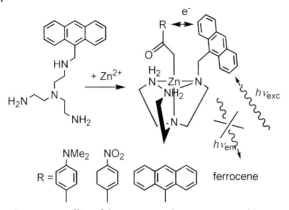

Figure 15.16 Effect of the anion coordination input on luminescent properties (see text).

LIGHT ABSORPTION

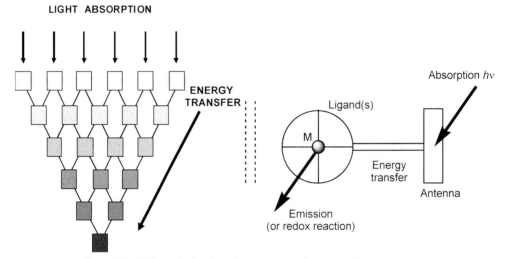

Figure 15.17 Schematic drawing of an antenna and antenna effect.

We have previously used in this chapter the term *"antenna"* or *"antenna effect"*. An antenna for light harvesting is a *multicomponent system* in which several chromophoric molecular species absorb the incident light and channel the excitation energy to a common acceptor component (Figure 15.17). The *chromophore* (C) is, thus, excited by a photon of given wavelength, and then transfers its energy to a *luminophore* (L) which, in turn, emits another photon of a different wavelength. These *chromophore–luminophore* (CL) complexes mimic the "antenna effect" of natural systems (photosynthetic light-harvesting antennae, for example). The most interesting systems are treated below. Other relevant systems, such as Ru-bpy derivatives can be found in Ref. [22].

15.3.2.1 Metallo-porphyrin Systems

Arrays containing porphyrin molecules are certainly the largest class of artificial antennae. Chromophore types closely related to chlorophyll derivatives are porphyrines and phtalocyanines. Although chlorophyll employs Mg^{2+}, artificial systems use Zn^{2+} as the most typical substitute. A review of cyclic porphyrin arrays as artificial photosynthetic antennae has been published recently [23].

Porphyrins with Equatorial Coordination Mode Di-, tri-, tetra-, and pentameric porphyrin arrays have been obtained by using several linkers between aryl groups of adjacent tetraarylporphyrin macrocycles. The array shown in Figure 15.18 has been studied with and without Zn in the central porphyrin. In this system efficient energy transfer from the peripheral Zn-containing units to the free-base core was observed (95–99%), without competitive electron-transfer reactions [24]. Giant porphyrin wheels, possessing large electron coupling, are novel models of light-harvesting photosynthetic antenna [25]. The wheel shown in Figure 15.19, and others very similar, exhibit an efficient excitation energy rate along the array.

Porphyrins with Apical Coordination Mode *Side-to-face* connection in porphyrin arrays is obtained when a porphyrin carrying suitable peripheral Lewis-base functions binds, by means of axial coordination, to the metal of the second porphyrin [26]. The stability constant reaches $10^{11}\,M^{-1}$ in nonpolar solvents. This motif can be applied to the formation of linear and macrocyclic species, both of which have been regarded as excellent light harvesting antennae. The macroring of Figure 15.20 can accommodate a tetrapodal ligand based on complementary and cooperative coordination [27]. It has been identified as the *best model* for the light-harvesting complexes of photosynthetic purple bacteria, which also adopt cyclic arrangements. The tetrapodal ligand makes the supramolecular structure more stable, without modification of the fluorescence. Circular structures are favored in

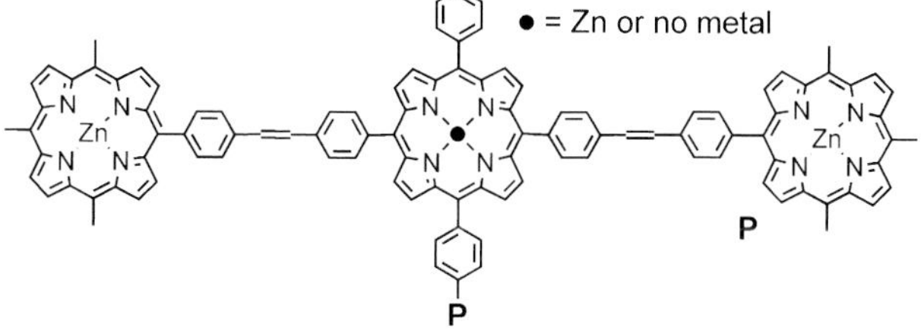

Figure 15.18 A pentameric porphyrin array used as antenna (one P part has been omitted in the drawing).

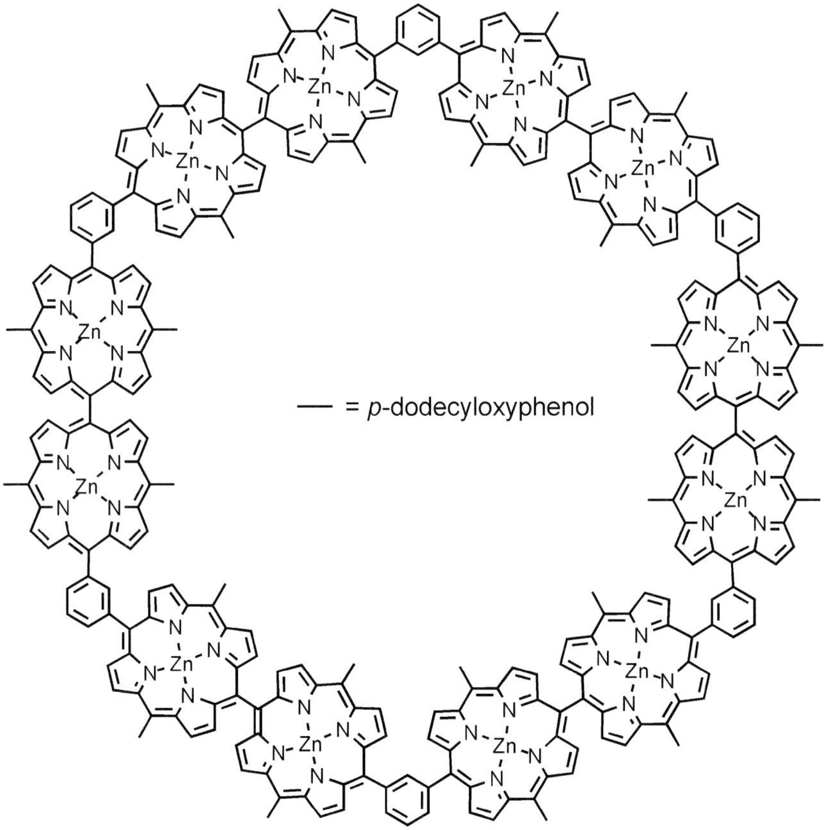

= *p*-dodecyloxyphenol

Figure 15.19 Giant discrete porphyrin wheel, possessing large electron coupling, as a model of light-harvesting photosynthetic antenna.

forming densely packed two-dimensional arrays in biological membranes. They are effective in absorbing light from any direction due to the isotropic nature of the circle and in transferring the energy to neighboring antenna molecules in the ring and, further, to other rings [28].

15.3.2.2 Dendrimers

Dendrimers have been studied in Chapter 5. Photoactive units can be directly incorporated or appended with covalent or coordination bonds in different regions of a dendritic structure and can also be noncovalently hosted in the cavities of a dendrimer or associated with the dendrimer surface.

Let us imagine a dendrimer with the [Ru(bpy)₃] moiety in the center, dimethoxi-benzene as branches and naphthalene as surface. A very efficient energy transfer process occurs, converting the very short-lived UV fluorescence of the aromatic

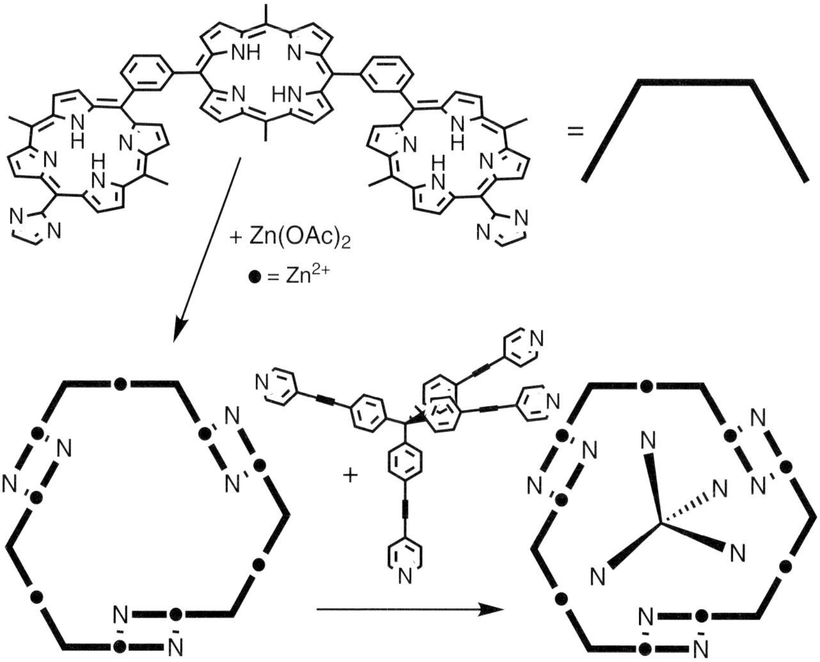

Figure 15.20 Antenna formed by porphyrins with apical coordination mode.

units of the wedges to the long-lived orange emission of the metal-based dendritic core. It should also be noted that in aerated solution the luminescence intensity of the dendrimer core is more than twice as intense as that of the $[Ru(bpy)_3]^{2+}$ parent compound, because the dendrimer branches protect the Ru-bpy-based core from dioxygen quenching.

Dendrimers with metal complexes in each branching center have been carefully investigated photophysically. For example, the $[\{(Zn)_7\}_4P]$ system, which incorporates 28 light absorbing porphyrin units into a dendritic scaffold with an energy-accepting core, mimics several aspects of natural light-harvesting systems [29].

15.3.2.3 Lanthanide Systems

For artificial systems the term 'antenna effect' was first used in the discussion of lanthanide ions surrounded by strongly absorbing ligands, in which the luminescence of the lanthanide ion was sensitized by excitation in the ligand-centered excited states. Since 1990 many complexes with very selective groups that improve the performance have been reported, as already commented in Section 15.2.3.3.

15.3.3
Photoinduced Charge Separation (Solar Energy Conversion)

Complementary to the *antenna effect* (photoinduced energy transfer) there is the simultaneous possibility of *photoinduced charge separation* (CS) as a fundamental step in photochemical energy conversion.

15.3.3.1 Photoinduced Energy Transfer Coupled to Charge Separation [30]

Great efforts have been made in relation to '*artificial photosynthesis*', where the goal is to mimic the green plants and other photosynthetic organisms that use sunlight to make high-energy chemicals. In the design of molecularly based systems for light-to-chemical energy conversion, this step is studied through the construction of two-, three- and multi-component systems (*dyads, triads, tetrads, pentads*, etc.) having suitable electron donor and acceptor moieties placed at specific positions on a charge-transfer chromophore. The theoretical scheme is shown in Figure 15.21. The performance of a triad (or tetrad, etc.) is measured by the following properties: (i) the lifetime of charge separation, (ii) the quantum yield of charge separation, depending on the competition between forward and back processes.

In a transition-metal coordination compound, the metal ion is usually that which absorbs light. A great variety of acceptor molecular components and donor units have been explored so far. Charge recombination (CR) is one of the great obstacles that must be overcome to have an efficient light-driven energy-storing reaction. Some important systems are:

- *Ru-bpy derivatives.* Very recently a $Mn_2^{II,III}Ru^{II}$-NDI triad (NDI = naphthalenediimide) (see Section 15.3.3.2) was reported to exhibit a charge-separated state lifetime in the millisecond range at room temperature [31]. Likewise, a new system with photoinduced long-lived charge-separated state ($\tau \sim 2$ μs) of [Ru(bpy)$_3$]–methylviologen with cucurbit[8]uril in aqueous solution, has been reported [32].

- *Porphyrins.* A simple example is the system involving porphyrin (chromophore)–quinone (quencher) dyad depicted in Figure 15.22A. The porphyrin complex

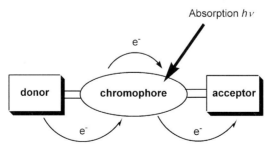

Figure 15.21 Scheme of a multi-component system (dyads, triads, tetrads, pentads, etc.) on a charge-transfer chromophore.

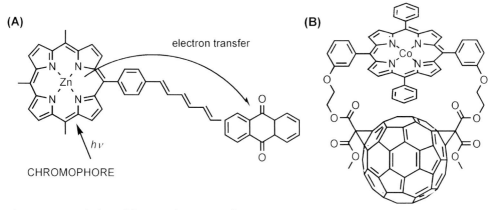

Figure 15.22 Dyads derived from porphyrin cromophores.

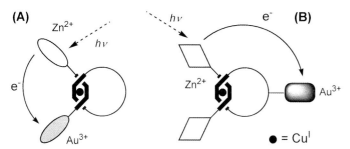

Figure 15.23 Charge separation in porphyrin-catenane derivatives.

transfers an electron to a nearby quinone (electron acceptor) following excitation by a photon. A more sophisticated system is the *trans*-2-linked C_{60}-cobalt(II) tetraphenylporphyrin diad (Figure 15.22B). Model calculations demonstrate that the fullerene moiety is easy to reduce. Many triads and tetrads combining the central porphyrin unit with carotene and ferrocene as donors and fullerene as acceptor have been reported. A description of the properties of many of these systems can be found in Ref. 33.

- *Catenanes and rotaxanes.* In 1993, Sauvage and coworkers prepared a rotaxane built around a central copper(I) bis(1,10-phen) complex with gold(III) and zinc(II) porphyrins acting as terminal stoppers (Figure 15.23) [34]. Zinc porphyrins are good electron donors and gold(III) porphyrins have good electron-accepting properties. Upon selective excitation of either porphyrin, rapid electron transfer occurs from the zinc porphyrin to the appended gold porphyrin. Electron transfer may proceed through a bond via the bridge between the two porphyrins or through space. These latter systems are photoreactive both with and without $M^+ = Cu^+$, Ag^+, Li^+, etc. in the center of the rotaxane. The coordination of M^+ enhances the rate of electron transfer between the porphyrins, indicating that the role of the M^+ is important in different perspectives. The same authors have recently published an interesting review on synthesis and photoinduced processes of porphyrin rotaxanes and catenanes [35].

15.3.3.2 Model of Photosynthesis

A key enzyme in green plant photosynthesis is Photosystem II (PSII). PSII carries out the general functions: light absorption, energy transfer, charge separation and charge stabilization. Therefore, the presence of both, an *acceptor side* and a *donor side* are necessary (Figure 15.24). A review of the chemistry of Photosystem II has been recently published [36] (see Chapter 17 for more details).

Over the past twenty years much effort has been devoted to studies of photoinduced electron transfer reactions from chlorophyll and analogs, e.g. to mimic the *acceptor side* of Photosystem II mainly with porphyrin–quinone (Figure 15.25A) and porphyrin–fullerene dyads [37]. The porphyrin (P) is a model for the less stable and synthetically less tractable chlorophyll.

In contrast, the attempts to mimic the *donor side* of Photosystem II have received less attention. In 1997, Akermark et al. prepared a simple model system, where a ruthenium(II)-bpy-type complex on illumination transfers an electron to an external acceptor (MV^{2+}) and then recovers an electron by internal ET from a coordinated Mn^{2+} complex that becomes Mn^{3+} (Figure 15.25B) [38]. The discovery that the tyrosine residue plays an important role in water oxidation has stimulated the synthesis of mimic systems for PSII that contain Ru(II)-bpy complexes connected to tyrosine and several ligands that can coordinate manganese.

To successfully mimic the global PSII system, very long-lived intermediate charge-separated states, stable on a time scale of seconds, are required. In 2005 Hammarstrom and coworkers reported a $Mn_2^{II,III}$-$Ru^{II}(bpy)_3$-NDI triad ($Mn_2^{II,III}$ = donor; NDI = naphthalenediimine acceptor, covalently linked to bpy) with the average lifetime of the $NDI\cdot^-$ radical of ca. $600\,\mu s$ at room temperature, which is at least 2 orders of magnitude longer than that for previously reported triads based on Ru-bpy photosensitizer [31].

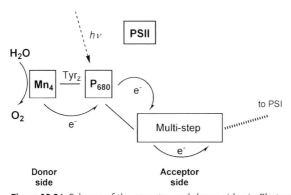

Figure 15.24 Scheme of the acceptor and donor sides in Photosystem II.

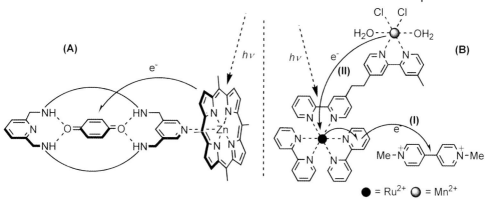

Figure 15.25 Models of the acceptor side (A) and the donor side (B) in Photosystem II.

15.4
Applications (Present and Future)

15.4.1
Modulation of Switches: Analytical Probes and Sensors

For some authors, switch and sensor are almost the same; for other authors the sensor is an application of a switch process. *Chemosensors* are molecules of abiotic origin that signal the presence of matter or energy (e.g. phenolphthalein). According to the nature of the output signal, chemosensors can be classified into different types, the so-called *fluorescent chemosensors* being the most interesting for this chapter.

15.4.1.1 Fluorescent (or Luminescent) Chemosensors [39]
Chemosensors incorporating a *fluorophore*, and a mechanism for communication between the parts are called *fluorescent chemosensors*. They are, in principle, systems that modify their light emission (enhancement or quenching) through chemical parameters, such as pH, complexation of metal ions, anions or molecules. The recognition event is converted into an optical signal expressed as a fluorescence enhancement or quenching of the fluorogenic fragment directly (*intrinsic chemosensors*) or indirectly linked via a spacer to the receptor unit (*conjugated chemosensors*).

Metal-ion Sensors In general, the majority of the fluorescent chemosensors possess a signalling unit sensitive to the binding to metal ions, leading to *chelation enhancement of the fluorescence*, (CHEF) or *chelation enhancement of the quenching* (CHEQ). Artificial optical sensors that respond selectively to biologically important metal ions (e.g. Li^+, Na^+, K^+, Mg^{2+} and Ca^{2+}) have attracted much attention for their potential use in chemical and biological applications. The detection of poisonous heavy metals is also of great importance in medicine. The most important groups

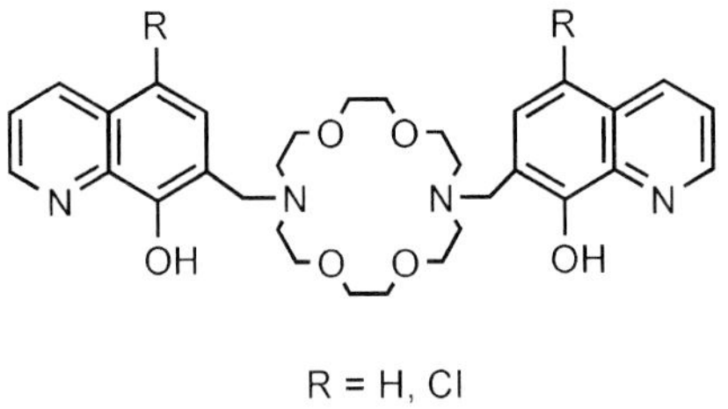

R = H, Cl

Figure 15.26 A fluorescent chemosensor for Mg^{2+}.

of ligands than act as metal-ion sensors are: crown ethers and derivatives, other macrocycles, cryptands and derivatives. Recently the so-called nanoporous polymers have been exhaustively studied. For the sake of brevity, we are going to study only two examples:

Magnesium ions, the most abundant divalent cation in cells, are required for many cell processes such as proliferation and cell death. Diaza-18-crown-6-hydroxyquinoline (Figure 15.26) derivatives are very good positive fluorescent sensors for Mg^{2+} in living cells [40]. Remarkably, fluorescence output is not significantly affected by other divalent cations, most importantly Ca^{2+}, or by pH changes within the physiological range. Furthermore, this probe is readily permeable to cells.

Cadmium ions. The azacryptand shown in Figure 15.27, has the fluorophore 7-nitrobenz-2-oxa-1,3-diazole attached, to give an integrated fluorophore-receptor configuration. The fluorophoric system, however, does not show any appreciable emission when excited, due to an efficient intramolecular photoinduced electron transfer (PET). But when a metal ion enters the cavity, the intramolecular PET is blocked, causing recovery of the fluorescence. Cd^{2+} *gives the highest quantum yield.* However, in the presence of coordinating ions such as Cl^-, N_3^- and SCN^- the metal ion comes out of the cavity, causing PET to take place once again and the fluorescence is lost. Thus, translocation of Cd^{2+} between the inside and outside of the cryptand cavity can lead to a reversible fluorescence on/off situation [41]. This signalling agent is, thus, a sensor of cations and anions. Other selective fluorescent sensors for imaging Cd^{2+} in living cells have been reported recently [42].

Other cations. A sensitized europium complex generated by micromolar concentrations of copper(I) has been reported very recently [43]. Sensors of Zn [44], Fe [45], Hg [46], and other metals have been reported. The design of sensor molecules that show cation-induced fluorescence enhancement by heavy and transition metal ions has been reviewed by Rurack [47].

Anion Sensors Many anions have an important biological and environmental relevance and efforts have been made recently to make anion-controlled luminescent chemosensors. Much of this effort stem from the work of Parker, who investigated the influence of iodide, chloride, bromide, hydrogencarbonate, hydrogenphosphate, sulfate, and lactate on Eu^{3+} and Tb^{3+} complexes with cyclen derivatives [48]. The photophysical properties of both Eu-L and Tb-L depend drastically on the nature of the anion in solution; in particular, adding two equivalents of

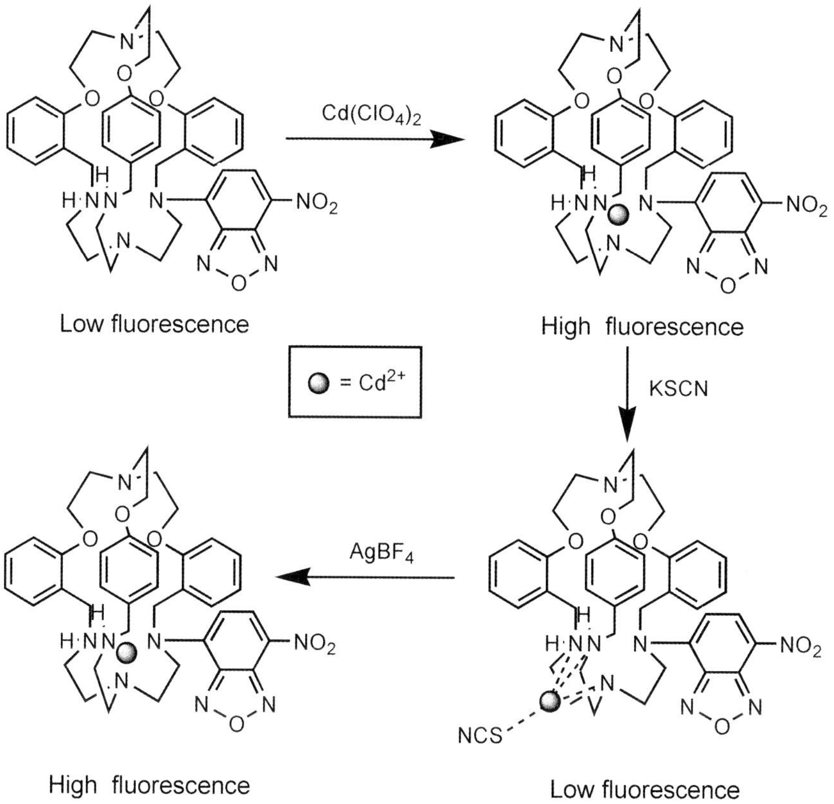

Figure 15.27 Azacryptand as fluorescent chemosensor for Cd^{2+}.

nitrate to a solution of [Eu-L](OTf)$_3$ results in an 11-fold enhancement of the metal-centered emission intensity. A pyrophosphate-selective fluorescent chemosensor at physiological pH (important in cancer research) has been reported recently [49]. Fluorescent probes for fluoride and superoxide are given in Refs. [50, 51] respectively.

Molecule Sensors   Molecule sensors for oxygen, NO, small molecules and biomolecules, etc. have been largely developed in past years. Let us describe two of them:

- *Oxygen sensors.* Luminescence-quenching-based oxygen sensors find application in a variety of areas from medicine to chemical and environmental analysis. Meso-substituted bisporphyrins metallated with Zn(II), Cu(II) or Pd(II) led to homobimetallic systems, [(M)$_2$(bisporphyrin)], with a cofacial structure. The different spacers placed in the meso position allow one to vary the intermacrocycle distance and the planar angle. This cofacial feature offers the possibility of having dioxygen molecules inside the cavity for a period of time, allowing dynamic (collisional) phosphorescence quenching to be more efficient [52]. Several methods for $^1O_2$ detection have also been developed [53].

- *Nitric oxide sensors.* Nitric oxide (NO) has captured the attention of chemists, biologists, and medical researchers. NO regulates both beneficial and harmful biological processes, depending on a variety of not yet fully delineated factors. At low concentration NO regulates vasodilatation in the circulatory system and long-term potentiation in the brain. In contrast, micromolar concentration of NO can trigger the formation of reactive nitrogen species, leading to carcinogenesis and neurodegenerative disorders but also providing a defense against invading pathogens.

Reviews of current strategies for detecting NO, by coupling transition metal-nitrosyl complexes with fluorescence signalling, can be found in Ref. [54].

15.4.2
Photomolecular-scale Machines [55]

Photomolecular machines are devices that undergo light-induced nanomechanical processes resulting in reversible changes in shape. The two most important systems are:

15.4.2.1 Systems Based on Photoisomerization Processes

Cis–trans photoisomerization reactions involving —N=N—, —C=N—, or —C=C— bonds are well known processes. In fact, our visual sense uses *trans–cis* photoisomerization of rhodopsin in the retina, resulting in an electric signal to the nervous systems through a multistep chemical reaction process. In general, they are extremely clean and reversible reactions, the prototypical case being the *cis–trans* isomerization of azobenzene with visible or UV light, respectively (Figure 15.28, inset).

One of the most fascinating devices has been named "*Light-driven open-close motion of chiral molecular scissors*" (Figure 15.28A) [56]. In these "scissors" the ferrocene is the pivot part, since the two cyclopentadienyl rings are parallel to each other and rotate freely, even at low temperature. The azobenzene is the driving group. Thus, the photoisomerization of the azobenzene unit induces the open–close motion of the "blade" parts (the two external phenyl rings). The two chiral isomers, *cis* and *trans*, were possible to synthesize and separate. When the *trans*-isomer is irradiated at 350 nm (UV light) the absorption spectrum changes indicate isomerization to the *cis*-isomer, in a molar ratio 10:90, approximately. On the other hand, upon irradiation with visible light ($\lambda > 400$ nm) a backward isomerization takes place to furnish a *cis/trans* isomer ratio of approximately 55/45.

Following this system, in 2006, Aida and coworkers reported that light-induced scissor-like conformational changes of one molecule, can give rise to mechanical twisting on a non-covalent bound guest molecule. With two Zn-porphyrin moieties

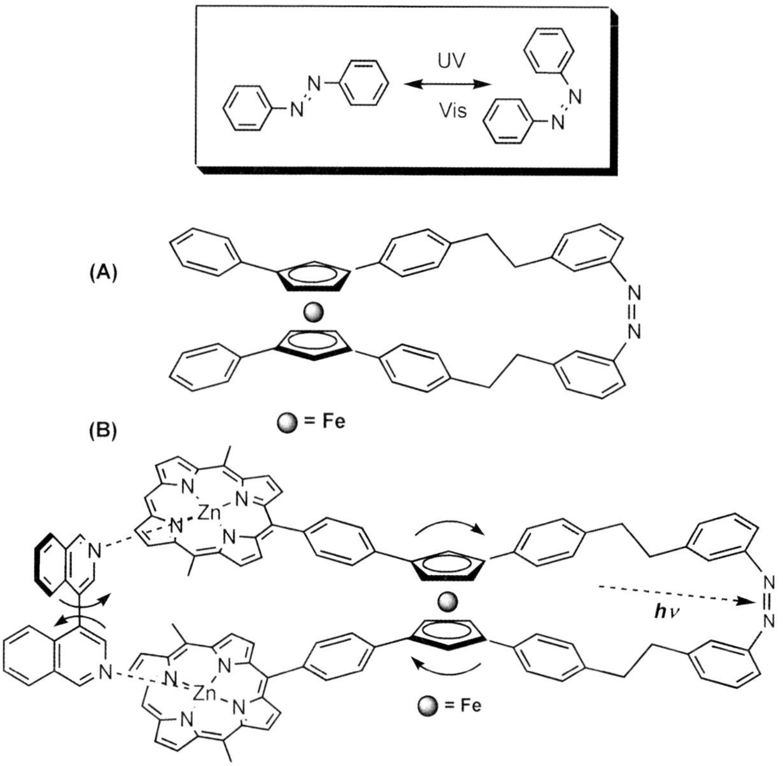

Figure 15.28 Molecular-scale machines based of photoisomerization processes.

added to ferrocene parts, it is possible to translate the scissor effect into intermolecular coupling of motion (Figure 15.28B) [57].

15.4.2.2 Systems Based on Photoinduced Electron Transfer (PET)

Amongst the most 'popular' and well-known nanomachines are the rotaxanes, called *molecular shuttles*, which incorporate two structurally different bipyridinium sites (stations **1** and **2** in Figure 15.29) [13, 55]. One of the bulky end-groups of the rotaxane's string is a Ru-bpy complex. This can absorb a photon of visible light and so form an excited state that donates an electron to the more easily reduced of the two bipyridinium sites (station **2**). Reduced bipyridinium is a very much poorer binding site for the ring, and a net flux of rings occurs to the unreduced station **1**. After $10\,\mu s$, however, as back electron transfer finally takes place, station **2** regains its stickiness. A next flux of rings occurs back from station **1**, the system returns to its original equilibrium, and a machine cycle has taken place. The process can be repeated for at least a thousand cycles. To generate net flows of rings between the stations one would have to switch the light source rapidly on and off.

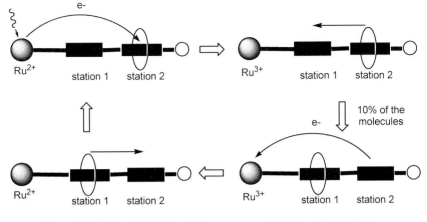

station 1 = 3,3'-dimethyl-4,4'-bipyridinium   station 2 = 4,4'-bipyridinium

ring (electron donor) = bis-*p*-phenylene[34]crown-10

Figure 15.29 A molecular-scale machine based on photoinduced electron transfer.

15.4.3
Splitting of Water upon Visible Light Irradiation: Hydrogen Liberation

One of the most important aspects of the photochemistry and photophysics is to achieve the possibility to harvest the solar energy in order to convert it to electric or chemical energy.

15.4.3.1 The Solar Spectrum and Energy Requirements

The nature of the solar spectrum imposes fundamental limits on the efficiency of any photochemical process. Most of the energy in sunlight appears as photons of relatively long wavelength; only a small portion lies in the UV region of the spectrum. Thus, to take full advantage of solar energy through photochemical and photophysical processes, several conditions must be fulfilled: (i) the absorption spectrum of the active chemical must coincide with the distribution spectrum of solar energy; (ii) the reaction potentially used to harvest solar energy must be endergonic, the greater the ΔG value of the corresponding thermal reaction, the greater the harvested energy; (iii) the inverse reaction must be easily controllable, that is, despite being exergonic, a sufficiently large activation energy must prevent it from occurring too rapidly in normal conditions; (iv) the quantum yield of the photochemical process must be high in order to have a great efficiency; (v) systems must be cheap, abundant, easy to manipulate, stable to water and air, etc. This has to apply to both the reactants and products of the reaction.

With these premises, many photochemical processes have been studied during the past forty or more years. Concerning the redox processes special attention has been paid to the splitting of water upon visible light irradiation:

$$H_2O\ (l) \rightarrow H_2(g) + \frac{1}{2}O_2(g)$$

The easy and cheap hydrogen liberation from water is, undoubtedly, one of the most exciting projects in chemistry. However, many problems arise when attempting to carry out this process either on the laboratory scale or, more importantly, on the industrial scale. The water splitting is a highly endergonic process with ΔG = 237.2 kJ mol^{-1}. This corresponds to the UV region below 300 nm, where the intensity of the solar radiation at the earth's surface is negligibly small. All these reasons explain why *photocatalytic water splitting* has received much attention since the 1970s.

15.4.3.2 The First Steps of the History: From the Beginning to the 1980s [58]

An excited state M* of a metal complex M may undergo oxidation or reduction on reaction with an electron acceptor A or donor D, respectively: M* + A $\rightarrow$ M$^+$ + A$^-$ or M* + D $\rightarrow$ M$^-$ + D$^+$. These two reactions, in principle, may be applied to the production of hydrogen,

$$A^-\ (M^-) + H^+ \rightarrow A\ (M) + \frac{1}{2}H_2$$

M is usually called a *sensitizer* (S). [Ru(bpy)$_3$]$^{2+}$ has been undoubtedly the main sensitizer in this research. By far the most popular reagent for *oxidative quenching* (or *relay*, R) of [Ru*(bpy)$_3$]$^{2+}$ is methylviologen (1,1'-dimethyl-4,4-bipyridinium, or "paraquat", abbreviated here as MV^{2+}).

$$[Ru*(bpy)_3]^{2+} + MV^{2+} \rightarrow [Ru(bpy)_3]^{3+} + MV^+$$

At this stage, at least theoretically, both the oxidant and reductant could be used to liberate hydrogen and oxygen from water: $E^0(Ru^{3+}/Ru^{2+})$ = 1.26 V; $E^0(Ru^{3+}/Ru^{*2+})$ = −0.86 V; $E^0(MV^{2+}/MV^+)$ = −0.46 V; $E^0(2H^+/H_2)$ = 0 V; $E^0(\frac{1}{2}O_2,2H^+/H_2O)$ = 1.229 V. Thus:

$$2MV^+ + 2H^+ \rightarrow 2MV^{2+} + H_2$$

$$2[Ru(bpy)_3]^{3+} + H_2O \rightarrow 2[Ru(bpy)_3]^{2+} + \frac{1}{2}O_2 + 2H^+$$

There are, however, several problems that are not easily overcome: (i) the two reactions need a catalyst; (ii) the back reaction, [Ru(bpy)$_3$]$^{3+}$ + MV$^+$ $\rightarrow$ [Ru(bpy)$_3$]$^{2+}$ + MV^{2+} must be suppressed; (iii) likewise it is necessary to suppress the reaction of O$_2$ with MV$^+$ to produce again MV^{2+}.

Concerning point (i) two electrons are needed for each water molecule but MV$^+$ only provides one at a time. To avoid this problem, the use of heterogeneous catalysts, particularly colloidal noble metals (Pt) and noble metal oxides (PtO$_2$), was developed. A catalyst for multielectron redox processes is essentially a 'charge pool', that is, a species capable of acquiring electrons (or holes) from a one-electron reducing (or oxidizing) species in a stepwise manner at constant potential and then delivering these electrons (or holes) to the substrate in a 'concerted' manner, to avoid the formation of high-energy intermediates. Under optimum conditions a quantum efficiency of $\phi \approx 0.35$ of hydrogen yield was reached.

Concerning point (ii) $[Ru(bpy)_3]^{3+}$ must be efficiently scavenged to prevent the back reaction. Amines, such as EDTA reduce $[Ru(bpy)_3]^{3+}$, leading ultimately to a degradation of EDTA into inert products (for this reason this additional reactant is called a *sacrificial donor*) (Figure 15.30). Other methods to reduce the back electron transfer have been reported, one of the best being the use of microemulsions or micelles of MV^{2+}, such as obtained with $C_{14}MV^{2+}$. The oxidized species has a pronounced hydrophilic character whereas it is strongly hydrophobic when reduced. This hydrophobic character impedes close contact between the species $C_{14}MV^+$ and the $[Ru(bpy)_3]^{3+}$ cation, decreasing by some two orders of magnitude the back reaction.

The report by Grätzel et al. of the simultaneous production of hydrogen and oxygen from water, using a two-catalyst system (Pt and RuO_2) therefore represented an advance (Figure 15.31). In the first experiment, individual particles of the two catalysts were suspended in solution. Later, the two catalysts, Pt and RuO_2, were codeposited in a common carrier i.e. colloidal TiO_2. With this cyclic system, the H_2 evolution can be sustained over long irradiation times: after two days of irradiation there was no noticeable decrease in the hydrogen generation rate. A series of black experiments was performed to confirm the above results: there is no H_2 formation in the absence of light and both the electron-relay and sensitizer have to be present.

15.4.3.3 The Second Step: The Use of Sensitizers and Semiconductors [58]

An alternative approach consists of the use of sensitizers and semiconductors, without the need of a relay, R. The general scheme is shown in Figure 15.32A. Noble metals as catalysts are necessary owing to their small overvoltage for hydro-

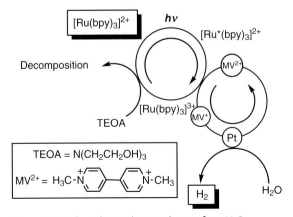

Figure 15.30 Photochemical H_2 production from H_2O.

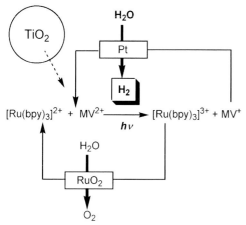

Figure 15.31 Photochemical H_2 and O_2 production from H_2O.

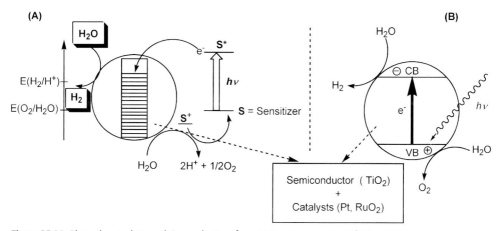

Figure 15.32 Photochemical H_2 and O_2 production from H_2O using a semiconductor.

gen generation from water. The main problem of these systems is, once again, the back reaction $H_2 + 1/2O_2$ giving H_2O. This has been overcome through the development of bifunctional redox catalysts, in which both Pt and RuO_2 catalysts are loaded onto the same carrier particle, such as colloidal TiO_2.

15.4.3.4 The Third Step: The Use of Only Semiconductors [59]
Finally, there is another possibility: light is absorbed *directly* by a colloidal particle of the semiconductor transferring an electron from the valence band to the conducting band. The advent of nanocrystalline semiconductor systems has rekindled interest in cells for water cleavage by visible light. Through appropriate catalysts, the electron is able to liberate H_2 and the positive hole to liberate O_2 from water. In this case there is no need for either sensitizers or relays (Figure 15.32B).

15.4.3.5 Grätzel's System

A low-cost tandem device that achieves the direct cleavage of water into hydrogen and oxygen by visible light was developed by Grätzel and coworkers [60]. A thin film of nanocrystalline WO_3 or Fe_2O_3 serves as the top electrode absorbing the blue part of the solar spectrum. The valence band holes (h^+) created by band gap excitation of the film oxidize water to oxygen, while the conduction band electrons are fed into the second photosystem consisting of the dye-sensitized nanocrystalline TiO_2 cell. The latter is placed directly under the WO_3 film, capturing the green and red parts of the solar spectrum that are transmitted through the top electrode. The photovoltage generated by the second photosystem enables the generation of hydrogen by the conducting band electrons. A photograph of such a cell is shown in Ref. [60].

15.4.3.6 Non-Ru-bpy Systems

Owing to its intrinsic importance, many researchers have tried alternative ways to split water photochemically. It is outside the scope of this chapter to review all these attempts. The interested reader can find some new developments in Ref. [61].

15.4.3.7 Dye-sensitized Solar Cells

An alternative approach to the photovoltaic device for the direct conversion of sunlight to electric energy is the dye-sensitized solar cells (DSC or DSSC), in which an irradiated sensitizer transfers an electron to a wide band-gap semiconductor electrode, leading to photocurrent. The oxidized sensitizer or dye is then regenerated using a solution redox couple that also serves to accept electrons from the external circuit (Figure 15.33) [62]. The voltage generated under illumination

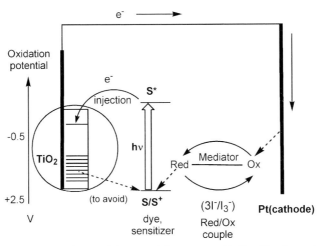

Figure 15.33 Scheme of the Grätzel dye-sensitized solar cell. Photoexcitation of the sensitizer (S) is followed by electron injection into the conduction band of a semiconductor oxide film. The dye molecule is regenerated by the redox system, which itself is regenerated at the counter electrode by electrons passed through the load.

corresponds to the difference between the Fermi level of the electron in the solid and the redox potential of the electrolyte. The prototype and most successful and extensively studied DSSC is the Grätzel cell (1991) [62], which is based on mesoscopic metal oxide, most notably TiO_2, thin films that are sensitized by surface-bound metal complexes (mainly ruthenium bipyridyl dyes) and function through the agency of a redox couple in the fluid electrolyte. The operational principles and description of the device can be found in Refs. [60, 62].

The low-cost and easy preparation makes DSSC one of the most promising photovoltaic cells for conversion of sunlight to electricity. The maximum conversion efficiency was achieved by using *cis*-di(thiocyanato)bis(2,2′-bipyridyl-4,4′-dica rboxylate)ruthenium(II) (referred to as N3) as a photosensitizer. Gratzel et al. reported incident photon-to-current conversion efficiencies (IPCE) of 80–85% with an overall conversion efficiency of 10%. However, the conversion efficiency of DSSCs is still lower than that of the silicon-based photovoltaic cells. The reader can find several reviews on metal complex sensitizers in dye-sensitized solar cells in Ref. [63].

15.4.3.8 The Future

An excellent viewpoint was published in 2004 by Turner [64]: "*Sustainable hydrogen production technologies that may affect hydrogen production in the future include photobiological and photoelectrochemical approaches. These systems produce hydrogen directly from sunlight and water, and offer the possibility of increasing the efficiency of the solar-to-hydrogen pathway and lowering the capital cost of the system . . . These systems might allow the use of seawater directly as the feedstock instead of high-purity water*".

In 2005, Nocera and coworkers stated: "*. . . when the water-splitting problem is posed in the simplest chemistry framework . . . the overall transformation is a multielectron process. The treatment of even a single-electron reaction represents such an important milestone of chemistry that it was worthy of Nobel Prizes . . .* (Marcus and Taube). *A similar understanding about multielectron redox reactions has yet to be realized.*" [65].

References

1 Balzani, V., Carassiti, V. (1970) *Photochemistry of Coordination Compounds*, Academic Press, London.

2 Porterfield, W.F. (1993) *Inorganic Chemistry. A Unified Approach*, Academic Press, New York.

3 Balzani, V., Credi, A., Venturi, M. Photochemistry and photophysics of coordination compounds: An extended view, *Coord. Chem. Rev.* 1998, *171*, 3.

4 Barakat, K.A., Cundari, T.R., Omary, M.A. Jahn-Teller Distortion in the Phosphorescent Excited State of Three-Coordinate Au(I) Phosphine Complexes, *J. Am. Chem. Soc.* 2003, *125*, 14228.

5 Lee, J., Kovalevski, A.Y., Novozhilova, I.V., Bagley, K.A., Coppens, P., Richter-Addo. G.B. Single- and Double-Linkage Isomerism in a Six-Coordinate Iron Porphyrin Containing Nitrosyl and Nitro Ligands, *J. Am. Chem. Soc.* 2004, *126*, 7180.

6 (a) Rachford, A.A., Petersen, J.L., Rack, J.J. Phototriggered sulfoxide isomerization in

[Ru(pic)$_2$(dmso)$_2$], *Dalton Trans.* 2007, 3245; (b) Butcher, D.P., Rachford, A.A., Petersen, J.L., Rack, J.J. Phototriggered S → O Isomerization of a Ruthenium-Bound Chelating Sulfoxide, *Inorg. Chem.* 2006, *45*, 9178.

7 Shibahara, T., Tsuboi, M., Nakaoka, S., Ide, Y. Photochromic Isomerization of a Dinuclear Molybdenum Complex with Ethylene-1,2-Dithiolate and Disulfur Ligands: X-ray Structures of the Two Isomers, *Inorg. Chem.* 2003, *42*, 935.

8 (a) Vos, J.G., Kelly, J.M. Ruthenium polypyridyl chemistry; from basic research to applications and back again, *Dalton Trans.* 2006, 4869; (b) De Cola, L., Belser, P., von Zelewsky, A., Vögtle, F. Design, synthesis and photophysics of ruthenium and osmium complexes through 20 years of collaboration, *Inorg. Chim. Acta.* 2007, *360*, 775.

9 (a) Fraysse, S., Coudret, C., Launay, J.P. Molecular Wires Built from Binuclear Cyclometalated Complexes, *J. Am. Chem. Soc.* 2003, *125*, 5880; (b) Son, S.U., Park, K.H., Lee, Y-S., Kim, B.Y., Choi, C.H., Lah, N.S., Jang, Y.H., Jang, D-J., Chung, Y.K. Synthesis of Ru(II) Complexes of N-Heterocyclic Carbenes and Their Promising Photoluminescence Properties in Water, *Inorg. Chem.* 2004, *43*, 6896.

10 Bünzli, J-C.G., Piguet, C. Taking advantage of luminescent lanthanide ions, *Chem. Soc. Rev.* 2005, *34*, 1048.

11 Ward, M.D. Transition-metal sensitised near-infrared luminescence from lanthanides in d–f heteronuclear arrays, *Coord. Chem. Rev.* 2007, *251*, 1663.

12 Faulkner, S., Matthews, J.L. (2003) *Fluorescent and Luminescent Complexes for Biomedical Applications*, Vol. 9 of *Comprehensive Coordination Chemistry*, 2nd edn., Wrad, M.D. (ed.), Elsevier, Oxford.

13 Balzani, V., Credi, A., Venturi, M. (2004) *Molecular Devices and Machines. A Journey into the Nano World*, Wiley-VCH, Weinheim.

14 Jukes, R.T.J., Adamo, V., Hartl, F., Belser, P., De Cola, L. Photochromic Dithienylethene Derivatives Containing Ru(II) or Os(II) Metal Units. Sensitized Photocyclization from a Triplet State, *Inorg. Chem.* 2004, *43*, 2779.

15 Engeser, M., Fabbrizzi, L., Licchelli, M., Sacchi, D. A fluorescent molecular thermometer based on the nickel(II) high-spin/low-spin interconversion, *Chem. Commun.* 1999, 1191.

16 Lazarides, T., Easun, T.L., Veyne-Marti, C., Alsindi, W.Z., George, M.W., Deppermann, N., Hunter, C.A., Adams, H., Ward, M.D. Structural and Photophysical Properties of Adducts of [Ru(bipy)(CN)$_4$]$^{2-}$ with Different Metal Cations: Metallochromism and Its Use in Switching Photoinduced Energy Transfer, *J. Am. Chem. Soc.* 2007, *129*, 4014.

17 Martínez, R., Ratera, I., Tárraga, A., Molina, P., Veciana, J. A simple and robust reversible redox–fluorescence molecular switch based on a 1,4-disubstituted azine with ferrocene and pyrene units, *Chem. Comm.* 2006, 3809.

18 Fabrizzi, L., Licchelli, M., Pallavicini, P., Parodi, L. Controllable Intramolecular Motions That Generate Fluorescent Signals for a Metal Scorpionate Complex, *Angew. Chem. Int. Ed. Eng.* 1998, *37*, 800.

19 Barboiu, M., Prodi, L., Montalti, M., Zaccheroni, N., Kyritsakas, N., Lehn, J-M. Dynamic Chemical Devices: Modulation of Photophysical Properties by Reversible, Ion-Triggered, and Proton-Fuelled Nanomechanical Shape-Flipping Molecular Motions, *Chem. Eur. J.* 2004, *10*, 2953.

20 Linke-Schaetzel, M., Anson, C.E., Powell, A.K., Buth, G., Palomares, E., Durrant, J. D., Balaban, T.S., Lehn, J-M. Dynamic Chemical Devices: Photoinduced Electron Transfer and Its Ion-Triggered Switching in Nanomechanical Butterfly-Type Bis(porphyrin)terpyridines, *Chem. Eur. J.* 2006, *12*, 1931.

21 Encinas, S., Bushell, K.L., Couchman, S.M., Jeffery, J.C., Ward, M.D., Flamigni, L., Barigelletti, F. Switching of the inter-component photoinduced electron- and energy-transfer properties of a Ru(II)-aza-crown–Re(I) complex; effects of changing temperature, and of incorporation of Ba^{2+} ion into the macrocyclic spacer between the chromophores, *J. Chem. Soc., Dalton Trans*, 2000, 1783.

22 Medlycott, E.A., Hanan, G.S. Designing tridentate ligands for ruthenium(II) complexes with prolonged room temperature luminescence lifetimes, *Chem. Soc. Rev.* 2005, *34*, 133.

23 Nakamura, Y., Aratani, N., Osuka, A. Cyclic porphyrin arrays as artificial photosynthetic antenna: synthesis and excitation energy transfer, *Chem. Soc. Rev.* 2007, *36*, 831.

24 Seth, J., Palaniappan, V., Johnson, T.E., Prathapan, S., Lindsey, J.S., Bocian, D.F. Investigation of Electronic Communication in Multi-Porphyrin Light-Harvesting Arrays, *J. Am. Chem. Soc.* 1994, *116*, 10578.

25 Hori, T., Aratani, N., Takagi, A., Matsumoto, T., Kawai, T., Yoon, M-C., Yoon, Z.S., Cho, S., Kim, D., Osuka, A. Giant Porphyrin Wheels with Large Electronic Coupling as Models of Light-Harvesting Photosynthetic Antenna, *Chem. Eur. J.* 2006, 1319.

26 Kobuke, Y. Artificial Light-Harvesting Systems by Use of Metal Coordination, *Eur. J. Inorg. Chem.* 2006, 2333.

27 Kuramochi, Y., Satake, A., Kobuke, Y. Light-Harvesting Macroring Accommodating a Tetrapodal Ligand Based on Complementary and Cooperative Coordinations, *J. Am. Chem. Soc.* 2004, *126*, 8668.

28 Hajjaj, F., Yoon, Z.S., Yoon, M-C., Park, J., Satake, A., Kim, D., Kobuke, Y. Assemblies of Supramolecular Porphyrin Dimers in Pentagonal and Hexagonal Arrays Exhibiting Light-Harvesting Antenna Function, *J. Am. Chem. Soc.* 2006, *128*, 4612.

29 Choi, M.S., Aida, T., Yamazaki, T., Yamazaki, I. Dendritic Multiporphyrin Arrays as Light-Harvesting Antennae: Effects of Generation Number and Morphology on Intramolecular Energy Transfer, *Chem. Eur. J.* 2002, *8*, 2668.

30 Alstrum-Acevedo, J.A., Brennaman, M.K., Meyer, T.J. Chemical Approaches to Artificial Photosynthesis. 2, *Inorg. Chem.* 2005, *44*, 6802.

31 Borgstrom, M., Shaikh, N., Johansson, O., Anderlund, M.F., Styring, S., Akermark, B., Magnuson, A., Hammarstrom, L. Light Induced Manganese Oxidation and Long-Lived Charge Separation in a $Mn_2^{II,II}$-$Ru^{II}(bpy)_3$-

Acceptor Triad, *J. Am. Chem. Soc.* 2005, *127*, 17504.

32 Sun, S., Zhang, R., Andersson, S., Pan, J., Akermark, B., Sun, L. The photoinduced long-lived charge-separated state of $Ru(bpy)_3$–methylviologen with cucurbit[8]uril in aqueous solution, *Chem. Comm.* 2006, 4195.

33 (a) Imahori, H. Creation of Fullerene-Based Artificial Photosynthetic Systems, *Bull.Chem. Soc. Jpn.* 2007, *80*, 621; (b) Winters, M.U., Dahlstedt, E., Blades, H.E., Wilson, C.J., Frampton, M.J., Anderson, H.L., Albinsson, B. Probing the Efficiency of Electron Transfer through Porphyrin-Based Molecular Wires, *J.Am. Chem. Soc.* 2007, *129*, 4291.

34 Chambron, J.C., Harriman, A., Heitz, V., Sauvage, J-P. Ultrafast Photoinduced Electron Transfer between Porphyrinic Subunits within a Bis(porphyrin)-Stoppered Rotaxane, *J. Am. Chem. Soc.* 1993, *115*, 6109.

35 Flamigni, L., Heitz, V., Sauvage, J-P. Porphyrin Rotaxanes and Catenanes: Copper(1)-Templated Synthesis and Photoinduced Processes, *Struct. Bond.*, 2006, *121*, 217.

36 McEvoy, J.P., Brudvig, G.W. Water-Splitting Chemistry of Photosystem II, *Chem. Rev.* 2006, *106*, 4455.

37 Gust, D., Moore, T.A., Moore, A.L. Mimicking Photosynthetic Solar Energy Transduction, *Acc. Chem. Res.* 2001, *34*, 40.

38 Sun, L., Hammarstrom, L., Norrby, T., Berglund, H., Davydov, R., Andersson, M., Borje, A., Korall, P., Philouze, C., Almgren, M., Styring, S., Akermark, B. Intramolecular electron transfer from coordinated manganese(II) to photogenerated ruthenium(III), *Chem. Commun*, 1997, 607.

39 Prodi, L. Luminescent chemosensors: from molecules to nanoparticles, *New J. Chem.* 2005, *29*, 20.

40 Farruggia, G., Iotti, S., Prodi, L., Montalti, M., Zaccheroni, N., Savage, P.B., Trapani, V., Sale, P., Wolf, F.I. 8-Hydroxyquinoline Derivatives as Fluorescent Sensors for Magnesium in Living Cells, *J. Am. Chem. Soc.* 2006, *128*, 344.

41 Bag, B., Bharadwaj, P.K. Attachment of an Electron-Withdrawing Fluorophore to a

Cryptand for Modulation of Fluorescence Signaling, *Inorg. Chem.* 2004, *43*, 4626.

42 Peng, X., Du, J., Fan, J., Wang, J., Wu, Y., Zhao, J., Sun, S., Xu, T. A Selective Fluorescent Sensor for Imaging Cd²⁺ in Living Cells, *J. Am. Chem. Soc.* 2007, *129*, 1500.

43 Viguier, R.F.H. Hulme, A.N. A Sensitized Europium Complex Generated by Micromolar Concentrations of Copper(I): Toward the Detection of Copper(I) in Biology, *J. Am. Chem. Soc.* 2006, *128*, 11370.

44 Hanaoka, K., Kikuchi, K., Kojima, H., Urano, Y., Nagano, T. Development of a Zinc Ion-Selective Luminescent Lanthanide Chemosensor for Biological Applications, *J. Am. Chem. Soc.* 2004, *126*, 12470.

45 Kikkeri, R., Traboulsi, H., Humbert, N., Gumienna-Kontecka, E., Arad-Yellin, R., Melman, G., Elhabiri, M., Albrecht-Gary, A.M., Shanzer, A. Toward Iron Sensors: Bioinspired Tripods Based on Fluorescent Phenol-oxazoline Coordination Sites, *Inorg. Chem.* 2007, *46*, 2485.

46 (a) Wu, D., Huang, W., Duan, C., Lin, Z., Meng, Q. Highly Sensitive Fluorescent Probe for Selective Detection of Hg²⁺ in DMF Aqueous Media, *Inorg. Chem.* 2007, *46*, 1538; (b) Praveen, L., Ganga, V.B., Thirumalai, R., Sreeja, T., Reddy, M.L.P., Varma, R.L. A New Hg²⁺-Selective Fluorescent Sensor Based on a 1,3-Alternate Thiacalix[4]arene Anchored with Four 8-Quinolinoloxy Groups, *Inorg. Chem.* 2007, *46*, 6277.

47 Rurack, K. Flipping the light switch 'ON' – the design of sensor molecules that show cation-induced fluorescence enhancement with heavy and transition metal ions, *Spectrochim. Acta. A.* 2001, *57*, 2161.

48 Parker, D., Williams, J.A.G. (2003) *Responsive Luminescent Lanthanide Complexes, Metal Ions in Biological Systems*, Sigel, A., Sigel, H. (eds.), Vol. 40, Marcel Dekker, New York.

49 Lee, H.N., Xu, Z., Kim, S.K., Swamy, K.M.K., Kim, Y., Kim, S.J., Yoon, J. Pyrophosphate-Selective Fluorescent Chemosensor at Physiological pH: Formation of a Unique Excimer upon Addition of Pyrophosphate, *J. Am. Chem. Soc.* 2007, *129*, 3828.

50 (a) Jose, D.A., Kar, P., Koley, D., Ganguly, B., Thiel, W., Ghosh, H.N., Das, A. Phenol- and Catechol-Based Ruthenium(II) Polypyridyl Complexes as Colorimetric Sensors for Fluoride Ions, *Inorg. Chem.* 2007, *46*, 5576; (b) Fillaut, J-L., Andries, J., Perruchon, J., Desvergne, J-P., Toupet, L., Fadel, L., Zouchoune, B., Saillard, J-Y. Alkynyl Ruthenium Colorimetric Sensors: Optimizing the Selectivity toward Fluoride Anion, *Inorg. Chem.* 2007, *46*, 5922; (c) Zhang, B-G., Xu, J., Zhao, Y-G., Duan, C-Y., Cao, X., Meng, Q-J. Host–guest complexation of a fluorescent and electrochemical chemsensor for fluoride anion, *J. Chem. Soc., Dalton Trans.*, 2006, 1271.

51 Maeda, H., Yamamoto, K., Kohno, I., Hafsi, L., Itoh, N., Nakagawa, S., Suzuki, K., Uno, T. Design of a Practical Fluorescent Probe for Superoxide Based on Protection-Deprotection Chemistry of Fluoresceins with Benzenesulfonyl Protecting Groups, *Chem. Eur. J.* 2007, *13*, 1946.

52 Faure, S., Stern, C., Guilard, R., Harvey, P.D. Synthesis and Photophysical Properties of Meso-Substituted Bisporphyrins: Comparative Study of Phosphorescence Quenching for Dioxygen Sensing, *Inorg. Chem.* 2005, *44*, 9232.

53 Song, B., Wang, G., Tan, M., Yuan. J. A Europium(III) Complex as an Efficient Singlet Oxygen Luminescence Probe, *J. Am. Chem. Soc.* 2006, *128*, 13442.

54 (a) Nagano, T., Yoshimura, T. Bioimaging of Nitric Oxide, *Chem. Rev.* 2002, *102*, 1235; (b) Lim, M.H., Lippard, S.J. Metal-Based Turn-On Fluorescent Probes for Sensing Nitric Oxide, *Acc. Chem. Res.* 2007, *40*, 41.

55 Credi, A. Artificial Molecular Motors Powered by Light, *Aust. J. Chem.* 2006, *59*, 157.

56 Muraoka, T., Kinbara, K., Kobayashi, Y., Aida, T. Light-Driven Open-Close Motion of Chiral Molecular Scissors, *J. Am. Chem. Soc.* 2003, *125*, 5612.

57 Muraoka, T., Kinbara, K., Aida, T. Mechanical twisting of a guest by a photoresponsive host, *Nature*, 2006, *440*, 512.

58 (a) Gratzel, M. Artificial Photosynthesis: Water Cleavage into Hydrogen and Oxygen by Visible Light, *Acc. Chem. Res.* 1981, *14*, 376; (b) Maverick, A.W., Gray, H.B. Solar Energy Storage Reactions Involving Metal Complexes, *Proc. ICCC, Toulouse*, 1981, 19.

59 McDevitt, J.T. Photoelectrochemical solar cells, *J. Chem. Educ.* 1984, *61*, 217.

60 Grätzel, M. Mesoscopic Solar Cells for Electricity and Hydrogen Production from Sunlight, *Chem. Lett.* 2005, *34*, 8.

61 (a) Esswein, A.J., Veige, A.S., Nocera, D.G. A Photocycle for Hydrogen Production from Two-Electron Mixed-Valence Complexes, *J. Am. Chem. Soc.* 2005, *127*, 16641; (b) Du, P., Schneider, J., Jarosz, P., Eisenberg, R. Photocatalytic Generation of Hydrogen from Water Using a Platinum(II) Terpyridyl Acetylide Chromophore, *J. Am. Chem. Soc.* 2006, *128*, 7726; (c) Zhang, J., Du, P., Schneider, J., Jarosz, P., Eisenberg, R. Photogeneration of Hydrogen from Water Using an Integrated System Based on TiO₂ and Platinum(II) Diimine Dithiolate Sensitizers, *J. Am. Chem. Soc.* 2007, *129*, 7726.

62 (a) O'Regan, B., Grätzel, M. A Low-Cost, High-Efficiency Solar-Cell Based On Dye-Sensitized Colloidal TiO_2 Films, *Nature*, 1991, *353*, 737; (b) Grätzel, M. Photoelectrochemical cells, *Nature*, 2001, *414*, 338; (c) Grätzel, M., Solar Energy Conversion by Dye-Sensitized Photovoltaic Cells, *Inorg. Chem.* 2005, *44*, 6841; (d) Robertson, N. Optimizing Dyes for Dye-Sensitized Solar Cells, *Angew. Chem. Int. Ed.* 2006, *45*, 2338.

63 (a) Polo, A.S., Itokazu, M.K., Iha, N.Y.M. Metal complex sensitizers in dye-sensitized solar cells, *Coord. Chem. Rev.* 2004, *248*, 1343; (b) Campbell, W.M., Burrell, A.K., Officer, D.L., Jolley, K.W. Porphyrins as light harvesters in the dye-sensitised TiO_2 solar cell, *Coord. Chem. Rev.* 2004, *248*, 1363; (c) Meyer, G.J. Molecular Approaches to Solar Energy Conversion with Coordination Compounds Anchored to Semiconductor Surfaces, *Inorg. Chem.* 2005, *44*, 6852.

64 Turner, J.A. Sustainable Hydrogen Production, *Science*, 2004, *305*, 972.

65 Dempsey, J.L., Esswein, A.J., Manke, D.R., Rosenthal, J., Soper, J.D., Nocera, D.G. Molecular Chemistry of Consequence to Renewable Energy, *Inorg. Chem.* 2005, *44*, 6879.

Bibliography

Special Issue on Photochemistry in *J. Chem. Educ.*, October 1983.

Hovarth, O., Stevenson, K.L. (1993) *Charge Transfer Photochemistry of Coordination Compounds*, VCH, New York.

Roundhill, D.M. (1994) *Photochemistry and Photophysics of Metal Complexes*, Plenum Press, New York.

Balzani, V., Jurius, A., Venturi, M., Campagna, S., Serroni. S. Luminescent and Redox-Active Polynuclear Transition Metal Complexes, *Chem. Rev.*, 1996, *96*, 759.

16
Crystal Engineering: Metal–Organic Framework (MOFs)

16.1
Coordination Polymers and Crystal Engineering

16.1.1
Coordination Polymers [1]

In the 1970s, Wells focused on the overall structures of inorganic compounds and abstracted crystal structures in terms of their topology (polyhedra or infinite networks) [2]. In the 1990s, Robson developed and extrapolated Wells' work into the realm of metal–organic compounds and coordination chemistry [3].

Coordination polymers, also known as metal-organic frameworks (MOFs), are metal–ligand compounds that extend 'infinitely' into one, two or three dimensions (1D, 2D or 3D) respectively via metal–ligand bonding. The ligand must be a bridging organic group. Infinite metal–ligand assemblies where the metal–organic connectivity is interrupted by '*pure inorganic*' bridges, such as OH^-, Cl^-, N_3^-, etc., are called *organic–inorganic hybrid materials*. This differentiation is, maybe, a pure formalism. However, for some authors, it is the *bridging organic ligand* that generates the large diversity in the topologies and possible properties of the metal–organic coordination networks [1].

16.1.2
Crystal Engineering

In previous chapters we have dealt with polynuclear complexes, focusing almost exclusively on discrete (thus '*molecular*') systems. We have seen that there are some principles and rules for synthesizing new molecules in a non-serendipitous manner. There is, thus, a logical '*molecular architecture*'. A further step is required to 'create' new materials: to pass from the molecular architecture level to the more complicated *crystal engineering* level. This will be the aim of this chapter.

Crystal engineering is making crystals by design [4]. This implies the ability to assemble molecular or ionic components into the desired architecture by engineering a target network of interactions. The difference in bonding types offers a

Coordination Chemistry. Joan Ribas Gispert
Copyright © 2008 WILEY-VCH Verlag GmbH & Co. KGaA, Weinheim
ISBN: 978-3-527-31802-5

practical way of differentiating target materials and hence synthetic strategies. For most scientists, crystal engineering is not only *making crystals by design* but also – and especially – *making crystals by design with specific and predicted properties* [1, 4, 5].

16.1.3
Strategies in Coordination Chemistry

In 1989, Hoskins and Robson demonstrated that networks with predictable architectures can be engineered by a related strategy based on the interaction of metals with ligands containing multiple sites of coordination [6]. Indeed, since the 1990s crystal engineering, especially network-based materials design, has emerged as a promising field of research. The recent launch of three new journals dedicated to this subject indicates its emergence as a major discipline in the physical sciences [7].

16.1.3.1 Node-and-Spacer Approach
Many novel crystal structures are achieved by serendipity but the researchers engaged in the field of crystal engineering are increasingly seeking different non-serendipitous *strategies*, based on the characteristics of the metallic centers and the ligands. According to Zaworotko and coworkers, "*The most common strategy that has thus far been applied in the context of the design of coordination polymers is to generate a coordination polymer network that is a simple extension of transition metal or metal cluster coordination chemistry. For this approach to become effective one requires exodentate bifunctional ligands to link metal moieties. The metal coordination environment therefore functions as the node which defines the overall network geometry of the coordination polymer*" [8].

This strategy is called the *building-block methodology*, or *node-and-spacer approach*. Nodes are metal ions with different *ideal topologies*, which tend to give *specific* geometries. Due to their different coordination connectivities and geometries, the choice of metal center mainly dictates the architecture of the resultant framework. For example, two-fold linear connectivity leads to 1D chains, three-fold connectivity leads to brickwall (T-shaped) or hexagonal networks, four-fold square-planar connectivities often afford 2D sheets, while four-fold tetrahedral and six-fold octahedral centers give corresponding 3D networks (diamond-like or Prussian-Blue-like, respectively). Such a strategy would, for example, be expected to afford any of the architectures illustrated in Figure 16.1, in which it is assumed, in all cases, that bridges are bidentate ligands with only two ligating positions (linear connectors).

The use of other tri- tetra- and polydentate ligands (connectors) can afford the same and/or other more complicated architectures. This usually requires multidentate ligands with two or more donor atoms. Such bridging ligands are called di-, tri-, tetra topic, depending on the number of donor sites (Chapter 2). Some typical connectors, according to their connectivity, are schematized in Figure 16.2.

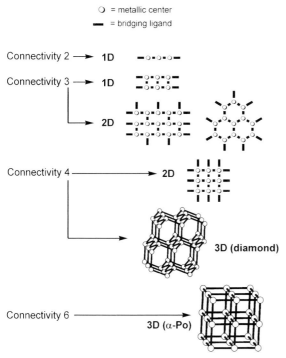

Figure 16.1 A network approach to crystal structure prediction.

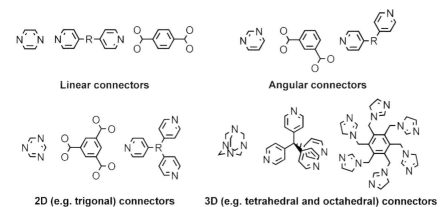

Figure 16.2 Some representative organic connectors and their connectivity.

Thus, judicious combination of a metal "node" and a ligand "spacer" is the most synthetically useful approach for producing predictable network architectures. However, the major limitation of this strategy is that the predictability of networks is subjective and cannot be generalized. Still, an accurate prediction of the overall crystal structure is usually not possible, as has been emphasized by Batten [9] in an article with a paradigmatic title: *"Glorious uncertainty. Challenges for network design"*. Batten finishes his article, saying: *"Despite these important challenges for the future, there is no doubt the modular, net-based approach is an essential part of crystal engineering . . . It has produced some spectacular and fascinating results, and will continue to do so for the foreseeable future. However, despite the increasing number of examples of true structure design, we should not be blind to the limitations and challenges of this approach. We can, at best, direct rather than predict structures, i.e. we can create the conditions under which our desired structure is possible, but we cannot be certain of its formation. This is nonetheless a considerable achievement, and its significance should not be underestimated. However, this area is still very much an experimental pursuit, with surprises and subtleties at every turn"*.

16.1.3.2 The Counter-anion/cation Influence

The influence of different anions/cations on the overall structure of MOFs is very important and has been extensively studied. For example, the reaction between ethylenediaminetetrapropionitrile and $AgNO_3$, $Ag(CF_3SO_3)$, or $AgClO_4$ yields a 1D chain, a 2D layer, or a boxlike 2D network, respectively [10].

16.1.4
General Analysis of Framework Structures

16.1.4.1 A Question of Nomenclature

The Wells notation [2] is usually used whenever possible. Let us assume a simple hexagonal (*honeycomb*) sheet, in which each node is equivalent. Because all nodes are 6-membered, it is classified as a *"uniform"* net, and the symbol (6,3) can be used: 6 is the number of nodes in the shortest circuit and 3 is the connectivity of each node.

The (n, p) notation of Wells can also be applied to 3D networks, as long as they are uniform nets (the shortest circuits are all of the same size) and all points are of the same connectivity. In the (n, p) notation, *n* is the size of smallest circuit, i.e. the number of nodes in the shortest path, and *p* is the connectivity of nodes. Many different nets are possible for each (n, p) combination and are designated (n, p)-a, (n, p)-b, etc. For example the (10,3)-a, (10,3)-b, (10,3)-c etc. 3D nets all have 3-connecting nodes, and all their shortest circuits are 10-membered rings. Since this designation does not uniquely define a net we also frequently find names based on a relation to a similar inorganic structure. Thus the most symmetric (10,3) net, the (10,3)-a is also called the $SrSi_2$ net.

If the net is not uniform (i.e. all nodes and/or all shortest circuits are not the same) then the (n, p) notation cannot be used, and the net has to be described using either the Schläfli symbol (sometimes complicated for chemists), or a common prototype (e.g. rutile), very well-known to chemists.

16.1.4.2 1D Framework Structures

There are no noticeable generalities in this kind of framework. Some of them will be treated in later Sections.

16.1.4.3 2D Framework Structures

The simplest 2D sheets, which comprise just one kind of regular polygon, are so-called regular *"tilings"* or *"tessellations"*. There are three common architectures which are based on triangles, squares or hexagons, since six triangles, four squares, and three hexagons meet at a node in a 2D network, giving angles of 60°, 90°, and 120°. The corresponding Wells notations are (6, 3), (4, 4), and (3, 6), respectively (Figure 16.3). An example of (6,3) is $[Cu^{I}(pyrazine)_{1.5}]^{+}$; examples of (4,4) are $[Co(4,4'-bpy)_2(H_2O)_2]^{2+}$ and $[Cd(4,4'-bipyridine-N,N'-dioxide)(NO_3)_2]$, (3,6) topologies are very rare. Indeed, the first example was reported in 2005, with $[Zn_3(1,4-BDC)_3(DEF)_2] \cdot DEF$ (1,4-BDC = 1,4-benzenedicarboxylate; DEF = diethyl-formamide). The metallic node is a trinuclear building block [11]. Surprisingly, this (3, 6) tessellation is very common in close-packed structures of many elements, but is very unusual in coordination polymers.

Loss of this symmetry results in alternative 2D networks. For example, 3-connected nodes with angles of 180°, 90° and 90° result in brickwall or herringbone architectures (Figure 16.4A) while those with angles of 90°, 135° and 135° give rise to networks such as that of Figure 16.4B. Similarly, 4-connected nodes with angles of 120°, 60°, 120° and 60° give rise to networks called Kagome lattices (Figure 16.4C). In all these cases, the networks are uninodal and the edges of the polygons in any one network are of identical length. Loss of these constraints leads to networks of more complex topology.

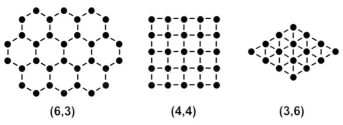

(6,3)   (4,4)   (3,6)

Figure 16.3 Three regular tessellations in 2D networks.

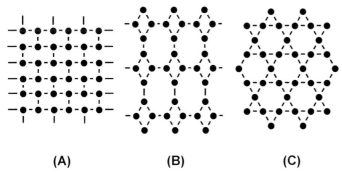

(A)   (B)   (C)

Figure 16.4 Some non-regular 2D networks.

16.1.4.4 **3D Framework Structures**

It appears that, for the majority of 3D MOF structures, there are well-known proto-types in metallic or binary inorganic solids. For example, diamond-related nets composed of tetrahedral nodes and linear linkers are the most common; a primitive cubic net (α-Po) based on octahedral nodes with linear linkers is also frequently observed. Other typical 3D structures have the following topologies: boracite, $CdSO_4$, CaB_6, feldspar, NbO, perovskite, Pt_3O_4, PtS, pyrite, quartz, rutile, sodalite, $SrSi_2$ (also known as (10, 3)-a net), $ThSi_2$ (also known as (10, 3)-b net), etc.

Note that the connectivity of the building blocks in any of these frameworks does not exceed six and, in general, coordination networks with a local connectivity higher than six are very rare. For example, MOFs with a fluorite-like framework have been reported only recently. We will comment on some of these high-connected frameworks later.

The large majority of 3D MOFs with uninodal net topology have tetrahedral geometry (43.3%), followed by octahedral (19.6%) and triangular (14.6%) geome-tries. Significantly fewer MOF structures have square (6.1%), mixed square and tetrahedral (8.7%) and pentagonal (trigonal bipyramidal and square pyramidal, 7.7%) geometry. In the case of tetrahedral geometry, 70% have the diamond topol-ogy, and only 9% the $SrAl_2$ topology. With square building blocks, 45% have the topology of NbO; 38% of $CdSO_4$ and 17% the so-called lvt net [12]. Other divisions can be found in this review.

Furthermore, layer-pillared 3D frameworks have been successfully synthesized from substrates of metal-carboxylate 2D sheets and organic amines or the reverse. For example, $[Zn(4,4'-bpy)(H_2O)_2]SiF_6$ is a 2D square-grid network. By dehydra-tion, $[Zn(4,4'-bpy)_2(SiF_6)]$ is formed, which is a cubic 3D network, in which the SiF_6 groups act as linear bridge between the layers [13].

16.2
MOFs with Polydentate Polypyridyl Derivatives

16.2.1
Pyrazine, 4,4′-Bipyridine and Similar Ditopic "Rigid" Derivatives

16.2.1.1 **One- and Two-dimensional Networks**

Some specific and different examples of 1D systems are shown in Figure 16.5.

2D examples with different coordination nodes are depicted in Figure 16.6A. The Cd atom has a distorted octahedral geometry with four pyridyl groups at the equatorial positions and two nitrate ions at the apical positions. With Cu^{2+} instead of Cd^{2+} the 2D network obtained is very similar but with square planar coordina-tion of copper(II) ions. Different 2D networks may be obtained by a *multicomponent self-assembly process* using two kind of ligands, such as in $[Cu(4,4'-bpy)(pyz)(H_2O)](PF_6)_2$, (Figure 16.6B).

The effect of the rigid 2,2′-, 3,3′-, and 4,4′ positional orientation of the nitrogen atoms in acetylene-bridged *N,N′*-bidentate ligand on the resultant structures has

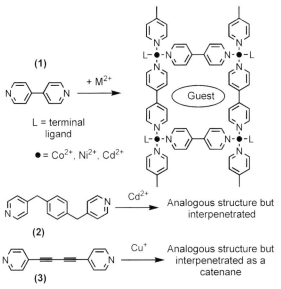

Figure 16.5 Some 1D networks from polypyridyl bridging ligands.

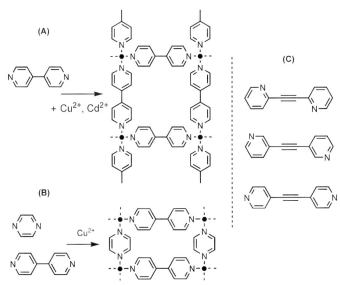

Figure 16.6 (A), (B) Some 2D networks from polypyridyl bridging ligands; (C) three isomeric polypyridyl connectors.

been studied (Figure 16.6C). With several divalent metal ions, a variety of structural architectures, ranging from sinusoidal and zigzag chain, to two-dimensional channel-type architectures have been reported [14].

16.2.1.2   Three-dimensional Networks

1. *Tetrahedral nodes.* Cu^+ and Ag^+ in $[CuL_2]^+$ and $[AgL_2]^+$ contain the diamond-type network with many linear rods (L), such as 4,4′-bipyridine, 4-cyanopyridyne, pyrazine, etc. The structure of $[Cu^I(4,4′\text{-bpy})_2]PF_6$ is a typical example. Tetrahedral ligand-nodes such as 4,4′,4″,4‴-tetracyanotetraphenylmethane can also give diamond networks with cationic tetrahedral nodes (Cu^+). These networks can be either interpenetrated or with the counterions occupying the cavities of the net.

2. *Octahedral nodes.* Given the ubiquity of octahedral metal environments, it is somewhat surprising that simple 3D octahedral polymers with such polypyridinic ligands have been largely unexplored. The reasons seem to be steric. $[Ag(pyrazine)_3](SbF_6)$ and $[Sc(4,4′\text{-bpy-}N,N′\text{-dioxide})_3]X_3$ (X = NO_3^-, $CF_3SO_3^-$, ClO_4^-) adopt the α-polonium-type structure [15].

3. *Layer-pillared* 3D frameworks. The 3D NaCl-type frameworks of $[M(NO_2)(pyz)_2]ClO_4$ (M = Co, Cu), consist of (4,4) sheets of metal-pyrazine connected by $\mu_{1,3}$-nitrito [16]. This structure is, formally, a hybrid network (see Section 16.8).

16.2.2
Tritopic (Trigonal) and Similar Polytopic Bridging Ligands

Trigonal molecular building blocks, with rigid or flexible backbones and various donor functionalities, are ideal for the assembly of layer honeycomb or 3D (10, 3) nets. Although not pyridyl donor ligands, 1,3,5-tricyanobenzene (TCB) and larger derivatives are paradigmatic examples. They react with Ag^+ to give $[Ag(CF_3SO_3)(TCB)]$, whose crystal structure is trigonal and consists of honeycomb sheets based on alternating 1,3,5-tricyanobenzene and Ag(I) units. Larger derivatives give nets that generate large void spaces, leading to interpenetrating sheets and, furthermore, the cavities (>15 Å in diameter) are filled by solvent molecules.

The more complex ligand hexakis(imidazol-1-ylmethyl)benzene (Figure 16.7) has a coordinating functional group small enough to allow six to assemble around an octahedrally coordination metal: with Cd^{2+} it gives the infinite α-Po-related structure (Figure 16.7).

16.2.3
Flexible Bis-pyridyl-like Derivatives

16.2.3.1   Simple Cases

1. The reaction of $Co(NO_3)_2 \cdot aq$ with 1,2-bis(4-pyridyl)ethane (Figure 16.8A) gives
   three different isomers: a linear chain of square boxes, a molecular bilayer and

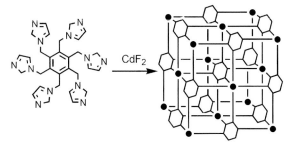

Figure 16.7 An α-Po-like structure with Cd²⁺ and the ligand shown in the figure.

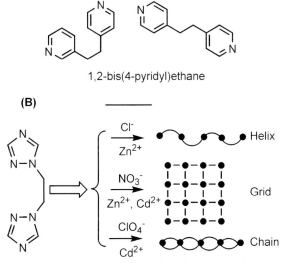

(A)

gauche   *anti*

1,2-bis(4-pyridyl)ethane

(B)

Figure 16.8 Flexible bis-pyridyl-like ligands and some different MOFs obtained from ligand B.

a molecular ladder, according to the crystallization medium (that is, solvent(s) and/or template) [17].

2. Using the flexible 1,2-bis(1,2,4-triazole-1-yl)ethane (Figure 16.8B) new helical chains, [ZnCl₂(L)] (Zn tetracoordinated), two-dimensional polymers [M(L)₂(H₂O)₂](NO₃)₂ (M = Zn, Cd) and a one-dimensional double-bridged chain, [Cd(L)₂(H₂O)₂](ClO₄)₂ have been reported (Zn and Cd hexacoordinated). Anions tune the self-assembly process to form different coordination networks (Figure 16.8B) [18].

16.2.3.2 Metal–organic Rotaxane Frameworks (MORFs)

The possibilities of different modes of entanglement of polyrotaxanes (Chapter 14) open new perspectives in an almost completely unexplored area of chemical

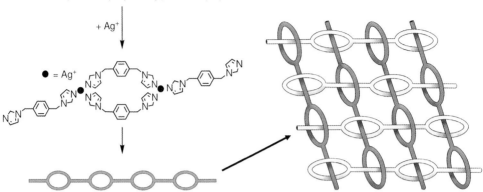

Figure 16.9 below shows the ligand and structure. Captions in image area:

1,4-bis(imidazole)-1-yl-methyl)benzene (bix)

● = Ag⁺

Figure 16.9 A 2D polyrotaxane with formula [Ag$_2$(bix)$_3$(NO$_3$)$_2$].

topology. A very peculiar and unprecedented two-dimensional polyrotaxane, [Ag$_2$(bix)$_3$(NO$_3$)$_2$] (bix, see Figure 16.9) has been reported [19]. In the 1D polymeric chain the three-fold connecting nodes are provided by the three-coordinated Ag⁺. Each polymeric chain is entangled through rotaxane interactions with an infinite number of other chains. This generates the 2D polyrotaxane sheet represented in Figure 16.9. The [Cd$_2$(bix)$_3$(SO$_4$)$_2$] system is very similar. The only difference is that the sulfates complete the coordination sphere of the Cd^{2+} centers, which display trigonal bipyramidal geometry [20].

16.2.3.3 Cucurbituril Networks: Influence of the Metal Center and the Bridging Ligands

1. With an *N,N′*-bis(3-pyridylmethyl)-1,4-diamino*pentane* "string". Formation of a pseudorotaxane by threading a cucurbituril "bead" with the flexible string, followed by reaction with AgNO$_3$ yields a novel polyrotaxane, the structure of which reveals the "beads" threaded onto a *helical* one-dimensional coordination polymer [21]. The scheme of this fantastic network is given in Figure 16.10.

2. With an *N,N′*-bis(3-pyridylmethyl)-1,4-diamino*butane* "string". With AgNO$_3$ it yields an unprecedented polyrotaxane in which cucurbituril beads are threaded on a 2D coordination polymer network (Figure 16.11) [22]. The 2D network consists of large edge-sharing chair-shaped hexagons with a Ag(I) ion at each corner. The 2D polyrotaxane network forms layers stacked on each other. When silver tosylate is used instead of silver nitrate *only* a 1D polyrotaxane coordination polymer is formed (Figure 16.11).

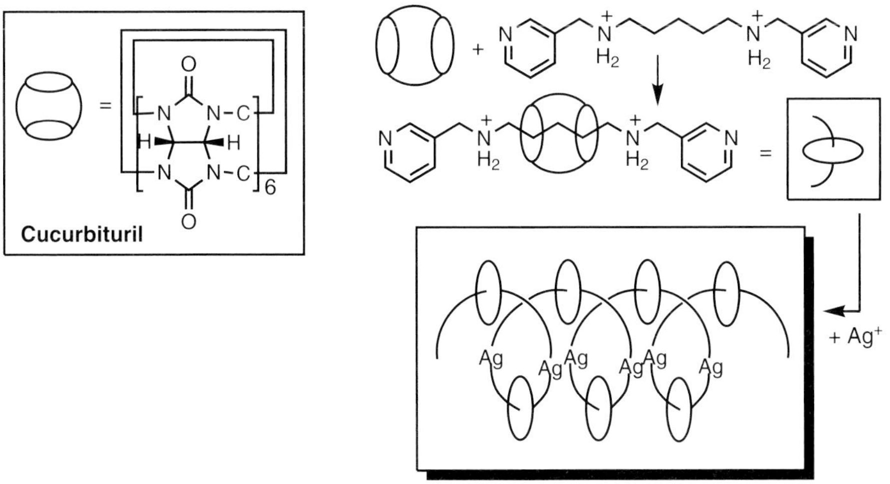

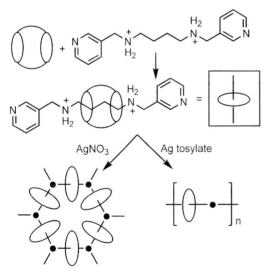

Figure 16.10 A helical network of Ag⁺ formed with cucurbituril units.

Figure 16.11 2D and 1D polyrotaxane networks formed with Ag⁺ and cucurbituril units.

16.3
MOFs with Carboxylate Linkers

In order of their number of $-COO$ groups, some relevant examples are the following:

CO_2 acts as a template in the fabrication of metal(III) formates $[M^{III}(HCOO)_3 \cdot 3/4CO_2 \cdot 1/4H_2O \cdot 1/4HCOOH]$ (M = Fe, Al, Ga, In). The metals have octahedral geometry and are linked by HCOO⁻ in the *anti–anti* style into a ReO₃ net (where Fe replaces Re and HCOO⁻ replaces O) [23].

The oxalate ion acts as a rigid bidentate ligand (Chapter 2) that can facilitate the formation of extended structures by bridging metal centers. The dimensionality ranges from zero to three. $[M^{II}_2(ox)_3]^{2-}$, $[M^IM^{III}(ox)_3]^{2-}$ and $[M^{II}M^{III}(ox)_3]^-$ are very good building blocks for the synthesis of 2D and 3D homo- and bi-metallic networks. Indeed, the formation of the 2D or 3D motif is strongly dependent on the choice of the template counterion: when the countercation is NR_4^+, PPh_4^+ and similar, 2D layers of hexagonal honeycomb oxalate-based networks are formed (Figure 16.12). In this case, building blocks of opposite chirality are alternately linked, which confines the bridged metal ions to lie within a plane. When the countercation is the chiral $[M^{II}(bpy)_3]^{2-}$ (M^{II} = Fe, Co, Ni), the metal-oxalato centers form 3D chiral frameworks, in which each $[M(ox)_{3/2}]$ subunit can be considered as a 3-connecting point, giving a (10,3)-a net with $[M(bpy)_3]^{2+}$ cations occupying the vacancies [24].

Benzene-1,4-dicarboxylate (terephthalic acid, tp) is a linker, which allows the formation of a variety of structures from discrete molecules to one- two- or three-dimensional networks. Long flexible aliphatic dicarboxylates have good conformational freedom, which manifests itself in the various connecting modes that yield novel frameworks. They also act as pillars for metal–oxygen layers to give higher dimensionality frameworks. Many compounds using $[OOC-(CH_2)_n-COO]^{2-}$ ions have been prepared and characterized, in the form of polynuclear (0D), chains (1D), or layers (2D) [25].

1,3,5-Benzenetricarboxylate (btc) (trimesic acid). Many multidimensional porous frameworks using this acid and various metal salts have been reported. The flexible tripodal *benzene-1,3,5-triacetic acid* can adopt at least two different conformations (*cis, cis, cis* and *cis, trans, trans*) upon coordination with metal ions. With Ag(I) and Cd(II) it forms 3D structures with open channels occupied by water molecules, $[Ag_3(bta)]\cdot1.5H_2O$ and $[Cd_3(bta)_2(H_2O)_7]\cdot5H_2O$ [26].

A review of cobalt and nickel polycarboxylate frameworks, focusing mainly on di-, tri-, tetra-, and hexacarboxylates has been published recently [27].

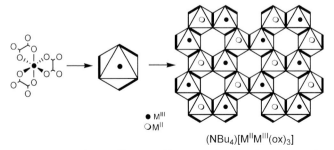

$(NBu_4)[M^{II}M^{III}(ox)_3]$

Figure 16.12 2D layers of hexagonal honeycomb oxalate-based networks (see text for details).

16.4
MOFs with Polynuclear Building Nodes

The introduction of metal clusters into MOFs may lead to new functional solid-state materials that possess fascinating structures and special properties. Furthermore, this strategy provides a promising pathway towards the generation of MOFs with highly connected nodes.

Yaghi and coworkers developed the use of metal-carboxylate clusters for the design of porous MOFs, in which chelating functional groups lock the metal ion in position and produce M–O–C–O–M entities referred to as *secondary building units* (SBUs) [28]. Those commonly occurring in metal carboxylates include: (i) the square paddle-wheel in $[M_2(carboxylate)_4]$, (ii) the octahedral 'basic zinc acetate' cluster and (iii) the trigonal prismatic oxo-centered trimers. All of them will be commented on in the following sections.

16.4.1
Dinuclear Building Nodes

Metal carboxylates tend to form dimers, the classic example being copper diacetate, which can be linked to form infinite arrays in the following ways:

1. *In the axial position* (Figure 16.13A). With bifunctional N-ligands, infinite 1D chains have been extensively reported, in which the bifunctional spacer links the paddle-wheels in their axial positions. A 2D honeycomb structure based on a paddle-wheel diruthenium complex, $[Ru_2^{II}(O_2CPh)_4]$ and triazine as a three connected node has been reported [29].

2. *In the equatorial position* (Figure 16.13B). Within this perspective the $[M_2(O_2C–)_4]$ (metal carboxylate paddle-wheel) can be considered one of the most paradigmatic square SBUs with regard to the disposition of the R groups. 2D copper(II) networks formed with this unit are known, in which the axial positions are occupied by solvent molecules. The metal–organic layers are further stacked to generate porous 3D structures. Guest-dependent cavities can be tailored with the aid of appropriate guest species. In addition, the nature and dimension of interstitial spaces can be modified using various axial ligands of SBUs.

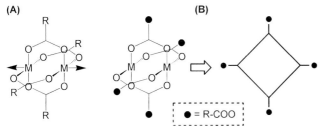

Figure 16.13 Scheme of dinuclear $[M_2(carboxylate)_4]$ building blocks (see text for the details on coordination).

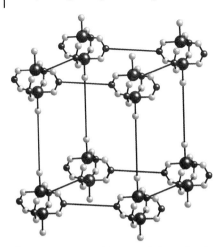

Figure 16.14 A α-Po-like network formed with [Cu₂(*trans*-1,4-cyclohexanedicarboxylate)₄] and 4,4′-bpy connectors.

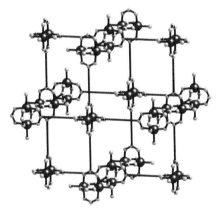

Figure 16.15 A NbO-like network formed with [Cu₂(*o*-bromobenzenedicarboxylate)₄] and 4,4′-bpy connectors.

However, frequently, 3D networks are formed, in which the dimers are linked via *both* the axial and equatorial positions, such as in the α-Po nets of copper(II)-*trans*-1,4-cyclohexanedicarboxylate frameworks with 4,4′-bpy and 4,4′-bpe (4,4′-bis(pyridyl)ethylene) linkers (Figure 16.14) [30] and in the NbO-like net obtained with *o*-Br-benzenedicarboxylic acid (Figure 16.15) [31]. Similar to the α-Po net is the interpenetrated pillared-grid with large pores based on 4,4′-biphenyldicarboxylate and 4,4′-bpy, which retains the same structure on desolvation [32].

In general, the ubiquitous [M₂(R–COO)₄L₂] SBU can give very different MOFs depending on the carboxylate spacer. For example, bdc (1,3-benzenedicarboxylate) allowed the synthesis of triangular lattices and of the first "nanoscale" Kagome lattice [33]; two isomeric [Zn₂(bdc)₂(dabco)] (bdc = 1,4-benzenedicarboxylate; dabco

= 1,4-diazabicyclo[2.2.2]octane), networks, a tetragonal phase and a pillared Kagome net topology, based on a [Zn_2(carboxylate)$_4$] paddlewheel unit, have been reported recently [34].

16.4.2
M_3O and M_4O Building Nodes

Interesting structures have been obtained from triangular oxocentered (M_3O) building units with vanadium, zinc, iron, chromium, copper, etc. For example, cationic and mixed-valent forms of [$M_3O(COO)_6$] have been linked by ditopic dicarboxylate to produce porous 3D MOFs. The carboxylate carbon atoms as the points-of-extension define the vertices of a trigonal prismatic building unit (Figure 16.16A). These [$M_3^{III}OL_3$(dicarbox)$_{6/2}$] · guest nets are currently labelled MIL-n (MIL standing for Materials of Institut Lavoisier) and are very important for their porosity (see below).

An exceptional 1-D MOF with unprecedented heptanuclear copper units, schematized as $Cu_3^{II}O$–Cu^{II}–$Cu_3^{II}O$ [35] and a unique (3,9)-connected net, [$Co_3(\mu$-OH)(pdc)$_3$] · H_2O (pdc = pyridine-3,5-dicarboxylate), in which the [$Co_3(\mu$-OH)(COO)$_6$] SBUs act as nine-connected nodes of a tricapped trigonal prism, and the pdc groups act as three connected nodes of trigonal geometry have been reported recently [36].

The octahedral basic zinc carboxylate [$Zn_4O(O_2C-)_6$] is the most paradigmatic SBU, acting as octahedral joints (i.e. C vertices) in the frameworks (Figure 16.16B). In 1999, Yaghi and coworkers reported the so-called MOF-5, [$Zn_4O(BDC)_3(DMF)_8$ (C_6H_5Cl)] (H_2BDC = 1,4-benzenedicarboxylic acid) [28a]. The structure may be derived from a simple cubic six-connected net, α-Po (Figure 16.1). This structure is exceptionally stable (up to 400 °C) and highly porous. It can be functionalized with different longer dicarboxylate groups and its pore size can be expanded with the long molecular struts. An isoreticular series of highly crystalline materials,

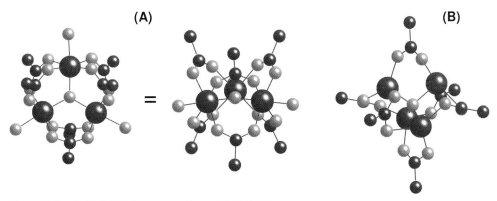

(A) = **(B)**

Figure 16.16 (A) M_3O SBU (two perspectives); (B) M_4O SBU (T_d). (large spheres = metal ions; small spheres: black, C; gray, O).

(named IRMOF-*n*) has been reported. The pore space can be varied from 3.8 to 28.8 Å, representing up to 91% of the crystal volume.

16.4.3
"Nanoscale" Building Nodes

High-nuclearity clusters can also be assembled into infinite coordination polymers. The $[Zn_5(\mu_3\text{-}O)_2(bpdc)_4(DMF)_2(EtOH)_2]\cdot$solvents (bpdc = 4,4′-biphenyldicarboxylate) is a novel 3D framework with an expanded diamondoid topology constructed from double connected pentanuclear SBUs linked by rigid bpdc ligands [37].

$Na_3[Co_6O(OH)(isophthalate)_6]\cdot H_2O$ (MIL-104) is a three-dimensional MOF with an unprecedented octahedral building unit [38]. The metal is octahedrally coordinated and six octahedra are associated to form a super-octahedron. This unusual hexanuclear unit has never been observed either in the solid state or in coordination chemistry. Each hexamer is joined by twelve bridging isophthalate ions to six neighboring units, with each metal unit serving as the octahedral node of a cuboidal net. The topology here is the same as that found for ReO_3, with $[Na_6Co_6O_{26}]$ units as a giant octahedron.

In 2006 a novel and unique coordination polymer based on a polynuclear core, $[Ni_6(OH)_6(1,4\text{-cdc})_3(H_2O)_6]\cdot 2H_2O$, was reported [39]. This compound adopts a 2D structure with a hexanuclear prismatic nickel cluster $[Ni_6(OH)_6(-CO_2)_3(H_2O)_6]$ as a building unit. Each "prismane" cluster is connected to six adjacent prismane clusters through six $-C_6H_{10}-$ groups of cdc ligands, forming a 2D layer structure (Figure 16.17).

The solvothermal reaction of 9,10-dicyanoanthracene and $ZnCl_2/NaN_3$ gave the MOF $[Zn_7(OH)_8(DTA)_3]\cdot H_2O$ (DTA = 9,10-ditetrazolanthracene), which presents a rare topological framework, formed by DTA^{2-} bridging unprecedented heptanu-

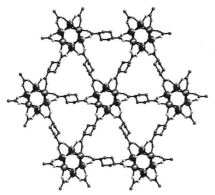

Figure 16.17 MOF derived from the hexanuclear tecton $[Ni_6(OH)_6(1,4\text{-cdc})_3(H_2O)_6]$ (1,4-cdc = 1,4-cyclohexanedicarboxylate).

clear spindle $[Zn_7(OH)_8]^{6+}$ clusters as SBUs, exhibiting strong luminescent emission with long lifetime [40].

Larger and more spectacular infinite networks from nanoscale metal–organic building blocks have been extensively reported in recent years. The final structures can be 1D, 2D or 3D (some of them as simple as the α-Po network). The reader can find some of them in Ref. [41].

16.4.4
Oligonuclear Complexes as Tectons

A large variety of high-dimensionality coordination polymers can be constructed by taking advantage of the high-flexibility of oligonuclear multimetallic nodes. The 3d–4f nodes are particular interesting since the metal ions interact selectively with various spacers. A review on this subject has been published recently [42].

16.5
Highly Connected Solid-state Materials

Highly connected materials remain scarce because the construction of such systems is severely hampered by the available number of coordination sites at the metal centers and the sterically demanding nature of organic ligands. Schröder and coworkers have termed *"non-natural"* these highly connected solids [43]. Two possibilities are known.

16.5.1
From Lanthanide Nodes [43]

The combination of high coordination number lanthanide metal centers and 4,4'-bipyridine-*N,N'*-dioxide and similar ligands, allows the formation of these *"non-natural"* frameworks. The organic ligands have flexible angular geometries at the O-donor, are complementary to hard lanthanide metal centers, and, therefore, form highly stable yet sterically flexible complex geometries. $La(CF_3SO_3)_3$, $La(NO_3)_3$ and $Yb(CF_3SO_3)_3$ react with 4,4'-bipyridine-*N,N'*-dioxide giving different polymeric structures based on networks of eight-coordinated Ln^{3+} nodes, linked by bridging L ligands.

16.5.2
From Cluster Nodes

The simplest eight-connected framework would be the fluorite-like network. In 2004 Kim and coworkers reported the *first* MOF of fluorite built with an eight-connecting tetranuclear cadmium cluster (twisted square prism) and a tetrahedral four-connecting ligand (H_4TCPM=(tetrakis(4-carboxyphenyl)methane):$[Cd_4(O_2CR)_8$ $(DMF)_4] = [Cd_4(TCPM)_2(DMF)_4]$ [44]. To understand the framework topology it is necessary to simplify the building blocks from which the 3D net is built. The $[Cd_4]$ cluster unit and TCPM ligand can be represented by cubic and tetrahedral units, respectively. The (4,8)-connected net has a strong resemblance to distorted fluorite (CaF_2) (Figure 16.18).

A 12-connected MOF was reported in 2005 [45]. The $[Cu_3(CN)(pdt)_2]$ (pdt = 4-pyridinethiolate) has a 3D twelve-connected coordination framework with Cu_6S_4 clusters as nodes, with a face-centered cubic topological framework. The connectors are the pyridine ring of pdt and the cyanide groups. Each Cu_6S_4 cluster is connected to twelve adjacent clusters via four cyanide groups and eight pyridine rings. The net, when idealized, simply corresponds to a cubic close packed array of spheres, namely a face-centered cubic lattice (Figure 16.19).

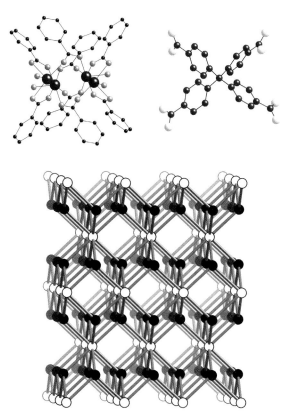

Figure 16.18 (A) (4, 8)-connected net, similar to distorted CaF_2: $[Cd_4(O_2CR)_8(DMF)_4]$ (see text for the explanation of the topology).

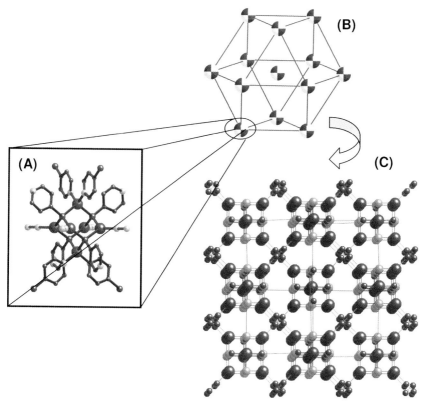

Figure 16.19 A 12-connected MOF, [Cu$_3$(pdt)$_2$(CN)] (pdt = 4-pyridinethiolate). The nodes are [Cu$_6$S$_4$] clusters (large black spheres = Cu).

16.6
Interpenetrating MOFs [46]

Basically, *interpenetrating* structures are ones in which two or more networks are not connected directly to each other, but could not be separated (in an imaginary sense) without breaking the bonds. *Interdigitation* occurs for 1D or 2D networks where each network intrudes on the space of the other, but nonetheless could be separated without breaking bonds.

When relatively large cavities are generated in a network, interpenetration of independent networks usually occurs, to mitigate against the stable open framework structure. It is generally accepted that longer ligands often favor the formation of interpenetrating structures. By far the most common interpenetrated 3D nets are the diamandoid (>40%) following by the α-polonium nets (ca. 20%). An analysis of the crystallographic structural database for interpenetrating metal–organic and inorganic 3D networks has been published by Proserpio and coworkers [47].

16.6.1
Topology of Interpenetration

- 1D ladders. Interpenetration of 1D networks results in dimensional increase, such as 1D → 2D parallel interpenetration. Ladders are the most common topology for interpenetrating 1D nets.

- 2D sheets. The two major classes of interpenetration seen with sheet structures are *parallel interpenetration*, in which each sheet is intimately entangled with a finite number of others (often two but occasionally more) whose mean planes are parallel (Figure 16.20A) and *inclined interpenetration*, in which the mean planes of the sheets are not all parallel and each sheet is penetrated by an infinite number of inclined ones. Parallel interpenetration usually results in 2D layers of interpenetrating networks whereas inclined interpenetration always generates an overall 3D structure (Figure 16.20B).

An example of parallel interpenetration concerns the tcm⁻ (tricyanomethanide) bridging ligand, in the [Cu^I(tcm)(4,4′-bpy)] network [48]. It contains puckered (4, 4) sheets composed of tetrahedral Cu$^+$ ions bridged by 2-connecting tcm⁻ and 4,4′-bpy. The sheets participate in parallel interpenetration in pairs (Figure 16.21).

A typical example of inclined interpenetration is [Zn(4,4′-bpy)$_2$(H$_2$O)$_2$](SiF$_6$) which contains two perpendicular and equivalent stacks of square-grid [Zn(4,4-

(A)   **(B)**

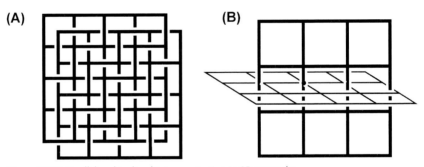

Figure 16.20 Parallel and inclined interpenetration in 2D networks.

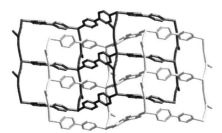

Figure 16.21 Example of parallel interpenetration of the [Cu^I(tcm)(4,4′-bpy)] network (tcm = tricyanomethanide ion).

bpy)$_2$(H$_2$O)$_2$]$^{2+}$ sheets, which interpenetrate so that any particular sheet has an infinite number of perpendicular ones with it [49].

- 3D nets. Interpenetration of the diamondoid networks is the most common form of interpenetration. This mode of interpenetration holds regardless of the number of interpenetrating nets in the whole structure (the number of interpenetrating nets range from 2 to 11). Little progress has been made with other four-connected networks: the first interpenetrating NbO-type network was reported in 1998 in the coordination polymer [Cu(bpye)$_2$(NO$_3$)$_2$] (bpye = 1,2-bis(4-pyridyl)ethane) [50]. Some relevant examples of other 3D interpenetrated types, with different n-connected nets, such as α-Po, PtS-type, CdSO$_4$, and SrAl$_2$-type have also been reported.

16.7
Porous Coordination Polymers

Open MOFs are widely regarded as promising materials for applications in catalysis, separation, gas storage and molecular recognition. Compared to conventionally used microporous inorganic materials such as zeolites, these "*organic*" structures have the potential for more flexible rational design, through control and functionalization of the pores.

Kitagawa et al. [51], divide the pore walls into three classes: (i) in the absence of special substituents in the ligand, a guest molecule can interact with the pore walls in the restricted space through dispersion forces; (ii) the introduction of hydrogen bonding sites (via amido, hydroxo, carboxy, amino, etc. substituents) can change the surface properties; (iii) incorporation of metal sites capable of forming coordinatively unsaturated metal centers, having an energy that is much greater than that of either dispersion forces or hydrogen bonds. These linkers have been named "*metalloligands*".

The following points should be taken into consideration when creating porous coordination polymers [52]: (i) it is impossible to synthesize compounds containing vacant space because nature dislikes vacuums. In other words, the pores will always be filled with some sort of guest or template molecules; (ii) large linkers, which extend the distance between nodes (connectors) of a framework, are often used for the preparation of large micropores. However, with such linkers interpenetration frequently occurs. Regulation of interpenetration is an important challenge in crystal engineering.

16.7.1
Interpenetration vs. Porosity: a Problem

Robson et al. demonstrated a way of avoiding interpenetration in a key early paper in the field of coordination polymers [6b]. Neutral frameworks Cd(CN)$_2$ and Zn(CN)$_2$ were found to exist as doubly interpenetrated diamondoid nets. However, a charged framework would also have to accommodate counterions. The framework salt $[N(CH_3)_4][Cu^I Zn^{II}(CN)_4]$ was therefore prepared and, as hoped, the bulky ammonium counterions present within the adamantoid cavities prevented interpenetration.

Certain aromatic molecules such as naphthalene, pyrene, and others can fill up the cavities in the MOF. These aromatic guests are in contact with each other through aromatic interactions, so that they could actually be described as forming a second, non-covalent, network, which interpenetrates the coordination polymer. If there are aromatic moieties, such as bpy, in the coordination polymer, short (face-to-face) contacts between the aromatic guest and the bpy help to avoid the interpenetration of MOFs.

It is also possible to design frameworks in which interpenetration is forbidden by the inherent dimensions of the framework itself. This relies on the fact that in order to interpenetrate, the size of the vertex or diameter of the polymer strand must be less than the size of the cavity. The use of bulkier vertices and the use of metal clusters and cages, as opposed to single metal ions, are therefore of interest as building units.

16.7.2
Types of Porous Structures

According to Kitagawa et al. porous coordination compounds can be classified into three categories, 1st, 2nd, and 3rd generation [53]. The 1st generation compounds have microporous frameworks, which are sustained only with guest molecules and show irreversible framework collapse on removal of the guest molecules. The 2nd generation compounds, which possess stable frameworks, reversibly lose and reabsorb guest species without undergoing a change in phase or morphology. The 3rd generation compounds exhibit dynamic structures which change their own frameworks in response to external stimuli, such as pressure, light, electric field, guest molecules, and so on.

Based on spatial dimensions there are four types of porous structures: 0D cavities (dots), channels (1D), layers (2D), and intersecting channels (3D).

16.7.2.1 Dots (0D Cavities)
In this case, guest molecules are unable to pass out of the cavities. An example is the interpenetrated 3D network of $[Zn(CN)(NO_3)(tpt)_{2/3}]$ (tpt = 2,4,6-tri(4-pyridyl)1,3,5-triazine) which provides a barrier impenetrable to even the smallest molecules (with the possible exception of H_2), effectively isolating each cavity from its neighbors and from the outside world. The cavities, sealed-off in this manner, are, however, exceptionally spacious. The cavity is large enough to accommodate ca. 18 solvent molecules [54].

16.7.2.2 Channels (1D Space)
One of the most rational methods is to insert pillar moieties between the layers of 2D compounds. For example, a very good pillar for [Cu(glutarate)] is 4,4'-bpy, giving corrugated sheets of metal-glutarate moieties that are pillared via axial

coordination of canted 4,4′-bpy ligands. The network possesses a topology related to the α-Po net. The resulting 3D network contains channels occupied by water molecules that form hydrogen bonded chains. The removal of the guest molecules does not influence the 3D network. Furthermore, once dehydrated, the network can reabsorb water molecules under various conditions. This complex represents a novel porous network generated in water and acts as a highly selective absorbent for water molecules [55].

16.7.2.3 Layers (2D Space)

A few 2D MOFs have been reported in which several guests can be incorporated between the layers. A typical example is $[Co(btc)_3py_2] \cdot 2/3py$ (btc = 1,3,5-benzenetricarboxylate) with btc acting as bidentate and monodentate. The metal centers have a position occupied by pyridine molecules. This arrangement generates extended stacked sheets. The anchored pyridine ligands hold these layers by mutual π-stacking between pyridines. In the remaining space between the sheets there are uncoordinated pyridine molecules. This material remains crystalline upon removal of the pyridine inclusions. Further heating reveals the absence of pyridine in the sample, but upon addition of pyridine the original X-ray pattern for the solid as synthesized is obtained [56].

16.7.2.4 Intersecting Channels (3D Space)

3D intersecting channels, which frequently occur in zeolites, are constructed by the interconnection of 1D channels from various directions. Paradigmatic examples derive from $[Zn_4O(carboxylate)_6] \cdot$ guest, previously discussed (Section 16.4.2) affording exceptionally rigid and highly porous structures with 3D intersecting channels. The use of different carboxylates (generally linear) yields the same type of frameworks with diverse pore size and functionalities.

16.7.2.5 Reversible Transformation

It is necessary to distinguish between reversible adsorption/desorption of free "solvent" molecules and adsorption/desorption of molecules linked to a metal ion. The first case is easier to study. A review of crystal-to-crystal structural transformations in MOFs was published in 2006 [57]. MOFs often contain metal ions that are coordinated to solvent molecules. These open structures commonly collapse on removal of the coordinating solvent molecules. Therefore the chance of finding MOFs that exhibit permanent porosity after removal of coordinating solvent molecules is rare [58]. However, it is generally believed that MOFs generate vacant coordination sites when coordination solvent molecules are removed from the metal ions; such MOFs have been anticipated to enhance the H_2 storage capability.

16.7.3
Functions of Porous Structures

Microporous *inorganic materials* such as zeolites find widespread application in heterogeneous catalysis, adsorption and ion-exchange processes. The novel microporous MOFs have the same potential with major control.

16.7.3.1 Gas Sorption or Storage

One of the most promising applications of MOFs is gas storage. The first report on gas adsorption at ambient temperature appeared in 1997 with the $[M_2(4,4'\text{-bpy})_3(NO_3)_4] \cdot xH_2O$ (M = Co, Ni, Zn) framework, which reversibly absorbs CH_4, N_2, and O_2 in the pressure range 0–36 atm without collapse of the crystal framework [59]. $[Cu(SiF_6)(4,4'\text{-bpy})_2]$ is able to absorb very large quantities of methane [60]. In fact, relative to framework weight, more methane is absorbed by this material than by zeolite 5A, which was taken as the optimum conventional zeolite for methane sorption.

In general, the bis-carboxylate ligands can form a range of multinuclear nodes with predefined geometries. By varying the length of the organic backbone it is possible to investigate the correlation of pore size with gas-adsorption behavior. Some examples are the following: (i) the so-called IRMOF-6 which uses the cyclobutane-derivatized benzenedicarboxylate linker, was found to be particularly effective in the uptake of methane, with capacity exceeding that of zeolite 5A. Many of the IRMOF-n frameworks have been investigated for their potential use as hydrogen and alkane storage media [61]; (ii) MIL-96 is a porous aluminium trimesate 3D structure constructed from a hexagonal network of 18-membered rings and μ_3-oxo-centered trinuclear units. It has been demonstrated that this product is able to absorb both CO_2 and CH_4 at room temperature and H_2 at 77 K [62]. Serre and coworkers have recently reported an interesting study on the role of solvent–host interactions that lead to very large swelling on MIL-n frameworks [63].

Hydrogen is a very important guest. Two new cubic zeotypic metal carboxylates, denoted MIL-100 and MIL-101 are built up from trimers of metal octahedra and di- or tricarboxylic acid, exhibit giant pores and unprecedented surface areas without any loss of crystallinity after water evacuation. They are good candidates for hydrogen storage [64]. Furthermore, they can adsorb and deliver the analgesic and anti-inflammatory drug, Ibuprofen [65].

Interpenetration can be utilized to strengthen the interaction between the gaseous molecules and the framework by an entrapment mechanism in which a hydrogen molecule is in close proximity with several aromatic rings from interpenetrating networks. One example is found with the $[Zn_4O(L)_3(dmf)_2]$ clusters (L = dibenzoic acid derivatives), giving interpenetrated 3D networks with large cubic cavities of approximately $19 \times 19 \times 19 \text{Å}^3$ and having a comparable hydrogen storage to that of purified single-walled carbon nanotubes [66].

Chemical reduction of MOFs as a method to enhance gas uptake and binding has been recently explored [67]. The $[Zn_2(NDC)_2(diPyNI)]$ (where NDC = 2,6-naphthalenedicarboxylate and dyPyNI = N,N'-di-(4-pyridyl)-1,4,5,8-naphthalenetetracarboxydiimide) is a very robust MOF, which retains 54% solvent-accessible void volume. It can be reduced, without destruction, with lithium metal in DMF as solvent. Remarkable enhancement is seen for N_2 and H_2 absorption (almost double) in the Li$^+$-doped MOF. N_2 absorption measurements suggest a reversible structural change, represented speculatively as framework displacement. Different lithium-doped MOFs, with the typical $[Zn_4O(OOC\text{–}R\text{–}COO)_6]$ connector coupled to different aromatic linkers (R) through the O–C–O common to each linker, present reversible H_2 storage at ambient temperature [68]. These –and other cases– suggest that the doping of MOFs with electropositive metals is a promising strategy for practical H_2 storage.

The number of publications on H_2 storage in porous MOFs *only* in 2007 has increased extraordinarily. Two reviews on hydrogen storage in metal–organic frameworks have been published recently: in 2005 [69] and in 2007 [70]. It seems, however, that, to date, no *practical* means for H_2 storage and transportation have yet been fully developed.

16.7.3.2 Exchange

Porous zeolites have cation-exchange properties as a result of their anionic frameworks. Porous coordination polymers, in contrast to zeolites, tend to have cationic frameworks, which are constructed from cationic metal ions and neutral bridging ligands. These MOFs accommodate counteranions in the cavities and, therefore, have anion-exchange properties. Anion exchange was first reported in 1990 with [{Cu(4,4',4'',4'''-tetracyanotetraphenylmethane}$BF_4 \cdot xC_6H_5NO_2$] [6b]. The structure contains a diamond-like cationic framework with large adamantine-like cavities occupied by disordered nitrobenzene and BF_4^- ions. This crystal readily undergoes anion-exchange with PF_6^- ions. A review on anion separation with MOFs has been published recently [71].

Quantitative exchange studies with a variety of small molecules have identified MOFs with specificity towards guest shape or functionality. For example, the [Zn(BDC)(4,4'-bpy)$_{0.5}$] (BDC = 1,4-benzenedicarboxylate) MOF has a microporous framework composed of paddle-wheel [Zn$_2$(COO)$_4$] units, bridged by the BDC ligands forming a 2D square grid. The 2D square grids are pillared by 4,4'-bpy molecules, to form a 3D framework. Two of these frameworks interpenetrate, giving 1D channels with a size and shape that are useful for the gas chromatographic separation of linear and branched alkanes [72].

16.7.3.3 Catalysis

To date, solid catalysts have been almost exclusively inorganic materials, the microporous zeolites being especially useful. Coordination polymers also show catalytic activities but, as indicated by Kitagawa and coworkers in 2004, "*their catalytic activities are largely unexplored*" [52]. However, catalysis is often cited as a desirable characteristic of metal–organic frameworks, and a justification for general research into these materials.

One of the first examples was reported in 1994, with [Cd(4,4'-bpy)$_2$(NO$_3$)$_2$], which contains a 2D network with cavities surrounded by 4,4'-bpy units, showing shape-specific catalytic activity for cyanosilylation of aldehydes [73]. This reaction is apparently promoted by the heterogeneous polymer since no reaction takes place with separated powdered Cd(NO$_3$)$_2$ and 4,4'-bpy.

In 2006, Hupp and coworkers reported a new microporous MOF, featuring chiral [Mn(salen)] struts, which is highly effective as an asymmetric catalyst for olefin epoxidation, yielding enantiomeric excesses that rival those of the free molecule analog [74]. The catalytic strut is a Mn-salen derivative incorporated in a very robust pillared paddlewheel structure containing pairs of zinc ions together with biphenyldicarboxylate (bpdc) as the second ligand. Framework confinement of the manganese salen entity enhances catalyst stability, imparts substrate-sized selectivity and permits catalyst separation. In the case of the homogeneous epoxidation catalysts, [Mn(salen)] is initially highly effective, but loses much of its activity after the first few minutes. After a few hours essentially all activity is lost. In contrast, the framework-immobilized catalyst exhibited much better reactivity and selectivity.

The $[Cu_3(btc)_2]$ (btc = benzene-1,3,5-tricarboxylate) is a rigid MOF with a zeolite-like structure and with free coordination sites on the Cu^{2+} ions. It is a highly selective Lewis acid catalyst for the isomerization of terpene derivatives [75].

The oxidation of CO to CO_2 in porous MOFs, such as [Cu(5-methylisophthalate)$(H_2O)]\cdot(H_2O)$ has been studied recently. The $[Cu(mipt)(H_2O)]$ MOF is formed by the typical paddle-wheel units, interconnected by bent mipt linkers to form an undulated 2D net. Each copper center has a square-pyramidal environment. The framework can be dehydrated/hydrated reversibly with a pronounced color change. The Lewis-acid coordination sites (H_2O) are on the interior of the channel walls, and thus copper sites are accessible for adsorption and catalytic conversion. Notably, the microporous framework remains intact after removal of guest molecules and after catalytic reaction [76].

Kitagawa and coworkers give in their review a table of microporous coordination polymers capable of catalytic activity [52].

16.8
Inorganic Hybrid Materials

We have left for the final part of this chapter a brief summary of three special 'inorganic' ligands, which generate many different hybrid crystal topologies, usually different than those derived from materials constructed by organic (usually long) linkers: the azido (N_3^-), the cyanido (CN^-), and polyoxometalates.

16.8.1
Metal-azido Coordination Polymers

Due to the various coordination modes and efficient magnetic pathway, the azido ligand has attracted considerable interest in the last decades, mainly for magnetic studies [77] (see Chapter 10). The coordination networks with *only* N_3^- bridging ligands present singly or, simultaneously, various bridging modes giving, thus, many different topologies. Unfortunately, so far, it is impossible to control the nature of the framework formed. There are many examples reported in the literature with azido and other bridging ligands. Two typical cases are: (i) a 2D grid-like Cu(II) complex with EO azido and pyrazine bridges, $[Cu(N_3)_2(pyz)]$ [78]. The 2D grid-like network consists of a 1D chain of Cu–pyz units connected by EO azido bridges; (ii) [Co$_{1.5}$(isonic)(N$_3$)(OH)] is the first metal azido complex with isonico-tinato as a bridging ligand, in which the overall structure exhibits a new (3,6) connected net topology. The complicated 3D network can be described in terms of sheets of [Co$_3$(OH)$_2$(N$_3$)$_2$]$^{2+}$ pillared by isonicotinate subunits [79].

16.8.2
Metal-cyanido Coordination Polymers

The most important series is, undoubtedly, the Prussian Blue derivatives, which are good candidates for new molecule-based magnets (see Chapter 10). The replacement of [Fe(CN)$_6$]$^{4-}$ by octahedral cluster analogs [Re$_6$X$_8$(CN)$_6$]$^{n-}$ (X = chalcide) or [Nb$_6$X$_{12}$(CN)$_6$]$^{n-}$ (X = halide) has led to the preparation of *expanded* Prussian-Blue-type frameworks with large cavities and sometimes interesting gas absorption properties as well as new magnetic materials [80].

A substantial modular substitution is the alteration of the key Prussian Blue building block, namely the octahedral cyanometallate [M(CN)$_6$]$^{n-}$, with non-octahedral cyanometallates [M(CN)$_x$]$^{p-}$. This has generated a series of other "*basic structural motifs*". For example, [Ag(CN)$_2$]$^-$ has coordination number 2. In this case the bridging units will be –NC–Ag–CN–. A typical example is [Cu(en)$_2$][Ag(CN)$_2$]$_2$. Four-coordinate square-planar [M(CN)$_4$]$^{2-}$ (M = Ni, Pd, Pt) building blocks were found to generate square-grid arrays (Hoffmann-clathrate type, for example) with transition metal cations. Tetrahedral cyanometallate units (Zn, Cd) favor the forma-tion of three-dimensional diamond-like networks. With seven- or eight-coordinate cyanometallates, such as [Mo(CN)$_7$]$^{4-}$ and [Mo/W(CN)$_8$]$^{4-}$, novel three-dimensional networks have been obtained. Let us emphasize, the recently reported sodalite-like framework based on octacyanomolybdate and neodymium with guest metha-nol molecules and neodymium octahydrate ions, [{Nd^{III}(CH$_3$OH)$_4$Mo^{IV}(CN)$_8$}$_3$] [Nd^{III}(H$_2$O)$_8$] · 8CH$_3$OH [81]. The anionic 3D [{Nd^{III}(CH$_3$OH)$_4$Mo^{IV}(CN)$_8$}$_3$]$^{3-}$ has a zeolite-like framework, which contains [Nd^{III}(H$_2$O)$_8$]$^{3+}$ ions and methanol mole-cules as guests in its channels. The repeating unit of the framework is a cage composed of eight 18-membered rings (Figure 16.22). All of the cage edges are Mo–CN–Nd linkages. The cage can be visualized as a truncated cube, with 14 neighbors sharing different faces. Even if distorted, it is still convenient to regard it as a β-cage, the primary building unit of sodalite (Na$_8$Cl$_2$[Al$_6$Si$_6$O$_{24}$]).

16.8.3
Polyoxometalates as Nodes

The polyoxometalates have been studied in Chapter 5. These clusters can act as nodes of promising MOFs. One of the strategies used for the design and synthesis of 3D networks with POMs is to employ a specific coordinated metal cation as a charge compensation agent and template. Some representative examples are: (i) The first zeolitic topological porous polyoxometalate, [Cu$_4$V$_{13}$^{IV}V$_5$^{V}O$_{42}$(NO$_3$)(1,2-dia minopropane)$_8$] · 10H$_2$O [82], (ii) the [Cu$_2$(bpy)$_2$(μ-ox)][M^{III}(OH)$_7$Mo$_6$O$_{17}$] (M = Al, Cr), with 1D chains constructed of alternating Anderson-type polyoxoanions and

(A)   (B)

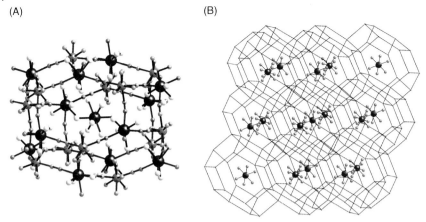

Figure 16.22 (A) The sodalite-like β-cage of the anionic framework [{Nd^{III}(CH$_3$OH)$_4$Mo^{IV}(CN)$_8$}$_3$]$^{3-}$ Mo = large black spheres, Nd = large gray spheres. (B) Zeolite-like framework (see text).

oxalate-bridged dinuclear copper complexes. Extensive hydrogen bonding plays an important role in the formation of the 3D networks [83], (iii) the unprecedented 3D network constructed by coordination polymer layers with three-connected Cu$^+$ nodes (Cl$^-$, 4,4′-bpy) and a Keggin type anion as "pillar" [PW$^{VI}_{10}$W$^V_2$O$_{40}$]$^{5-}$. Thanks to the Cu-bpy long sticks, the individual network is porous, giving a fivefold interpenetrating network [84].

References

1 (a) Janiak, C. Engineering coordination polymers towards applications, *Dalton Trans.* 2003, 2781; (b) James, S.L. Metal-organic frameworks, *Chem. Soc. Rev.* 2003, *32*, 276; (c) Brammer, L., Developments in inorganic crystal engineering, *Chem. Soc. Rev.* 2004, *33*, 476; (d) M.W. Hosseini, Molecular Tectonics: From Simple Tectons to Complex Molecular Networks, *Acc. Chem. Res.* 2005, *38*, 313.

2 (a) Wells, A.F. (1984) *Structural Inorganic Chemistry*, 5th edn., Oxford University Press, Oxford; (b) Wells, A.F. (1977) *Three-dimensional Nets and Polyhedra*, Wiley, New York.

3 Robson, R., Abrahams, B.F., Batten, S.R., Gable, R.W., Hoskins, B.F., Liu, J.P. Supramolecular Architecture, ed. T. Bein, *ACS Symp. Ser.* 1992, *499*, 256.

4 (a) Braga, D. Crystal engineering, Where from? Where to? *Chem. Comm.* 2003, 2751; (b) Sharma, C.V.K. Crystal engineering – where do we go from here? *Cryst. Growth Des.*, 2002, *2*, 465.

5 Braga, D., Brammer, L., Champness, N.R. New trends in crystal engineering, *CrystEngComm.* 2005, *7*, 1.

6 (a) Hoskins, B.F., Robson, R., Infinite Polymeric Frameworks Consisting of Three-dimensionally Linked Rod-like Segments, *J. Am. Chem. Soc.* 1989, *111*, 5962; (b) Hoskins, B.F., Robson, R. Design and Construction of a New Class of Scaffolding-like Materials Comprising Infinite Polymeric Frameworks of 3D-Linked Molecular Rods. A Reappraisal of the Zn(CN)$_2$ and Cd(CN)$_2$ Structures and the Synthesis and Structure of the Diamond-Related Frameworks [N(CH$_3$)$_4$]

[Cu^{I}Zn^{II}(CN)$_4$] and CU^I[4,4′4″,4‴ tetracyanotetraphenylmethane]BF$_4$.xC$_6$H$_5$NO$_2$, *J. Am. Chem. Soc.* 1990, *112*, 1546.

7 These three Journals are: *Cryst. Eng.* (1998); *CrystEngComm* (1999) and *Cryst. Growth Des.* (2001).

8 Power, K.N., Hennigar, T.L., Zaworotko, M.J. Crystal structure of the coordination polymer [Co(bipy)$_{1.5}$(NO$_3$)$_2$] · CS$_2$ (bipy=4,4′-bipyridine), a new motif for a network sustained by "T-shape" building blocks, *New J. Chem.* 1998, *22*, 177.

9 Batten, S.R. Glorious uncertainty -- challenges for network design, *J. Solid State Chem.*, 2005, *178*, 2475.

10 Min, K.S., Suh, M.P. Silver(I)-Polynitrile Network Solids for Anion Exchange: Anion-Induced Transformation of Supramolecular Structure in the Crystalline State, *J. Am. Chem. Soc.* 2000, *122*, 6834.

11 Williams, C.A., Blake, A.J., Hubberstey, P., Schröder, M. A unique example of a 3^6 tessellated 2-D net based on a trinuclear zinc(II)-1,4-benzenedicarboxylate framework, *Chem. Comm.* 2005, 5435.

12 Ockwig, N.W., Delgado-Friedrichs, O., O'Keeffe, M., Yaghi, O.M. Reticular Chemistry: Occurrence and Taxonomy of Nets and Grammar for the Design of Frameworks, *Acc. Chem. Res.* 2005, *38*, 176.

13 Subramanian, S., Zaworotko, M.J. Porous Solids by Design: [Zn(4,4′-bpy)$_2$(SiF$_6$)]$_n$ · xDMF, a Single Framework Octahedral Coordination Polymer with Large Square Channels, *Angew. Chem. Int. Ed.* 1995, *34*, 2127.

14 Zaman, M.B., Smith, M.D., Ciurtin, D.M., zur Loye, H-C. New Cd(II)-, Co(II)-, and Cu(II)-Containing Coordination Polymers Synthesized by Using the Rigid Ligand 1,2-Bis(3-pyridyl)ethyne (3,3′-DPA), *Inorg. Chem*, 2002, *41*, 4895.

15 Long, D-L., Hill, R.J., Blake, A.J., Champness, N.R., Hubberstey, P., Wilson, C., Schröder, M. Anion Control over Interpenetration and Framework Topology in Coordination Networks Based on Homoleptic Six-Connected Scandium Nodes, *Chem. Eur. J.* 2005, *11*, 1384.

16 Liu, T., Chen, Y-H., Zhang, Y-J., Wang, Z-M., Gao, S. NaCl-Type Frameworks of

[M(pyrazine)$_2$NO$_2$]ClO$_4$ (M = Co, Cu), the First Examples Containing μ$_{1,3}$-Nitrito Bridges Showing Antiferromagnetism, *Inorg. Chem.* 2006, *45*, 9148.

17 Hennigar, T.L., MacQuarrie, D.C., Losier, P., Rogers, R.D., Zawototko, M.L. Supramolecular Isomerism in Coordination Polymers: Conformational Freedom of Ligands in [Co(NO$_3$)$_2$(1,2-bis(4-pyridyl)ethane)$_{1.5}$]$_n$, *Angew.Chem. Int. Ed.* 1997, *36*, 972.

18 Yi, L., Yang, X., Lu, T., Cheng, P. Self-assembly of right-handed helical infinite chain, one- and two-dimensional coordination polymers tuned via anions, *Cryst. Growth Des.*, 2005, *5*, 1215.

19 Hoskins, B.F., Robson, R., Slizys, D.A. An Infinite 2D Polyrotaxane Network in Ag$_2$(bix)$_3$(NO$_3$)$_2$ (bix = 1,4-Bis(imidazol-1-ylmethyl)benzene), *J. Am. Chem. Soc.* 1997, *119*, 2952.

20 Carlucci, L., Ciani, G., Proserpio, D.M. Parallel and inclined (1D → 2D) interlacing modes in new polyrotaxane frameworks [M$_2$(bix)$_3$(SO$_4$)$_2$] [M = Zn(II), Cd(II); Bix=1,4-bis(imidazol-1-ylmethyl)benzene], *Cryst. Growth Des.*, 2005, *5*, 37.

21 Whang, D., Heo, J., Kim, C-A., Kim, K. Helical polyrotaxane: cucurbituril "beads" threaded onto a helical one-dimensional coordination polymer, *Chem. Commun.* 1997, 2361.

22 Whang, D., Kim, K. Polycatenated Two-Dimensional Polyrotaxane Net, *J. Am. Chem. Soc.* 1997, *119*, 451.

23 Tian, Y-Q., Zhao, Y-M., Xu, H-J., Chi, C-Y. CO$_2$ Template Synthesis of Metal Formates with a ReO$_3$ Net, *Inorg. Chem.* 2007, *46*, 1612.

24 (a) Clemente-Leon, M., Coronado, E., Gómez-García, C.J., Soriano-Portillo, A. Increasing the Ordering Temperatures in Oxalate-Based 3D Chiral Magnets: the Series [Ir(ppy)$_2$(bpy)][M^{II}M^{III}(ox)$_3$] · 0.5H$_2$O (M^{II}M^{III} = MnCr, FeCr, CoCr, NiCr, ZnCr, MnFe, FeFe); bpy = 2,2′-Bipyridine; ppy = 2-Phenylpyridine; ox = Oxalate Dianion), *Inorg.Chem.* 2006, *45*, 5653; (b) Decurtins, S., Schmalle, H.W., Schneuwly, P., Ensling, J., Gütlich, P. A Concept for the Synthesis of 3-Dimensional Homo- and Bimetallic Oxalate-Bridged Networks [M$_2$(ox)$_3$]$_n$, Structural, Mossbauer, and

Magnetic Studies in the Field of Molecular-Based Magnets, *J. Am. Chem. Soc.* 1994, *116*, 9521.

25 Rao, C.N.R., Natarajan, S., Vaidhyanathan, R. Metal Carboxylates with Open Architectures, *Angew. Chem. Int. Ed.* 2004, *43*, 1466.

26 Zhu, H-F., Fan, J., Okamura, T., Zhang, Z-H., Liu, G.X., Yu, K-B., Sun, W-Y., Ueyama, N. Metal-Organic Architectures of Silver(I), Cadmium(II), and Copper(II) with a Flexible Tricarboxylate Ligand, *Inorg. Chem.* 2006, *45*, 3941.

27 Guillou, N., Livage, C., Férey, G. Cobalt and Nickel Oxide Architectures in Metal Carboxylate Frameworks: From Coordination Polymers to 3D Inorganic Skeletons, *Eur. J. Inorg. Chem.* 2006, 4963.

28 (a) Li, H., Eddaoudi, M., O'Keeffe, M., Yaghi, O.M. Design and synthesis of an exceptionally stable and highly porous metal-organic framework, *Nature*, 1999, *402*, 276; (b) Rosi, N.L., Kim, J., Eddaoudi, M., Chen, B., O'Keeffe, M., Yaghi, O.M. Rod Packings and Metal-Organic Frameworks Constructed from Rod-Shaped Secondary Building Units, *J. Am. Chem. Soc.* 2005, *127*, 1504.

29 Furukawa, S., Ohba, M., Kitagawa, S. Rational synthesis of a two-dimensional honeycomb structure based on a paramagnetic paddlewheel diruthenium complex, *Chem. Comm.* 2005, 865.

30 Chen, B., Fronczek, F.R., Courtney, B.H., Zapata, F. α-Po nets of copper(II)-trans-1,4-cyclohexanedicarboxylate frameworks based on a paddle-wheel building block and its enlarged dimer, *Cryst. Growth Des.* 2006, *6*, 825.

31 Eddaoudi, M., Kim, J., O'Keeffe, M., Yaghi, O.M. $Cu_2[o\text{-}Br\text{-}C_6H_3(CO_2)_2]_2(H_2O)_2 \cdot (DMF)_8(H_2O)_2$: A Framework Deliberately Designed To Have the NbO Structure Type, *J. Am. Chem. Soc.* 2002, *124*, 376.

32 Pichon, A., Fierro, C.M., Nieuwenhuyzen, M., James, S.L. A pillared-grid MOF with large pores based on the $Cu_2(O_2CR)_4$ paddle-wheel, *CrystEngComm.* 2007, *9*, 449.

33 Moulton, B., Lu, J., Hajndl, R., Hariharan, S., Zaworotko, M.J. Crystal

Engineering of a Nanoscale Kagomé Lattice, *Angew. Chem. Int. Ed.* 2002, *41*, 2821.

34 Chun, H., Moon, J. Discovery, Synthesis, and Characterization of an Isomeric Coordination Polymer with Pillared Kagome Net Topology, *Inorg. Chem.* 2007, *46*, 4371.

35 Ferrer, S., Aznar, E., Lloret, F., Castiñeiras, A., Liu-González, M., Borras, J. One-Dimensional Metal-Organic Framework with Unprecedented Heptanuclear Copper Units, *Inorg. Chem.* 2007, *46*, 372.

36 Zhang, X-M., Zheng, Y-Z., Li, C-R., Zhang, W-X., Chen, X-M. Unprecedented (3, 9)-connected $(4^2 \cdot 6)_3(4^6 \cdot 6^{21} \cdot 8^9)$ net constructed by trinuclear mixed-valence cobalt clusters, *Cryst. Growth Des.* 2007, *7*, 980.

37 Fang, Q-R., Zhu, G-S., Jin, Z., Xue, M., Wei, X., Wang, D-J., Qiu, S-L. A novel metal-organic framework with the diamondoid topology constructed from pentanuclear zinc-carboxylate clusters, *Cryst. Growth Des.* 2007, *7*, 1035.

38 Livage, C., Guillou, N., Chaigneau, J., Rabu, P., Drillon, M., Ferey, G. A Three-Dimensional Metal-Organic Framework with an Unprecedented Octahedral Building Unit, *Angew. Chem. Int. Ed.* 2005, *44*, 6488.

39 Chen, J., Ohba, M., Zhao, D., Kaneko, W., Kitagawa, S. Polynuclear core-based nickel 1,4-cyclohexanedicarboxylate coordination polymers as temperature-dependent hydrothermal reaction products, *Cryst. Growth Des.* 2006, *6*, 664.

40 Li, J-R., Tao, Y., Yu, Q., Bu, X-H. A pcu-type metal–organic framework with spindle $[Zn_7(OH)_8]^{6+}$ cluster as secondary building units, *Chem. Comm.* 2007, 1527.

41 (a) McManus, G.J., Wang, Z., Zaworotko, M.J. Suprasupermolecular chemistry: Infinite networks from nanoscale metal-organic building blocks, *Cryst. Growth Des.*, 2004, *4*, 11; (b) Luo, F., Batten, S.R., Che, Y., Zheng, J-M. Synthesis, Structure, and Characterization of Three Series of 3d-4f Metal-Organic Frameworks Based on Rod-Shaped and (6,3)-Sheet Metal Carboxylate Substructures, *Chem. Eur. J.* 2007, *13*, 4948; (c) Zhang, J-J., Zhou, H-J., Lachgar, A. Directed Assembly of Cluster-Based

Supramolecules into One-Dimensional Coordination Polymers, *Angew. Chem. Int. Ed.* 2007, *46*, 4995; (d) Zou, W-Q., Wang, M-S., Li, Y., Wu, A-Q., Zheng, F-K., Chen, Q-Y., Guo, G-C., Huang, J-S. Unprecedented (3,10)-Connected 2-D Metal-Organic Framework Constructed from Octanuclear Cobalt(II) Clusters and a New Bifunctional Ligand, *Inorg. Chem.* 2007, *46*, 6852.

42 Andruh, M. Oligonuclear complexes as tectons in crystal engineering: structural diversity and magnetic properties, *Chem. Comm.* 2007, 2565.

43 Long, D-L., Hill, R.J., Blake, A.J., Champness, N.R., Hubberstey, P., Proserpio, D.M., Wilson, C., Schröder, M. Non-Natural Eight-Connected Solid-State Materials: A New Coordination Chemistry, *Angew. Chem. Int. Ed.* 2004, *43*, 1851.

44 Chun, H., Kim, D., Dybtsev, D.N., Kim, K. Metal-Organic Replica of Fluorite Built with an Eight-Connecting Tetranuclear Cadmium Cluster and a Tetrahedral Four-Connecting Ligand, *Angew. Chem. Int. Ed.* 2004, *43*, 971.

45 Zhang, X-M., Fang, R-Q., Wu, H-S. A Twelve-Connected Cu_6S_4 Cluster-Based Coordination Polymer, *J. Am. Chem. Soc.* 2005, *127*, 7670.

46 (a) Batten, S.R., Robson, R. Interpenetrating Nets: Ordered, Periodic Entanglement, *Angew. Chem. Int. Ed.* 1998, *37*, 1460; (b) Batten, S.R. Topology of interpenetration, *CrystEngComm.* 2001, *3*, 67.

47 Baburin, I.A., Blatov, V.A., Carlucci, L., Ciani, G., Proserpio, D.M. Interpenetrating metal-organic and inorganic 3D networks: a computer-aided systematic investigation. Part II. Analysis of the Inorganic Crystal Structure Database (ICSD), *J. Solid State Chem.* 2005, *178*, 2452.

48 Batten, S.R., Hoskins, B.F., Robson, R. Interdigitation, Interpenetration and Intercalation In Layered Cuprous Tricyanomethanide Derivatives, *Chem. Eur. J.* 2000, *6*, 156.

49 Gable, R.W., Hoskins, B.F., Robson, R. A new type of interpenetration involving enmeshed independent square grid sheets. The structure of diaquabis-(4,4′-

bipyridine)zinc hexafluorosilicate, *J. Chem. Soc., Chem. Commun.* 1990, 1677.

50 Power, K.N., Hennigar, T.L., Zaworotko, M.J. X-Ray crystal structure of {Cu[1,2-bis(4-pyridyl)ethane]$_2$(NO$_3$)$_2$}$_n$: the first example of a coordination polymer that exhibits the NbO 3D network architecture, *Chem. Commun.* 1998, 595.

51 Kitagawa, S., Noro, S., Nakamura, T. Pore surface engineering of microporous coordination polymers, *Chem. Commun.* 2006, 701.

52 Kitagawa, S., Kitaura, R., Noro, S. Functional Porous Coordination Polymers, *Angew. Chem. Int. Ed.* 2004, *43*, 2334.

53 Kitagawa, S., Kondo, M. Functional micropore chemistry of crystalline metal complex-assembled compounds, *Bull. Chem. Soc. Jpn*, 1998, *71*, 1739.

54 Batten, S.R., Hoskins, B.F., Robson, R. Two Interpenetrating 3D Networks Which Generate Spacious Sealed-Off Compartments Enclosing of the Order of 20 Solvent Molecules in the Structures of Zn(CN)(NO$_3$)(tpt)$_{2/3}$.solv (tpt = 2,4,6-tri(4-pyridyl)-1,3,5-triazine,solv = ~3/4C$_2$H$_2$Cl$_4$.3/4CH$_3$OH or ~3/4CH$_3$Cl$_3$.1/3CH$_3$OH), *J. Am. Chem. Soc.* 1995, *117*, 5385.

55 Rather, B., Zaworotko, M.J. A 3D metal-organic network, [Cu$_2$(glutarate)$_2$(4,4′-bipyridine)], that exhibits single-crystal to single-crystal dehydration and rehydration, *Chem. Comm.* 2003, 830.

56 Yaghi, O.M., Li, G., Li, H. Selective Binding And Removal Of Guests In A Microporous Metal-Organic Framework, *Nature*, 1995, *378*, 703.

57 Halder, G.J., Kepert, C.J. Single crystal to single crystal structural transformations in molecular framework materials, *Aust. J. Chem.* 2006, *59*, 597.

58 A recent example of reversibility can be found in: Suh, M.P., Cheon, Y.E., Lee, E.Y. Reversible Transformation of Zn(II) Coordination Geometry in a Single Crystal of Porous Metal-Organic Framework [Zn$_3$(ntb)$_2$(EtOH)$_2$] · 4EtOH, *Chem. Eur. J.* 2007, *13*, 4208.

59 Kondo, M., Yoshimoti, T., Seki, K., Matsuzaka, H., Kitagawa, S. Three-Dimensional Framework with Channeling Cavities for Small Molecules: {[M$_2$(4,4′-

bpy)$_3$(NO$_3$)$_4$] · xH$_2$O}$_n$ (M = Co, Ni, Zn), *Angew. Chem. Int. Ed.* 1997, *36*, 1725.

60 Noro, S., Kitagawa, S., Kondo, M., Seki, K. A New, Methane Adsorbent, Porous Coordination Polymer [{CuSiF$_6$(4,4'-bipyridine)$_2$}$_n$], *Angew. Chem. Int. Ed.* 2000, *39*, 2081.

61 Zhang, L., Wang, Q., Wu, T., Liu, Y-C. Understanding Adsorption and Interactions of Alkane Isomer Mixtures In Isoreticular Metal-Organic Frameworks, *Chem. Eur. J.* 2007, *13*, 6387.

62 Loiseau, T., Lecroq, L., Volkringer, C., Marrot, J., Ferey, G., Haouas, M., Taulelle, F., Bourrelly, S., Llewellyn, P.L., Latroche, M. MIL-96, a Porous Aluminum Trimesate 3D Structure Constructed from a Hexagonal Network of 18-Membered Rings and μ$_3$-Oxo-Centered Trinuclear Units, *J. Am. Chem. Soc.* 2006, *128*, 10223.

63 Serre, C., Mellot-Draznieks, C., Surblé, S., Audebrand, N., Filinchuk, Y., Férey, G. Role of solvent-host interactions that lead to very large swelling of hybrid frameworks, *Science*, 2007, *315*, 1828.

64 Latroche, M., Surblé, S., Serre, C., Mellot-Draznieks, C., Llewellyn, P.L., Lee, J-H., Chang, J-S., Jhung, S.H., Férey, G. Hydrogen Storage in the Giant-Pore Metal-Organic Frameworks MIL-100 and MIL-101, *Angew. Chem. Int. Ed.* 2006, *45*, 8227.

65 Horcajada, P., Serre, C., Vallet-Regí, M., Sebban, M., Taulelle, F., Férey, G. Metal-Organic Frameworks as Efficient Materials for Drug Delivery, *Angew. Chem. Int. Ed.* 2006, *45*, 5974.

66 Kesanli, B., Cui, Y., Smith, M.R., Bittner, E.W., Bockrath, B.C., Lin, W. Highly Interpenetrated Metal-Organic Frameworks for Hydrogen Storage, *Angew. Chem. Int. Ed.* 2005, *44*, 72.

67 Mulfort, K.L., Hupp, J.T. Chemical Reduction of Metal-Organic Framework Materials as a Method to Enhance Gas Uptake and Binding, *J. Am. Chem. Soc.* 2007, *129*, 9604.

68 Han, S.S., Godard III, W.A. Lithium-Doped Metal-Organic Frameworks for Reversible H$_2$ Storage at Ambient Temperature, *J. Am. Chem. Soc.* 2007, *129*, 8422.

69 Rowsell, J.L.C., Yaghi, O.M. Strategies for Hydrogen Storage in Metal-Organic Frameworks, *Angew. Chem. Int. Ed.* 2005, *44*, 4670.

70 Lin, X., Jia, J., Hubberstey, P., Schröder, M., Champness, N.R. Hydrogen storage in metal-organic frameworks, *CrystEngComm.* 2007, *9*, 438.

71 Custelcean, R., Moyer, B.A. Anion Separation with Metal-Organic Frameworks, *Eur. J. Inorg. Chem.* 2007, 1321.

72 Chen, B., Liang, C., Yang, J., Contreras, D.S., Clancy, Y.L., Lobkovsky, E.B., Yaghi, O.M., Dai, S. A Microporous Metal-Organic Framework for Gas-Chromatographic Separation of Alkanes, *Angew. Chem. Int. Ed.* 2006, *45*, 1390.

73 Fujita, M., Kwon, Y.J., Washizu, S., Ogura, K. Preparation, Clathration Ability, and Catalysis of a Two-Dimensional Square Network Material Composed of Cadmium(II) and 4,Y-Bipyridine, *J. Am. Chem. Soc.* 1994, *116*, 1151.

74 Cho, S-H., Ma, B., Nguyen, S.T., Hupp, J.T., Albrecht-Schmitt, T.E. A metal–organic framework material that functions as an enantioselective catalyst for olefin epoxidation, *Chem. Comm.* 2006, 2563.

75 Alaerts, L., Séguin, E., Poelman, H., Thibault-Starzyk, F., Jacobs, P.A., De Vos, D.E. Probing the Lewis Acidity and Catalytic Activity of the Metal-Organic Framework [Cu$_3$(btc)$_2$] (BTC=Benzene-1,3,5-tricarboxylate), *Chem. Eur J.* 2006, *12*, 7353.

76 Zou, R-Q., Sakurai, H., Han, S., Zhong, R-Q., Xu, Q. Probing the Lewis Acid Sites and CO Catalytic Oxidation Activity of the Porous Metal-Organic Polymer [Cu(5-methylisophthalate)], *J. Am. Chem. Soc.* 2007, *129*, 8402.

77 (a) Ribas, J., Escuer, A., Monfort, M., Vicente, R., Cortés, R., Lezama, L., Rojo, T. Polynuclear Ni(II) and Mn(II) azido bridging complexes. Structural trends and magnetic behavior, *Coord. Chem. Rev.* 1999, *193-195*, 1027; (b) Ribas, J., Escuer, A., Monfort, M., Vicente, R., Cortés, R., Lezama, L., Rojo, T., Goher, M.A.S. (2001) *Magnetism: Molecules to Materials*, Volume II., J.S. Miller, M. Drillon (Eds.), Chapter 9. Wiley-VCH, Weinheim.

78 Dong, W., Ouyang, Y., Liao, D-Z., Yan, S-P., Cheng, P., Jiang, Z-H. Synthesis, structure and magnetic property of a 2D grid-like copper(II) complex with end-on azido and pyrazine bridges, *Inorg. Chim. Acta*, 2006, *359*, 3363.

79 Liu, F.-C., Zeng, Y-F., Jiao, J., Bu, X-H., Ribas, J., Batten, S.R. First Metal Azide Complex with Isonicotinate as a Bridging Ligand Showing New Net Topology: Hydrothermal Synthesis, Structure, and Magnetic Properties, *Inorg. Chem.* 2006, *45*, 2776.

80 Zhang, J., Lachgar, A. Superexpanded Prussian-Blue Analogue with $[Fe(CN)_6]^{4-}$, $[Nb_6Cl_{12}(CN)6]^{4-}$, and $[Mn(salen)]^+$ as Building Units, *J. Am. Chem. Soc.* 2007, *129*, 250.

81 Wang, Z-X., Shen, X-F., Wang, J., Zhang, P., Liu, Y-Z., Nfor, E.N., Song, Y., Ohkoshi, S. Hashimoto, K., You, X-Z. A Sodalite-like Framework Based on Octacyanomolybdate and Neodymium with Guest Methanol Molecules and Neodymium Octahydrate Ions, *Angew. Chem. Int. Ed.* 2006, *45*, 3287.

82 Xu, Y., Nie, L.-B., Zhu, D., Song, Y., Zhou, G.-P., You, W-S. Two new three-dimensional porous polyoxometalates with typical ACO topological open frameworks: $\{[(Cu_4V_{13}V_5O_{42})$-V-IV-O-V(NO_3)(C_3H_{10}N_2)(8)] \cdot 10H(_2)O\}(n)$ and $\{[(Cu4V_{12}V_6O_{42})$-V-IV-O-V(SO_4)(C_3H_{10}N_2)(8)] \cdot 10H(2)O\}(n)$, *Cryst. Growth Des.* 2007, *7*, 925.

83 Cao, R., Liu, S., Xie, L., Pan, Y., Cao, J., Ren, Y., Xu, L. Organic-Inorganic Hybrids Constructed of Anderson-Type Polyoxoanions and Oxalato-Bridged Dinuclear Copper Complexes, *Inorg. Chem.* 2007, *46*, 3541.

84 Fan, L., Xiao, D., Wang, E., Li, Y., Su, Z., Wang, X., Liu, J. An unprecedented fivefold interpenetrating network based on polyoxometalate building blocks, *Cryst. Growth Des.* 2007, *7*, 592.

Bibliography

Tiekink, E.R.T., Vittal, J.J. (Eds.) (2006) *Frontiers in Crystal Engineering*, Wiley, Chichester (England)

17
Biocoordination Chemistry: Coordination Chemistry and Life

17.1
Introduction

An important part of bioinorganic chemistry involves studying the role played by metal ions in biological processes. Wieghardt, in the foreword of the book *Concepts and Models in Bioinorganic Chemistry* [1], reiterates the statement made by Wood (University of Minnesota, 1975): *"If you think that biochemistry is the organic chemistry of living systems, then you are misled; biochemistry is the coordination chemistry of living systems"*.

Bioinorganic chemistry studies various aspects related to the role of metal ions in living beings, for example: the ions which are found in living beings, the structure of the complexes they form, their function, and the mechanism of the reactions they produce. However, there is another complementary but highly important issue which must also be taken into account, namely, the effect of external metal ions in terms of both their toxicity and their potential use as drugs, as well as the question of how they can be used as diagnostic agents.

It is often very difficult to study metal centers in biological molecules. Therefore, one of the aims of bioinorganic chemistry is to compare the properties of the metal center of a biological system with those of small coordination compounds that are easier to characterize. Comparing the properties of the two systems can provide highly useful information about the type of ligands involved, the coordination number, geometry, the oxidation state of the metal, etc.

The present chapter does not aim to give a comprehensive account, or even an overview, of bioinorganic chemistry, but rather explores the application of the subject matter from earlier chapters to this topic.

17.1.1
Biodistribution of Metal Ions

An obvious limitation to the biodistribution of metal elements is that the process is clearly restricted by the availability of such metal elements in the environment. Availability itself is limited by ambient *aqueous* conditions. The first cells were formed in water and, consequently, they incorporated the dissolved elements that were found there in the greatest concentration. The limiting equilibrium conditions are governed in part by the pH and the redox potential in particular local environments. Assuming a pressure of 1 atm and a temperature of 0 to 30 °C, it is possible that some thirty elements could be taken up by cells at a reasonably low energy cost. The availability of metal elements due to these factors has changed over time, as shown in Table 17.1.

At present we are aware of thirteen metal ions which are essential for plants and animals. Four of these (Na^+, K^+, Mg^{2+} and Ca^{2+}) are present in large quantities and are known as the *bulk metals*; the remaining nine, which are present in small quantities, are the d-block elements (Table 17.2) and are known as *trace elements*. It is this latter group which we are concerned with in biocoordination chemistry. Their definition varies in scope according to different authors. In general, the definition of trace elements with regard to the human body involves a daily requirement of fewer than 25 mg. Strictly speaking, elements should be called essential only if, on the one hand, their total absence in the organism causes severe damage and, on the other, their biological function is fully understood.

For several authors [2], trace metals may be divided into two subgroups. Iron, copper and zinc form one group (Fe > Zn > Cu), while the remaining six metals are termed *ultra-trace elements*, being present in exceedingly small concentrations. Of these ultra-trace elements only five (Mn, Mo, Co, Ni and V) have been identified

Table 17.1 Changes over time in the estimated available concentration of metal ions in the sea.

Metal ion	Original condition/M	Aerobic conditions/M	Metal ion	Original condition/M	Aerobic conditions/M
Na^+	$>10^{-1}$	$>10^{-1}$	Co^{2+}	$<10^{-13}$	$\sim10^{-11}$
K^+	$\sim10^{-2}$	$\sim10^{-2}$	Ni^{2+}	$<10^{-12}$	$<10^{-9}$
Mg^{2+}	$\sim10^{-2}$	$>10^{-2}$	Cu	$<10^{-20}$ (Cu^+)	$<10^{-10}$ (Cu^{2+})
Ca^{2+}	$\sim10^{-3}$	$\sim10^{-3}$	Zn^{2+}	$<10^{-12}$	$<10^{-8}$
V	$\sim10^{-7}$	$\sim10^{-7}$ (VO_4^{3-})	Mo	$<10^{-10}$ (MoS_4^{2-})	$\sim10^{-7}$ (MoO_4^{2-})
Mn^{2+}	$\sim10^{-7}$	$\sim10^{-9}$	W	$\sim10^{-9}$ (WS_4^{2-})	$\sim10^{-9}$ (WO_4^{2-})
Fe	$\sim10^{-7}$ (Fe^{2+})	$\sim10^{-17}$ (Fe^{3+})	H^+	pH $\sim$ 7	pH $\sim$ 8

Taken from Ref. [1].

Table 17.2 Important d-elements (in bold case) in biocoordination chemistry.

Ti	**V**	**Cr**	**Mn**	**Fe**	**Co**	**Ni**	**Cu**	**Zn**
Zr	Nb	**Mo**	Tc	Ru	Rh	Pd	Ag	Cd
Hf	Ta	W	Re	Os	Ir	Pt	Au	Hg

as forming metalloenzymes. The remaining metal, Cr, has been reported as being essential in glucose metabolism in higher mammals.

Cu is found in processes that involve oxygen in some way or another, and this suggests that copper became incorporated into biological systems at a time when the atmosphere was already oxidizing and the copper could be found in the form Cu^{2+}, which is more soluble in water than Cu^+. Copper is, thus, a relatively "modern" element, having become bioavailable with the advent of an oxygen atmosphere. The case of iron is different. Fe^{2+} compounds are much more soluble than those of Fe^{3+}, and it can thus be deduced that it began to take part in biological processes during the reducing phase of the formation of living beings. Molybdenum is a rare element in the earth's crust but is quite soluble at pH 7 as MoO_4^{2-}; it is thus rather abundant in sea water and has therefore been found as an essential element in many organisms.

It is noteworthy that elements such as Si, Al or Ti, which are prominent as mineral components, play only a marginal role in the biosphere. One reason for this is that, in general, the physiological conditions for living processes include pH values of about 7 in aqueous solution. Under these conditions, the aforementioned elements in their usual high oxidation state exist as nearly insoluble oxides or hydroxides and are therefore not bioavailable.

17.1.2
Biocoordination Compounds

The properties of a given coordination compound are highly sensitive to the type of ligand, the coordination number and the geometry. Similarly, the functions of biological coordination compounds (active center or reaction center) will depend greatly on the metal ion that is present at the reaction center and on the ligands with which it is coordinated, as well as on the surrounding geometry.

17.1.2.1 Types of Reactions
As occurs in chemistry in general, the reactions which take place in living beings can be classified into two broad groups:

1. Redox reactions. In these reactions the metal center must have more than one oxidation state that can easily be reached, and in most cases this involves Fe and Cu ions. Other important ions that take part in these processes include Mo (activation of N_2), Mn (oxidation of water during photosynthesis), and Ni (hydrogenases).

   The activation of small molecules with large bond energies requires the action of a catalyst. The ability of transition metal centers to simultaneously accept or donate electron charge (π-back bonding) allows organisms to carry out energetically and mechanistically difficult reactions under physiological conditions such as: (i) the reversible uptake, transport, storage and activation–reduction of O_2; and (ii) the fixation of molecular N_2 and its conversion in NH_3.

2. Acid–base reactions. The majority occur in processes of hydrolysis. These processes require ions with a 2+ oxidation state, for example, Zn^{2+} or Mg^{2+} (and, to a lesser extent, Ni^{2+} and Ca^{2+}).

17.1.2.2 Knowledge of the Active Site: Crystallographic, Spectroscopic and Other Techniques

It is generally believed that the function of a biological system depends upon its structure. For several decades single X-ray diffraction has served as the final arbiter for biological structure determination, although more recently synchrotron radiation, which requires smaller crystals, has been used. Other structural techniques such as EXAFS have been used to determine the structure of metal-biomolecules as, despite having poor precision, they do not rely on single crystals being obtained. The crystalline form, however, is not the one pertaining to physiological environments *in vivo* and so the question remains as to whether the structures in solution and in the crystal are really the same. Powerful NMR techniques are available for the determination of structures in solution. When metal ions are paramagnetic, EPR spectra are of fundamental importance in determining the oxidation states of active centers, and even in monitoring the mechanism of the processes. In the particular case of Fe, Mössbauer spectra prove highly useful and effective.

In processes involving O_2, those techniques which enable one to observe the strength of the O—O bond have been widely used, a good example being Raman spectra (or Raman resonance). Other widely used techniques include electron spectra (visible and UV), as the band intensity enables a distinction to be made between d–d bands and charge transfer (CT) bands (see Chapter 9). Finally, many aspects of the reactions which take place in redox processes can be clarified by studying the redox potentials of systems.

17.2
Biological Ligands and Their Environment

Many simple ligands come from the environment itself: H_2O and ligands derived from H_2O such as OH^-, O^{2-} bridge, etc. Given that during the evolutionary process the first environment in which compounds were formed with metal ions was reducing and rich in H_2S and derivatives, it is not surprising that there are many substances with sulfido bridges.

The two most important classes of bioligands are *proteins* with amino acid side chains usable for coordination, and biosynthesized *macrocyclic ligands*. A general outline of the role played by metal-biomolecules is given in Figure 17.1.

17.2.1
Proteins

Proteins, including enzymes, consist of α-amino acids which are connected via peptide bonds. However, it should be noted that only a small number of the amino

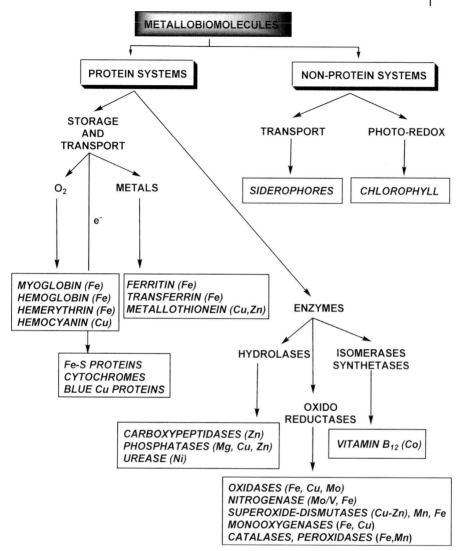

Figure 17.1 Scheme of metallobiomolecules.

acids that form proteins have atoms which can act as ligands (Table 17.3) (Figure 17.2).

O-donor ligands prefer hard metal ions, N-donors prefer intermediate metal ions, and S-donors prefer soft metal ions. As occurs in coordination chemistry that is not related to biological environments, the carboxylato ligand–very important in biocoordination–can be coordinated in many different ways (see Chapter 2).

Table 17.3 Amino acids able to form proteins in living beings.

Amino acid	Donor atom	Type of ligand	
histidine[a]	N	imidazole	N—His
cysteine[a]	S	thiol	Cys—S$^{-[b]}$
methionine	S	thioether	Met—S—CH$_3$
glutamate[a]	O	carboxylate	Glu—COO$^-$
aspartate[a]	O	carboxylate	Asp—COO$^-$
tyrosine	O	phenol	Ph—O$^{-[b]}$

a The most frequent.
b Deprotonated upon coordinating with the metal ion.

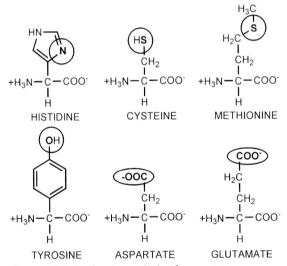

Figure 17.2 Natural amino acids that form proteins.

17.2.2
Macrocyclic Tetrapyrrole Ligands (Porphyrinoids)

The most important macrocyclic ligands in living beings are the tetrapyrroles (see Chapter 2). The tetrapyrroles are at least partially unsaturated tetradentate macrocyclic ligands which, in their deprotonated forms, can tightly bind even substitutionally labile divalent metal ions. The natural porphyrinoids, such as the green pigments of plants and the red pigment of human and animal blood, are intensely colored compounds that are indispensable in all spheres of life. Accordingly, they have been called the "pigments of life". Metalloporphyrinoids are used as cofactors in a multitude of enzymes in order to achieve transport and storage of electrons, of O_2, and of other oxide gases (e.g. NO), to collect and convert solar energy, and to use and activate O_2, methyl groups and other organic moieties in complex enzyme reactions.

As regards natural metalloporphyrinoids the redox inactive Mg^{2+} is used in chlorophyll, which is directly involved in photosynthesis. Furthermore, the redox active Fe, Ni, and Co centers are found in another set of natural metalloporphyrinoids, the biochemical function of which exploits the availability of several oxidation states.

Biosynthesis provides the natural porphyrinoid ligands with remarkable structures, spanning a variety of ring sizes, degree of ligand unsaturation, and topologies of π-conjugation. They are roughly circular, macrocyclic ligands with a planar array of four nucleophilic nitrogens, which provides a coordination hole with a diameter of about 4 Å. Therefore, the porphyrinoids are able to bind any of the transition metal ions at their center in a kinetically and thermodynamically inert way.

Some characteristics of these metalloporphyrinoid systems should be considered: one or two axial coordination sites are generally available, thus opening a further dimension in their chemistry. Axial coordination is highly important in terms of protecting the complex. If, for example, one of the axial positions is occupied by an amino acid of the protein, a smaller molecule such as water may coordinate in the remaining free position. The size of the macrocyclic cavity and its flexibility is also important. These two factors enable the selection of the metal and favor or hinder its reactivity. The main general examples are shown in Figure 17.3.

1. The heme group, which consists of an iron center and a substituted porphyrin ligand.

2. Chlorophylls contain (otherwise labile) Mg^{2+} as the central ion and partially hydrogenated and substituted porphyrin ligands (chlorin).

3. Cobalamins contain cobalt and a partially conjugated corrin which contains one ring member less than the porphyrin macrocycles.

4. In 1980 a porphyrinoid nickel complex, "factor 430" (coenzyme F_{430}), was isolated from methane-producing microorganisms and characterized. It constitutes a good candidate for "first hour catalyst" in biochemistry [3].

porphyrin   chlorin   corrin F_{430}

Figure 17.3 Four typical porphyrinoid macrocyclic ligands.

17.2.2.1 **Heme Group**

There are many biological systems in which the reaction center is formed by a Fe^{2+} or Fe^{3+} ion coordinated to a porphyrin ligand. This Fe-porphyrin unit is known as a *heme* group. This active center may have various functions, which will be studied later. The reaction center is similar in all these systems as the porphyrin ligand occupies four coordination positions and can thus coordinate one or two ligands in the axial position. In addition to small structural variations the main factor which affects the reactivity of the metal center is whether or not there are different ligands coordinated to the Fe in the axial positions. For example, depending on which ligands are present in the axial position, the Fe-porphyrin complex may present either a high-spin or low-spin electron configuration.

When there is only one ligand in the axial position, this will be an amino acid from the protein chain and thus the Fe is pentacoordinated. The position that remains vacant enables the coordination of other molecules, for example, O_2 (see below). When the Fe-porphyrin system has two ligands in axial positions, there are two possible situations which may arise:

1. One of the ligands is an amino acid from the protein chain and the remaining position is occupied by a water molecule. In this case the system continues to be high spin and the water shows labile coordination, thus enabling interaction with other species. Consequently, it may act as a catalyst (enzyme).

2. The two axial positions are occupied by amino acids from the protein chain. Here, the electron configuration of the Fe is low spin. In these cases the metal center is unable to interact with any external reagent and thus loses any catalytic ability. The only reaction in which it can participate is electron transfer, as the Fe^{3+}/Fe^{2+} process. The Fe^{3+}/Fe^{2+} reduction potential is highly sensitive to the ligands in the axial position and to the R groups of the aromatic ring (see cytochromes).

In systems with high spin (haemoglobin) the Fe^{3+}/Fe^{2+} redox potential is small and positive (of the order of +0.17 V). In contrast, in those cases when the system acts as a catalyst (enzymes) the potentials are negative. For example, catalases and peroxidases have E^0 between –0.17 and –0.43 V. In such cases the reduction of Fe^{3+} is not at all favorable; indeed, the process that takes place is the oxidation of Fe^{3+}, which moves to oxidation states 4+ or 5+.

Given the importance of systems with a heme group they will be studied in greater depth later.

17.2.2.2 **Macrocyclic Ligands derived from Non-heme Porphyrins**

Slight modifications in the ring of the heme group cause the systems to have very different properties (Figure 17.3). The following are of particular interest:

1. Modification of the aromatic system, leading to less conjugation. Chlorophyll is the pigment responsible for capturing light in the process of photosynthesis. When the chlorophyll is excited by light, electron transfer takes place, but these electrons come from the porphyrin ligand rather than the metal. Although one

might, therefore, think that the presence of the Mg ion is not necessary, experimental data show this not to be the case: the Mg coordinates a water molecule in the vacant axial position and this enables the formation of hydrogen bonds between different chlorophylls, which are thus piled up in a staggered structure that makes them highly adapted to the absorption of sunlight.

2. Vitamin B_{12} (cyanocobalamin) is a Co(III) compound with a macrocyclic ligand similar to the porphyrins but with a couple of differences: (i) it has one carbon less in the ring, which means the cavity is smaller and (ii) it has a lower π system, which means the ring is more flexible.

B_{12} coenzymes provide the best examples of physiologically-relevant bioorganometallic chemistry, as attached to one of the axial positions there is either a CH_3 group (as in methylcobalamin) or a CH_2-adenosyl group, as in the B_{12} coenzyme (adenosyl-cobalamin). Methylcobalamin participates in biomethylation reactions such as that of homocysteine ($HS-CH_2-CH_2-CH(NH_2)COOH$) into methionine ($CH_3-S-CH_2-CH_2-CH(NH_2)COOH$). Adenosyl-cobalamin is very effective in inducing rearrangement (1,2-shift) with migrating groups. In order to catalyze the two processes the Co(III) is reduced and then re-oxidized. In the methyl transfer reaction the redox process over the cobalt ion is Co(III)/Co(I), whereas in the reordering process with coenzyme-B_{12}, the reaction takes place via radicals and the oxidation states of the cobalt are Co(III)–Co(II).

3. The only nickel-containing porphyrinoid with a known biological function is the nickel coenzyme F_{430}. The ligand has the same cavity as porphyrin, but as with the chlorin ring the π system is less delocalized, giving the ring greater flexibility. Its name comes from the absorption band that is visible at 430 nm. Although its mechanism is not clearly understood it would seem to be an organometallic species with a CH_3 group attached to Ni, which can shift through the oxidation states Ni(I)–Ni(II)–Ni(III). This system is responsible for biological methane formation.

17.3
Systems that Interact with O_2

17.3.1
O_2 Carriers

Three classes of O_2 transport proteins are known: the hemoglobin–myoglobin family, hemerythrins (marine invertebrate), and hemocyanin (molluscs and arthropods).

17.3.1.1 Myoglobin and Hemoglobin

In myoglobin and hemoglobin the iron(II) is in a square-pyramidal environment, the four square-planar coordination sites being associated with the porphyrin ring and the axial with a histidine from the protein chain (Figure 17.4). The

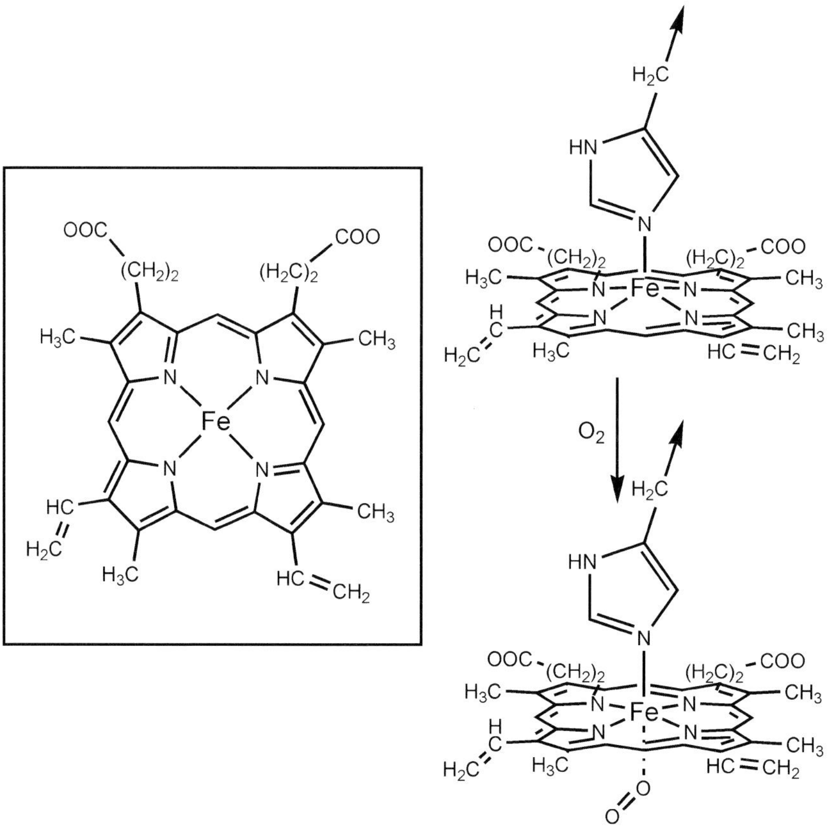

Figure 17.4 Active center in myoglobin and haemoglobin.

five-coordinate iron is high-spin. These species are colored whether or not they have oxygen coordinated; their color may change somewhat from red (without oxygen) to red–purple (with oxygen coordinated). This phenomenon is due to the characteristic spectrum of the porphyrin ligand, whose bands can be attributed to the system's π transitions. The spectrum of this ligand is modified upon coordination with Fe and also with an axial group, as this changes the value of Δ.

Taking into consideration the low solubility of the oxygen in water, one needs a species (such as hemoglobin) with high affinity for oxygen in order to increase its concentration in water, together with another species with low affinity (such as myoglobin) so that oxygen can readily be made available where it is needed, for example, within a muscle. Myoglobin consists of a protein with a molecular weight of just less than 18 000 and has only one iron-porphyrin unit, a *heme* (Figure 17.4). Hemoglobin, however, has four heme units, each with its own protein chain, and these are intermeshed to give a single molecule. The unusual oxygen-affinity properties of hemoglobin arise from cooperative behavior between the four heme groups, a cooperation that must surely be mediated by electrostatic interactions between the protein chains. Oxygen release from hemoglobin to myoglobin is also determined by the cooperative factor as well as by pH (protonation or deprotonation of amino acid residues may well change the preferred geometry of the peptide chains).

Oxygen (π-acid ligand) is adsorbed by attachment to iron atoms in myoglobin and hemoglobin; however, it is not uniquely σ or π bonded, but rather at an intermediate, angular orientation, either 115° (oxy-Mb) or ~156° (oxy-Hb) (Figure 17.4). Myoglobin or hemoglobin may also coordinate with other molecules such as CO or CN^- as they are also π-acid ligands, although in these cases the angle formed is approximately 180°.

Oxygen enters the sixth site, completing an octahedral coordination around the iron(II) and giving low-spin, $(t_{2g})^6$, $S = 0$. The reduction of the ionic radius of iron(II), high-spin to low-spin, is sufficient to enable the iron(II) – which is initially a bit too large to fit in the porphyrin hole and so sits slightly outside the ring and towards the histidine ligand – to drop more or less into place in the ring in oxyhemoglobin. The consequent tug on the histidine connecting the iron(II) to the porphyrin presumably contributes to the cooperative oxygenation effect seen in hemoglobin. It is worth mentioning that in a recent computational study, the different electronic states of dioxygen and heme iron have been calculated and the lowest energy route for the binding process has been identified [4].

Oxyhemoglobin and oxymyoglobin do not give any EPR signal, which could be attributed to Fe^{2+}(l.s.; $S = 0$)–O_2 or to Fe^{3+}(l.s.; $S = \frac{1}{2}$)–O_2^-($S = \frac{1}{2}$) with antiferromagnetic interaction. There is scientific evidence in support of both theories: the most likely scenario is that since O_2 is a π-acid ligand there will be an intermediate situation between Fe(II)–O_2 and Fe(III)–O_2^-. For comparison, a peak at 1103 cm^{-1} has been reported for oxymyoglobin in Raman spectroscopy, quite close to the value of 1145 cm^{-1} reported for the O_2^- anion.

17.3.1.2 Non-heme Carriers: Hemerythrin and Hemocyanin (Figure 17.5)

The non-heme O_2 carriers, hemerythrin (Hr) and hemocyanin (Hc), have only been found in invertebrates. Hr and Hc possess binuclear iron and copper centers, respectively. The uptake of O_2 by either Hr or Hc is accompanied by two-electron rearrangement of the dinuclear center to produce bound peroxide. This rearrangement is reversible because when the O_2 is released the electron density (charge) is redistributed, leading to the starting form.

Deoxy-Hr contains two high-spin Fe(II) ions that are weakly AF coupled through an OH^- ion and two carboxylates from glutamate and aspartate, $[Fe(\mu\text{-}O)(\mu\text{-}COO)_2]$ core. The terminal N-ligands are five histidine groups (3 + 2). Molecular O_2 binds to the pentacoordinated iron(II) at the vacant position site. Electrons are then transferred from the Fe(II) ion to generate a binuclear Fe(III) site with bound peroxido, which accepts the proton from the OH bridge to facilitate a μ-oxido bridge formation (Figure 17.5). Thus, the oxy-Hr active site is best described as a (μ-oxido)(μ-carboxylato)diiron(III) complex with a terminal (hydrogen)peroxido ligand. The presence of the Fe–O–Fe motif is consistent with the strong AF coupling. This protein shows a very small cooperative effect, which depends on the

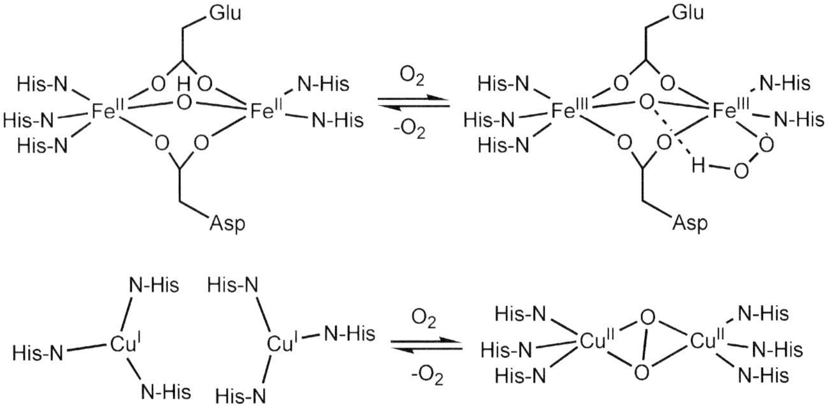

Figure 17.5 Scheme of hemerythrin and hemocyanin O_2 carriers.

interactions between the subunits. The iron atoms are barely modified by the oxidation and the shift from the OH^- bridge to the O^{2-} bridge, without any change in the spin state (high spin before and after oxidation).

In the structure of hemocyanin, each copper ion is coordinated by three histidine residues. The peroxido group is formed after uptake of O_2 by the dinuclear copper protein, passing from Cu(I) to Cu(II). This feature gives the "blue blood" common to arthropods and molluscs; the intense blue color is due to a charge transfer band. Once oxidized, the two Cu(II) centers are strongly AF coupled, yielding a diamagnetic complex (no EPR signal). The strong magnetic interaction suggests that the peroxido group is coordinated via its two oxygen atoms to the two copper ions. This feature has been corroborated by Raman spectra as well as by X-ray crystal determination (Figure 17.5). In this case, the cooperative effect is important because, prior to oxidation, the two copper atoms are far away, whereas after the oxidation process there is a peroxido bridge, which leads to a noticeable decrease in the Cu–Cu length. This induces a strong displacement of the amino acids linked to them, and greatly affects the subunit interaction, thus creating important cooperative effects.

17.3.2
Activation of O_2

There are two kinds of enzymes that insert oxygen atoms from molecular O_2 to the inert substrates: *monooxygenases* (insertion of one oxygen atom to the substrate and reduction the other oxygen atom to water) and *dioxygenases* (insertion of two oxygen atoms). It is interesting to note that monooxygenases have a similar core to the O_2-carriers. The new species formed are very powerful thermodynamic oxidizing agents, much more so than O_2 itself.

17.3.2.1 **Hemoglobin-like Systems**
The most widely studied enzyme is the cytochrome P_{450}, which catalyzes the reaction:

$$R_3C-H + O_2 \rightarrow R_3C-OH$$

The name of this enzyme comes from the characteristic absorption band at 450 nm of the CO adduct. This enzyme presents an axial position occupied by S-Cys$^-$ and a vacant position that enables coordination and activation of the O_2 molecule. The iron is low spin. The redox potential of the Fe^{3+}/Fe^{2+} system is negative, which favors oxidation from Fe^{2+} to Fe^{3+}:

$$Porphyrin-Fe(II) + O_2 \rightarrow Porphyrin-Fe(III)-O_2^-$$

A new supply of electrons (from another electron-transport protein) is required to lead to the formation of an intermediate $Fe^{III}-O_2^{2-}$ or $Fe^{III}-O_2H^-$. Very recently, resonance Raman has allowed the detection of the (hydrogen)peroxido intermediate, by means of the direct observation of the structure-sensitive internal vibrational modes of the (Fe–OOH) fragment [5]. The remaining two electrons required to break the O–O (peroxido) bond come from the complex itself, the presence of an intermediate in the form $Fe^V=O$ being detected. This intermediate can be stabilized in various ways: (i) by the electron donor (cys–S$^-$) ligand in the axial position and (ii) by the porphyrin ligand which has two negative charges. A radical mechanism has been proposed based on the possible resonant species:

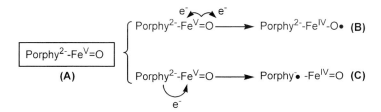

The **B** form would explain the breaking of the C–H bond:

$$Fe^{IV}-O^{\bullet} + C-H \rightarrow Fe^{IV}-O-H + C^{\bullet} \rightarrow Fe^{3+} + C-O-H$$

17.3.2.2 Hemerythrin-like Systems

These non-heme iron proteins typically bind iron using only histidine and carboxylato ligands. Soluble methane monooxygenase (sMMO) is one of the best known examples. This enzyme, a multiproteic complex, is employed by microorganisms which use CH_4 as a source of carbon and energy, the latter being provided through the first step of methane oxidation: CH_3OH. sMMO has been found to be capable of oxidizing a wide range of carbon-containing substrates. These include halogenated aliphatic compounds such as trichloroethylene, a significant groundwater pollutant.

The active center is similar to that of hemerythrin (O_2 transport system; see above). The oxygen activation is thus carried out by dinuclear complexes of Fe(II)–Fe(II) that coordinate the O_2, producing an irreversible redox process:

$$Fe(II)-Fe(II) + O_2 \rightarrow Fe(III)-Fe(III)-OOH^-$$

The reduced form is colorless as it is a Fe(II) complex without colored ligands, whereas the oxidized form, Fe(III)–peroxido, is colored due to a Fe(III)–OOH$^-$ CT band.

The proposed mechanism assumes that O_2 binds to the reduced diiron(II) form of the enzyme, yielding a diiron(III) peroxido adduct. The [O–O] bond is broken through oxidation from Fe(III) to Fe(IV), and Fe^{IV}=O species are possibly formed, similar to what occurs with the intermediate of the cytochrome P_{450} reaction. The reaction of these high-valent centers with substrate yields a diiron(III) state that can be reduced by the cellular machinery to the diiron(II) state, thus completing the catalytic cycle. Many structural models have already been reported [1].

17.3.2.3 Hemocyanin-like Systems

Tyrosinase (monophenol monooxigenase) is one of the most important *copper monooxygenase enzymes*, and is ubiquitous in bacteria, fungi, plants and animals. Tyrosinase is an enzyme that catalyzes the oxidation of phenols (such as tyrosine) to 1,2-dihydroxy substituted aromatic systems (catechols). This involves, therefore, the insertion of one oxygen atom (monooxygenase enzymes). Tyrosinase can also oxidize catechols to *o*-quinones. This reaction is not only essential for the synthesis of melanine and catecholamine neurotransmitters (dopamine, noradrenaline, adrenaline) (Figure 17.6A) but also for the microbial degradation of aromatic compounds in the environment.

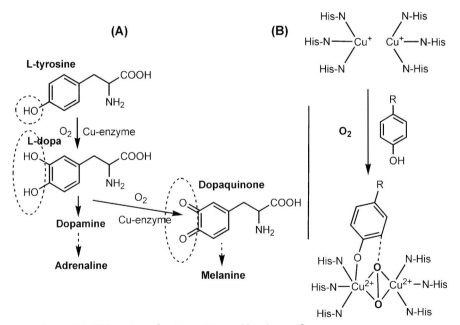

Figure 17.6 (A) Tyrosinase functions, (B) possible scheme of the reaction in the Cu$^+$–Cu$^+$ center.

Tyrosinases have been isolated from a wide variety of plant, animal and fungi species and studied. It has been suggested that there is no common tyrosinase protein structure occurring across all species. However, all tyrosinases have in common a dinuclear complex of copper(I) coordinated to three histidines with, probably, a labile water molecule on each copper. When it reacts with O_2 a $[Cu_2^{II}(\mu\text{-}O_2)]^{2-}$ species is formed, as in the hemocyanin O_2-carrier system. The labile site occupied by water is substituted by the phenol and the C(aromatic)–H bond is thus close to the peroxido bridge, leading to the rearrangement which converts C–H into C–OH (Figure 17.6B). Structurally and functionally similar to tyrosinase is the catechol-oxidase which converts catechols to the corresponding o-quinones. Tyrosinases and catechol oxidases are collectively termed polyphenol oxidases.

17.3.3
Other Fe-Heme Enzymes: Peroxidases and Catalases

Hydrogen peroxide can be produced as an intermediate in oxygen reduction during respiration; only about 80% of the dioxygen taken up by breathing is *completely* reduced. Cellular detoxification mechanisms include enzymes such as peroxidases and catalases. Peroxidases use H_2O_2 to oxidize different kinds of substrates, S (S = cytochrome c, Mn^{2+}, Cl^-, RNH_2, etc.). There are numerous compounds such as fatty acids, amines, phenols and toxins that can serve as substrates for peroxidases. Catalases disproportionate the hydrogen peroxide into H_2O and O_2, the first step of this reaction being the reduction of the H_2O_2, that is, the same as in the case of peroxidases.

$$\text{Peroxidases: } H_2O_2 + S_{(red)} \rightarrow 2H_2O + S_{(ox)}$$

$$\text{Catalases: } H_2O_2 + H_2O_2 \rightarrow 2H_2O + O_2$$

In the catalase case, therefore, the $S_{(red)}$ is the H_2O_2 and $S_{(ox)}$ the O_2. In most peroxidases and catalases the reaction center is a Fe-heme group, closely related to cytochrome P_{450}: the iron center shows a vacant coordination site in the native enzyme that seems an obvious place for the interaction between the enzyme and H_2O_2 (or OOH^-) to take place.

The most important difference is that in cytochrome P_{450} the O_2 is coordinated with Fe(II), whereas in peroxidases and catalases the ligand that coordinates with the iron is the peroxido (O_2^{2-}). Therefore, the oxidation state of the iron must be Fe(III), a similar intermediate (porphyrin–Fe^{3+}–OOH^-) being formed in both enzymes. As in the case of the cytochrome P_{450} (Section 17.3.2.1), two electrons from the complex itself are required to break the O–O bond of the peroxido group, leading to the formation of a $Fe^{IV}=O$ intermediate, in accordance with EXAFS measurements and Raman resonance. The catalytic cycle as it is presently believed to occur is shown schematically in Figure 17.7.

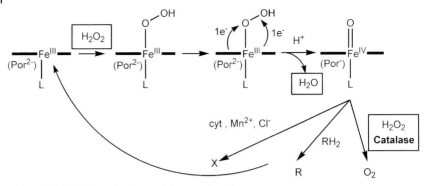

Figure 17.7 Catalytic cycles in peroxidases and catalases.

17.4
Electron Transfer

A large series of biomolecules was designed by nature to regulate electron transfer (ET) in all living systems. Among these are redox-active compounds such as flavins, quinones and nicotinamide cofactors, and a series of metalloproteins such as ferredoxins, cytochromes and blue copper proteins, in which a metal ion enables specific redox functions of the protein. Theoretical considerations are the same as those reported in Chapter 12 for electron transfer in general. For example, long-range electron tunnelling in cytochrome c and other proteins is a viable mechanism for biological ET [1].

17.4.1
Cytochromes

There are a number of different cytochromes according to the groups present in the porphyrin ring, and within each group the ligands in the axial position may vary. For these reasons, cytochromes with a wide range of E^0 can be found in nature. As mentioned previously, cytochromes participate in electron transport as the hexacoordinated Fe limits their role to electron transfer. Cytochrome c, for example, exists either in a diamagnetic Fe^{2+} form (the reduced state) or in a low-spin Fe^{3+} form (the oxidized state). In both states, the porphyrin-bound iron center is hexacoordinated, in a pseudooctahedral environment, via axial binding of histidine and methionine residues. The one-electron redox process is accompanied by only a small restructuring of the protein structures in the vicinity of the active site. Given that the values of E^0 are small and changeable, the processes are favorable. Cytochromes present different redox potentials according to the group in the axial position: between −0.1 and −0.4 V if there are two *N*-histidine groups; and between −0.1 and +0.4 V if there is one *N*-histidine ligand and another of *S*-methionine. The presence of a soft ligand (*S*-methionine) favours reduction and therefore yields greater reduction potentials.

17.4.2
Fe—S Proteins

Most iron–sulfur centers in proteins are involved in electron transfer at a typically negative redox potential. They are believed to be among the oldest metal-containing proteins, and are widely distributed throughout anaerobic, aerobic, and photosynthetic bacteria, as well as in mammals, plants and fungi. Approximately 1% of the iron content in mammals is present in the form of Fe–S proteins. Many features suggest that they might have played an important role very early in evolution, i.e. in the absence of free O_2.

A characteristic feature of iron–sulfur proteins is the coordination of iron with protein-bound cysteinate sulfur (cys-S^-) and, in polynuclear Fe–S centers, with "inorganic" acid-labile sulfido (S^{2-}) ligand. The coordination of the cys-S^- group produces a red color due to charge transfer bands. Unlike systems with a porphyrin ligand, the Fe ion in these Fe–S proteins is only found in its "normal" oxidation states: 2+ and 3+. The nuclearity may vary, although in all cases the Fe has a pseudotetrahedral environment (high-spin). In this protein family only one electron is exchanged, regardless of the number of metal centers; they present an oxidized and a reduced form. The Fe–S centers can be divided according to their nuclearity:

1. Rubredoxins are small redox proteins which occur in certain bacteria and contain just one iron center, tetrahedrally surrounded by the sulfurs of four cysteine ligands, $[Fe(S-Cys)_4]^{-/2-}$ (Figure 17.8A). The oxidized species (Fe^{3+}) is $(e)^2(t_2)^3$ and shows the EPR spectrum corresponding to a $S = 5/2$, with a component $g > 2.0$. The reduced form (Fe^{2+}) presents no EPR spectrum in standard conditions ($S = 2$). The value of E^0 is –0.06 V.

2. [2Fe–2S] centers (ferredoxins) (Figure 17.8B). The most abundant have $E^0 < 0$. The schematic formula is $[Fe_2(S-Cys)_4(\mu-S)_2]^{n-}$. The bridging sulfido ligands can readily be displaced by acid media and are thus called labile sulfides. The [2Fe–2S] center is particularly common in chloroplasts (spinach leaves, for example).

   Although the oxidized form gives no signal in EPR spectroscopy (strong AF coupling), the reduced form does. It can thus be deduced that electron transport occurs via the reaction Fe(III)—Fe(III)$\leftrightarrow$Fe(III)—Fe(II). EPR and Mössbauer spectroscopy suggest a localized description with fixed valences Fe(II) and Fe(III) for reduced [2Fe–2S] center in proteins, as in model complexes.

   Within this group of proteins there are some systems that contain centers with unusual spectroscopic properties and relatively high redox potentials ($E^0 > 0$). These [2Fe–2S] systems are termed "Rieske" centers. The Rieske proteins contain two markedly different iron centers and present fewer sulfido ligands coordinated with Fe. It has been suggested that one of the Fe has two Cys-S^- and the other two N–Hist, $[Fe_2(S-Cys)_2(N-His)_2(\mu-S)_2]^{n-}$. As occurs in the previous case, only the reduced form gives an EPR signal. The fact that they have an E^0 of different sign is due to the difference in the terminal ligands. The coordinated N–His, in place of the (softer) Cys–S⁻, may stabilize the oxidized form.

3. [3Fe–4S] centers, such as $[Fe_3(\mu\text{-}S)_4(S\text{–}Cys)_3]^{n-}$ (Figure 17.8C). The reduction potential is negative: $E^0 < 0$. The oxidized form ($3Fe^{3+}$) gives an EPR signal with a value of $g \approx 2.0$, but the reduced form gives none. The Mössbauer spectrum of the oxidized form presents the signal corresponding to Fe(III), whereas the reduced form gives a signal corresponding to Fe(III) and an intermediate between Fe(III) and Fe(II) (delocalized mixed valence).

4. [4Fe–4S] centers (Figure 17.8D). The most common and stable Fe–S centers are of the [4Fe–4S] type, with a distorted cubane-like structure. The prototype is $[Fe_4(\mu\text{-}S)_4(S\text{–}Cys)_4]^{n-}$. [4Fe–4S] centers participate in many complex biological redox reactions such as photosynthesis, respiration, and N_2 fixation. Two kinds of these proteins are found: with low potential, $E^0 < 0$ and with high potential, $E^0 > 0$. As in previous cases, EPR spectroscopy is highly useful for determining the oxidation states of the iron ions. For the low potential proteins, the reduced form gives an EPR signal with $g \approx 2.0$, whereas the oxidized form shows none. The redox process is 2Fe(III)–2Fe(II) → Fe(III)–3Fe(II). For the high potential iron–sulfur proteins (HiPIP), the oxidized form gives an EPR signal at $g \approx 2.0$, whereas the reduced form gives no signal in the spectrum. The redox process is 3Fe(III)–Fe(II) → 2Fe(III)–2Fe(II). Therefore, in both cases the most stable form is 2Fe(III)–2Fe(II). Indeed:

$$3Fe(III)\ Fe(II) \xrightarrow{\ E^0 > 0\ } 2Fe(III)\ 2Fe(II) \xrightarrow{\ E^0 < 0\ } Fe(III)\ 3Fe(II)$$

The oxidized and reduced forms in both cases give a single signal in the Mössbauer spectrum, which is indicative of delocalized mixed valence (oxidized

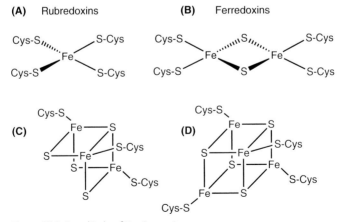

(A) Rubredoxins   **(B)** Ferredoxins

(C)   **(D)**

Figure 17.8 Four kinds of Fe–S proteins.

form: $Fe^{2.5+}$). This implies that a smaller variation in bond length, and electron transfer can thus occur more easily.

17.4.3
Blue Copper Proteins

The blue copper proteins form a series of mononuclear copper compounds that are involved in long-range electron transfer reactions. The redox potentials are between 0.2 and 0.4 V (more positive than the E^0 for $Cu^{2+}/Cu^+(aq)$, 0.153 V). The compounds are called "blue" because of their intense color, about 400 times more intense that than of most copper(II) salts. The blue color is due to a charge-transfer transition (LMCT) band around 620 nm, and not to the d–d transitions. The blue band also persists when the Cu^{2+} is replaced by Ni^{2+} or Co^{2+}, consistent with a low energy ligand-to-metal charge-transfer assignment.

Some of them have appropriate names such as *azurin* (participation in bacterial photosynthesis) or *plastocyanin* (participation in plant photosynthesis). In 1978 the plastocyanin X-ray crystal structure was identified and revealed, in particular, that the copper is coordinated by a cysteine thiol group, a methionine thioether group, and two histidine imidazole groups [6].

An unusual aspect of their EPR spectrum is a very small hyperfine coupling $A_{\parallel}$ value ($\sim 60 \times 10^{-4} \, cm^{-1}$), which reflects a reduced interaction of the unpaired electron with the Cu nucleus. This is due to the short Cu^{II}–S distance (~ 2.1 Å), reflecting a strong covalency for the Cu–S bond. The geometries of the copper center in both oxidized and reduced forms are nearly identical, thus minimizing reorganization and facilitating very efficient (fast) electron transfer.

17.5
Electron Transport and Enzyme Activity

Many enzymatic processes are of the redox type and require electron transport systems in addition to the catalytic center. There are two cell processes of this type which take place inside membranes: cell respiration (mitochondrial membrane) and photosynthesis (thylakoidal membrane). Another process that combines electron transport with enzymatic activity is nitrogen fixation by means of bacteria.

17.5.1
Cell Respiration: Cytochrome Oxidase

In the *mitochondrial membrane* there are a series of proteins and molecules that are involved in the process of electron transport: Fe–S proteins, cytochromes and quinones. As they have similar redox potentials, the electron can pass from one center to another until reaching the cytochrome *c*, this being the source of the electrons for cytochrome oxidase, which catalyzes the four-electron reduction of O_2 to H_2O. Cytochrome oxidase couples the energy released from the reduction of

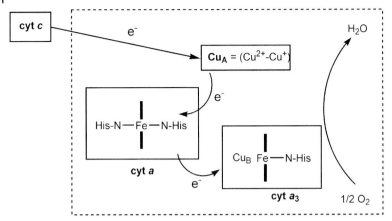

Figure 17.9 Coordination compounds in cytochrome oxidase (see text).

O_2 to proton transport, generating a trans-membrane proton gradient (Figure 17.9).

Cytochrome oxidase is one of the most important, but also one of the most complex, metalloproteins and contains up to thirteen different subunits. In fact, the metal content of each unit has not yet been fully defined. The most recent studies suggest that the main subunit of the monomeric protein contains the following coordination compounds (Figure 17.9):

1. A dinuclear Cu complex, $[Cu_2(\mu\text{-}S_{Cys})_2]$ core (Cu_A), whose oxidized form shows an EPR spectrum with seven lines in the $g_{\parallel}$, due to hyperfine coupling. This finding is indicative of a delocalized mixed valence centre (Cu^+–Cu^{2+});

2. One separated cytochrome (cyt *a*) with low-spin iron(III), two axial histidine ligands, low porphyrin symmetry and a high redox potential;

3. The catalytic center (cyt a_3-Cu_B), comprising a pentacoordinated cytochrome (high spin) and a tricoordinated Cu.

The electrons coming from the cytochrome *c* go to the dinuclear Cu_A, from where they are transferred to cyt *a* and then on to the reaction center (cyt a_3-Cu_B).

When the catalytic center is in the reduced form, porphyrin–Fe(II)–Cu(I) ($S = 2$, EPR silent) can coordinate the O_2 (π-acid ligand) as shown in Figure 17.10. The species $Fe^{3+}\cdots O_2^{2-}\cdots Cu^{2+}$ gives no EPR signal as the magnetic coupling between the metal ions is AF ($S = 2$). The addition of two new electrons from the Cu_A and cyt *a* centers reduces the Cu^{2+} and Fe^{3+}; finally, the addition of two new electrons enables the formation of H_2O from O_2^{2-} and regenerates the initial complex, Fe^{2+}–Cu^+. Unfortunately, no satisfactory full structural details are known.

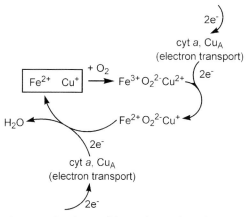

Figure 17.10 Scheme of the catalytic cycle in the cytochrome oxidase.

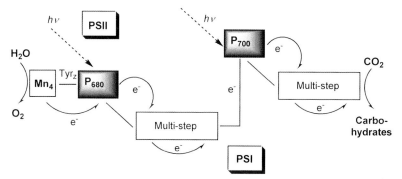

Figure 17.11 Outline of the two photosystems, PSI and PSII.

17.5.2
Photosynthesis

Photosynthesis in the thylakoidal membrane involves two processes, the oxidation of H_2O to O_2 and the reduction of CO_2, to give rise to carbohydrates. These two processes take place in two different photosystems, each one with a different chlorophyll. Photosynthesis runs on visible and near-infrared photons. The light is collected by antenna systems (see Chapter 15), i.e. pigment molecules, such as carotene and porphyrin-derived chromophores. Electronic excitation in the antenna migrates from chromophore to chromophore and ultimately to a reaction center, where it is converted into chemical energy in the form of charge separation. An outline of the process is given in Figure 17.11. It is important to say, from the outset, that the complete details of water oxidation are yet to be fully understood.

The PSI system receives light and, subsequently, an electron is transferred to the neighboring molecule. There must therefore be an electron transport system between PSI and CO_2: this involves Fe–S proteins, among other groups.

A key enzyme in green plant photosynthesis is photosystem II (PSII). This is the only photosynthetic reaction center that can use water as an electron source, and it thereby provides plants (and, indirectly, the entire biosphere) with an unlimited source of electrons. PSII carries out the general reactions: light absorption, energy transfer, charge separation and charge stabilization.

The key player in water oxidation is a triad composed of a multimer of chlorophylls (named P_{680}), a redox active amino acid (named tyrosine-z), and a Mn-cluster composed of four Mn ions (termed oxygen evolving center: OEC). The Mn cluster can be in several oxidation states, named S-states; from S_0 (lowest oxidation state) to S_4 (highest oxidation state). The latter form (S_4) is capable of oxidizing the water molecule.

Upon illumination, P_{680} in PSII is excited. It becomes strongly reductant and transfers an electron to the acceptor systems (Figure 17.11). The oxidized primary donor (P_{680}^+ cation) is one of nature's most oxidizing species and reaches a potential of +1.2 V. P_{680}^+ is rapidly reduced by tyrosine-z, which is located about 12 Å away. The main function of Tyr_z is to act as a very fast electron donor to P_{680}^+, according to the S-state in which the cluster of four manganese ions (OEC) is present. The reaction which takes place is: $Tyr–O–H \rightarrow Tyr–O^\bullet + H^+ + e^-$ (EPR active). For the regeneration of $Tyr–O^\bullet$ it must capture electrons coming from the Mn cluster. The multi-step electron transfer systems (Figure 17.11) contain quinones, cytochromes and blue copper proteins (plastocyanine).

17.5.3
Nitrogen Fixation: Nitrogenase [7]

Inert atmospheric nitrogen (N_2) is transformed to ammonium ions by microorganisms. The [Fe–S] clusters of nitrogenase have proved a seductive enigma for more than a generation of chemists, given the ease with which they catalyze the reduction of dinitrogen to ammonia under conditions milder than any known in the laboratory. Nitrogenases catalyze the reduction of N_2 to NH_3, according to the following reaction:

$$N_2 + 6H^+ + 6e^- \rightarrow 2NH_3$$

The nitrogenases exist in three forms: molybdenum-nitrogenase, vanadium-nitrogenase and iron-nitrogenase. Due to the thermodynamic stability of the N_2 molecule, its reduction requires a large amount of energy and six electrons per N_2 at a physiologically quite negative potential of less than −0.3 V. Fortunately, N_2 is a π-acid ligand: the N_2 can receive electronic density from the metal ions in its π^* MOs, weakening the triple bond N–N. Thus, N_2 complexation and fixation requires d_π electron-rich metal centers. The low redox potential and the necessarily high reactivity of nitrogenase enzymes further require the absence of competing – i.e.

related but better coordinating–molecules such as O_2 (there are mechanisms that include iron-containing proteins which serve as O_2 sensors).

Furthermore, nitrogenases possess an intrinsic hydrogenase activity which leads to the production of H_2 during the biological N_2 fixation reaction. The molecule of CO, a potent inhibitor of N_2 reduction, does not inhibit the production of H_2.

The most widely studied molybdenum-dependent form of nitrogenase is a complex system which contains a [4Fe–4S] cluster, and two very special systems, called the P clusters and the FeMo cofactors (Figure 17.12). The two latter clusters are compact assemblies consisting of cuboidal halves fused through a central atom with a μ_6 bridging modality that is found nowhere else in biology.

The P cluster has a structure in which two Fe_4S_3 units are bridged by an uncommon μ_6-S atom and by two μ-S-Cys residues. All iron sites possess distorted tetrahedral FeS_4 coordination. The FeMo-co is built from Fe_4S_3 and $MoFe_3S_3$ fragments bridged by three μ-S atoms and one μ_6-X atom, whose most likely identity is N [8]. The seven iron sites have distorted tetrahedral geometry, while the Mo atom is six-coordinated with distorted octahedral MoS_3NO_2 coordination. The Mo is located at one corner of this latter cluster and seems to be well away from the site of likely action in the nitrogen fixation process. Six of the seven iron atoms are coordinated with three S ligands and to a central N atom through a weak bond. This is a very unusual situation which must surely be connected with the coordination of N_2 as the first step of the fixation process.

As regards the actual mechanism underlying the transformation of N_2 into NH_3, knowledge remains at the level of hypotheses. Inside the protein the polynuclear P cluster may be reduced to the all-Fe^{2+} form and, presumably, regulate the low-potential transfer of electrons to the FeMo cofactor. In contrast to the electron-transport Fe–S proteins, which can only exchange one electron, the P cluster may pass through various oxidation states and accumulate up to eight electrons.

The currently accepted thesis suggests that the P cluster accepts electrons from the [4Fe–4S] cluster and transfers them to the FeMo-co, which acts as the site for N_2 binding, activation and subsequent reduction to NH_3 [7]. The coordinative saturation of the Mo center and the existence of V-nitrogenase or Fe-nitrogenase, where V or Fe replace the Mo ion, indicate that the Fe ion of the central cavity is better than Mo for the coordination of N_2. Recent theoretical calculations assume

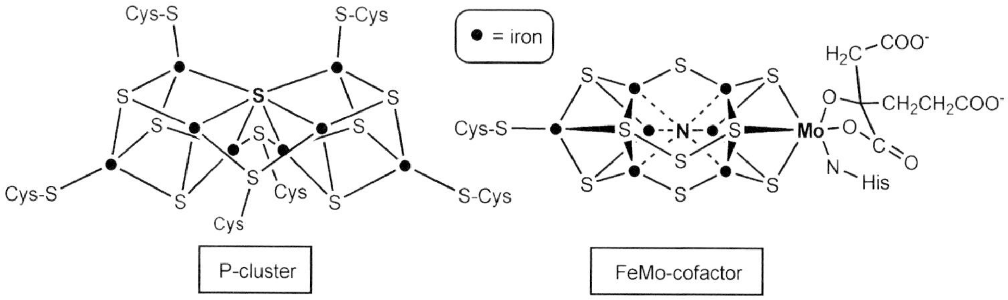

Figure 17.12 P cluster and the "FeMo cofactor" in nitrogenase.

than the central N-atom is exchangeable nitrogen, which thus participates in the process of NH_3 formation [9]. Vanadium-nitrogenases – rather than Mo-nitrogenases – have been found in free-living micro-organisms, and at lower temperatures they are more efficient than their Mo-counterparts. Another difference is the production of small amounts of hydrazine.

17.6
Medicinal Coordination Chemistry

The field of medicinal inorganic (coordination) chemistry can be roughly divided into two main streams of applications: therapeutic and diagnostic. The common principles that underlie new developments in medicinal inorganic chemistry derive from the understanding of both the coordination chemistry and the metabolism of metal ions.

17.6.1
Therapeutic Agents: Some Examples

One of the seminal developments in the understanding of metal ions in human health and disease was the success of applying coordination chemistry to metal-overload disorders, such as Wilson's disease (copper overload) and thalassaemia (iron overload). Wilson's disease, a formerly lethal genetic disorder, is characterized by copper accumulation in tissue that gives rise to physiological abnormalities. A first successful treatment of the disorder was reported in 1948 when British antiLewisite (BAL) was proposed as a chelating agent to remove excess copper. The formula of BAL is $CH_2SH–CHSH–CH_2OH$. The treatment of iron overload works on a similar principle: chelating agents with a high affinity for iron were first developed during the 1950s.

Mercury chelation therapy has the same aim. Mercury compounds are known to be highly toxic; indeed, some of them, such as dimethylmercury $[Hg(CH_3)_2]$ or methylmercury cation $[Hg(CH_3)]^+$, are toxic to such an extent that they are considered "supertoxic". However, relatively little is known about the molecular mechanism of their toxicity. Dimethylmercury, being apolar, readily crosses cell membranes, while the methylmercury cation, which has a polar part (enabling it to move closer) and an apolar part, is even more able to permeate cell membranes and easily enters the nucleus. Mercury has a high affinity for both thiols (the old name for thiols is "mercaptan", which is derived from "mercury-capture") and selenols.

Mercury poisoning has been treated using chelation therapy that generally uses one or both of two drugs: *meso*-dimercaptosuccinic acid (DMSA) and dimercapto-propanesulfonic acid (DMPS). While it is clear that treatment with these agents can remove mercury from the body, recent studies have found that they do not act as true chelating agents, being equivalent to monothiols. Instead, they form complexes with two mercury ions.

A very different – and more important – case is that of cisplatin targeted toxicity. The serendipitous discovery of the anti-tumor activity of *cis*-[PtCl$_2$(NH$_3$)$_2$] during the mid 1960s revolutionized medicinal inorganic chemistry. Before the introduction of cisplatin treatment into cancer chemotherapy, testicular cancer had a >90% fatality rate, but today it is a mostly curable (<5% mortality) disease. The elucidation of the mechanism of action of platinum-based chemotherapeutics, and the development of ways to mitigate the toxic side effects, has been the focus of intense research efforts. Carboplatin, in which two chlorides are replaced by cyclobutane-1,1-dicarboxylato, was shown to be less toxic and was introduced into clinical use just ten years after cisplatin. Differences in toxicity appear to derive mostly from variation in the kinetics and thermodynamic properties of the complexes, with carboplatin more than 100-fold more resistant to aquation, and with a much longer half-life of decomposition ($t_{1/2}$ = 30 h, vs. 2 h for cisplatin).

The anti-tumor activity of *cis*-[PtCl$_2$(NH$_3$)$_2$] is based on its ability to coordinate with DNA. Since the concentration of Cl$^-$ ions is much higher outside the cell than inside it, the *cis*-[PtCl$_2$(NH$_3$)$_2$] is not easily aquated in the exterior and, by being neutral, readily passes through the cell membrane. Inside the cell it is slowly hydrolyzed, thus liberating the two Cl$^-$ ligands and yielding the intermediate *cis*-[Pt(H$_2$O)$_2$(NH$_3$)$_2$]$^{2+}$ which, due to its positive charge, can more easily approach the DNA chain (which has exterior polar phosphate groups). The presence of the NH$_3$ ligands enables the formation of hydrogen bonds and the Pt(II) ion can be coordinated with two neighbouring nitrogenated bases of the DNA chain. These nitrogenated bases have N-donor atoms with an affinity for Pt(II). The coordination of the Pt(II) compound with DNA gives rise to a certain flexion (or distortion) of the DNA chain of the tumor cell, and this change in structure seems to be responsible for the anti-tumor activity of this type of complex.

17.6.2
Diagnostic Agents: Some Examples

The most widely used diagnostic agent is the gadolinium-based MRI (magnetic resonance imaging) contrast enhancement. MRI is a very powerful non-invasive diagnostic technique characterized by excellent three-dimensional visualization. An MRI image is based on the relative intensity of the signal from water protons in tissues of higher affinity for the contrast agent compared to background signals from surrounding tissues. As signal intensity (in tumors, for example) is increased with shorter relaxation times, image enhancement is achieved for as long as the contrast agent stays in the particular tissue. The optimal properties of MRI contrast agents are: maximum relaxivity, such as paramagnetic molecular species; minimum toxicity; minimum osmolality (neutral better than charged); rapid excretion (renal faster) and high tissue specificity.

Current MRI contrast agents are in the form of either paramagnetic complexes or magnetic nanoparticles. Paramagnetic complexes are usually Gd^{3+} or Mn^{2+}. The first report of a Gd(III) complex for MRI contrast enhancement was with an

N-methylglucosamide of [Gd^{III}(DTPA)(H$_2$O)]$^{2-}$ (Magnevist) (DTPA = diethylenetri-aminepentaacetic acid) (first published in 1973; first used in 1988) (Figure 17.13). Contrast agent-enhanced MRI now accounts for more than one-third of all clinical diagnostic scans. In all of these the presence of H$_2$O labile molecules is very important. Paramagnetic Gd^{3+} is particularly well-suited to enhance the proton magnetic effect as it has seven unpaired electrons in half-filled 4f orbitals, and a symmetric electronic ground state (^{8}S). Appropriate ligands, such as polyamino-carboxylate analogs of DOTA, tetrakis(carboxymethyl-1,4,7,10-tetraazacyclododec-ane, (Figure 17.13) can protect against the inherent toxicity of inorganic Gd^{3+} by providing thermodynamic stability and neutral charge, and also by contributing to more specific tissue targeting and enhancing intact renal excretion. Thus, *in vivo* tolerance and predictability can be ensured through appropriate ligand choice.

Manganese-enhanced MRI which uses Mn^{2+} ion as a contrast agent is applicable to animals only, owing to the toxicity of Mn^{2+} when it accumulates excessively in tissues and despite the increasing appreciation of this technique in neuroscience research. Very recently, MnO nanoparticles without this inconvenience have been reported and used [10]. Furthermore, MnO nanoparticles functionalized by recep-tor antibodies selectively target cell surfaces of breast cancer [10].

MRI for sensing tissue pH (an important goal in the diagnosis and aetiology of kidney disease and cancer) has been developed. In these cases, the relaxivity of the

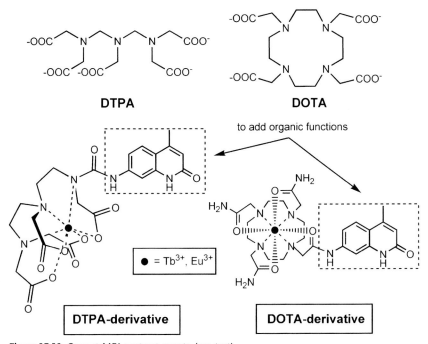

Figure 17.13 Current MRI contrast agents (see text).

complex should respond to changes in pH over the extremes of tissue pH (pH ≈ 5–8) and, ideally, the relaxivity of the complex should not be sensitive to endogenous metal ions, anions or proteins. Some Gd-DOTA derivatives have been proven to be nearly ideal for sensing tissue pH [11].

In Chapter 15 a number of lanthanide systems were discussed in terms of their role as immunoassay probes. Hence, novel cocktails of heterometallic Tb^{3+}/Eu^{3+} complexes have very recently been designed for ratiometric optical probes [11]. DTPA and DOTA derivative ligands (Figure 17.13) are able to sensitize both Tb^{3+} and Eu^{3+} emission, and serve as a suitable scaffold for the design of antenna chromophores, which could readily sensitize a Tb^{3+}/Eu^{3+} cocktail. Chemical derivation of these ligands (by means of adding ester, amide and carboxylic acid groups) has been shown to alter the degree to which they sensitize Tb^{3+} and Eu^{3+}, so it should be possible to design probes that exhibit ratiometric luminescent responses to chemical/enzymatic transformations. Satisfactory results have been obtained, for example, with enzymatic probes (esterase activity) [12].

References

1 Kraatz, H-B., Metzler-Nolte, N. (Eds.) (2006) *Concepts and Models in Bioinorganic Chemistry*, Wiley-VCH, Weinheim.

2 Fenton, D.E. (1995) *Biocoordination Chemistry*, Oxford Chemistry Primers, Oxford.

3 Kaim, W., Schwederski, B. (1996) *Bioinorganic Chemistry: Inorganic Elements in the Chemistry of the Life. An Introduction and Guide*, Wiley, Chichester.

4 Ribas-Ariño, J., Novoa, J.J. The mechanism for the reversible oxygen addition to heme. A theoretical CASPT2 study, *Chem. Comm.* 2007, *30*, 3160.

5 Mak, P.J., Denisov, I.G., Victoria, D., Makris, T.M., Deng, T., Sligar, S.G., Kincaid, J.R. Resonance Raman Detection of the Hydroperoxo Intermediate in the Cytochrome P450 Enzymatic Cycle, *J. Am. Chem. Soc.* 2007, *129*, 6382.

6 Colman, P.M., Freeman, H.C., Guss, J.M., Murata, M., Norris, V.A., Ramshaw, J.A.M., Venkatappa, M.P. X-ray crystal structure analysis of plastocyanin at 2.7 Å resolution, *Nature*, 1978, *272*, 319.

7 Lee, S.C., Holm, R.H. The Clusters of Nitrogenase: Synthetic Methodology in the Construction of Weak-Field Clusters, *Chem. Rev.* 2004, *104*, 1135.

8 Einsle, O., Teczan, F.A., Andrade, S.L.A., Schmid, B., Yoshida, M., Howard, J.B., Rees, D.C. Nitrogenase MoFe-Protein at 1.16 Å Resolution: A Central Ligand in the FeMo-Cofactor, *Science* 2002, *297*, 1696.

9 Huniar, U., Ahlrichs, R., Coucouvanis, D. Density Functional Theory Calculations and Exploration of a Possible Mechanism of N_2 Reduction by Nitrogenase, *J. Am. Chem. Soc.* 2004, *126*, 2588.

10 Na, H.B., Lee, J.H., An, K., Park, Y.I., Park, M., Lee, I.S., Nam, D-H., Kim, S.T., Kim, S-H., Kim, S-W., Lim, K-H., Kim, K-S., Kim, S-O., Hyeon, T. Development of a T_1 Contrast Agent for Magnetic Resonance Imaging Using MnO Nanoparticles, *Angew. Chem. Int. Ed.* 2007, *46*, 5397.

11 Kalman, F.K., Woods, M., Caravan, P., Jurek, P., Spiller, M., Tircso, G., Kiraly, R., Brucher, E., Sherry, A.D. Potentiometric and relaxometric properties of a gadolinium-based MRI contrast agent for sensing tissue pH, *Inorg. Chem.* 2007, *46*, 5260.

12 Tremblay, M. S., Halim, M., Sames, D. Cocktails of Tb^{3+} and Eu^{3+} Complexes: A General Platform for the Design of Ratiometric Optical Probes, *J. Am. Chem. Soc.* 2007, *129*, 7570.

Appendix 1: Tanabe–Sugano Diagrams (Chapter 8)

d^2 Tanabe-Sugano Diagram

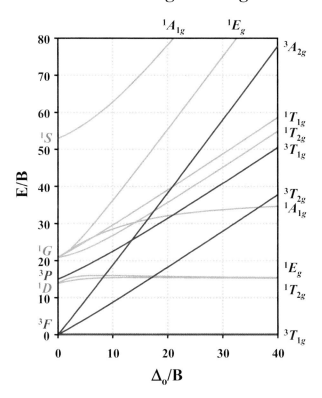

Coordination Chemistry. Joan Ribas Gispert.
Copyright © 2008 WILEY-VCH Verlag GmbH & Co. KGaA, Weinheim
ISBN: 978-3-527-31802-5

d^3 Tanabe-Sugano Diagram

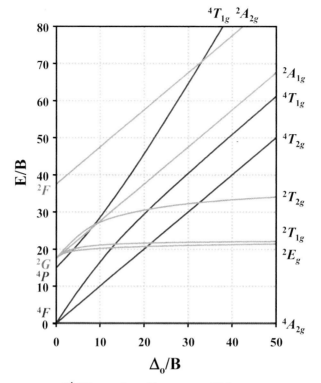

d^4 Tanabe-Sugano Diagram

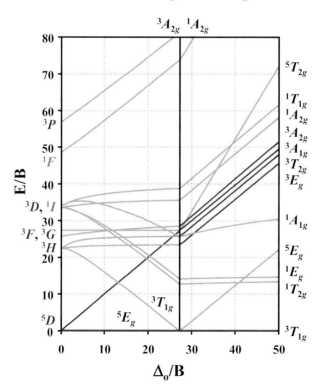

d^5 Tanabe-Sugano Diagram

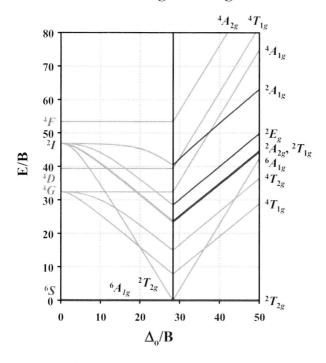

d^6 Tanabe-Sugano Diagram

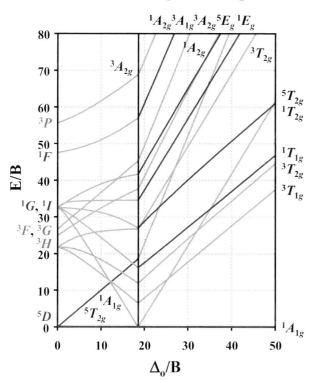

d^7 Tanabe-Sugano Diagram

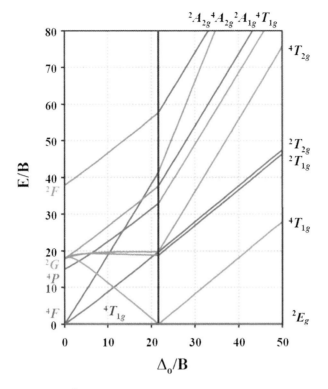

d^8 Tanabe-Sugano Diagram

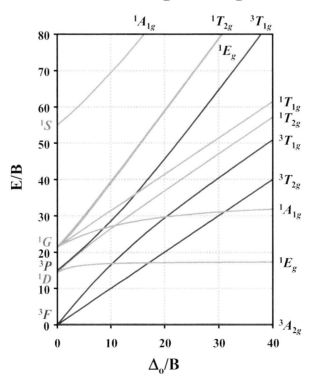

Appendix 2: Definitions and Units in Molecular Magnetism (Chapter 10)

A survey of the recent literature reveals certain confusion regarding the units used for physical magnitudes in the field of molecular magnetism when compared with the units used in physics for the same magnitudes. Indeed, there are two main systems used in magnetochemistry for electromagnetic units, which can be termed, "rational" (the SI system) and "irrational or unrationalized" (the CGS-EMU system) respectively [1]. This poses some difficulties, which prompted the publication in 2005 of the "Practical Guide to Measurement and Interpretation of Magnetic Properties (IUPAC Technical Report) [1]. The authors of this report suggested '*the use of quantities that lead to values independent of the two systems*'", such as the effective Bohr magneton number μ_{eff}, in an attempt to circumvent these problems.

Following the recommendations in the above mentioned guide [1], χ (molar susceptibility) may be converted from the first system (SI) into the second (CGS-EMU) using only the factor 4π, with no dimensions involved, i.e. $\chi = 4\pi\chi^{(ir)}$. Similarly, the magnetic field strengths are related as follows; $H^{(ir)} = 4\pi H$, whereas the units are related by Oersted $= 10^3/4\pi\,A\,m^{-1}$. Likewise, to avoid confusion, in the IUPAC Technical Report [1] it is recommended that one uses, for example, in graphs $B_0 = \mu_0 H$ (loosely called "magnetic field") with conversion factor $10^{-4}\,T/G$ (T = Tessla, G = Gauss; μ_0 being the permeability of vacuum).

What is actually used in papers from the area of molecular magnetism? Since the early times of the discipline, the units used have been almost exclusively those of the CGS-EMU system. In relation to this, Olivier Kahn states in his book [2], "*The SI is the legal, but legality is not science. Indeed, this system is particularly inappropriate in molecular magnetism …*"

With regard to all the above we should clarify a few aspects for the purposes of this book. Strictly speaking, the unit of magnetic field strength in the CGS-emu system is the Oersted. However, as most authors in this field, we will express H in Gauss, which is the unit of the magnetic induction (B) created by a magnetic field H. In the "unrationalized" CGS-EMU system, B and H are "taken as being dimensionally equivalent". In general, and except for cases where extreme precision is needed, B and H may be taken as equal since the value of the magnetic permeability *of air* is nearly unity.

Coordination Chemistry. Joan Ribas Gispert
Copyright © 2008 WILEY-VCH Verlag GmbH & Co. KGaA, Weinheim
ISBN: 978-3-527-31802-5

In order to illustrate the current state of things, the literature on molecular magnetism appearing in the most important chemistry journals has been scanned from January to September of 2007, for the magnetochemical unit system used (see Table below). The journals examined were Angewandte Chemie; Chemistry: a European Journal; Chemical Communications; Dalton Transactions; European Journal of Inorganic Chemistry; Inorganic Chemistry, and Journal of the American Chemical Society. These seven are the journals that publish the best articles in this field.

Units	H (G or Oe)		B, or $B_0 = \mu_0 H$ (T or mT)	Mixed H and $\mu_0 H$	Indefinite	Total
	explicit	*implicit*				
Total	93	48	24	12	6	183
% of total	77.0		13.1	6.6	3.3	100

In 77 % of the papers the CGS-EMU system is used explicitly or implicitly, with Gauss or Oersted used with similar frequency. Only 13.1% of the reports use the SI system or that recommended in 2005 [1]. Interestingly, 6.6% of articles mix the use of H and $\mu_0 H$ in their graphs. In the latter group, all papers are authored by teams composed of both chemists and physicists. In 3.3% of papers it is quite difficult to determine the system used. In this survey, neither reports that are exclusively theoretical nor spin crossover papers have been considered, since these do not mention, or very rarely, the magnitudes H or B.

In conclusion, it is not the aim of this book to instruct on which is the appropriate system used. Instead, as mentioned in the Introduction, my true wish is to stimulate (both in Professors and Students) reading of the specialized literature on the topic. Since most of the quoted papers in Chapters 10 (Molecular Magnetism) and 11 (EPR) use almost exclusively the CGS-EMU system or at most, mix the use of this system with the recommendations provided by IUPAC [1], I have decided to consistently use the CGS-EMU system, even while aware that this is not the "rational" system but the "unrationalized" one.

References

1 S. Hatscher, H. Schilder, H. Lueken, W. Urland, Practical Guide to Measurement and Interpretation of Magnetic Properties (IUPAC Technical Report), *Pure Appl. Chem.* 2005, *77*, 497.

2 O. Kahn (1993) *Molecular Magnetism*, VCH Publishers, New York.

Appendix 3 (Chapter 10)

χ_m formulae for some homodinuclear d^n–d^n complexes

1/2 - 1/2 $\quad f(x) = \dfrac{2\exp(x)}{1 + 3\exp(x)}$

$$\chi_M = \frac{Ng^2\beta^2}{kT} \times f(x)$$

1 - 1 $\quad f(x) = \dfrac{2\exp(x) + 10\exp(3x)}{1 + 3\exp(x) + 5\exp(3x)}$

$$x = J/kT$$

3/2 - 3/2 $\quad f(x) = \dfrac{2\exp(x) + 10\exp(3x) + 28\exp(6x)}{1 + 3\exp(x) + 5\exp(3x) + 7\exp(6x)}$

2 - 2 $\quad f(x) = \dfrac{2\exp(x) + 10\exp(3x) + 28\exp(6x) + 60\exp(10x)}{1 + 3\exp(x) + 5\exp(3x) + 7\exp(6x) + 9\exp(10x)}$

5/2 - 5/2 $\quad f(x) = \dfrac{2\exp(x) + 10\exp(3x) + 28\exp(6x) + 60\exp(10x) + 110\exp(15x)}{1 + 3\exp(x) + 5\exp(3x) + 7\exp(6x) + 9\exp(10x) + 11\exp(15x)}$

χ_m formulae for some heterodinuclear d^n–$d^{n'}$ complexes

$S_A \quad S_B$

1/2 1 $\quad \chi_M = \dfrac{Ng^2\beta^2}{kT} \cdot \dfrac{1 + 10\exp(1.5x)}{4 + 8\exp(1.5x)}$

1/2 3/2 $\quad \chi_M = \dfrac{Ng^2\beta^2}{kT} \cdot \dfrac{2 + 10\exp(2x)}{3 + 5\exp(2x)}$

1/2 2 $\quad \chi_M = \dfrac{Ng^2\beta^2}{kT} \cdot \dfrac{10 + 35\exp(2.5x)}{8 + 12\exp(2.5x)}$

1/2 5/2 $\quad \chi_M = \dfrac{Ng^2\beta^2}{kT} \cdot \dfrac{10 + 28\exp(3x)}{5 + 7\exp(3x)}$

$$x = J/kT$$

Coordination Chemistry. Joan Ribas Gispert
Copyright © 2008 WILEY-VCH Verlag GmbH & Co. KGaA, Weinheim
ISBN: 978-3-527-31802-5

Index

Coordination Chemistry. Joan Ribas Gispert
Copyright © 2008 WILEY-VCH Verlag GmbH & Co. KGaA, Weinheim
ISBN: 978-3-527-31802-5